Lecture Notes in Computer Science

Lecture Notes in Computer Science

Vol. 88: Mathematical Foundations of Computer Science 1980. Proceedings, 1980. Edited by P. Dembiński. VIII, 723 pages. 1980.

Vol. 89: Computer Aided Design - Modelling, Systems Engineering, CAD-Systems. Proceedings, 1980. Edited by J. Encarnacao. XIV, 461 pages. 1980.

Vol. 90: D. M. Sandford, Using Sophisticated Models in Resolution Theorem Proving.
XI, 239 pages. 1980

Vol. 91: D. Wood, Grammar and L Forms: An Introduction. IX, 314 pages. 1980.

Vol. 92: R. Milner, A Calculus of Communication Systems. VI, 171 pages. 1980.

Vol. 93: A. Nijholt, Context-Free Grammars: Covers, Normal Forms, and Parsing. VII, 253 pages. 1980.

Vol. 94: Semantics-Directed Compiler Generation. Proceedings, 1980. Edited by N. D. Jones. V, 489 pages. 1980.

Vol. 95: Ch. D. Marlin, Coroutines. XII, 246 pages. 1980.

Vol. 96: J. L. Peterson, Computer Programs for Spelling Correction: VI, 213 pages. 1980.

Vol. 97: S. Osaki and T. Nishio, Reliability Evaluation of Some Fault-Tolerant Computer Architectures. VI, 129 pages. 1980.

Vol. 98: Towards a Formal Description of Ada. Edited by D. Bjørner and O. N. Oest. XIV, 630 pages. 1980.

Vol. 99: I. Guessarian, Algebraic Semantics. XI, 158 pages. 1981.

Vol. 100: Graphtheoretic Concepts in Computer Science. Edited by H. Noltemeier. X, 403 pages. 1981.

Vol. 101: A. Thayse, Boolean Calculus of Differences. VII, 144 pages. 1981.

Vol. 102: J. H. Davenport, On the Integration of Algebraic Functions. 1–197 pages. 1981.

Vol. 103: H. Ledgard, A. Singer, J. Whiteside, Directions in Human Factors of Interactive Systems. VI, 190 pages. 1981.

Vol. 104: Theoretical Computer Science. Ed. by P. Deussen. VII, 261 pages. 1981.

Vol. 105: B. W. Lampson, M. Paul, H. J. Siegert, Distributed Systems – Architecture and Implementation. XIII, 510 pages. 1981.

Vol. 106: The Programming Language Ada. Reference Manual. X, 243 pages. 1981.

Vol. 107: International Colloquium on Formalization of Programming Concepts. Proceedings. Edited by J. Diaz and I. Ramos. VII, 478 pages. 1981.

Vol. 108: Graph Theory and Algorithms. Edited by N. Saito and T. Nishizeki. VI, 216 pages. 1981.

Vol. 109: Digital Image Processing Systems. Edited by L. Bolc and Zenon Kulpa. V, 353 pages. 1981.

Vol. 110: W. Dehning, H. Essig, S. Maass, The Adaptation of Virtual Man-Computer Interfaces to User Requirements in Dialogs. X, 142 pages. 1981.

Vol. 111: CONPAR 81. Edited by W. Händler. XI, 508 pages. 1981.

Vol. 112: CAAP '81. Proceedings. Edited by G. Astesiano and C. Böhm. VI, 364 pages. 1981.

Vol. 113: E.-E. Doberkat, Stochastic Automata: Stability, Nondeterminism, and Prediction. IX, 135 pages. 1981.

Vol. 114: B. Liskov, CLU, Reference Manual. VIII, 190 pages. 1981.

Vol. 115: Automata, Languages and Programming. Edited by S. Even and O. Kariv. VIII, 552 pages. 1981.

Vol. 116: M. A. Casanova, The Concurrency Control Problem for Database Systems. VII, 175 pages. 1981.

Vol. 117: Fundamentals of Computation Theory. Proceedings, 1981. Edited by F. Gécseg. XI, 471 pages. 1981.

Vol. 118: Mathematical Foundations of Computer Science 1981. Proceedings, 1981. Edited by J. Gruska and M. Chytil. XI, 589 pages. 1981.

Vol. 119: G. Hirst, Anaphora in Natural Language Understanding: A Survey. XIII, 128 pages. 1981.

Vol. 120: L. B. Rall, Automatic Differentiation: Techniques and Applications. VIII, 165 pages. 1981.

Vol. 121: Z. Zlatev, J. Wasniewski, and K. Schaumburg, Y12M Solution of Large and Sparse Systems of Linear Algebraic Equations. IX, 128 pages. 1981.

Vol. 122: Algorithms in Modern Mathematics and Computer Science. Proceedings, 1979. Edited by A. P. Ershov and D. E. Knuth. XI, 487 pages. 1981.

Vol. 123: Trends in Information Processing Systems. Proceedings, 1981. Edited by A. J. W. Duijvestijn and P. C. Lockemann. XI, 349 pages. 1981.

Vol. 124: W. Polak, Compiler Specification and Verification. XIII, 269 pages. 1981.

Vol. 125: Logic of Programs. Proceedings, 1979. Edited by E. Engeler. V, 245 pages. 1981.

Vol. 126: Microcomputer System Design. Proceedings, 1981. Edited by M. J. Flynn, N. R. Harris, and D. P. McCarthy. VII, 397 pages. 1982.

Voll. 127: Y. Wallach, Alternating Sequential/Parallel Processing. X, 329 pages. 1982.

Vol. 128: P. Branquart, G. Louis, P. Wodon, An Analytical Description of CHILL, the CCITT High Level Language. VI, 277 pages. 1982.

Vol. 129: B. T. Hailpern, Verifying Concurrent Processes Using Temporal Logic. VIII, 208 pages. 1982.

Vol. 130: R. Goldblatt, Axiomatising the Logic of Computer Programming. XI, 304 pages. 1982.

Vol. 131: Logics of Programs. Proceedings, 1981. Edited by D. Kozen. VI, 429 pages. 1982.

Vol. 132: Data Base Design Techniques I: Requirements and Logical Structures. Proceedings, 1978. Edited by S.B. Yao, S.B. Navathe, J.L. Weldon, and T.L. Kunii. V, 227 pages. 1982.

Vol. 133: Data Base Design Techniques II: Proceedings, 1979. Edited by S.B. Yao and T.L. Kunii. V, 229–399 pages. 1982.

Vol. 134: Program Specification. Proceedings, 1981. Edited by J. Staunstrup. IV, 426 pages. 1982.

Vol. 135: R.L. Constable, S.D. Johnson, and C.D. Eichenlaub, An Introduction to the PL/CV2 Programming Logic. X, 292 pages. 1982.

Vol. 136: Ch. M. Hoffmann, Group-Theoretic Algorithms and Graph Isomorphism. VIII, 311 pages. 1982.

Vol. 137: International Symposium on Programming. Proceedings, 1982. Edited by M. Dezani-Ciancaglini and M. Montanari. VI, 406 pages. 1982.

Vol. 138: 6th Conference on Automated Deduction. Proceedings, 1982. Edited by D.W. Loveland. VII, 389 pages. 1982.

Vol. 139: J. Uhl, S. Drossopoulou, G. Persch, G. Goos, M. Dausmann, G. Winterstein, W. Kirchgässner, An Attribute Grammar for the Semantic Analysis of Ada. IX, 511 pages. 1982.

Vol. 140: Automata, Languages and programming. Edited by M. Nielsen and E.M. Schmidt. VII, 614 pages. 1982.

Vol. 141: U. Kastens, B. Hutt, E. Zimmermann, GAG: A Practical Compiler Generator. IV, 156 pages. 1982.

Lecture Notes in Computer Science

Edited by G. Goos and J. Hartmanis

185

Mathematical Foundations of Software Development

TAPSOFT

Proceedings of the International Joint Conference on Theory and Practice of Software Development (TAPSOFT)

Berlin, March 25–29, 1985

Volume 1:
Colloquium on Trees in Algebra and Programming (CAAP' 85)

Edited by Hartmut Ehrig, Christiane Floyd, Maurice Nivat and James Thatcher

Springer-Verlag
Berlin Heidelberg New York Tokyo

Editors

H. Ehrig
C. Floyd
Technische Universität Berlin
Fachbereich 20, Informatik, SWT FR 5–6
Franklinstr. 28/29, D-1000 Berlin 10, FRG

M. Nivat
L.I.T.P., U.E.R. de Mathématiques
Université de Paris
2, Place Jussieu, F-75251 Paris Cedex, France

J. Thatcher
Math. Department IBM, T.J.W. Research Center
Yorktown Heights, NY 10598, USA

CR Subject Classification (1982): F, F3, F4, E1, D2

ISBN 3-540-15198-2 Springer-Verlag Berlin Heidelberg New York Tokyo
ISBN 0-387-15198-2 Springer-Verlag New York Heidelberg Berlin Tokyo

Printing and binding: Beltz Offsetdruck, Hemsbach/Bergstr.
2145/3140-543210

PREFACE

TAPSOFT is an international Joint Conference on Theory and Practice of Software Development. The idea for TAPSOFT originated when it was suggested that the 1985 annual Colloquium on Trees in Algebra and Programming (CAAP) should be held in Berlin. In view of the desired interaction between theory and practice, it was decided to supplement CAAP with a corresponding Colloquium on Software Engineering (CSE), and an advanced seminar connecting both parts. The overall aim of the conference is, to bring together theoretical computer scientists and software engineers (researchers and practitioners) with a view to discussing how formal methods can usefully be applied in software development.

TAPSOFT is being held from March 25-29, 1985 at the Technical University of Berlin. It is organized by the Technical University of Berlin, the Gesellschaft für Informatik, and the European Association for Theoretical Computer Science. The general organizers are Hartmut Ehrig (TU Berlin), Christiane Floyd (TU Berlin), Maurice Nivat (Université de Paris VI) and Jim Thatcher (IBM Research, Yorktown Heights).

TAPSOFT comprises three parts:

- Advanced Seminar on the Role of Semantics in Software Development

  The aim of this advanced seminar is to bring together leading experts in the fields of formal semantics and software engineering so as to enable them to present their own work and views on the role of semantics in software development, and to provide a forum for discussions between seminar speakers and all other conference participants.

  The seminar consists of invited lecturers and a panel discussion chaired by W. Turski and entitled "Formalism - or else?". Each lecture is followed by a brief response, giving a critical appraisal, and by a general discussion.

  The invited speakers are

J. Backus (USA)	H.D. Mills (USA)
M. Broy (Germany)	U. Montanari (Italy)
R.M. Burstall (UK)	P. Naur (Denmark)
A.P. Ershov (USSR)	D.L. Parnas (Canada)
J.J. Horning (USA)	J.C. Reynolds (USA)
C.B. Jones (UK)	D. Scott (USA)

The invited lectures are arranged in three sections:

Concepts of Semantics with a View to Software Development
The Role of Semantics in Language Design
The Role of Semantics in the Development of Software Systems

with a considerable overlap between the concerns of the contributions to the different sections.

- Colloquium on Trees in Algebra and Programming (CAAP '85)

  The previous Colloquia on Trees in Algebra and Programming were held in France and Italy as autonomous conferences. CAAP '85 is integrated into the TAPSOFT conference.

  Following the CAAP tradition, papers accepted for CAAP '85 cover a wide range of topics in theoretical computer science. In line with the theme of the TAPSOFT conference, special emphasis is given in the CAAP '85 program to problems arising in software development.

  The selected papers are organized in six sections:

  Algorithms and Combinatorics
  Rewriting
  Concurrency
  Graph Grammars and Formal Languages
  Specifications
  Semantics and Data Types.

  The program committee for CAAP '85 consists of:

A. Arnold (France)	M.C. Gaudel (France)
G. Ausiello (Italy)	H.-J. Kreowski (Germany)
E. Blum (USA)	B. Mahr (Germany)
W. Brauer (Germany)	U. Montanari (Italy)
R. Cori (France)	M. Nivat (France)
M. Dauchet (France)	G. Plotkin (UK)
H.D. Ehrich (Germany)	G. Rozenberg (Netherlands)
H. Ehrig (Chair, Germany)	E. Wagner (USA)

- Colloquium on Software Engineering (CSE)

  CSE focusses on the relevance of formal methods to software development. It tries to answer the following questions:

  - Can notions of software engineering be clarified with the help of formal concepts?

- Can concepts of formal semantics be applied in practical software development? Which supporting tools are needed?
- What experiences have been gained using formal methods and how are they related to the claims of their proponents?
- What alternatives to formal methods can be proposed?

These questions are taken up by researchers presenting concepts together with their realisation, and by practitioners reporting on industrial experience with methodical approaches. This session is supplemented by invited lectures given by experts from leading computer and software companies.

The selected papers are presented in the sections listed below:

Concepts and Methods in Software Development

Tools and Environments

Rigorous Approaches to Programming

Abstract Data Types in Software Development

Views of Concurrency

Industrial Experience

The program committee for CSE consists of:

C. Floyd (Chair, Germany)	P. Löhr (Germany)
C. Haenel (Germany)	P. Naur (Denmark)
P. Henderson (UK)	M. Sintzoff (Belgium)
H.-J. Hoffmann (Germany)	J. Thatcher (USA)
C.B. Jones (UK)	W. Turski (Poland)
G. Kahn (France)	H. Weber (Germany)
W. Koch (Germany)	J. Witt (Germany)
C.H.A. Koster (Netherlands)	

The TAPSOFT conference proceedings are published in advance of the conference in two volumes. The first volume includes the first two sections of invited papers for the Advanced Seminar and the final versions of 19 papers from CAAP '85, selected from a total of 54 papers. One additional selected paper from CAAP '85 was withdrawn by the authors because of an error in the proofs. The second volume includes the third section of invited papers from the Advanced Seminar, the final versions of 20 papers from CSE, selected from a total of 62 papers, together with three invited papers on industrial experience.

We would like to extend our sincere thanks to all program committee members and referees of CAAP '85 and CSE, as well as to the subreferees for CAAP '85 listed below for their care in reviewing and selecting the submitted papers:

E. Astesiano, M. Bellia, G. Berthelot, B. Biebow, M. Bidoit, P. Boehm, L. Bonsiepen, L. Bradley, G. Brebner, B. Buchberger, H. Carstensen, C. Choppy, H. Cohen, A. Corradini, B. Courcelle, E. Dahlhaus, P. Degano, V. Diekert, K. Drosten, J. Engelfriet, W. Fey, P. Flajolet, L. Fribourg, M. Gogolla, V. Goltz, U. Grude, J. Gruska, A. Habel, H. Hansen, K.-P. Hasler, D. Haussler, C. Kirchner, H. Kirchner, W. Kowalk, K.-J. Lange, G.A. Lanzarone, M. Lightner, U. Lipeck, S. Maaß, M. Main, C. Montangero, R. de Nicola, H. Oberquelle, R. Orsini, P. Padawitz, F. Parisi-Presicce, J.C. Raoult, M. Simi, F. Simon, W. Struckmann, R. Valk, B. Vauquelin, M. Venturini Zilli, G. Violal-Naquet, F. Voisin, H. Wagner, E. Welzl, K.-J. Werner, A. Wilharm and K. Winklmann.

We gratefully acknowledge the financial support provided by the following institutions and firms:

Deutsche Forschungsgemeinschaft
Senator für Wirtschaft und Verkehr, Berlin
IBM Europe
Arthur Andersen & Co., Hamburg
Softlab GmbH, München
Cray-Research GmbH, München
Siemens AG, München
Epsilon GmbH, Berlin

We wish to express our gratitude to the members of the Local Arrangements Committee: Wilfried Koch, Bernd Mahr, Ulrike Niehaus and Christoph Oeters and to all members of the Computer Science Department of the TU Berlin who helped in the organization, in particular: G. Ambach, P. Bacon, H. Barnewitz, M. Bittner, D. Fähndrich, W. Fey, A. Habel, H. Hansen, K.-P. Hasler, K. Kautz, W. Köhler, R. Kutsche, M. Löwe, H. Pribbenow, M. Reisin, K. Schlicht, H. Wagner and K. Wohnhas. Without their help the conference would not have been possible.

Finally, we would like to thank the Springer Verlag, in particular Mrs. I. Mayer for her friendly cooperation in preparing the proceedings.

Berlin, March 1985

Hartmut Ehrig
Institut für Software
und Theoretische Informatik
Technische Universität Berlin

Christiane Floyd
Institut für Angewandte
Informatik
Technische Universität Berlin

Maurice Nivat
U.E.R. de Mathematiques
Université de Paris VI

James Thatcher
IBM T.J. W. Research Center
Yorktown Heights

CONTENTS OF VOLUME 1 page

COLLOQUIUM ON TREES IN ALGEBRA AND PROGRAMMING

page

ALGORITHMS AND COMBINATORICS

REWRITING

CONCURRENCY

CONTENTS OF VOLUME 2

INTRODUCTION

Although the proceedings of the TAPSOFT conference are published in two volumes, there is, hopefully, no clear distinction between theoretical and practical contributions to this conference. In fact, there are several papers in both volumes which try to bridge the gap between the theory and practice of software development. Most of the authors, as well as the program committee members, are aware that there is such a gap, and most of us have already started to build bridges, but it is awfully difficult to reach the other side. I remember an excellent lecture by Jim Thatcher in Nyborg last year, where he explained his own difficulties in trying to bridge gaps within a big computer company, where the practical side was not only represented by people working in software development, but also by the management of the company.

Two years ago we began a seminar at the Technical University of Berlin to try and study the role of formal methods in software development. The participants were people from theoretical computer science working in the field of algebraic specifications, and people from software engineering partly using algebraic and partly using informal techniques for software development. Even in this small group of people, especially for Christiane Floyd and myself, it was most difficult to understand each other's terminology and main aims. But all of us were convinced that it would be most desirable to understand the aims of the other group and to launch joint projects.

The main joint project started at that time was the organization of this TAPSOFT conference. To begin with, we only wanted to combine the Colloquium on Trees in Algebra and Programming (CAAP '85) with a corresponding Colloquium on Software Engineering (CSE). It was Maurice Nivat who convinced us that we should have a third part, an Advanced Seminar on the Role of Semantics in Software Development, and Jim Thatcher, who enthusiastically took up these initial ideas which eventually led to this joint conference on Theory and Practice of Software Development.

After discussing the possible integration of CAAP into TAPSOFT with Maurice Nivat, Max Dauchet, André Arnold, and Giorgio Ausiello at CAAP '83 in L'Aquila, we were all convinced that combining CAAP with more practical aspects of computer science might help to bridge the gap between theory and practice. We all hope that at this 10th Colloquium on Trees in Algebra and Programming in Berlin 1985 people from theory and practice are listening to, and learning from, each other. This is only possible if all participants, especially the speakers, try to explain their ideas and results in simple terms and avoid the temptation to present too many of the technical details given in their papers.

However, I am aware that not all of the papers in this volume and not all of the corresponding lectures at the TAPSOFT conference will focus on bridging the gap between theory and practice. Actually, there is also a great need for the development of new theories in a mathematically precise and consistent way, without looking all the time at practical applications. Working in theoretical computer science, we have to make use of all kinds of mathematical theories, constructions and results in order to obtain a powerful theory from the mathematical point of view. But it is equally important that it should also be powerful from the point of view of computer science. This means the theory should be helpful in solving problems in practical computer science.

At this conference we want to focus principally on problems in software development. What are the main problems? How can they be stated precisely? Which of them can be better solved using formal methods and languages? Let us take a look at one specific question: In what way can formal specification methods and languages improve the software development process? Before we can try to answer this question, we have to make sure we know what kind of improvements we have in mind. What are the desirable concepts for software development, and how can these concepts be realized? It is difficult to get more than some informal ideas from people in software engineering in answer to these questions. Even basic notions such as "requirements specification", "design specification", "interface of a module" or "user interface" seem, as yet, to have no precise meaning in software engineering. They are interpreted differently in different papers, and even one and the same author offers a variety of interpretations.

For a theoretician, a simple solution in this case is to define all these notions as he pleases and start constructing his theory. Although this sometimes seems to be necessary, in most cases it turns out to be the wrong way: Theory and practice then diverge. It seems to be much better, although sometimes frustrating, to keep in contact with people in software engineering during the development of the theory. Similarly, it is advisable for people in software development to remain in contact with the users during the development of the software system in order to integrate this properly into the user environment.

In addition to the problems mentioned above, there are even bigger ones when considering the development of software for distributed systems, including aspects of nondeterminism and concurrency. On the one hand, there are powerful mathematical theories for concurrency and nondeterminism, using techniques from topology, modal logic, algebra and category theory. On the other hand, there are complex languages including concurrency, like ADA, and even more complex computer systems and networks which have been installed, although nobody is able to understand these systems in detail or even prove their correctness. Certainly, the need for interaction between people in theory and practice in such fields is even more important than in other areas which are at present better known.

Last but not least, several of these problems, especially the development of software which may be used for the control of critical processes or military purposes, should also be discussed with respect to their social impact.

We hope that this conference also constitutes a lively forum for discussions of this kind.

Berlin, March 1985 Hartmut Ehrig

Specification and Top Down Design of Distributed Systems

Manfred Broy

Fakultät für Mathematik und Informatik
Universität Passau
Postfach 25 40
8390 Passau

Abstract

Stream-processing functions provide an excellent semantic model for the abstract representation of systems of nondeterministic concurrent communicating agents. Based on this model a formalism for the specification of such functions is suggested. This way a fully modular, compositional methodology for the specification and the design of distributed systems and their components is derived. Concepts of correctness are defined and rules of inference are discussed that help to transform such specifications into a network of communicating agents. A combinatorial ("functional") notation for the sequential and parallel composition as well as feedback for those agents is introduced.

1. Introduction

For the top-down design of concurrent communicating systems a specification formalism is an indispensible requisite. Only if one is able to give modular specifications, i.e. abstract specifications also for the subcomponents of a distributed system, then the decomposition of the system can be done properly and the subcomponents can be developed and verified separately.

In this paper we are not interested in the internal structure of (components of) concurrent communicating systems ("internal behaviour") that could be described for instance by event structures (cf. [Winskel 82]) but rather in the "input/output" behaviour of these systems ("extensional behaviour").

In the following a simple language formalism is suggested that can be used as a formal framework for the specification, design and verification of concurrent, communicating systems and their components. It is based on a semantic model for concurrent, communicating systems and is an attempt to combine ideas and concepts from denotational semantics for concurrent systems (cf. [Broy 83b]), programming logic (such as temporal logic, cf. [Pnueli 77]), algebraic specifications (cf. [Broy, Wirsing 82]), program transformation (cf. [Broy 84d]), and functional (multi-)programming (cf. [Broy 81]).

2. A Specification Formalism

In this section a specification formalism is introduced that allows to specify communicating agents with a finite set of input lines and a finite set of output lines. We start by giving some examples, then give a formal syntax and finally define the semantics.

2.1. Specifications of Communicating Agents

A communicating agent has n input lines and m output lines, where m and n are arbitrary natural numbers (including 0). On every input line a finite or infinite sequence of data is transmitted to the agent and on every output line a finite or infinite sequence of data is generated by the agent. The input lines and output lines have internal (local) names that are used in a predicate for expressing the relationship between the input and output.

2.1.1. First Examples for Specifications of Agents

We start with the example of a very simple agent store which nevertheless shows the full power of the method:

agent store = **input stream data** d, **stream bool** b, **output stream data** r,

first b ⟹ r = store(**rest** d, **rest** b),
¬ **first** b ⟹ r = **first** d & store(d, **rest** b) **end**

This specification defines an agent with 2 input lines and 1 output line:

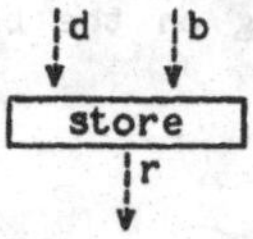

The identifiers d, b, r for the input lines and output lines are only internal ("bound") names and not relevant to the outside. The operator & puts an element in front of a sequence, **first** s returns the first element of a sequence s, **rest** s returns the sequence s without the first element.

According to the specification taking the sequences

d = 1 & 2 & 3 & 4 & ...
b = false & false & true & false & true & false & ...

as input implies the output sequence

r = 1 & 1 & 2 & 3 & ...

The agent store may be seen as a memory cell where b can be interpreted as the read/write command sequence: If the input on input line b is true, then a new data value on the input line d is stored, if the input on input line b is false, then this is interpreted as a read command: the current input on d is copied as output.

As a second example the agent schedule is specified:

agent schedule =
input stream data a, **stream data** b, **stream bool** s, **output stream data** r,

first s ⇒ r = **first** a & schedule(**rest** a, b, **rest** s),
¬ **first** s ⇒ r = **first** b & schedule(a, **rest** b, **rest** s) **end**

The agent schedule receives three input streams: two data streams and one stream of booleans. The data streams are merged according to the boolean values in the third input stream. With the input

a = 0 & 2 & 4 & ...
b = 1 & 3 & 5 & ...
s = true & true & false & false & true & false & ...

the agent schedule produces

r = 0 & 2 & 1 & 3 & 4 & 5 & ...

Another example is the agent switch:

agent switch = **input stream data** s, **stream bool** b, **output stream data** r1, r2,

∀ **stream data** d1, d2: (d1, d2) = switch(**rest** s, **rest** b) ⇒
(**first** b ⇒ (r1 = **first** s & d1 ∧ d2 = r2)) ∧
(¬ **first** b ⇒ (d1 = r1 ∧ r2 = **first** s & d2)) **end**

The agent switch produces two output streams by sending the input on its input line s either to the left or to the right output line depending on the boolean input in its input line b. One may prove

switch(s, b) = (d1, d2) ⇒ schedule(d1, d2, b) = s.

An agent even may have no input lines at all. An example reads:

agent onestream = **output stream nat** r, r = 1 & r **end**

The agent onestream produces an infinite stream of 1's:

r = 1 & 1 & 1 & ...

One may write also agents that perform arithmetic operations on its input streams. A simple example is:

agent addstream = **input stream nat** a, **stream nat** b, **output stream nat** r,

r = (**first** a + **first** b) & addstream(**rest** a, **rest** b) **end**

This agent adds the sequences of input streams elementwise; if

a = 0 & 2 & 4 & ...
b = 1 & 3 & 5 & ...

then one obtains

r = 1 & 5 & 9 & ...

Agents may also be specified based on other agents

agent natstream =
output stream nat r, r = 0 & addstream(r, onestream) **end**

The agent natstream produces the infinite stream of the natural numbers:

r = 0 & 1 & 2 & ...

So far the specified agents looked more or less deterministic, i.e. uniquely specified. They precisely describe functions from the tuple of input streams to the tuple of output streams.

2.1.2. Nondeterministic Agents

Of course, nondeterministic agent specifications may be written in the formalism, too.

agent infinite = **output stream bool** r, r = true & r ∨ r = false & r **end**

The agent infinite has two possible output streams:

r = true & true & ... or r = false & false & ...

The possibility of writing nondeterministic agent specifications leads to the possibility (and the problems) of using nondeterministic expressions. For instance with the specification above the agent infinite produces nondeterministically one of the two infinite sequences. This leads to the problem of nondeterministic terms and their meaning. Using nondeterministic terms in equations brings a number of complications. Writing

t1 = t2

for nondeterministic terms t1 and t2 what does that mean? That t1 and t2 must stand for the same set of possible values? For showing some of the subtle differences consider:

agent any = **output stream bool** r, r = true & any() ∨ r = false & any() **end**

The agent any produces any infinite sequence of boolean values. Another even more famous and more important example is the agent merge:

agent merge = **input stream data** a, **stream data** b, **output stream** r

∃ **stream bool** s: s = any() ∧ r = schedule(a, b, s) **end**

The agent merge is highly nondeterministic. For instance with the infinite streams

a = true & true & ...
b = false & false & ...

the term merge(a, b) may stand for **any** sequence of boolean values.

We take here a very liberal viewpoint for equations between nondeterministic terms, i.e. terms in which nondeterministic agents occur. With every agent a set of stream processing functions is associated. An equation between terms using agents is fulfilled, if for every occurrence of an agent identifier one may find a stream-processing function in the given set associated with the agent identifier, such that the equations are fulfilled.

After having given a number of simple examples the syntax and the semantics of the specification language is now formally defined.

2.2. Syntactic Form of an Agent Specification

In this section a syntax for specifications of communicating agents is given in BNF-style notation.

Syntax

<system> ::= <agent-spec>*

<agent-spec> ::= {**rec**} **agent** <agent-id> =
{ **input** <dec-tuple> } { **output** <dec-tuple> }
<formula> **end**

<dec-tuple> ::= { <dec> , }*

<dec> ::= {**stream**} <sort> <id> { , <id> }*

<formula> ::= <equation> | (<formula>) | true | false |
<formula> { ∧ | ∨ | ⇒ | , }^{1} <formula> |
¬ <formula> |
{∃ | ∀ }^{1} <dec-tuple> <dec> : <formula> |
{ ○, ◇, □ }^{1} <id> { , <id> }* : <formula>

<equation> ::= <exp> = <exp>

<exp> ::= <id> | { **first** | **rest** | **isempty** }^{1} | <exp> & <exp> |
<function> ({ <exp> { , <exp> }*}) |
if <exp> **then** <exp> **else** <exp> **fi** |
(<exp> { , <exp> }*)

<function> ::= <agent-id> | <primitive function>

In this syntax it is assumed

- a set <id> of identifiers for data or streams of data,
- a set <agent-id> of identifiers for agents,
- a set <primitive functions> of identifiers for primitive (given) agents.

Of course a number of context conditions (such as well-formedness or type-correctness of terms) have to be presupposed for ensuring that an agent specification is meaningful. For convenience and lack of space context conditions are not given explicitly.

2.3. Semantics of Agent Specifications

In this section a semantic model is introduced, based on an algebra of primitive data and functions. The semantic model consists essentially of stream processing functions. Then the agent specification formalism is related to this semantic model.

2.3.1. Streams and Stream-processing Functions

As one of the most fundamental domains for communicating programs one may consider the domain of streams (cf. [Broy 81]). Given a flat domain $A^{\perp}$ (i.e. a set A, with $\perp \notin A$, $A^{\perp} = A \cup \{\perp\}$, ordered by $a1 \sqsubseteq a2$ iff $a1 = a2 \vee a1 = \perp$ for $a1, a2 \in A^{\perp}$) the domain STREAM(A) of streams over A is defined by

$$\text{STREAM}(A) = (A^{*} \times \{\perp\}) \cup A^{*} \cup A^{\infty}$$

where A^{*} denotes the finite sequences (words) over A, and A^{∞} denotes the infinite sequences (words) over A. For $s1, s2 \in \text{STREAM}(A)$ a partial ordering is defined by

$$s1 \sqsubseteq s2 \text{ iff } s1 = s2 \text{ or } \exists\, s3 \in A^{*}, s4 \in \text{STREAM}(A) : s1 = s3 \circ \langle\perp\rangle \wedge s2 = s3 \circ s4$$

Here "$\circ$" denotes the usual concatenation where for $s \in A^{\infty}$ we define $s \circ s' = s$; for all $a \in A$ by $\langle a\rangle$ we denote the one-element sequence consisting of a. By ε we denote the empty sequence.

With these definitions (STREAM(A), $\sqsubseteq$) forms an algebraic domain where $(A^{*} \times \{\perp\}) \cup A^{*}$ is the set of finite elements, A^{∞} the infinite ones, $A^{*} \times \{\perp\}$ the partial ones and $A^{*} \cup A^{\infty}$ the total ones. The stream $\langle\perp\rangle$ represents the least element.

Streams can be used for representing the sequence of communications of a program, for instance the output on a specific channel. Of course for nonterminating programs the output may be an infinite sequence.

For communicating programs one has to distinguish two forms of nontermination: nontermination with infinite output and nontermination without any further output. The first is represented by an infinite stream, the second by a finite stream ending with the $\perp$-symbol. In this case one may speak of divergence.

The following four basic functions are used on streams:

$$\begin{aligned} ap &: A^{\perp} \times \text{STREAM}(A) \rightarrow \text{STREAM}(A) \\ rest &: \text{STREAM}(A) \rightarrow \text{STREAM}(A) \\ first &: \text{STREAM}(A) \rightarrow A^{\perp} \\ isempty &: \text{STREAM}(A) \rightarrow \mathbb{B}^{\perp} \end{aligned}$$

defined by

$$ap(a, s) = \begin{cases} \langle a\rangle \circ s & \text{if } a \in A,\ s \in \text{STREAM}(A) \\ \langle\perp\rangle & \text{otherwise} \end{cases}$$

For ap(a, s) we often write a&s. Note that ap is a nonstrict function: the result of applying ap may be different from the least element $\langle\bot\rangle$ even if the second argument is $\langle\bot\rangle$. However, if the first argument is the least element of the domain (is $\bot$) then the result is the least element (is $\langle\bot\rangle$). The function ap is left-strict. The stream a&s can be seen as a sequence of communicated data (for instance output). As soon as a is $\bot$, then there cannot be any defined output afterwards: $\bot$&s = $\langle\bot\rangle$. So the definition of ap mirrors the simple fact of communicating processes, that after divergence there cannot be any further output.

Note that STREAM(A) is not closed with respect to concatenation since for streams $s \in A^* \times \{\bot\}$ and $s' \in A^* \setminus \{\varepsilon\}$: $s \circ s' \notin$ STREAM(A).

Let $a \in A$, $s \in$ STREAM(A), then the functions rest, first, isempty are defined by:

rest(a&s) = s, rest(ε) = rest($\langle\bot\rangle$) = $\langle\bot\rangle$,

first(a&s) = a, first(ε) = first($\langle\bot\rangle$) = $\bot$,

isempty(a&s) = false, isempty(ε) = true, isempty($\langle\bot\rangle$) = $\bot$.

One simply proves:

<u>Lemma:</u> The functions ap, rest, first and isempty are monotonic and continuous. □

For obvious reasons streams can be considered as one of the most fundamental domains when dealing with systems of communicating processes. For procedural concurrent programs with shared memory one may consider streams of states, for processes with explicit communication primitives one can think of streams of communication actions (cf. [Broy 83a, 84a, 84b]).

For giving meaning to agent specifications a fixed set DATA of atomic data objects is assumed. For writing examples we assume $\mathbb{N} \subseteq$ DATA and true, false $\in$ DATA. In a more complete framework, one may assume some abstract data type specification method for specifying the atomic data objects.

The set D of all objects on which agents operate is defined by

$$D =_{def} DATA^{\bot} \cup STREAM(DATA)$$

An agent with m input lines and n output lines is a continuous mapping. The set of those continuous mappings is defined by:

$AGENT^m_n = [D^m \rightarrow D^n]$

The set of all agents is defined by

$AGENT = \{f \in [D^m \rightarrow D^n]: n, m \in \mathbb{N}\}$

Having fixed the data universe D and the universe of agents $AGENT^m_n$ now meaning can be assigned to specifications.

2.3.2. Assigning Meaning to Agent Specifications

For assigning meaning to agent specifications the well-known technique of environments is used: Two kinds of environments are needed, data environments and agent environments.

$ENV =_{def} (\langle id \rangle \rightarrow D)$

$AENV =_{def} (\langle agent\text{-}id \rangle \rightarrow P(AGENT) \setminus \{\emptyset\})$

An agent environment associates with every agent identifier a set of agents. An environment for identifiers associates with every identifier a stream or a data object.

Note that for very particular reasons (cf. concluding remarks) we have chosen to associate with every agent identifier a set of continuous stream processing functions instead of using functions mapping (tuples of) streams into sets of (tuples of) streams.

As usual the updating of environments σ is denoted by $\sigma[d/x]$:

$$\sigma[d/x](y) = \begin{cases} d & \text{if } x = y \\ \sigma(y) & \text{otherwise} \end{cases}$$

The meaning of an expression is defined by the semantic function

$B: \langle exp \rangle \rightarrow AENV \rightarrow P(ENV \rightarrow D)$

We write $B_\sigma[E]$ for $B(E)(\sigma)$. Note that due to the fact that agent identifiers stand for sets of stream processing functions, this particular semantic model is chosen.

Let $f \in \langle agent\text{-}id \rangle$

$$B_\sigma[f(E_1, \ldots, E_n)] =_{def} \{ h: ENV \rightarrow D: \exists\, g \in \sigma[f], h_1 \in B_\sigma[E_1], \ldots, h_n \in B_\sigma[E_n]: \forall\, e \in ENV: h(e) = g(h_1(e), \ldots, h_n(e)) \}$$

$$B_\sigma[\textbf{rest}\ E] =_{def} \{ h: ENV \rightarrow D: \exists\, g \in B_\sigma[E]: \forall e \in ENV: h(e) = rest(g(e)) \}$$

$$B_\sigma[\textbf{first}\ E] =_{def} \{ h: ENV \rightarrow D: \exists\, g \in B_\sigma[E]: \forall e \in ENV: h(e) = first(g(e)) \}$$

$$B_\sigma[\textbf{isempty}\ E] =_{def} \{ h: ENV \rightarrow D: \exists\, g \in B_\sigma[E]: \forall e \in ENV: h(e) = isempty(g(e)) \}$$

$$B_\sigma[E1 \;\&\; E2] =_{def} \{ h: ENV \to D: \exists\, h1 \in B_\sigma[E1], h2 \in B_\sigma[E2]: \forall e \in ENV: h(e) = ap(h1(e), h2(e)) \}$$

$$B_\sigma[x] =_{def} \{h: ENV \to D: e \in ENV: h(e) = e(x)\} \qquad \text{for } x \in \text{<id>}$$

$$B_\sigma[\textbf{ if } E0 \textbf{ then } E1 \textbf{ else } E2 \textbf{ fi }] =_{def} \{ h: ENV \to D: \exists\, h0 \in B_\sigma[E0], h1 \in B_\sigma[E1], h2 \in B_\sigma[E2]: \forall e \in ENV: h(e) = if(h0(e), h1(e), h2(e)) \}$$

where if : $D^3 \to D$ is defined by

$$if(d0, d1, d2) = \begin{cases} d1 & \text{if } d0 = true \\ d2 & \text{if } d0 = false \\ \bot & \text{otherwise} \end{cases}$$

$$B_\sigma[(E_1, \ldots, E_n)] =_{def} \{ h: ENV \to D: \exists\, h_1 \in B_\sigma[E_1], \ldots, h_n \in B_\sigma[E_n]: \forall e \in ENV: h(e) = (h_1(e), \ldots, h_n(e)) \}$$

The meaning of formulas is defined by the semantic function

$$M: \text{<formula>} \to AENV \to P(ENV \to \mathbb{B})$$

We write $M_\sigma[H]$ for $M(H)(\sigma)$.

$$M_\sigma[E1 = E2] =_{def} \{ h: ENV \to \mathbb{B}: \exists\, g1 \in B_\sigma[E1], g2 \in B_\sigma[E2]: \forall e \in ENV: h(e) = (g1(e) \equiv g2(e)) \}$$

For boolean expressions E we write often just E instead of E = true.

$$M_\sigma[H1 \wedge H2] =_{def} \{ h: ENV \to \mathbb{B}: \exists\, h1 \in M_\sigma[H1], h2 \in M_\sigma[H2]: \forall e \in ENV: h(e) = (h1(e) \wedge h2(e)) \}$$

We often write "," instead of "$\wedge$". Analogous definitions are assumed for the remaining logical connectives.

$$M_\sigma[\ \forall\, \mathbf{m}\ x: H\ ] =_{def} \{h: ENV \to \mathbb{B}: \exists\, h0 \in M_\sigma[H]\ \forall e \in ENV: h(e) = \forall d \in R: h0(e[d/x]) \}$$

where R denotes the subset of D that is indicated by the sort **m**.

$$M_\sigma[\ \exists\, \mathbf{m}\ x: H\ ] =_{def} \quad M_\sigma[\neg\ \forall\, \mathbf{m}\ x: \neg\, H]$$

In agent specifications agent identifiers stand for (sets of) stream processing functions. Every time a stream identifier occurs in a specification another instantiation may be taken, i.e. another stream processing function out of the set of stream processing functions can be chosen.

Given a formula H and an agent-environment σ and a data environment e we say H is <u>valid</u> for σ and e and write

$(\sigma, e) \models H$

if

$$\forall \sigma 0 \in AENV: \sigma 0 \leq \sigma \Rightarrow \exists h \in M_{\sigma 0}[H]: h(e) = true$$

Here $\sigma 0 \leq \sigma$ is used in the sense of pointwise subset relation.

The choice of this particular definition of validity is essentially motivated by concepts of implementation and program design as explained in the following sections: An agent may be used in networks of communicating agents and may occur several times. For each occurrence one arbitrary function out of the set of allowed functions may be chosen. However, the formula even holds if one uses an implementation of the resp. agent (i.e. a subset from the set of possible agents).

If H does not contain free data identifiers, the validity is independent of e and we write

$$\sigma \models H$$

In the following we use $\overline{x}$ (and $\overline{y}$ resp.) as abbreviation for tuples of declarations, i.e. for phrases of the syntactic unit <dec-tuple>. Let $\overline{x}$ stand for

$$s_1\ x_1,\ \ldots,\ s_n\ x_n$$

where the s_i specify the sort of the identifiers x_i. Let furthermore x (and y resp.) stand for $(x_1, \ldots, x_n)$. Now it is defined, under which circumstances an agent specification is fulfilled by an environment. Given an agent specification

agent f = **input** $\overline{x}$, **output** $\overline{y}$, H **end**

where we assume that in H only agent identifiers and the identifiers from $\overline{x}$ and $\overline{y}$ occur, we say for a given agent environment σ "σ fulfils the specification for f" if

$$\sigma \models \forall \overline{x}: \exists \overline{y}: y = f(x) \wedge H$$

Given a family DEF of definitions of the agent, $f_1, \ldots, f_k$

agent f_1 = ... H_1 **end** ... **agent** f_k = ... H_n **end**

an agent environment

$$\sigma: \{f_1, \ldots, f_k\} \rightarrow P(AGENT)$$

is called <u>consistent</u> and we write $\sigma \models$ DEF if σ fulfils all the agent specifications for $f_1, \ldots, f_k$. The semantics of the family of agent definitions then is given by the sets of all agent environments that fulfil the specification, i.e. by

$$\{\sigma: \{f_1, \ldots, f_k\} \rightarrow P(AGENT): \sigma \models DEF\}$$

This way every family of agent specifications f_i is semantically defining a set of environments, i. e. a set of families (mappings from identifiers to sets) of sets of stream-processing functions.

In the context of equational ("algebraic") specifications of families of agents similar questions arise as for the algebraic specification of abstract (data) types in hierarchies. One may introduce notions like persistency, hierarchy completeness (sufficient completeness), or hierarchy consistency in (algebraic) specifications of agents.

2.4. Temporal Logics

A highly developed notation and calculus for reasoning on (sets of) sequences is (linear time) temporal logic. Temporal logic mainly has been advocated and used for reasoning about complete concurrent systems cooperating via shared memory. Since agents are functions on sequences temporal logic should provide a framework also for agent specifications. Since one has to reason in agent specifications about several sequences in one formula, the temporal logic framework has to be slightly generalized. Three temporal operators are introduced.

Let s be a stream identifier and H be a predicate where s is used as a stream. Then we define

$$\circ\ s_1, \dots, s_k: H \ =_{def}\ H[(\mathbf{rest}\ s_1)/s_1, \dots, (\mathbf{rest}\ s_k)/s_k]$$

$$\Box\ s_1, \dots, s_k: H \ =_{def}\ \forall i \in \mathbb{N}: H[(\mathbf{rest}^i\ s_1)/s_1, \dots, (\mathbf{rest}^i\ s_k)/s_k]$$

$$\Diamond\ s_1, \dots, s_k: H \ =_{def}\ \exists i \in \mathbb{N}: H[(\mathbf{rest}^i\ s_1)/s_1, \dots, (\mathbf{rest}^i\ s_k)/s_k]$$

Here $\mathbf{rest}^i$ is assumed to by defined by

$$\mathbf{rest}^0\ s = s, \quad \mathbf{rest}^{i+1}\ s = \mathbf{rest}(\mathbf{rest}^i\ s)$$

By H[E/s] we denote the expression that is obtained by replacing all occurrences of s in H by the expression E.

In terms of our semantic definitions the meaning of the temporal operators is given by:

$$M_\sigma[\circ\ s_1, \dots, s_k: H] =_{def} \{h: ENV \to \mathbb{B}: \exists\ h0 \in M_\sigma[H]\ \forall e \in ENV: h(e) = h0(e[rest(s_1)/s_1, \dots, rest(s_k)/s_k])\}$$

$$M_\sigma[\Box\ s_1, \dots, s_k: H] =_{def} \{h: ENV \to \mathbb{B}: \exists\ h0 \in M_\sigma[H]\ \forall e \in ENV: h(e) = \forall\ i \in \mathbb{N}: h0(e[rest^i(s_1)/s_1, \dots, rest^i(s_k)/s_k])\}$$

$$M_\sigma[\Diamond\ s_1, \dots, s_k: H] =_{def} \{h: ENV \to \mathbb{B}: \exists\ h0 \in M_\sigma[H]\ \forall e \in ENV: h(e) = \exists\ i \in \mathbb{N}: h0(e[rest^i(s_1)/s_1, \dots, rest^i(s_k)/s_k])\}$$

With these temporal operators one can give specifications without any use of "recursion". For instance the example onestream then reads

agent onestream = **output stream nat** r, $\Box$ r: **first** r = 1 **end**

Temporal formulas therefore stand for infinite formulas and thus can be seen to allow often a more explicit way of specifications. For some of the examples of section 2 one can obtain more explicit specifications by temporal logics:

agent infinite = **output stream bool** r: ($\Box$ r: **first** r) $\vee$ ($\Box$ r: $\neg$ **first** r) **end**

agent addstream = **input stream nat** a, **stream nat** b, **output stream nat** r:

$\Box$ a, b, r: **first** r = **first** a + **first** b **end**

agent any = **output stream** r, $\Box$ r: (**first** r $\vee$ $\neg$ **first** r) **end**

For other examples such as the agent merge or the agent schedule, more explicit (nonalgebraic) specifications using temporal logic are less obvious. An appropriate, more powerful temporal logic for specifying communicating agents without algebraic (recursive) equations seems one of the interesting questions to be looked at.

Possible candidates for such an extended temporal logic are for instance versions of the until operator, the (iterated) combine operator, and the fixed point operator for logical formulas as suggested in [Barringer et al. 85]. However, it seems not clear so far, whether an algebraic style of specifications or such a more logic-oriented style of specifications is more appropriate. For the moment I prefer a formalism which supports both styles.

3. Agent Specifications Describing Algorithms

The formalism introduced so far was explained as a specification tool. Nevertheless agent specifications of particular syntactic forms can be seen as programs. Such specifications are called algorithmic agents.

3.1. Algorithmic Agents

An algorithmic agent is an agent specification of the following syntactic form:

agent f = **input** $\overline{x}$, **output** $\overline{y}$, $\exists$ $\overline{z}$: $H_1 \wedge \ldots \wedge H_j$ **end**

The set X of identifiers of $\overline{x}$ are called input ports, the set Y of identifiers in $\overline{y}$ are called the output ports, the set Z of identifiers in $\overline{z}$ are called the internal ports.

The formulas H_i, $1 \leq i \leq j$ are assumed to be of the form

$$(a_i = T^{(i)}_1 \vee \ldots \vee a_i = T^{(i)}_p)$$

where a_i is called the left-hand side of the clause H_i and the a_i are tuples of output ports and internal ports, i.e. $a_i \in (Y \cup Z)^p$ with some $p \in \mathbb{N}$. The $T^{(i)}_k$ are assumed to be expressions. Every identifier in $Y \cup Z$ must occur in exactly one left-hand side a_i. A system of agents is called algorithmic, if all agents of the system are algorithmic.

Algorithmic agent specifications can be trivially translated into applicative multiprograms written in programming languages such as AMPL (cf. [Broy 81]) for which an operational (reduction) semantics is available. This way algorithmic specifications define algorithms. In a system of agent specifications it is said that "the agent f uses the agent g" if

(1) g occurs in the body of the specification of agent f, or

(2) g is used by an agent h that is used by f.

If an agent f uses itself, then it is called recursive agent specification. If an algorithmic agent specification contains equations with stream identifiers occurring on both the left-handside and the right-handside of equations between streams then we speak of recursive (algebraic) equations for streams.

Every algorithmic agent defines a network (a data flow graph), if we take $X \cup Y \cup Z$ as arcs and introduce a node for every clause H_i. The arcs have as sources the nodes of the clauses where they appear on the lefthand side. If the agent specification is recursive, then the defined network is infinite, if the agent specification is not recursive but contains recursive stream equations, then the network is cyclic. If the algorithmic specification is neither recursive nor is containing recursive stream equations, then the network is a finite, acyclic (directed) graph. Examples are given below.

4. Design Issues

After having introduced a particular language for the specification of agents, in this section a number of design issues are treated that give some insights, how the language for specifying agents can be used.

4.1. A Combinatorial/Functional Notation for Agents

So far the specifications are written using (internal) identifiers for data objects. In this section a functional style notation is introduced that allows to combine given agent specifications by sequential composition, parallel composition and feedback.

In the following we assume the two agent specifications to be given:

agent a1 = **input** $\overline{x1}$, **output** $\overline{y1}$, H1 **end**

agent a2 = **input** $\overline{x2}$, **output** $\overline{y2}$, H2 **end**

where a1 has n1 input lines and m1 output lines and a2 has n2 input lines and m2 output lines. We assume, that $\overline{x1}$, $\overline{x2}$, $\overline{y1}$, $\overline{y2}$ are pairwise disjoint w.r.t. the (sets of) identifiers they contain.

The parallel composition of the two agents a1 and a2 is written by

a1 ∥ a2

where a3 = a1 ∥ a2 has n1 + n2 input lines and m1 + m2 output lines and it is specified by

agent a3 = **input** $\overline{x1}$, $\overline{x2}$, **output** $\overline{y1}$, $\overline{y2}$, $H1 \wedge H2$ **end**

For simplicity let us assume that now m1 = n2. The <u>sequential composition</u> of the two agents a1 and a2 is written by

a1·a2

where a4 = a1·a2 is an agent specification with n1 input lines and m2 output lines. It is defined by

agent a4 = **input** $\overline{x1}$, **output** $\overline{y2}$, $\exists$ $\overline{x2}$, $\overline{y1}$: $x2 = y1 \wedge H1 \wedge H2$ **end**

The <u>feedback</u> of an agent is defined by

C^i_j a

where the agent a is assumed to be given by the specification (with $1 \le i \le n$, $1 \le j \le m$)

agent a = **input** $s_1\ x_1, \ldots, s_n\ x_n$, **output** $r_1\ y_1, \ldots, r_m\ y_m$, H **end**

and the agent a5 = C^i_j a has n-1 input lines and m output lines. It is defined by the agent specification

agent a5 = **input** $s_1\ x_1, \ldots, s_{i-1}\ x_{i-1}, s_{i+1}\ x_{i+1}, \ldots, s_n\ x_n$,
output $r_1\ y_1, \ldots, r_m\ y_m$, $\exists\ s_i\ x_i$: $x_i = y_j \wedge H$ **end**

Note that the composition of algorithmic agents always leads to algorithmic agents again.

The agent store may be turned into an algorithmic agent (assuming algorithmic agents for switch and schedule) by:

agent store = **input stream data** d, **stream bool** b, **output stream data** r,

$\exists$ **stream data** s, z: (z, r) = switch(s, b), s = schedule(d, r, true & b) **end**

This agent store may be represented by the data flow diagram

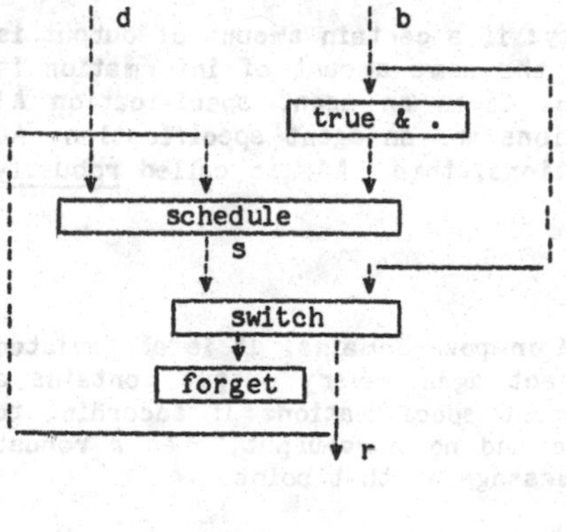

It can be also expressed by

$$C_1^2\ ((I \parallel I \parallel a) \cdot (\text{schedule} \parallel I) \cdot \text{switch} \cdot (\text{forget} \parallel I))$$

where I is the identity function and the agents forget and a are the trivial agents specified by

agent forget = **input stream data** d, true **end**
agent a = **input stream bool** b, **output stream bool** b1, b2,
b1 = false & b, b2 = b **end**

4.2. Correctness of Agents

During a design process of a distributed system a sequence or even a tree of agent specifications are produced. The final agent specification should be a program, i.e. an algorithmic agent, that must be correct w.r.t. the initial agent specification. In the most restrictive approach one could define correctness in the following way: An agent specification A1 is totally correct w.r.t. an agent specification A0, iff the set of functions denoted by A1 is identical to the set of functions denoted by A0. However, obviously this approach is too restrictive. More liberal notions of correctness have to be used and can be used much more flexible in the design process.

4.2.1. Partial Correctness

Partial correctness of concurrent, communicating systems is a so-called safety property: a program is partially correct (w.r.t. to some requirement specification) iff it never produces wrong results. This does not exclude that it diverges always immediately and therefore does not produce any results. Given an agent specification A1, that defines a set F1 of stream processing functions, and a requirement specification A0, that defines a set F0 of stream processing functions, A1 is called partially correct w.r.t. A0, iff

$$\forall f1 \in F1\ \exists f0 \in F0: f1 \sqsubseteq f0$$

This formula defines a preordering on sets that is also used in powerdomains. It is abbreviated by $F1 \sqsubseteq_E F0$. Accordingly for a partially correct agent every output stream is a prefix of an output stream included in the requirement specification.

4.2.2. Robust Correctness

Robust correctness is a typical liveness property: if a certain amount of output is guaranteed by the specification, then at least the same amount of information is to be guaranteed by a robust correct program. Given an agent specification A1 defining a set F1 of stream processing functions and an agent specification A0 defining a set F0 of stream processing functions, then A1 is called robustly correct w.r.t. A0, iff

$$\forall f1 \in F1\ \exists f0 \in F0: f0 \sqsubseteq f1$$

This is a preordering on sets that is also used in powerdomains. It is abbreviated by $F0 \sqsubseteq_M F1$. Accordingly for a robustly correct agent every output contains a prefix of some output included in the requirement specification. If according to the specification there may be some divergence and no moreoutput, then a robust correct implementation may include some error message at that point.

4.2.3. Correct Implementations

During the design process of a program for a given specification A0 typically particular design decisions are taken: specific algorithmic solutions are envisaged and certain nondeterministic alternatives are excluded. Accordingly for an agent specification A1 defining the set of functions F1 and the agent specification A0 defining the set of agents F0 the agent A1 is called a <u>(correct) implementation</u> of A0 if

$F1 \subseteq F0$

This definition reflects a very essential point of view of nondeterministic concurrent programs: a correct implementation may not include all nondeterministic possibilities but restrict itself to certain subsets, i.e. by choosing a particular scheduling strategy. Immediately one obtains the following lemma:

<u>Lemma</u>: If A1 is a correct implementation of A0, then

- A1 is partially correct w.r.t. A0
- A1 is robustly correct w.r.t. A0 □

Note that the converse statement does <u>not</u> hold: robust and partial correctness do not imply the correctness of an implementation.

4.3. Recursively Defined Algorithmic Agents

For recursively defined algorithmic agent specifications the semantic definitions are very liberal: the semantics is not restricted to least (defined) fixed points but considers the class of all fixed points. Trivially if one restricts the meaning of a recursively defined algorithmic agent by transition to a least (defined) fixed point semantics, then this represents a correct implementation of the given agent.

An algorithmic agent may trivially be translated into an applicative language for multiprocessing like AMPL (cf. [Broy 83b]). Then, however, a particular fixed point theory is used, while in the specification all fixed points are included. This restriction can be also expressed in our specification language by prefixing algorithmic specifications by **rec**. The system of agents

rec agent f_1 = ... E_1 **end** ... **rec agent** f_n = ... H_n **end**

has the meaning σ_{rec}, where

$\sigma_{rec} \in \{\sigma: \{f_1, ..., f_n\} \rightarrow P(AGENT)\}$

is the $\subseteq$-least ($\subseteq$ taken pointwise) function which is a $\sqsubseteq_M$-least and $\sqsubseteq_{EM}$-least fixed point of the equations E_i with

$\sigma_{rec} \models \forall\ \overline{x_i}\ \exists\ \overline{y}: y = f_i(x) \land H_i$

for $1 \leq i \leq n$. Here $\sqsubseteq_M$ is the preordering defined in the previous section. The preordering $\sqsubseteq_{EM}$ is defined by $\sqsubseteq_E \cap \sqsubseteq_M$.

The mathematical soundness and consistency of this definitions is implied by [Broy 84c].

Trivially, if one prefixes a given family DEF of agent specifications by rec, then one obtains a correct implementation of DEF: the least fixed points in the sense above provide a subset of the set of fixed points. Therefore the final step of a program development, viewing an algorithmic specification as a program, is trivially correct.

4.4. Transformations of Agent Specifications

The mathematical semantics of agent specifications and the definition of correct implementations clearly define what may be called a (partially, robustly) correct transformation step for a specification: the transformed specification must be a (partially, robustly) correct implementation of the initial program. On this basis a calculus for transformation rules can be developed. It comprises essentially the rules of inference of predicate logic, the algebraic axioms of streams and the basic rules of the used programming constructs.

Note, however, that due to the existence of nondeterministic agent specifications all laws of predicate logic where textual copying is involved have to be reconsidered and if necessary modified. Applying rules of inference where agent identifiers are copied may lead to incorrect implementations. Consider for instance the formula

$$y = f(x) \land z = y$$

which is essentially different from

$$y = f(x) \land z = f(x)$$

if f is a nondeterministic agent.

Now a simple example for the transformation of an agent specification is given. It essentially shows a development starting with recursive agents towards recursive stream equations.

Example: From recursion on functions to recursion on streams.
The agent sums0 computes the partial sums of a stream:

agent sums0 = **input stream nat** s, **output stream nat** r, r = help(s, 0) **end**

agent help = **input stream nat** s, **nat** n, **output stream nat** r,
r = n & help(**rest** s, n + **first** s) **end**

The agent sums0 or more precisely the agent help is defined recursively. It can be graphically represented by the diagram

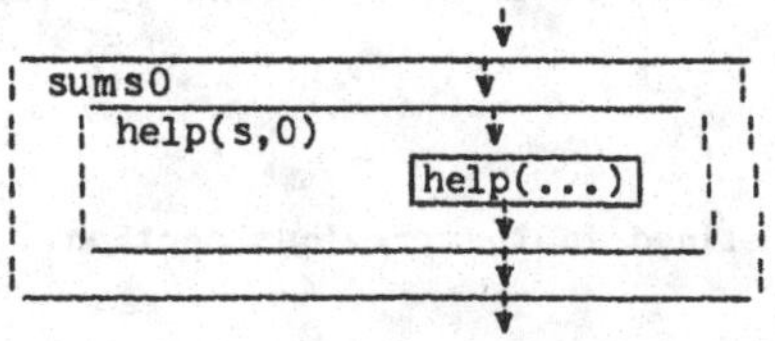

The agent sums0 is extensionally equivalent to the agent sums1:

agent sums1 = **input stream nat** s, **output stream nat** r, r = 0&addstream(r, s) **end**

The agent sums1 is defined by recursion on streams. It can be seen as defining a data flow diagram:

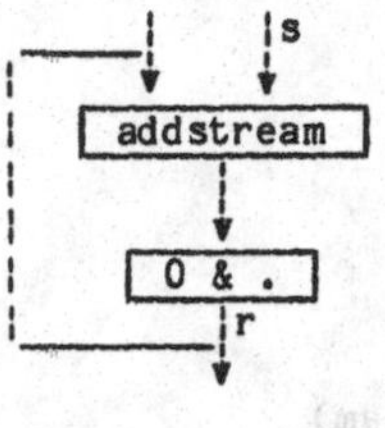

The agents sums0 and sums1 may also be represented by the term

C_1^1 (addstream · zero)

where zero is defined by

agent zero = **input stream nat** s, **output stream nat** r, r = 0 & s **end**

end of example

The example may be generalized to the following development rule: If an agent contains an equation with streams r, s of the form (with some given agent f):

(*) r = f(s)

and one can prove the formulas (with some given agent g):

(1) f(s) = g(f(s), s)

(2) ∀ s: r1 = g(r1, s) ∧ r2 = g(r2, s) ⇒ r1 = r2

then (*) can be replaced by the equation

r = g(r)

The condition (1) states the requirement that f(s) is a fixed point of g, the requirement (2) in addition requires that g has an unique fixed point.

<u>Example</u>: (Continued)

For verifying the applicability conditions (1) and (2) in our example above, one has to prove

(1) help(s, n) = n & addstream(help(s, n), s)

The condition (1) can be proved by structural induction on s:

(i) if s = <⊥>, then (and similarly for s = ε):

help(s, n) =
n & help(**rest** s, n + **first** s) =
n & ⊥ & help(...) =
n & <⊥> =
n & ⊥ & addstream(...) =
n & addstream(...)

(ii) if s = x & sn (where x ∈ ℕ) and (1) holds for sn:

help(s, n) =
n & help(sn, n + x) = (ind. hypoth.)
n & n+x & addstream(help(sn, n+x), sn) = (def addstream)
n & addstream(n & help(sn, n+x), x&sn) = (def help, s)
n & addstream(help(s, n), s)

For infinite s the continuity of the involved functions proves the result. Finally one has to prove that the equation

r = n & addstream(r, s)

has an unique fixed point for any s. Again this can be proved by induction on the length of s:

(i) if s = <⊥> then the equation implies

addstream(r, <⊥>) = <⊥>, and hence r = n & <⊥>

(ii) if for s the solution of the equation above is unique, then with x ∈ ℕ

r = n & addstream(r, x&s)

one gets

r = n & (**first** r + x) & addstream(**rest** r, s)

which now can be replaced by the equations

r = n & t, t = (n + x) & addstream(t, s)

According to our induction hypothesis t is uniquely determinated for finite streams s. For infinite streams the uniqueness of the fixed point of the equation follows from a continuity argument: the fixed point depends continuously on s.

end of example

Of course, the transformation rule above is only one simple example for transformations of agent specifications. It can be seen immediately that there is a rich class of transformation rules for agent specifications. Many of the transformation rules for sequential programs (cf. [Broy 84d]) can be adapted also for agent specifications.

4.5. Verification of Agent Specifications

Proving for two given agents A0, A1 that A1 is a (partially, robustly) correct implementation of A0 represents the classical case of verification. After resp. renaming of the input and output identifiers of A1, one basically has to show that the specifying predicate of A1 implies the specifying predicate of A0. Formally this again can be done by transforming A0 to A1 or more precisely deducing A0 from A1. Moreover one may also think of particular calculi for the verification of agents. For instance calculi dealing just with particular aspects of communicating agents may be developed such as calculi for partial or robust correctness.

4.6. Structured Programming with Agent Specifications

Trivially many different specifications can be given for one set of stream processing functions. Apart from different ways of writing equivalent predicates and renaming of local identifiers one could use different ways of structuring a specification.

A complete system of agent specifications consists of a family of agent specifications where for all occurring agent identifiers agent specifications are provided. The relation "agent a_i is used in agent a_j" then defines a quasiordering, and therefore induces a partial ordering on a quotient structure on the set of agents. This quasiordering is called the <u>macro structure of the system</u>.

Every agent itself contains some structure: the way how the specifying predicate is written. Especially for algorithmic agents this form of the predicate, the number of local identifiers is interesting, because it defines a finite network of communicating agents. This structure is called the <u>micro structure of (the components of) the system</u>.

Clearly the micro structure of an agent may again be seen as the macro structure of a (sub-)system with some further micro structure. Due to the use of recursive agents, such a transition from macro to micro views may be done up to infinity.

An important transformation rule that connects the micro structure with the macro structure is the unfold/fold rule for agents: Given an agent

agent a0 = **input** $\overline{x0}$, **output** $\overline{y0}$, E0 **end**

and an occurrence of the agent a0 in an equation

(*) t1 = a0(t2)

then one may "unfold" the agent specifications, i.e. one may replace the equation (*) by the formula

(**) $\exists\ \overline{x0}, \overline{y0}$: t1 = y0 $\wedge$ x0 = t2 $\wedge$ E0

Of course it is assumed that name clashes do not appear.

Even the reverse of the rule above is possible: the formula (**) may be replaced by the equation (*), provided (**) is occurring outside of the body of the agent a0. If a0 may be graphically represented by a finite data flow network

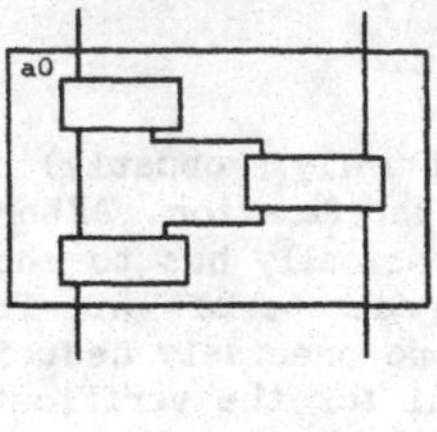

and an agent a uses a0:

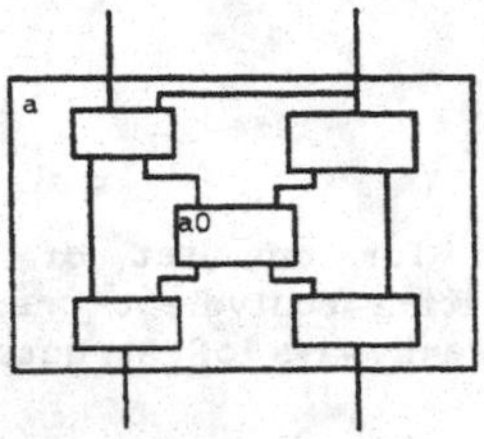

then by the rule one obtains for the agent a:

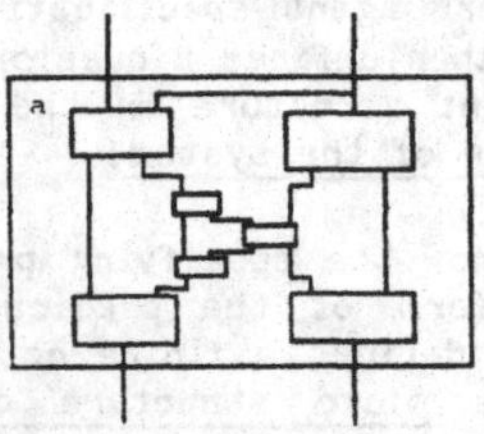

This way algorithmic agents are transformed into again algorithmic agents.

5. Application to Other Models for Concurrency

Specifications of concurrent communicating agents are based on the model of stream processing functions. This is essentially a model with implicit buffering message passing. Other important models are handshaking message passing or shared memory systems. An immediate question is, whether the specification method for agents can be applied for the other models, too.

The answer is rather simple: there exist semantic models both for shared memory as well as for handshake communication in terms of stream-processing functions (cf. [Broy 83, 84a, 84b]). Therefore one may simply write agent specifications, that specify stream-processing functions that represent the semantics of programs written in those programming languages according to those semantic models.

Example: Specification of producer and consumer

According to the semantic model for shared memory as given in [Broy 84a] components of a system of parallel programs working on some shared memory the access to which is protected by conditional critical regions may semantically be modelled by (sets of) functions mapping streams of states to streams of states and "rejects". For instance the specification of the producer/consumer problem may read as follows (here x, y, z, q denote programming variables):

```
agent producer = input stream state s, output stream(state ∪ {REJECT}) r,

      first r = (first s)[x0/x] ∧
      □ r, s: ○ r, s: first r = (first s)[next((first s)(x))/x,
                                app((first s)(q),(first s)(x))/q]     end

agent consumer = input stream state s, output stream (state ∪ {REJECT}) r,

      first r = (first s)[y0/y] ∧
      □ r, s: ○ r, s:
           ¬ isempty((first s )(q))  ⇒ first r = first s[top(q)/z,
                                                      pop(q)/q,
                                                      consume(y, top(q))/y],
           isempty((first s )(q))  ⇒ r = REJECT                 end
```

Here the notation s[d/x] is used as the notation for updating the state s by replacing the value for the identifier x by d.

Procedural programs that fulfil this specification then read:

```
producer:
  x := x0;
  while true do await true then q:= app(q, x) endwait;
               x := next(x)                                  od

consumer:
  y := y0;
  while true do await ¬ isempty(q) then q, z := pop(q), top(q) endwait;
               y := consume(y, z)                            od
```

Here the specification does not look very convincing: It is textually larger and (since less familiar) seems harder to understand than the programs. However a tuned notation for the specification might change this view. Writing x\` for the input value of a program variable (i.e. for (**first** s)(x)) and x' for its output value (i.e. for (**first** r)(x)) and dropping r, s from temporal formulas one obtains: sm-agents (shared memory agents). It is just consequent to add the information which variables are assumed to be shared:

```
sm-agent producer ≡ shared q,
      x' = x0 ∧ (□  ○  (x' = next(x`) ∧ q' = app(q`, x'))      end

sm-agent consumer ≡ shared q,
      y' = y0 ∧   □   ○ ((¬ isempty(q`) ⇒ (z' = top(q`) ∧
                                             q' = pop(q`) ∧
                                             y' = consume(y`, z')) ∧
                       (isempty(q`)  ⇒ REJECT))          end
```

end of example

In a similar way one may develop tuned notations for CSP-like programs or CCS-like programs by using the stream-based semantic models for CSP/CCS as given in [Broy 83a, 84a].

6. Concluding Remarks

The approach of agent specifications reflects an attempt to bring together distinct research directions: algebraic (equational) specifications, denotational models for concurrent, communicating systems, temporal logic, functional programming and program development by transformations.

The presented approach to the specification of concurrent communicating systems includes a number of design decisions that should be shortly recapitulated in the end and some justification should be given, too.

The chosen semantic model are sets of continuous functions mapping tuples of streams to tuples of streams. Why tuples of streams are considered is obvious. Considering sets of functions instead of relations or set-valued functions might be less clear (for a more detailed justification see for instance [Broy 81]). It is done for avoiding some subtle problems in connection with definitions of streams by fixed points over nondeterministic functionals. Whether the restriction to continuous functions is actually always appropriate, however, seems less clear.

The chosen logical framework is basically algebraic: All properties are specified by equations. The inclusion of temporal logic can just be seen as a notational variant being appropriate since sequence-like structures like streams are used. May be it is important to point out that the logic is two-valued in spite of the existence of partial functions (in the disguise of total functions with $\perp$ as result) and of nondeterministic agents.

The chosen concept of validity of specifications represents a very subtle point. The validity of formulas containing nondeterministic terms can be defined in several ways (by several modalities). Our choice represents a compromise between universal validity (a formula has to be valid for all nondetermistic alternatives) and existential validity (a formula has to be valid for only one nondeterministic alternative). It is justified by the interpretation of formulas as specifying networks of nondeterministic agents.

May be that the most important properties of a specification method are not only the underlying theoretical concepts but more pragmatic issues such as readability, tractability, support for structuring, possibilities of visual aids and machine support. In this light the presented approach seems rather attractive. The essential principle that every family of agents can be also considered (and formally specified) as a network and vice versa, the integration of pure specification constructs and algorithmic views into one framework let expect that the given approach can be further developed into a flexible and practically helpful tool.

What has been presented in the previous section rather can be seen as a first attempt to develop a modular specification and design method for communicating concurrent systems than a fully worked out methodology. Much remains to be done until such an approach actually will work practically.

References

[Barringer et al. 85]
H. Barringer, R. Kuiper, A. Pnueli: A Compositional Temporal Approach to a CSP-like Language. (Unpublished manuscript)

[Broy 81]
M. Broy: A Fixed Point Approach to Applicative Multiprogramming. In: M. Broy, G. Schmidt (eds.): Theoretical Foundations of Programming Methodology, Reidel Publ. Comp. 1982, 565-623

[Broy 83a]
M. Broy: Denotational Semantics of Communicating Processes based on a Language for Applicative Multiprogramming. IPL 17:1 1983, 29-38

[Broy 83b]
M. Broy: Fixed Point Theory for Communication and Concurrency. In: D. Bjørner (ed.): IFIP TC2 Working Conference on Formal Description of Programming Concepts II, Garmisch, June 1982, Amsterdam-New York-Oxford: North Holland Publ. Company 1983, 125-147

[Broy 83c]
M. Broy: Applicative Real Time Programming. In: R.E.A. Mason (ed.): Information Processing 83, 259-264

[Broy 84a]
M. Broy: Semantics of Communicating Processes. Information and Control (to appear)

[Broy 84b]
M. Broy: Denotational Semantics of Concurrent Programs with Shared Memory. In: M. Fontet, K. Mehlhorn (eds.): STACS 84, Lecture Notes in Computer Science 166, Berlin-Heidelberg-New York: Springer 1984, 163-173

[Broy 84c]
M. Broy: On the Herbrand Kleene Universe of Nondeterministic Computations. In: M.P. Chytil, V. Koubeck (eds.): Mathematical Foundations of Computer Science 1984. Lecture Notes in Computer Science 176, Berlin-Heidelberg-New York-Tokyo: Springer 1984, 214-222

[Broy 84d]
M. Broy: Algebraic Methods for Program Construction: The Project CIP. In: P. Pepper (ed): Program Transformation and Programming Environments. NATO ASI Series. Series F: 8. Berlin-Heidelberg-New York-Tokyo: Springer 1984, 199-222

[Broy, Wirsing 82]
M. Broy, M. Wirsing: Partial Abstract Types. Acta Informatica 18, 1982, 47-64

[Dennis 74]
J.B. Dennis: First Version of a Data Flow Procedure Language. In B. Robinet (ed.): Colloque sur la Progammation, Lecture Notes in Computer Science 19, Berlin-Heidelberg-New York: Springer 1974, 362-367

[Hehner 84]
E.C.R. Hehner: Predicative Programming Part I+II. CACM 27:2 (1984) 134-151

[Hoare et al. 81]
C.A.R. Hoare, S.D. Brookes, A.W. Roscoe: A Theory of Communicating Sequential Processes. Oxford University Computing Laboratory, Programming Research Group, Technical Monograph PRG-21, Oxford 1981

[Kahn, MacQueen 77]
G. Kahn, D. MacQueen 77: Coroutines and Networks of Parallel Processes. In: Proc. of the IFIP Congress 77, Amsterdam:North-Holland 1977, 994-998

[Keller 78]
R. M. Keller: Denotational Models for Parallel Programs with Indeterminate Operators. In: E. J. Neuhold (ed.): Formal Description of Programming Concepts. Amsterdam: North-Holland 1978, 337-366

[Milner 80a]
R. Milner: A Calculus of Communicating Systems. Lecture Notes in Computer Science 92, Berlin-Heidelberg-New York: Springer, 1980

[Plotkin 76]
G. Plotkin: A Powerdomain Construction. SIAM J. Computing 5, 1976, 452-486

[Pnueli 77]
A. Pnueli: The Temporal Logic of Programs. Proc 18th FOCS, Providence, 1977, 46-57

[Pratt 78]
V.R. Pratt: Process Logic. 6th Popl, 1979, 83-100

[Smyth 78]
M. Smyth: Power Domains. J. CSS 16, 1978, 23-36

[Winskel 82]
G. Winskel: Event Structure Semantics of CCS and Related Languages. In: M. Nielsen, E. M. Schmidt (eds.): ICALP 82 Lecture Notes in Computer Science 140, Berlin-Heidelberg-New York 1982, 561-576

SPECIFICATION LANGUAGES FOR DISTRIBUTED SYSTEMS

by Pierpaolo Degano and Ugo Montanari

Dipartimento di Informatica, University of Pisa

Corso Italia 40, I-56100 Pisa, Italy

Abstract

Requirements of specification languages for distributed systems are considered, and a two level approach based on a kernel metalanguage and many application-oriented extensions is advocated. The method is applied to some models developed by the authors, organized in a tree-like refinement structure.

1. Introduction

Emphasis on the specification phase within the software life cycle has been suggested as a remedy against the many inconveniencies of presently available programming methodology. A more structured approach and a complete documentation of the design decisions taken in all the phases from requirements to coding should enable an easy modification of the resulting software product for maintenance and re-use. It has been also suggested that the main loop of the software life cycle be closed on an executable version of the program, completely specified but still not optimized, rather similar to a detailed specification /BCG83/. Fully optimized versions should be derivable either manually or semiautomatically but should not be the "main documents" of the software product.

Improvements on programming methodology in the specification phase demand suitable specification languages. They should be formally defined, executable, and easy to use. In this field close collaboration between computer scientists on one side and software engineers on the other is badly needed. The situation is even more demanding in the case of concurrent distributed systems, where many theoretical problems are still open and where our intuition is often inadequate. A number of recent workshops /Spec79, 83, 84/ are evidence of the interest in specification languages for sequential and concurrent systems.

This work has been partially supported by CNR - Progetto Finalizzato Informatica, Obiettivo Cnet.

2. Requirements for a specification language.

In a specification language we can distinguish two aspects. The first is related to its semantic definition and the second refers to its flexiblity and usability in a number of practical situations. The two aspects are somewhat conflicting, since the former leads to elementary and orthogonal constructs, while the latter tends to require the contemporary presence of many, overlapping constructs and languages, specialized for levels of abstraction and fields of application.

From a practical point of view, the presence of several languages in the specification of a large system will probably be a fact of life and it will be necessary to cope with it. In principle, it might be enough to formalise all the specification languages using "the" mathematical language. However, this approach might lead to serious practical problems, since if the specification languages employed can be interfaced only at a very primitive level (e.g. set theory) it will not be easy to reason and prove properties about the programs so specified. Furthermore, it will be almost impossible to efficiently implement all the languages used to specify the system in order to obtain early prototypes.

The above conflict can be solved by means of a two-level approach. First, a basic metalanguage should be introduced containing all the concepts needed, organized in an elementary way but well studied and easy to understand; second, several specification languages should be defined in terms of the metalanguage using few, simple extension mechanisms. A specification might become executable by providing an implementation of the metalanguage and interpreters (translators) for every language towards the metalanguage.

A good example of the above two-level approach is the Pebble specification language /BuLa84/, where the kernel is a version of typed lambda calculus and the mechanisms for modularization and data abstractions are provided as applications. The semantics of Pebble is operational, and is given in the style of SOS /Plot83/ using labelled transition systems.

Since we are interested in specification languages for concurrent, distributed systems, more concepts must be embedded in the metalanguage. Necessary constructs include at least primitives for describing nondeterminism, concurrency, synchronization and communication. An important step towards a metalanguage for concurrency is Milner's asynchronous and synchronous CCS /Miln80, 83/. It has rather few constructs, a rich and elegant theory and an operational semantics defined in terms of labelled transition systems.

A construct for parallel composition is available in CCS. However, its meaning can be expressed in terms of other operations, and thus it is not primitive in this sense. In fact, CCS, as well as all the models based on interleaving, describes the fact that a set of events may occur concurrently (independently from each other) by saying that they may occur in any order. In this way a total ordering among the possibly spatially separated and causally independent events is imposed.

Although this level of detail is adequate for many applications, according to a number of researchers it is insufficient to describe all those aspects of distributed, concurrent systems that have practical interest, e.g. fairness. Models have been proposed which use partial orderings to explicitely describe the fact that events may take place concurrently. Among these we mention the pioneering work on Petri nets /Bram83/ and Cosy by Lauer et al. /LTS79/. Also the authors have followed this approach defining models which will be used in the sequel as a case study.

A major motivation for formally specifying a system is to be able to prove properties about it. Providing a satisfactory proof system is not a simple matter, especially for an operationally defined metalanguage. A widely proposed approach takes a temporal logic as a starting point /Roev84/.

Once a satisfactory metalanguage has been designed, among the most needed extensions we mention those providing the ability of structuring and composing pieces of specifications. The features required concern parametrization, modularity and abstraction.

The use of a specification language for a sizable system is greatly improved by the availability of suitable tools. Many of the considerations valid for standard programming environments also apply to specification language tools. In particular, the syntax- or semantic-driven techniques used for adapting generic tools (like editors, type checkers, interpreters, debuggers, etc.) to a particular language, should be more convenient in a specification language environment, due to the hopefully simpler structure of specification languages themselves.

We already mentioned the convenience for a specification language to be executable in a reasonably efficient way. Of course the use of high parallelism and/or special purpose machine architectures may be of great help. However, we believe that the practical possibility of executing specifications, e.g. for early prototyping, must be a specific concern in designing them, since otherwise the inefficiencies can easily, and hopelessly, increase exponentially.

Finally, we mention the so-called human engineering issue. If a specification language has to be used at all in practice, it must require only a limited knowledge by the programmer of the deep theoretical issues involved in its definition. It must also be intuitively appealing and should use all the technically available expedients (e.g. sophisticated graphics) for achieving an easy interaction with the user (see for instance the documents of the ESPRIT project Graspin /GRAS84/).

3. A basic model and its refinements

In this section we follow the methodological guidelines surveyed in the previous section by presenting in some detail several models developed by the authors. All formalisms consider concurrency a basic, irreducible concept and are based on partial orderings. The models are organized in a tree-like refinement structure, starting from a basic model and adding to it independent features to describe different aspects of distributed concurrent systems. These features are meant as a kernel of a more elaborated specification language, which should combine them in a usable way.

3.1. Concurrent histories

In this section we introduce our basic model, which is intended as Level 1. of our structured presentation.

We give an informal introduction to the notion of concurrent history and to the semantics of a set Z of atomic histories. A detailed definition can be found in /DeMo84a, b/.

Let A be a countable set of observable actions and let E be a countable set of process types containing an element 0 called termination. Sets A and E are disjoint.

A concurrent history h in H_{fin} is a triple $h=(S,l,\leq)$ where:

S is a finite set of subsystems;

l is a labelling function

$l:S \rightarrow A \cup E$ and

$\leq$ is a partial ordering relation on S.

The subsystems with labels in A are called events, while those with labels in E are called process states or simply processes.

We require that events never be minimal nor maximal, the processes always be minimal or maximal, but not both, elements of $\leq$. Processes which are minimal are called heads, and processes which are maximal are called tails. Thus processes are partitioned into heads and tails. Two histories are called disjoint if their sets of subsystems are disjoint.

Two histories $h_1=(S_1,l_1,\leq_1)$ and $h_2=(S_2,l_2,\leq_2)$ are isomorphic iff there is a bijective mapping

$g:S_1 \rightarrow S_2$ such that

$l_1(s)=l_2(g(s))$ and

$s_1 \leq_1 s_2$ iff $g(s_1) \leq_2 g(s_2)$.

We define an associative nondeterministic partial replacement operation on histories. Given two disjoint histories h_1, h_2 and a history h, we write h_1 before h_2 gives h iff h can be obtained by the following procedure. A possibly empty subset S^h_2 of the head processes of h_2 is matched against a subset S^t_1 of the tail processes of h_1, and corresponding processes are identified. Of course two processes can match only if their labels are identical. If the subset S^h_2 is the whole set of the head processes of S_2, the operation is called full replacement or simply replacement, and set S^t_1 is called rewritable.

The relation $\leq_1 \cup \leq_2$ is then made transitively closed and the processes in the matching set $S^h_2=S^t_1$ are erased. Note that we have

$S=(S_1-S^h_1)\cup(S_2-S^t_2)$.

In Fig. 1 we see an example of replacement. Here a,b,c are in A, and E_1, E_2 are in E. Partial orderings are depicted through their Hasse diagrams, growing downwards. Processes (events) are represented as boxes (circles).

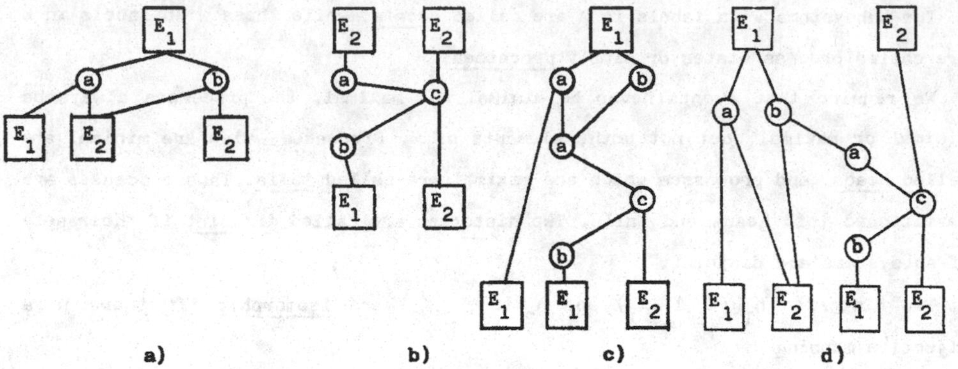

<u>Fig. 1.</u> Four concurrent histories h_1 (in a)), h_2 (in b)), h_3 (in c)) and h_4 (in d)) such that h_1 <u>before</u> h_2 <u>gives</u> h_3, and h_1 <u>before</u> h_2 <u>gives</u> h_4.

A history is <u>atomic</u> if either

i) there are no events and each head is smaller in the partial order than all the tails, or

ii) there is exactly one event greater than all heads and smaller than all tails.

An atomic history represents a single synchronization, either unobservable or observable.

We introduce a linear representation (up to isomorphism) for atomic histories.

i) $M_1 \dashrightarrow M_2$

ii) $M_1 \overset{a}{\dashrightarrow} M_2$

where M_1, M_2 are multisets of process types and a is an observable action.

Let Z be a set of disjoint histories.

A <u>computation on Z</u> is a finite or infinite sequence $D = \{h_i\} = (h_0, h_1 \ldots)$ such that

$$h_i \text{ \underline{before} } r_i \text{ \underline{gives} } h_{i+1} \qquad i=0,1,\ldots$$

where h_0, r_i $(i=0,1,\ldots)$ are disjoint atomic histories isomorphic to histories in Z.

As an example consider a computation with

$h_0 = E_1 \longrightarrow E_3, E_4$

$r_0 = E_3 \xrightarrow{a} E_1, E_5$

$r_1 = E_4 \xrightarrow{b} E_6, E_2$

$r_2 = E_5, E_6 \longrightarrow E_2$

where h_3 is the history in Fig 1a). Notice that all replacements are full. Note also that histories r_0 and r_1 might be interchanged obtaining the same h_3.

The histories belonging to a computation on Z are called <u>derivable from Z</u>. Furthermore, the <u>result</u> of a finite computation is its last element, if it does contain no rewritable set. The <u>result</u> of an infinite computation is its limit, if any, in a suitable metric space. Actually, in /DeMo84a/ (where replacement is required to be full) four distances on finite histories are defined, and the limits are obtained through standard topological completions. Remarkably enough, the non terminating computations converging in the four resulting complete metric spaces enjoy interesting liveness properties.

The four properties are: vitality (every running process will eventually produce an observable event), global fairness (a synchronizable set of processes will eventually run), local fairness (a process which is repeatedly ready to run, possibly with different partners, will eventually run), partial deadlock freedom (every non-terminated process will eventually run).

Moreover, the limits in the metric spaces are directly characterized in terms of their structural properties.

The proposed approach proves to be fruitful: a nondeterministic universal scheduler is defined which is capable of generating all and only computations being convergent in a given metric. This scheduler can be used in the four cases above, thus keeping only those computations which have the desired liveness property. The metric is a parameter of the scheduler, which is thus independent of the particular liveness property under consideration.

Our notion of concurrent history expresses in an abstract fashion the idea of concurrent computation. In fact, events not related by the partial ordering ≤ are meant to be concurrent. On the other hand, the above given notion of computation is purely sequential, and, since every computation step generates at most one event, a computation induces a well-founded total ordering on the events of its last element or of its result. We call it <u>generation ordering</u>.

The following theorem given in /DeMo84b/ links the sequential and concurrent notions of computation.

Theorem 3.1 The generation orderings induced by all computations on Z having the same history h=(S,1,≤) as last element or as result, are exactly those well-founded total orderings compatible with (i.e. larger than or equal to, in the set theoretical sense) the partial ordering ≤ of h.

In /DeMo84b/ a construction is presented for generating a most simplified Labelled Event Structure (LES) starting from a set Z of atomic histories. This LES defines the semantics of Z. Labelled Event Structures are models of nondeterministic concurrent computations described in the literature /CFM82/, /Wins82/. Our notion of simplification is based on an abstraction homomorphism having the property that, given a LES, a unique most simplified homomorphic LES always exists.

3.2. Petri nets

The notion of computation on a set Z of atomic histories defined in the previous section is a general language-independent framework for describing behaviours of concurrent programs. The set Z can be defined independently. If Z is finite, it can be given explicitly. We call this Level 1.1.. In this case a set Z is equivalent to a transition Petri net /JaVa79/ plus a partial function mapping the transitions of the net into observable actions. The places of the net correspond to the process types, while every transition is associated to an atomic history in Z. More precisely, immediate antecedent (successor) places of the transition correspond to heads (tails), and an event exists iff the partial function above is defined. In fig. 2 we see an example of this equivalence.

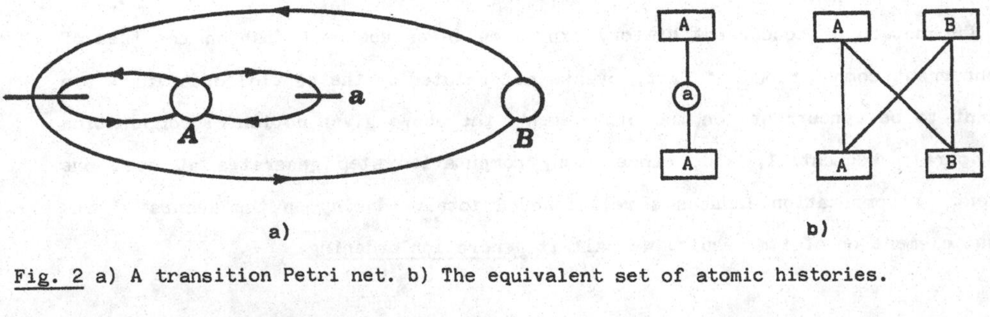

Fig. 2 a) A transition Petri net. b) The equivalent set of atomic histories.

3.3. Synchronization

So far, an atomic history involving many processes is understood as a single, indivisible move. On the other hand, it can often be conveniently seen as the composition, through a synchronization mechanism, of a separate move for each process involved. This consideration brings us to Level 1.2.. For the sake of symplicity, from now on we consider only atomic histories of type ii). Each possible move is described by a labelled production. The left member of a production contains a single process, and the label is called a communication protocol. The synchronization mechanism is represented by an associative commutative partial function f, mapping, if defined, any multiset of communication protocols into an observable action. Function f, called synchronization function, makes it possible to specify which productions can be composed in a single transition rule.

A production is a triple

$$A \xrightarrow{p} M$$

where A is a process type, M is a multiset of process types and p is a protocol.

Given a set W of productions and a synchronization function f, the atomic histories in Z are derivable by the following inference rule

$$\frac{A_1 \xrightarrow{p_1} M_1, \ldots, A_n \xrightarrow{p_n} M_n}{A_1, \ldots, A_n \xrightarrow{f(p_1, \ldots, p_n)} M_1 \cup \ldots \cup M_n}$$

In Fig. 3 we see a computation step where the atomic history used is generated by the above rule.

An example of a synchronization mechanism described in the literature which fits our present schema is that used for synchronization trees /Miln80/. Milner, however, does not distinguish between observable actions and communication protocols, since the observer himself is considered to be a process. Furthermore, the communication mechanism is intended to be based on message passing, and thus actions come in pairs like $a\bar{a}$, where an element represents the envoy of the message and the other its reception.

$A \xrightarrow{a'} A$

$A \xrightarrow{b'} A$

$B \xrightarrow{b''} B$

a)

$f(a')=a$

$f(b',b'')=b$

b)

$$\frac{A \xrightarrow{b'} A\ ,\ B \xrightarrow{b''} B}{AB \xrightarrow{b} AB}$$

c)

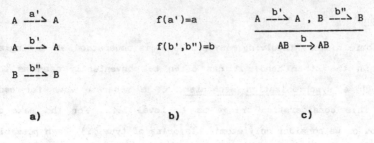

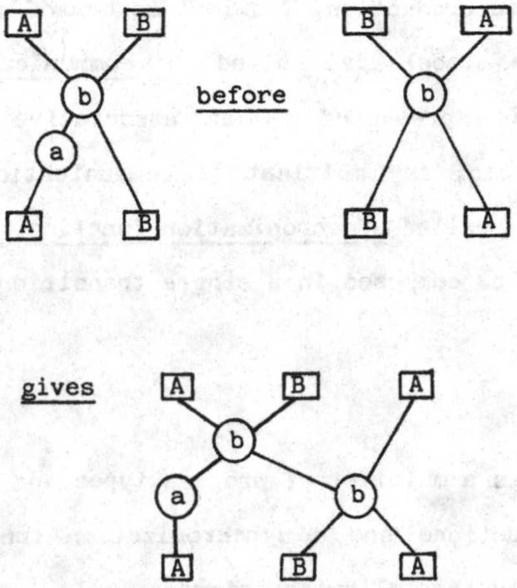

d)

Fig. 3 a) Three productions (notice the half-arrow).

b) The synchronization function.

c) The derivation of an atomic history.

d) A step in a computation.

To adapt synchronization trees to our framework, we assume that the observer can exchange a set of protocols $a'_1, \bar{a}'_1, a'_2, \bar{a}'_2, \ldots$. Thus we have,

$f(a'_1)=a_1,\ f(\bar{a}'_1)=\bar{a}_1, \ldots$

$f(a'_1,\bar{a}'_1)=\tau\ ,\ f(a'_2,\bar{a}'_2)=\tau\ , \ldots$

where τ represents Milner's unobservable action.

Conversely, the "intersection" synchronization mechanism proposed by Hoare /BHR84/ does not fit completely. This mechanism forces all processes which are present in the system to agree on the same protocol, i.e.

$f_n(a',\ldots,a')=a,\ f_n(b',\ldots,b')=b,\ldots$

where the synchronization function f_n applies iff there are exactly n processes in the system. Notice that this synchronization mechanism requires access to the global state of the system, and we exclude this possibility.

3.4. CCS

In the previous section we were able to define our set Z of atomic histories in terms of a given set of productions and a synchronization function. It is also possible to define directly Z using suitable inference rules, obtaining Level 1.3.. Here we define Z_{CCS} for Milner's CCS. It is possible to prove that there is a direct correspondence between atomic histories in Z_{CCS} and CCS derivation steps (/DDM84/). Recall that the concrete syntax of pure CCS terms is as follows.

$E ::= x \mid NIL \mid \mu E \mid E\backslash\alpha \mid E[\phi] \mid E+E \mid E|E \mid rec\ x.E.$

We introduce a notion of move

$$I_1 \xrightarrow[I_3]{L} I_2$$

which generalizes Milner derivation relation

$$E_1 \xrightarrow{\mu} E_2$$

and which is essentially an atomic history, according to the definition given in the Section 3.1. The elements I_1, I_2 and I_3 in a move are finite sets of grapes, i.e. of terms defined as follows.

$G ::= E \mid id|G \mid G|id \mid G\backslash\alpha \mid G\ [\phi]$

where E denotes a CCS term and α and $[\phi]$ have the same meaning as in E.

Intuitively speaking, a grape represents a suitable subterm of a CCS term, together with its access path. Given a move $I_1 \xrightarrow[I_3]{L} I_2$, the grapes in I_1, I_2, I_3 are called head, tail and idle grapes, respectively.

A CCS term can be decomposed into a set of grapes by the function dec, here defined by structural induction.

$dec(NIL) = \{NIL\}$

$dec(x) = \{x\}$

$dec(\mu E) = \{\mu E\}$

$dec(E\backslash\alpha) = dec(E)\backslash\alpha$

$dec(E[\phi]) = dec(E)[\phi]$

$dec(E_1+E_2) = \{E_1+E_2\}$

$dec(E_1|E_2) = dec(E_1)|id \cup id|dec(E_2)$

$dec(rec\ x.E) = \{rec\ x.E\}$

Here the application of a syntactic constructor to a set of grapes is defined as applying the constructor to all grapes in the set, e.g.

$I\backslash\alpha = \{G\backslash\alpha \mid G \text{ in } I\}$.

Notice that the decomposition stops when an action, a sum or a recursion in encountered. We have for instance:

(3.1) $dec((((rec\ x.\alpha x+\beta x)|rec\ x.\alpha x+\gamma x)|rec\ x.\bar{\alpha}x)\backslash\alpha) =$
$\{(((rec\ x.\alpha x+\beta x)|id)|id)\backslash\alpha,\ ((id|rec\ x.\alpha x+\gamma x)|id)\backslash\alpha,\ (id|rec\ x.\bar{\alpha}x)\backslash\alpha\}$.

It is easy to see that function dec is injective, and thus full information about E is contained in dec(E). It is not surjective instead, and a set of grapes which is the decomposition of a CCS term is called <u>complete</u>.

Going back to our definition, the last element of the move relation is a <u>synchronization term</u> L, namely a term defined as follows.

$L ::= \mu \mid id|L \mid L|id \mid L|L \mid L\backslash\alpha \mid L[\phi]$

where μ, α and ϕ have the same meaning as in CCS terms.

Intuitively, L brings information about the internal communication of the move. We define by structural induction on L a function syn, which embodies the synchronization algebra of CCS.

$syn(\mu) = \mu$

$syn(id|L) = syn(L|id) = syn(L)$

$syn(L_1|L_2) =$ <u>if</u> $syn(L_1) = \overline{syn(L_2)}$ <u>then</u> τ <u>else</u> synerror

$syn(L\backslash\alpha) =$ <u>if</u> $syn(L)=\alpha$ <u>or</u> $syn(L)=\bar{\alpha}$ <u>then</u> synerror <u>else</u> syn(L)

$syn(L[\phi]) =$ <u>if</u> syn(L)=synerror <u>then</u> synerror <u>else</u> $\phi(syn(L))$

Notice that if a subterm of L evaluates to synerror, the whole L evaluates to synerror.

We are now ready to define our move relation using axioms and inference rules in direct correspondence with those of Milner derivation relation.

Let

$$I_1 \xrightarrow[I_2]{L} I_3$$

be the least relation which satisfies

<u>Act</u>. $\{\mu E\} \xrightarrow[\emptyset]{\mu} \mathrm{dec}(E)$

<u>Res</u>. $I_1 \xrightarrow[I_3]{L} I_2$ <u>implies</u> $I_1\backslash\alpha \xrightarrow[I_3\backslash\alpha]{L\backslash\alpha} I_2\backslash\alpha$

<u>Rel</u>. $I_1 \xrightarrow[I_3]{L} I_2$ <u>implies</u> $I_1[\phi] \xrightarrow[I_3[\phi]]{L[\phi]} I_2[\phi]$

<u>Sum</u>.1) $I_1 \xrightarrow[I_3]{L} I_2$ <u>implies</u> $\{E_1+E\} \xrightarrow[\emptyset]{L} I_2 \cup I_3$

2) <u>and</u> $\{E+E_1\} \xrightarrow[\emptyset]{L} I_2 \cup I_3$

<u>where</u> $\mathrm{dec}(E_1) = I_1 \cup I_3$

<u>Com</u>.1) $I_1 \xrightarrow[I_3]{L} I_2$ <u>implies</u> $I_1|id \xrightarrow[I_3|id \cup id|\mathrm{dec}(E)]{L} I_2|id$

2) <u>and</u> $id|I_1 \xrightarrow[id|I_3 \cup \mathrm{dec}(E)|id]{L} id|I_2$

3) $I_1 \xrightarrow[I_3]{L} I_2$ <u>and</u>

$I'_1 \xrightarrow[I'_3]{L'} I'_2$ <u>implies</u> $I_1|id \cup id|I'_1 \xrightarrow[I_3|id \cup id|I'_3]{L|L'} I_2|id \cup id|I'_2$

<u>Rec</u>. $I_1 \xrightarrow[I_3]{L} I_2$ <u>and</u>

$I_1 \cup I_3 = \mathrm{dec}(E_1[\mathrm{rec}\ x.E_1/x])$ <u>implies</u> $\{\mathrm{rec}\ x.E_1\} \xrightarrow[\emptyset]{L} I_2 \cup I_3$

The intuitive meaning of the move relation

$$I_1 \xrightarrow[I_3]{L} I_2$$

is that the grapes in I_1 become the grapes in I_2 with internal communication L, while the grapes in I_3 stay idle. It is easy to see by induction that $I_1 \cup I_3$ and $I_2 \cup I_3$ are complete sets of grapes.

Therefore, given a move we can syntetize out of it a <u>head</u> term E_1 and a <u>tail</u> term E_2 such that $dec(E_1) = I_1 \cup I_3$ and $dec(E_2) = I_2 \cup I_3$. We can now shortly comment about our axiom and rules.

In axiom <u>Act</u>., a single grape is rewritten as a set of grapes, since the firing of the action makes explicit the (possible) parallelism of E. A move generated by rule <u>Sum</u>.1) can be understood as consisting of two steps. Starting from the singleton $\{E_1+E\}$ a first step discards alternative E and decomposes E_1 into $I_1 \cup I_3$; a second step (the condition of the inference rule) rewrites I_1 as I_2 and leaves I_3 idle. The net effect of the two steps, however, is to rewrite the singleton $\{E_1+E\}$ into the set $I_2 \cup I_3$, with no idle grape.

Rule <u>Com</u>.1) (<u>Com</u>.2)) can be read as follows. If we have a move where I_1 is rewritten as I_2 and I_3 stays idle, we can add in parallel to I_1, I_2 and I_3 to the right (to the left), a complete set of grapes dec(E), which stay idle. <u>Com</u>.3) is the synchronization rule: note that encoding the composition of L and L' into L|L' may permit to use different synchronization algebras.

As an example we show the proof of both a move and the corresponding Milner derivation.

$\{\alpha E'\} \xrightarrow[\emptyset]{\alpha} \{E'\}$, where $E' = rec\ x.\alpha x+\beta x$, by <u>Act</u>.;

$\{\alpha E'+\beta E'\} \xrightarrow[\emptyset]{\alpha} \{E'\}$, by <u>Sum</u>. 1);

$\{E'\} \xrightarrow[\emptyset]{\alpha} \{E'\}$, by <u>Rec</u>.;

i) $\{E'|id\} \xrightarrow[\{id|E''\}]{\alpha|id} \{E'|id\}$, where $E'' = rec\ x.\alpha x+\gamma x$, by <u>Com</u>.1);

$\{E\} \xrightarrow[\emptyset]{\bar{\alpha}} \{E\}$, where $E = rec\ x.\bar{\alpha}x$, by <u>Act</u>. and <u>Rec</u>.;

$\{(E'|id)|id,\ id|E\} \xrightarrow[\{(id|E'')|id\}]{(\alpha|id)|\bar{\alpha}} \{(E'|id)|id,\ id|E\}$, by i) and <u>Com</u>.3)

(3.2) $\{((E'|id)|id)\backslash\alpha, (id|E)\backslash\alpha\} \xrightarrow[\{((id|E'')|id)\backslash\alpha\}]{((\alpha|id)|\bar{\alpha})\backslash\alpha} \{((E'|id)|id)\backslash\alpha, (id|E)\backslash\alpha\}$.

Let us follow now the corresponding derivation in pure CCS.

$\alpha E' \xrightarrow{\alpha} E'$ where E'=rec $x.\alpha x+\beta x$

$\alpha E'+\beta E' \xrightarrow{\alpha} E'$

$E' \xrightarrow{\alpha} E'$

i) $E'|E'' \xrightarrow{\alpha} E'|E''$ where E''=rec $x.\alpha x+\gamma x$

$E \xrightarrow{\bar{\alpha}} E$ where E=rec $x.\bar{\alpha}x$

$(E'|E'')|E \xrightarrow{\tau} (E'|E'')|E$

$((E'|E'')|E)\backslash\alpha \xrightarrow{\tau} ((E'|E'')|E)\backslash\alpha$.

Note that if we write (3.2) as $I_1 \xrightarrow[I_3]{L} I_2$ and the latter derivation as $E_1 \xrightarrow{\tau} E_2$ we have $dec(E_1)=dec(E_2)=I_1 \cup I_3=I_2 \cup I_3$ by Example 3.1 and, according to the definition, $syn(((\alpha|id)|\bar{\alpha})\backslash\alpha)=\tau$.

In general it is possible to prove /DDM84/ that if $E_1 \xrightarrow{a} E_2$ then $I_1 \xrightarrow[I_3]{L} I_2$ with $syn(L)=a$, $dec(E_1)=I_1 \cup I_3$, $dec(E_2)=I_2 \cup I_3$, and viceversa.

We now translate moves to histories. Given a move

$$m = I_1 \xrightarrow[I_3]{L} I_2$$

with $syn(L)\neq synerror$, its head (tail) grapes in I_1 (I_2) can be interpreted as head (tail) processes of an atomic history, having one event labelled by $syn(L)$. Idle grapes are simply forgotten. However, since the sets of head and tail processes of a history are by definition non intersecting, while I_1 and I_2 may have non empty intersection, it is necessary to make "new" copies of I_1 and I_2. The same construction takes care of the fact that all histories in a set Z of atomic histories must be disjoint.

In Fig. 4 we see a picture of the history corresponding to the move in (3.2).

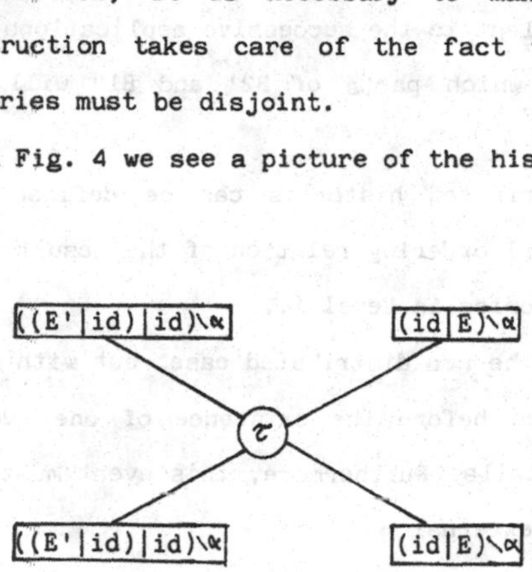

Fig.4 A picture of the history corresponding to the move (3.2).

3.5. Spatial distribution

In this section we describe Level 1.2.1., thus refining Level 1.2. of Section 3.3.. We enrich a history with a spatial structure by introducing a set of ports N and a function c specifying for every subsystem s the tuple of ports to which s is connected.

Formally, a distributed history is thus a quintuple $(N,S,c,l,\leq)$, where (N,S,c) is a (multiple, hyper-)graph, and $(S,l,\leq)$ is a history.

The replacement operation can be extended to deal with the spatial aspects by using a construction analogous to one developed in the algebraic theory of graph grammars /Ehri83/. More precisely, in the algebraic approach a production

$$p=(B1 \xleftarrow{b1} K \xrightarrow{b2} B2)$$

consists of three graphs B1, K and B2 and two graph homomorphisms b1 and b2. Graphs B1 and B2 represent the usual left and right members (B1 will be replaced by B2), while the "interface" graph K and the two homomorphims b1 and b2 are used for defining the "embedding", i.e. the connections of B2 with the rest of the graph. A distributed history h is analogous to a production P if we identify B1 and B2 with the heads and tails of h. Furthermore, if we assume that K consists only of ports and that b1 and b2 are injective, the role of K, b1 and b2 (i.e. specifying corresponding ports) can be played by those ports of the history connected to both heads and tails. These ports are called external ports.

In the algebraic theory, given two productions p' and p", a new production

$$p=p'*_R p''$$

can be constructed, whose application is equivalent to the successive applications of p' and p" /Ehri83/. The graph R specifies which parts of B2' and B1" will be identified.

The replacement operation between distributed histories can be defined in exactly the same way. The events and the partial ordering relation of the result are of course obtained as for non distributed histories in Level 1..

Atomic histories are also defined as in the non distributed case, but with one additional constraint. In fact we require as before the existence of one event greater than all heads and smaller then all tails. Furthermore, this event must be connected to all and only ports which are not external.

Non external ports of atomic distributed histories are called synchronization ports. The intended meaning of the above constraint is that all ports which are deleted and created by the transition are clearly affected by the event. Viceversa, a synchronization on some port, being a point-like, completely centralized transition, must involve all processes connected to that port. Thus external ports, which represent the boundary with the external world, cannot be loci of synchronization.

In Fig. 5 we see a set Z of three distributed, atomic histories and a three-step computation. Notice that ports are represented as variables and atomic histories are written as usual in linear form. Note also that in the distributed histories in Fig. 5b) and c) port y_1 is external, while in d) there are no external ports.

Also for distributed histories the problem arises of generating atomic histories through a synchronization mechanism. This is possible using algebraic constructions like the gluing star /Corr83, CDM84, BFH84/.

A formalism called GDS, essentially consistent with the one outlined here, is described in detail in /CaMo83, DeMo83, DeMo84b/.

$$A(x),B(x,y) \xrightarrow{l(x,x')} A(x'),B(x',y)$$
$$B(x,y),A(y) \xrightarrow{r(y,y')} B(x,y'),A(y')$$
$$A(z) \xrightarrow{a} A(z)$$

a)

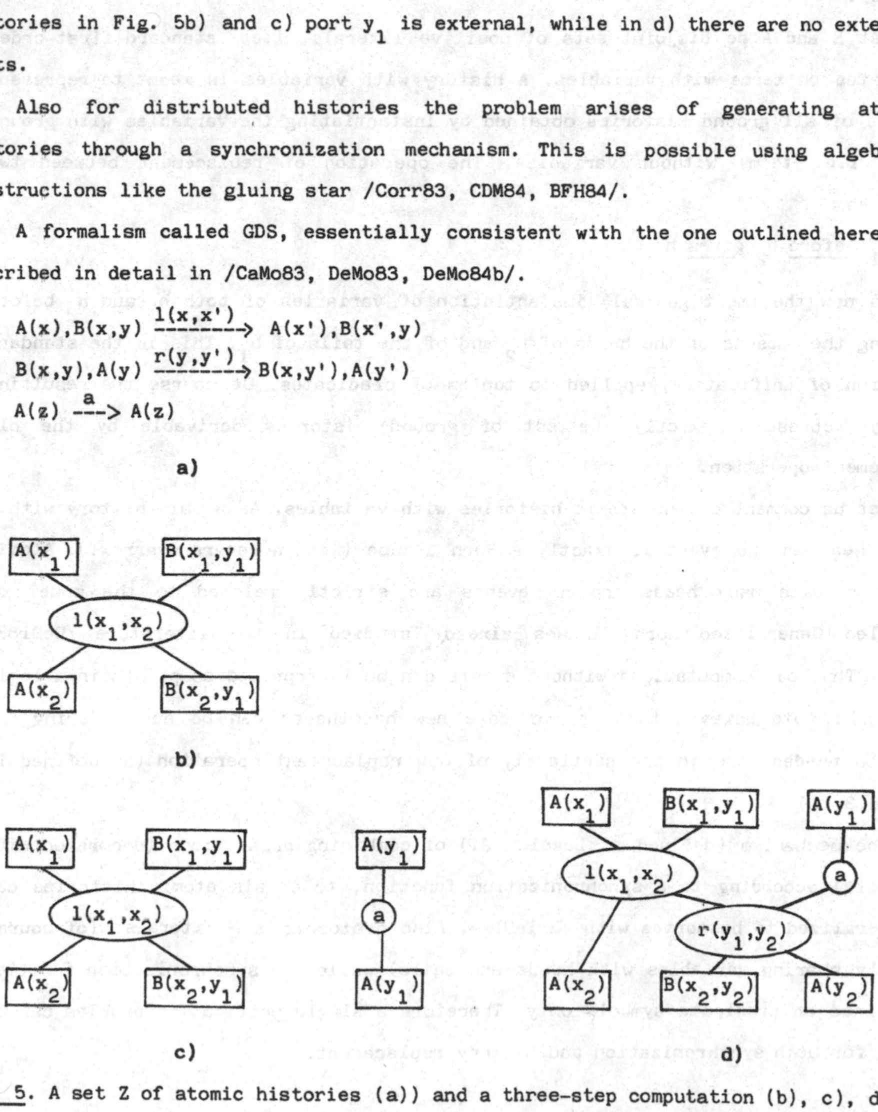

Fig. 5. A set Z of atomic histories (a)) and a three-step computation (b), c), d)).

3.6. Abstract data types and logic programming

In this section, we refine the model of Level 1.2. of Section 3.3. in a different direction. So far the alphabets E of process types and A of observable actions were unstructured, yet infinite. Thus no explicit notion of data type was available. In the present Level 1.2.2. we parametrize the symbols of both alphabets with (tuples of) terms of an Herbrand universe. Logic programming axiomatizations of data in terms of Horn clauses can be given by using an enriched version of the productions of Section 3.3., thus providing a well-known tool for defining abstract data types. For a comprehensive reference on logic and logic programming see /Robi79/.

Let E and A be disjoint sets of positive literals, i.e. standard first-order predicates on terms with variables. A history with variables is meant to represent the set of all ground histories obtained by instantiating the variables with ground terms, i.e. terms without variables. The operation of replacement between two histories

h_1 before h_2 gives h

permits now the (most general) instantiation of variables of both h_1 and h_2 before matching the subsets of the heads of h_2 and of the tails of h_1. This is the standard operation of unification, applied to tuples of predicates. Of course the resulting history represents exactly the set of ground histories derivable by the old replacement operation.

Let us comment on the atomic histories with variables. An atomic history with a single head and no event is exactly a Horn clause (with a reversed arrow!). Atomic histories with many heads and no events are strictly related to the model of so-called Generalized Horn Clauses already studied in the literature /DeDi83, FLP84/. Thus our computations without events can be interpreted as proofs in a logic framework. Note however that in our case new hypotheses can be added during the proof as needed, due to the partiality of our replacement operation (as defined in Section 3.1.).

The mechanism (defined at Level 2.2.) of combining productions (decorated with protocols) according to a synchronization function, to obtain atomic histories can be generalized to histories with variables. Also protocols are literals (of course possibly sharing variables with heads and tails) while the synchronization function is defined on predicate symbols only. Therefore a single unification problem can be set up for both synchronization and history replacement.

A similar mechanism has been defined in the literature by Monteiro /Mont84/. It uses essentially the CCS communication mechanism and generates histories without events.

As a general comment, in our approach we emphasize the description of the temporal evolution of concurrent distributed systems (as a partial ordering of events) because we aim at having a process-oriented semantics rather than an input-output semantics. Along this line we can rely also on our results on semantics and properties of infinite computations /DeMo84a, b/.

In Fig. 6 we see an example of a simple system consisting of a producer P and a consumer C connected by a queue Q. The producer generates a random natural number by iterating a successor operation s. When communication with the queue takes place, the current number is enqueued and the counter is reset. The queue can output its front value by communicating with the consumer, provided the latter is in a ready state. Eventually the consumer prints the value. The queue is implemented as a pair of stacks. When the output stack is empty, it receives the reversed content of the input stack. Note that the atomic histories handling the stacks do not produce events, and thus the observable behaviour of Q is only that of a queue. It may be interesting to examine the results of the infinite computations in this example (with heads P,Q and C) when different metrics are chosen to define the limits. It is easy to see that all computations are vital and thus have a result in out first metric space. When local fairness is required (namely our third metric is used) the producer cannot count forever, all the computations are also partial deadlock free and thus the limits have no tails. Note that in this case producer P realizes a choice operation, i.e. generates an unbound natural number still guaranteeing termination.

We conclude this section by showing the whole of our level structure:

Level 1. Concurrent histories

- Level 1.1. Petri nets
- Level 1.2. Productions - synchronization
  - Level 1.2.1. Spatial structure - GDS
  - Level 1.2.2. Abstract data types - logic programming
- Level 1.3. Inference rules of CCS.

$$P(n) \xrightarrow{count} P(s(n))$$

i) $$P(n) \xrightarrow{read1(n)} P(0)$$

ii) $$Q(S_1,S_2) \xrightarrow{read2(n)} Q(cons(n,S_1),S_2)$$

$$Q(cons(n,S_1),nil) \longrightarrow R(cons(n,S_1),nil)$$

$$R(cons(n,S_1),S_2) \longrightarrow R(S_1,cons(n,S_2))$$

$$R(nil,S_2) \longrightarrow Q(nil,S_2)$$

iii) $$Q(S_1,cons(n,S_2)) \xrightarrow{write1(n)} Q(S_1,S_2)$$

iv) $$C(ready) \xrightarrow{write2(n)} C(full(n))$$

$$C(full(n)) \xrightarrow{print(n)} C(ready)$$

$$f(read1,read2)=read$$

$$f(write1,write2)=write$$

i)+ii) $$P(n),Q(S_1,S_2) \xrightarrow{read(n)} P(0),Q(cons(n,S_1),S_2)$$

iii)+iv) $$Q(S_1,cons(n,S_2)),C(ready) \xrightarrow{write(n)} Q(S_1,S_2),C(full(n))$$

a)

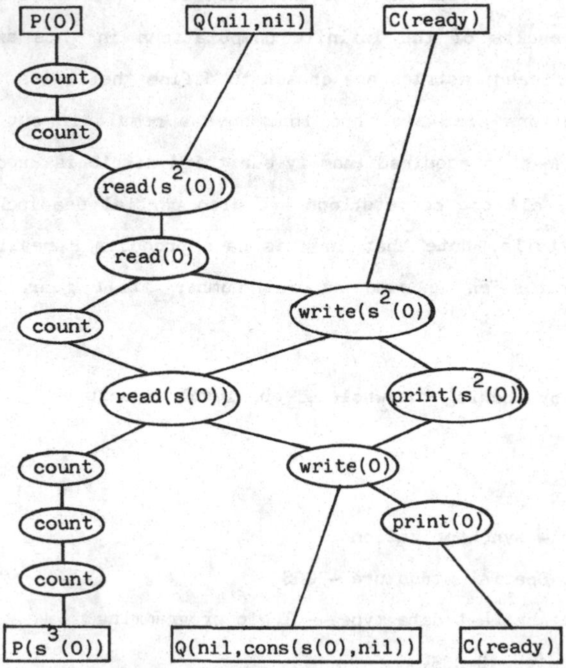

b)

Fig.6. Productions, synchronization function and atomic histories (a)), and a derived history (b)) for a producer-consumer example.

Acknowledgements

The authors wish to thank Sonia Raglianti and Silvia Giannini for having accurately typed this manuscript in an incredibly short time.

References

/BCG83/ Balzer,R., Cheatham,T.E. and Green,C., Software Technology in the 1990's: Using a New Paradigm, Computer 16, 11 (1983), 39-45.

/BFH84/ Boehm,P., Fonio,H., and Habel,A. Amalgamation of Graph Transformations with Applications to Synchronization, to appear in Proc. CAAP'85, Berlin, March 25-29, 1985, Springer-Verlag, Berlin.

/BHR84/ Brookes,S.D., Hoare,C.A.R., and Roscoe,A.W. A Theory of Communicating Sequential Processes, J. ACM 31, 3 (1984), 560-599.

/BuLa84/ Burstall,R., and Lampson,B. A Kernel Language for Abstract Data Types and Modules, Proc. Symp. on Semantics of Data Types (G. Kahn, D.B. MacQueen, and G. Plotkin Eds), Sophia Antipolis (France), June 1984, LNCS, 173, Berlin, 1984, pp. 1-50.

/Bram83/ Brams, G.W. Réseaux de Petri: Théorie et Pratique, Vol. I and II, Masson, Paris, 1983.

/CaMo83/ Castellani,I., and Montanari,U. Graph Grammars for Distributed Systems, Proc. 2nd Int. Workshop on Graph-Grammars and their Applications to Computer Science (H. Erigh, M.A. Nagel, and G. Rozenberg Eds), LNCS, 153, Springer-Verlag, Berlin 1983, pp. 20-83.

/CDM84/ Corradini,A., Degano,P., and Montanari,U. Specifying Highly Concurrent Data Structure Manipulation, to appear in Proc. ACM Int. Symp. on Computing, Firenze, March 1985, North-Holland.

/CFM82/ Castellani,I., Franceschi,P., and Montanari,U. Labeled Event Structures: A Model for Observable Concurrency, Proc. IFIP TC 2 - Working Conference: Formal Description of Programming Concepts II (D. Bjørner Ed), Garmisch - Partenkirchen. 1982. North-Holland. Amsterdam, 1983, pp. 383-400.

/Corr84/ Corradini,A. Aspetti Modellistici ed Implementativi di GDS, un Formalismo per la Specifica di Sistemi Distribuiti, Master Thesis, Computer Science Dept., Univ. of Pisa, Pisa, April 1984.

/DDM84/ Degano,P., De Nicola,R., and Montanari,U. Partial Ordering Derivations for CCS, submitted for publication.

/DeDi83/ Degano,P., and Diomedi,S. A First Order Semantics of a Connective Suitable to Express Concurrency, Proc. 2nd Logic Programming Workshop (L.M. Pereira Ed.), Albufeira (Portugal), 1983, pp. 506-517.

/DeMo83/ Degano,P., and Montanari,U. A Model for Distributed Systems Based on Graph Rewriting, Note Cnet 111, Computer Science Dept., Univ. of Pisa, Pisa, 1983.

/DeMo84a/ Degano,P., and Montanari,U. Liveness Properties as Convergence in Metric Spaces, Proc. 16th Annual ACM SIGACT Symposium on Theory of Computing, April 30 - May 2, 1984, Washington, DC, pp. 31-38.

/DeMo84b/ Degano,P., and Montanari,U. Distributed Systems, Partial Orderings of Events, and Event Structures, Lecture Notes of the 1984 International Summer School on Control Flow and Data Flow - Concepts of Distributed Programming (M. Broy Ed.), LNCS, Springer-Verlag, Berlin, 1984.

/Ehri83/ Ehrig,H. Aspects of Concurrency in Graph Grammars, Proc. 2nd Int. Workshop on Graph-Grammars and their Applications to Computer Science (H. Ehrig, M.A. Nagel, and G. Rozemberg Eds), LNCS, 153, Springer-Verlag, Berlin, 1983.

/FLP84/ Falaschi,M., Levi,G., and Palamidessi,C. A Synchronization Logic: Axiomatics and Formal Semantics of Generalized Horn Clauses, Info. and Co. 60 (1984), 36-69.

/GRAS84/ ESPRIT Pilot Project 125 - GRASPIN, Foundation Study for a Personal Software Development Workstation Prototype, Pilot Phase Project Report, October 1984.

/JaVa80/ Jantzen,M., and Valk,R. Formal Properties of Place/Transition Nets, In: Net Theory and Applications (W. Brauer Ed.), LNCS, 84, Springer-Verlag, Berlin 1979, pp. 165-212.

/LTS79/ Lauer,P.E., Torrigiani,P., Shields,M.W. COSY: A Specification Language based on Path Expressions, Acta Informatica, 12 (1979), 109-158.

/Miln80/ Milner,R. A Calculus of Communicating Systems, LNCS, 92, Springer-Verlag, Berlin 1980.

/Miln83/ Milner,R. Calculi for Synchrony and Asynchrony, TCS 25 (1983), 267-310.

/Mont84/ Monteiro,L. A Proposal for Distributed Programming in Logic, In: Implementations of PROLOG (J.A. Campbell Ed.), Ellis Horwood, Chichester, 1984, pp. 329-340.

/Plot83/ Plotkin,G.D. An Operational Semantics for CSP, Proc. IFIP TC 2-Working Conference: Formal Description of Programming Concepts II (D. Bjørner Ed.), Garmisch-Partenkirchen, June 1982, North-Holland, Amsterdam 1983, pp. 199-223.

/Robi79/ Robinson,J.A. Logic: Form and Function, Edinburgh University Press, 1979.

/Roev84/ de Roever, W.-P. The Quest for Compositionality - A Survey of Assertion--Based Proof Systems for Concurrent Programs, to appear in Proc. IFIP Working Conference on the Role of Abstract Models in Information Processing, Vienna, January 30 - February 1, 1985.

/Spec79/ Abstract Software Specifications, 1979 Copenhagen Winter School Proc. (D. Bjørner Ed.), LNCS, 86, Springer-Verlag, Berlin, 1980.

/Spec83/ STL/SERC Workshop on Analysis of Concurrent Systems, September 12-16, 1983, Cambridge (U.K.).

/Spec84/ Combining Specification Methods, Proc. 1984 Workshop on Formal Software Development, May 20-26, 1984, Nyborg (DK).

/Wins82/ Winskel,G. Event Structure Semantics for CCS and Related Languages, Proc. 9th ICALP (M. Nielsen and E.M. Schmidt Eds), LNCS, 140, Springer-Verlag, Berlin 1982, pp. 561-576.

Semantically Based Programming Tools (Summary)

William L. Scherlis and *Dana S. Scott*

Department of Computer Science
Carnegie-Mellon University
Pittsburgh, Pennsylvania 15213, USA

December 1984

The design and development of semantically based programming tools is a goal whose fulfillment is unquestionably many years off. The vision has been articulated in various forms by many researchers (including us in IFIP 83), but our impression now is that the expression of the vision provides little more than spiritual support and that there is an urgent need to distinguish shorter term goals along the way to mechanized tools.

While there does indeed seem to be some consensus that a notion of program derivation or "meta-program" is required (and there are indeed many informal examples of derivations in the literature), little has been said concerning the formal structure of these objects and how we will need to operate on them. What distinguishes derivations from mere proofs? What are the basic derivation steps? What deductive support is required? How can we assess the value of proposed derivation structures? Also, while much has been said concerning the functionality of the proposed tools, it is safe to say that we are still quite uncertain of their intended behavior and mode of interaction. What will be the correct balance of responsibility between programmer

---

This research was supported in part by the Office of Naval Research under Contract N0014-84-K-0415 and in part by the Defense Advanced Research Projects Agency (DoD), ARPA Order No. 3597, monitored by the Air Force Avionics Laboratory under Contract F33615-81-K-1539. The views and conclusions contained in this document are those of the authors and should not be interpreted as representing the official policies, either expressed or implied, of the Defense Advanced Research Projects Agency or the U.S. Government.

and tool and how will they interact? Can we realistically expect to build tools that will capture enough knowledge that they will become useful in practice?

This brief paper is intended to draw attention to several of the shorter-term issues and problems. Our recent focus has been on ths construction of a system for experimentation, so the bulk of our comments here will be on this aspect. Even though there are substantial theoretical problems still to be overcome, it is possible now to make progress towards experimental systems. The second section of the paper contains some remarks concerning programmer interaction with semantically based tools (what we have been calling inferential programming systems).

1. Towards experimental semantically based programming tools.

While a number of implemented systems have been built for experimenting with interactive theorem proving, program transformation, heuristically guided programming, and the like, we are still far from having a satisfactory framework for experimental inferential programming tools. The ML and LCF systems are perhaps the best examples currently available, but they are not well suited to the large scale syntactic manipulations that will be required. More importantly, ML is not as well able to support, as compared with Lisp, the (relatively) large scale system development we are undertaking.

We evaluated a number of possible architectures for prototype inferential programming systems, and it became immediately clear that the overall system design would continue to elude us until a number of conceptual and engineering questions had been settled. But we did find that in all of our scenarios, certain basic system functions were required above and beyond those provided in conventional programming environments. The Ergo Support System (ESS) is being created in order to provide an environment in which a diversity of implementation paths can be explored—an environment that will support the construction of prototype semantically based tools.

In our present view, the Support System includes four basic components, (1) an object and type management facility we call the *datafile*, (2) a set of *language*

tools, including a syntactic processor for supporting experimentation with programming and logical languages, (3) a basic *deductive facility* for supporting inference and simplification of expressions, and (4) an *interaction and view management* facility that supports communication between the system and human users. As the Support System evolves, other components will be added. (The Support System has been under active development since Spring 1984; it is being implemented in a Common Lisp environment.) The initial applications of the Support System will be experimentation with mechanizing several existing program derivations in order to better understand what kind of deductive support is needed.

While the Support System development will soon provide a useful set of facilities, a longer term goal is the development of metalanguage for controlling the support system and, ultimately, expressing knowledge about programming. Three requirements for the metalanguage are a suitable type framework for organizing objects, language for accessing the support system, and a means of easily expressing deductive knowledge. This deductive metaphor for computation is especially vivid in program manipulation systems, and a useful metalanguage should have a strong deductive flavor. Again, the best existing model for this language and its relationship to the system is ML.

Datafile. A principal source of complexity in large systems is the number and variety of the objects they manipulate and the lack of a uniform mechanism for managing the objects. This will be particularly true for program manipulation systems, in which it will be necessary to manipulate objects such as program text, program specifications, facts about programs, domain-specific facts, proofs of facts, documentation text, compiled code, program transformations, program derivations, program derivation patterns (for representing heuristic knowledge), and even mail messages and window descriptions for displays. Adding to the complexity is the fact that most of these objects, considered abstractly, will need a variety of representations appropriate to different computational needs (and appropriate to different storage media such as files, displays, and run-time storage). Ordinary software engineering considerations dictate that a uniform means—beyond typing—be developed for managing the many objects, kinds of objects, and representations.

In the Ergo Support System, the *datafile* facility serves in this information management role. The datafile functions as a combined file system and database by providing a means of naming, storing, classifying, and retrieving objects and collections of objects.

A major function of the datafile is classification and search management. The language for naming objects and collections of objects must, in order to permit fast search, be more expressive than that of an ordinary file system. But, in order to retain flexibility, we have avoided strong commitment to particular database organizations. Two basic mechanisms are provided—*classes*, which are special objects that represent sets of objects, and a simple set-theoretic language of *retrieval expressions* (which themselves are also objects) for describing classes. The datafile keeps track of selected retrieval expression objects and their values during incremental updates, thus enabling rapid search of the classes described even when the datafile state has changed.

Besides naming and search support, the datafile has a rudimentary object typing mechanism. The types are simply uninterpreted symbols; thus, there is no commitment in the support system to a particular ontology or organization of types. But this rudimentary typing system does provide special support for multiple representations of objects. Program fragments, for example, are most usually represented internally as annotated abstract syntax trees, but on a user's display or in an externally available file, they are best represented as concrete-syntactic text, and in an internal text file the best structure may be Lisp-like lists. The datafile helps keep track of the multiple instances and representations and maintain consistency as appropriate—allowing programs that use the datafile to manipulate objects without regard to representation.

Syntactic Tools. The major objects of attention in programming tools are, of course, programs. While it would be dangerous to make commitments to any specific languages of programs, support of a generic sort can be provided for manipulation of programs and other complex objects. The Support System provides sophisticated tools for compiling grammar specifications into parsers, unparsers, formatters, matchers, and so on. These tools provide a uniform means for associating concrete syntax with objects such as programs, specifications, transformations, assertions, and so on. Although syntax may appear to be a trivial problem, it is essential that it be handled gracefully even in prototype semantically based systems, in which so many different languages must be used; the user interface should not be the limiting factor to the success of experimental systems.

Reasoning Facility. The bulk of the computational activity in a semantically based tool is deduction. We are pursuing in our implementation a general approach that allows most of the formal reasoning activity to be described in a uniform framework, including proving assertions about programs, making simplifications to formulas and programs, applying transformations, and carrying out computations on virtual machines. This deductive framework would be the mechanism that carries out the fundamental program derivation steps and interacts with the user to plot strategy. LCF is the major antecedent of our design, though there will be important differences. For example, we are interested in directly manipulating proof objects and abstractions on proof structure.

Our recent efforts have been concentrated on building or collecting an implemented base of decision procedures that will facilitate inference in theories such as simple recursive data types and Presburger arithmetic, and on integrating these with a simple rewriting and pattern matching facility. We are not yet at the point of approaching substantive issues in this aspect of the implementation.

But, on the basis of the informal program derivations that exist in the literature, it can be argued that in most circumstances only a modest inference capability is required for most derivation steps. Of course, there are cases where substantial intellectual jumps must be made; in the first prototype tools these jumps will need to be made by the user.

Interaction Facility. In developing mechanisms by which users can interact with semantically based tools, perhaps most essential to recognize is the overwhelming complexity and size of the objects that will be manipulated. In order to evoke a decision, request advice, or simply inform a user concerning a particular object, such as a theorem to be proved or a program derivation step or a program specification, it is usually necessary for the tool to shield the user from the full extent of detail contained in the object being considered—and instead to present some view (i.e., projection or abstraction) of the complete object. Each such view presents a single aspect of the object; a collection of these views can provide a more natural and useful portrayal for a given interaction than a single monolithic representation. The use of views for manipulating objects allows internal object representations to include a wealth of supplementary information.

2. User Interaction with an Inferential Programming System.

If we are to develop useful tools, we must take a realistic attitude towards the way in which programmers will interact with semantically based environments. We must accept, for example, the fact that the vast majority of programming steps will likely involve the revision of previously made decisions, either because requirements have changed or in order to explore further a space of possible designs.

An example of inferential programming activity. Consider the sorts of abstract types that might be used in the implementation of a simple compiler. For certain types, such as syntax trees or run-time stack frames, choice of representation has a significant effect on the ultimate compiler design. In both of these cases the objects are used in so many different ways that, once the necessary specialized operations are included in the type interface, there would arguably be little representational structure left to hide. The author of a code generator, for example, will need to know the full details of stack frame layout in order to generate tolerably efficient code sequences for block entry and so on. It could then be argued that there is no point to making the abstract type explicit in the code since the intended abstraction boundary must be so consistently violated.

In the program derivation framework, uses of abstractions and the definitions of their representations need not coexist within individual programs (i.e., at individual derivation steps), but are instead spread over the derivation. This idea has been illustrated in a number of derivations in the literature. In these examples, instances of abstract operations appearing in early derivation steps are replaced in later steps by the corresponding operations on the representation, which, in still later steps, may be optimized based on the context in which they appear.

How, then, would we use an inferential programming system to help maintain a program containing abstract types in this regime? Ideally, every user interaction would result in addition of structure to the derivation under construction, either through commitment, simplification, or whatever. But, since semantically based tools are to be used by human programmers, we must allow for backtracking and revision of earlier decisions. For example, it may be necessary to revise various commitment steps along the way from specification to implementation, or it may be necessary to revise the specification itself (if it exists yet—derivations need not be constructed in any particular order).

Inferential programming steps. This brings us to our point. In interacting with an inferential programming tool, a user can make two kinds of steps. One would hope that most steps are *elaborative* steps, in which commitments or simplifications are made, augmenting an existing partial derivation. But there will always be some number of *lateral* steps, in which functional changes or design changes are made.

Attempts to work out this kind of scenario in greater detail inevitably lead to the question of what is the "right" notion of program derivation. Our view, and this is somewhat of a philosophical commitment, is that this is primarily a foundational question—as distinct from an analytical one. Analysing our programming experience will provide direction, but our purpose is not to build tools that imitate this experience.

Rather, our intent is to formulate notions of program derivation (together with criteria for their adequacy) through investigations into semantics and transformations.

3. Conclusions.

It is one of our fundamental theses that major improvements in software engineering practice will come about only through the development and use of software development tools. This thesis is based on our belief that formal methods will ultimately have a far more profound effect on software engineering productivity than management based methods, programming language design, or fast hardware. But we also believe that formal methods, by their nature, are suitable for practical use only in mechanized systems. This is not to say that *everything* is to be automated; the point is that the actual formal steps must be automated even if most of the guidance for their use is to come from the user. This is the same kind of observation that prompted the developers of LCF to introduce the ML type structure to protect the notion of "theorem," together with formal deductive structure that defines it, from the surrounding heuristic apparatus. It must be expected that heuristics will be under continual development, but a deductive system is fragile and will change only infrequently.

Acknowledgements. We thank Penny Anderson, Scott Dietzen, Paola Giannini, and Carl Gunter for careful readings of this paper on short notice.

From Function Level Semantics to Program Transformation and Optimization

John Backus
IBM Research Laboratory
San Jose, California 95193

ABSTRACT The software crisis results from our disorderly concepts of "program". These make programming an art, rather than an engineering discipline. Such a discipline would at least require that we have stocks of useful off-the-shelf programs and collections of standard theorems that can be applied repeatedly. We have neither.

Mathematical systems are often distinguished by a set of operations that (a) map a set of entities into itself, (b) have simply understood results, and (c) obey a set of strong algebraic laws. Neither conventional programs nor "object level" functional programs are entities belonging to such a system. The standard operations on conventional programs violate (b) and (c); object level functional programs normally employ lambda abstraction as their program building operation and it violates (a) and (c). Other problems of these program concepts are reviewed.

Function level programs are the entities of just such a mathematical system: programs are built by program-forming operations having good algebraic properties. Hence they are the subject of a large number of general theorems, theorems that are applicable in practice. We give examples. Function level programs also have the possibility of providing solutions to many of the other problems reviewed.

The paper reviews the function level FP system of programs, sketches a function level semantics for it, and from the equations of that semantics develops some moderately general results concerning linear, recursively defined functions, one concerning recursion removal. It then discusses other general, directly applicable results in the literature and shows that they are essentially function level results and are best presented and recognized in that form.

The final section is about optimization; it shows how some FP programs can be transformed into others that run as fast as Fortran programs. It introduces "Fortran constructs" into FP, pure functions that have an obvious corresponding Fortran-like program. It exhibits a number of function level identities for these constructs and shows how these can be used to convert inefficient FP programs into efficient Fortran-like ones.

1. Introduction

The primary reason for the software crisis is that programming has failed to become an engineering discipline. One mark of such a discipline is a collection of standard constructs and theorems. Thus electronics has a set of standard circuits and standard theorems that are used over and over by practicing engineers to design a new device and verify that it will work correctly.

In contrast, programs are not constructed from a collection of off-the-shelf programs; each program is basically written anew from the beginning. There are few general theorems that can applied directly to a given program to help prove it correct. Typically each program is proved correct, if at all, also from the beginning.

Thus programming is a low level business, one that is guided by many helpful principles, but which has not reached a level of elegance and power in which we can accumulate reusable programs and generally applicable theorems.

This software crisis is basically caused by the fact that we have chosen the wrong concepts of "program" to work with. Let us make a brief survey of the difficulties that attach to two basic kinds of program concepts.

1.1 Problems with the von Neumann concept of "program"

The von Neumann concept of a program as a mapping of one store into another has the following basic problems:

- **Programs depend on storage plans; programs to be combined must have a common storage plan, hence programs cannot be built from existing, independently written, off-the-shelf programs. Instead they are typically built from a set of subprograms that have all been planned and written together, de novo.**

- **The combining forms used to put existing program fragments together to form larger ones do not have good mathematical properties [they do not obey a strong set of algebraic laws]. Further, the most powerful of these, while-do and its analogs, lack the most important property of good combining forms: if you know what programs p and q do, then you should know in the most simple and direct way what combine(p,q) does.**

- **Programs are concerned with a low level of detail and with their "environment". Programming is still primarily the art of combining words to form a single new word and then arranging to repeat this process [a really "loopy" way to work!]. This low level style encourages a concern with computer efficiency that is humanly very inefficient. [Instead of relying on general theorems to transform inefficient high level programs into more complex efficient ones.]**

As a result of these defects programming has made little progress toward becoming an efficient engineering discipline over the last quarter century. Their combined effect is that programming is a terribly repetitive business, a black art. Beyond principles and techniques we have very few things in our profession of lasting usefulness: few reusable programs, few general theorems [and the statement of these is usually so complex that it is difficult to recognize an instance of one]. For a further discussion of the problems flowing from the von Neumann concept of "program" see [Backus 81a].

One example of this lack of generality and reusability is our treatment of "housekeeping operations". These consist of "control structures" to repeat a set of statements, together with certain index variables scattered throughout those statements [thus a housekeeping operation is not just a "for statement" but includes all the occurrences of the index variable]. As a result, these ubiquitous operations in our work are not reusable and compact operations, but comprise bits of information scattered over a program text. Thus housekeeping operations, which lie at the center of our business, cannot be defined, cannot be reused, cannot be transferred from one program to another, but must always be drearily redeveloped from scratch in each new program.

The area of reasoning about and transforming programs represents another example of the lack of generality in conventional programming. There are a great many papers in our literature giving examples of transformations and correctness proofs. Nevertheless, we still seem to be a long way from having a powerful and generally accepted body of knowledge that really simplifies the task of reasoning about programs, even very ordinary kinds of programs. We have few general theorems that are immediately useful to a programmer for transforming his program or proving it correct. [Although this situation is beginning to improve, at least in the area of functional programming.] What we do have is a large collection of <u>principles</u> that one may struggle to apply to an individual program; thus, for example, [Manna 74] contains a number of important theorems that justify various kinds of inductive proofs.

In the absence of general theorems, some authors propose that programs be developed starting from a predicate that is to be true when the program ends. The development process is illustrated by many examples and principles that may help one to develop a desired program. This approach is described in David Gries' book on structured programming, <u>The Science of Programming</u>. See also [Dijkstra 76, Hoare 69, Mills 75].

Structured programming is quite successful [it gives many helpful techniques to try, and deals with many difficult and subtle problems], but, like all conventional programming, it is demanding and infuriatingly repetitive -- with practice, one acquires technique but no lasting general results -- each problem is a new challenge. It is as if we were to solve many quadratic equations, treating each as a new problem and cleverly using various factoring techniques -- instead of solving the general equation once and for all, and then applying that solution to each equation.

1.2 Problems with functional, "object level" concepts of "program"

There has been a movement in recent years to work with functional programs [that map objects into objects, rather than stores into stores]. This approach has many advantages, although it has to overcome a number of problems before it becomes the obvious choice for everyday use. Even here, however, the most common notion of program is the "object level" one and this has problems. It is based on the use of variables that range over "objects" [the things programs transform]. In "object level" programming, building a new program from given ones involves applying the given programs to objects or object variables until the "result object" is built; the new program is then obtained by abstracting the object variables. Here are some of the problems:

- Unlike conventional programs, which use standard combining forms [e.g., composition, if-then-else, while-do], these programs tend to rely on lambda abstraction as the principal tool for building programs. Lambda abstraction is not a combining form. It combines a variable and an expression [with free variables] to form either a program [=function] or another expression. Its properties are syntactic rather than algebraic. Thus it combines two entities, neither of which is a program, to form a program; it fails to build programs from simpler ones and it fails to provide an algebra of programs on which to base general results about programs.

- Since these programs are not built by combining forms, reasoning about them concerns operations on objects and the mathematics of objects. This leads to theorems about objects but does not easily lead to general theorems about programs.

To have general theorems about programs one needs programs that are built by applying operations to existing programs, and these operations [combining forms, functionals, program-forming operations] must have attractive algebraic properties. This is the fundamental structure of many mathematical systems. Why are we in computer science so reluctant to build our basic entity, the program, in this way?

As in the case of conventional programs, work with object level programs has produced few general results, for the reasons sketched above. For example, one promising approach to transforming object level programs is that of [Burstall & Darlington 77]. Its authors give a set of transformation rules for transforming recursively defined functions at the object level. But again, as in the structured programming approach, there are few general results; instead there are lots of examples and strategies. These may be helpful when one is confronted with a new problem, but a lot of ingenuity is needed to define the right auxiliary function [the "eureka" step in the process]. Some powerful results are obtained by this approach. The problem is to try to make them more general and reusable.

1.3 Function level programs

The function level view of programs is one in which programs are constructed from existing ones by the use of program-forming operations (PFOs), operations that have two properties:

- If G(f,g) is a program built from programs f and g by the program-forming operation G, and if one knows how f and g behave, then one knows very simply and directly how the new program G(f,g) behaves.

- There is a powerful set of algebraic laws relating PFOs, particularly those that equate an expression in which one PFO is the main connective with another expression in which a different PFO is the main connective.

This way of building things -- by the application of operations that a) produce clearly understood results, and b) have strong algebraic properties -- distinguishes many of our most useful mathematical systems. The function level view asks that programs be built this way, that programming, by adopting this concept of program, become a mathematical system, the kind of system that history has proven to be extremely useful.

The function level approach seeks to bring this mathematical viewpoint to programming -- one that has heretofore been lacking -- just as the development of abstract data types has brought mathematics to bear on the subject of types. Originally we thought of data types as sets and we were concerned with the structure [or representation] of the objects belonging to a given type. Later we understood that we were really more interested in the abstract properties of the objects, not their structure or representation. And these properties are given by the algebraic structure of the operations on the objects, as determined by the laws they are required to satisfy. It is these algebraic laws that enable us to reason about data types, to prove general theorems about any type whose operations obey certain laws.

The function level view of programs seeks to do for programming what abstract data types have done for types, what algebra has done for arithmetic. It seeks to focus our attention on the operations on programs and on their algebraic structure, rather than on the structure and representation of the programs themselves. It seeks to have us view a program as the result of an operation, not as a syntactic entity whose structure is determined by punctuation -- semicolons and **begin - ends**, nor as a predicate transformer, nor as the syntactic result of substituting an argument for a variable in some expression.

In addition to their own algebraic structure, the program-forming operations [= combining forms = functionals] give a hierarchical structure to programs built with them. The operation that constructs a program is its main connective. Each of the programs put together by the main connective is itself built by various secondary connectives, and so on down, giving a tree structure to the entire program, with program-forming operations at the nodes and its constituent programs at the leaves. The importance of structure in conventional programs has long been recognized; we should now become aware of its importance in functional ones.

Just as we can now prove general theorems about objects by utilizing the algebraic properties of the operations on those objects [e.g., prove that all quadratic equations have a general solution], so we can prove general theorems about programs by the same means. But to do so, programs must be the result of operations, operations that have desirable algebraic properties. This is exactly what the function level view demands.

In contrast to this goal, the program-forming operations (PFOs) of von Neumann programs do not have desirable algebraic properties, and object level functional programs are not constructed by PFOs at all.

In summary, the fundamental goal of the function level view is to emphasize the mathematical structure of programs rather than the textual one. This view also offers solutions to the other difficulties discussed above, with the following possibilities:

1) The ability to construct a new program from off-the-shelf ones, since functional programs do not depend on storage plans.
2) Combining forms such that, if you understand the programs that are to be combined, then you very easily understand the resulting combined program.
3) Programs that are not concerned with low level details or with the names, size, or storage arrangement of their operands.
4) The ability to use powerful algebraic laws and theorems to convert simple but inefficient programs into complex, efficient ones. This can lessen the need to write efficient programs. [Our concern with efficiency is a major cause of inefficiency in programming.]
5) General "housekeeping" operations [data rearrangements] that can be defined and used over and over again.
6) A technology for reasoning about and transforming programs in which general theorems can be proved once and for all and easily applied in practice.

In this paper we hope to give indications for the validity of some of the above advantages of the function level view. We exhibit and use combining forms with the claimed properties; we use reusable, compact housekeeping operations. We exhibit compact programs that have a great disregard for efficiency and then turn them into very efficient ones. By developing a general result and applying it in examples, by examining the general results of others and restating them in function level form, and by using function level equations to optimize function level programs, we hope to show that this style offers a uniform, helpful technology for reasoning about, transforming, and optimizing programs.

1.4 Outline of the paper

Section 2 offers a brief review of the FP language [Backus 78, 81b] and some examples of function level equations for its primitives and combining forms that might make up a semantic description. Section 3 reviews some earlier general results, then develops some new ones leading to a moderately general recursion removal theorem. Section 4 presents a survey of examples of general results by some authors and their application, and exhibits them in function level form.

Section 5 presents a strategy for optimizing certain kinds of FP programs. It introduces some new "constructs" into FP -- pure functions with obvious counterparts as conventional programs -- and some equations involving them. It shows how these can be used to form identity functions for the arguments of functions to be optimized -- identity functions that describe the "shape" of those arguments. And it shows the use of these constructs and equations in optimizing some examples. These examples show how inefficient FP "housekeeping" operations, copying, and the computation of intermediate results can be eliminated by a relatively simple strategy to produce programs that run as fast as Fortran programs.

Section 6 presents some conclusions.

2. Review and semantics of FP

2.1 Review of FP

Readers familiar with FP may skip this section or use it as a reference for understanding the examples of later sections.

We adopt most of the notation and primitives of the FP system of [Backus 78] -- the few deviations are noted below with "(*)". Thus the set O of objects contains ⊥

["bottom"], the set of atoms, the empty sequence <>, and the sequence $<x_1,\ldots,x_n>$ of non-bottom objects x_i [where $<\ldots,\bot,\ldots> = \bot$]. The set of atoms typically include numbers, symbols or identifiers, and T and F for "true" and "false".

An object expression is either an object or a sequence of object expressions or an application F:E, where F is a function expression and E is an object expression.

A function f or respectively, a [function expression] is either

a) a primitive function [primitive function symbol], or

b) $G(f_1,\ldots,f_n)$, where G is one of the combining forms {composition, condition, construction, left insert, right insert, apply-to-all}, the f_i are functions [function expressions], and n corresponds to the arity of G, or

c) ~x, for some object x [object expression], where ~x is the function that is everywhere x except at $\bot$. Such a function is called a constant function. Or,

d) sel(n) or selr(n), for some integer [object expression with integer value] $n \geq 1$. Such functions are called selector functions and select the nth element of a sequence starting from the left end -- sel(n) -- or the right end -- selr(n)]. When it is clear that a selector function is intended, we write, e.g., "3" in place of "sel(3)". Or,

e) a defined function [defined function symbol], for which a unique definition exists [see definitions below].

A function definition has the form

def f = E

where f is an unused function symbol and E is a function expression that may involve f. [To see that the strict, bottom-up semantics of FP constitute a safe computation rule to compute the least fixed point of this recursive equation, refer to [Williams 82].] We may also give "extended" definitions [see 2.3 below for their description].

Each function is strict and maps a single object into a single object. We write f:x for the result of applying f to x. A function f is defined at x if $f:x \neq \bot$.

In the following examples of primitive FP functions and combining forms, we give object level descriptions of them so the reader may have an informal understanding of their behavior without first having to become familiar with the function level style. However, the function level equations of the next section indicate the way we propose to formally describe their semantics. [But we shall only indicate this in this paper.]

2.1.1 Examples of primitive FP functions. Here we give some FP primitive functions with brief explanations. These descriptions are merely to indicate the "object level" effect of these functions, their semantics are ultimately to be given by "function level" equations, some of which appear in 2.5 - 2.8 below.

Function	Description, examples
1,2.. 1r,2r..	Selector functions. 1:<A,B,C> = A, 2:<A,B,C> = B, 2:<A> = ⊥, 1r:<A,B,C> = C, 2r:<A,B,C,D> = C [We use the integers to represent selector functions when there is no confusion about whether we mean a function or an integer; when there is, we use the PFO sel that converts an integer n into the corresponding selector function. Thus in the above we could write sel(n):x rather than n:x.]
tl, tlr	Tail and tail right. tl:<A,B,C> = <B,C>, tlr:<A,B,C> = <A,B>
al, ar	Append left and append right. al:<A,<B,C>> = <A,B,C>, ar:<<A,B>,C> = <A,B,C>
distl, distr	distribute left, distribute right. distl:<A,<B,C>>=<<A,B>,<A,C>>, distr:<<A,B>,C>=<<A,C>,<B,C>>
trans	Transpose. trans:<<a,b>,<c,d>,<e,f>> = <<a,c,e>,<b,d,f>>
id	Identity. id:x = x for all objects x.
+, -, *, /	The arithmetic functions. +:<2,3> = 5, *:<3,5> = 15, etc.
add1, sub1	add1:3 = 4, sub1:7 = 6
and, not, eq	and:<T,T> = T, and:<T,F> = F, not:F = T, eq:<A,<B>> = F
atom	atom:A = T, atom:F = T, atom:<A,B> = F, atom:3.14 = T (defined for every non-bottom object)
null	null:<> = T, null:<A,B> = F, null:A = ⊥ (defined only on sequences)

2.1.2 Examples of function expressions using combining forms. Here are some function expressions involving functions p, f, g, $f_1,\dots,f_n$ showing the notation for the combining forms and the functions they build. In these examples, if no value is given for a function f at some object x, then f:x = ⊥.

Expression	Description, examples
f•g	**Composition** of f and g. f•g:x = f:(g:x)
p -> f; g	**Condition** of p [predicate], f and g. Read "if p then f else g". (p -> f; g):x = f:x if p:x = T = g:x if p:x = F We write p -> f; q -> g; h for p -> f; (q -> g; h) etc.
[$f_1,\dots,f_n$]	**Construction** of $f_1,\dots,f_n$.

$[f_1,\ldots,f_n]:x = \langle f_1:x,\ldots,f_n:x\rangle$
$[]:x = \langle\rangle$

/f — **Right insert of f.** (*)
$/f:\langle x,y\rangle = y$ if $x = \langle\rangle$ (*)
$= f:\langle 1:x, /f\cdot\langle tl:x,y\rangle\rangle$ if $x = \langle x_1,\ldots,x_n\rangle$ (*)

\f — **Left insert of f** (*)
$\backslash f:\langle x,y\rangle = y$ if $x = \langle\rangle$ (*)
$= f:\langle \backslash f:\langle tlr:x,y\rangle, lr:x\rangle$ if $x = \langle x_1,\ldots,x_n\rangle$ (*)

@f — **Apply-to-all of f.**
$@f:\langle\rangle = \langle\rangle$
$@f:\langle x_1,\ldots,x_n\rangle = \langle f:x_1,\ldots,f:x_n\rangle$

(a f b) — Infix notation for $f\cdot[a,b]$. $f + g = +\cdot[f,g]$

2.2 Specification of semantics: object level vs. function level

The semantics of the original FP language [Backus 78] was specified at the object level. That is, for each function denoted by an expression E and each object x the result of applying E to x [the object E:x] was specified by a collection of rules. From these object level semantics a set of function level laws was derived that expressed various algebraic properties of the primitive functions and of the combining forms [= functionals = PFOs].

Instead, one could have specified the function level laws relating the primitive functions and PFOs. For example, to partly describe the semantics of append left [al], instead of giving the object level equation: for all objects x, $y_1,\ldots,y_n$,

$$al:\langle x,\langle y_1,\ldots,y_n\rangle\rangle = \langle x,y_1,\ldots,y_n\rangle ,$$

that expresses the equality of two objects, one could give the corresponding "lifted" function level equation: for all functions x, $y_1,\ldots,y_n$

$$al\cdot[x,[y_1,\ldots,_n]] = [x,y_1,\ldots,y_n] ,$$

that expresses the equality of two functions. Not only can one obtain from this the object level equation, but its function level form is very useful in reasoning about programs: one has a general law in which one can substitute any functions for the variables of the law and then replace the resulting function expression on the left by that on the right [or vice versa]. On the other hand, the object level version is useful only for reasoning about objects.

2.3 Extended definitions in FP

Many people find it hard to read "proper" FP definitions like the following one for an iterative factorial function f, where factorial = $f\cdot[id,\bar{~}1]$,

```
7)    f = eq0•1 -> 2; f•[subl•1, *•[1,2]]
```

whereas they have no difficulty with the object level equation

```
8)    f:<x,y> = eq0:x => y; f:<subl:x, *:<x,y>>
```

which is to hold for all objects x and y [here "=>" denotes the object level operation if-then-else]. To make function level equations as easy to read as (8), object level equations can always be "lifted" [Backus 81c] to corresponding function level ones by doing the following

a) Change all object variables to function variables, thus x will range over all functions instead of all objects.
b) Change all pointed brackets < >, to square ones [ ].
c) Change all applications to compositions, i.e. change ":" to "•".
d) Change => to ->.

If we do this to (8) we get the function level equation that holds for all functions x and y

```
9)    f•[x,y] = eq0•x -> y; f•[subl•x, *•[x,y]]
```

that serves as readably to define f as (8) and has the additional advantage that it is a general law about the defined function f. We will call such a definition lifted from the object level an <u>extended definition</u> [Backus 81b] and use them freely. If we note that [1,2] is the identity on pairs, then for pairs, by substitution for x and y we get

```
      f = f•[1,2] = eq0•1 -> 2; f•[subl•1, *•[1,2]]
```

which is the original proper definition (7). We have found that extended definitions are by far the best form to use in defining a function, for readability, whereas proper definitions are the best to use for reasoning about [transforming, calculating with] the defined function. We will make use of this fact later.

2.4 FP semantics: soundness and completeness

There is the question of the soundness and completeness of any semantic description of a language. This question has not been completely answered for any treatment of the semantics of the FP language of [Backus 78], however this has now been done [Halpern, Williams, Wimmers, Winkler 85] for an FP system [let me call it FP_{84}] much richer than the one we deal with here. The FP system we treat here is almost identical to that described in [Backus 78]: all functions are strict, the sequence constructor is strict, all sequences are of finite length, and there are no functions of higher type than functionals.

None of the above properties hold for FP_{84}. That is, FP_{84} is an untyped system in which functions at all levels exist and are first class entities of the system: sequences of functions can be formed and other functions can be applied to them, thus "apply:<f,g>" is a meaningful expression in FP_{84}, whereas it is not in FP. Infinite sequences are admissible constructions.

Halpern, Williams, Wimmers, Winkler give the operational semantics of FP_{84} in terms of rewrite rules. They also define an independent denotational semantics. They prove that the rewrite rules preserve the denotational "meaning" of an expression [soundness]. They then prove that if the "meaning" of an expression is a finite object [the "meaning" of an expression may be a function or an infinite sequence], then the rewrite rules are sufficient to reduce the expression to that object [completeness].

2.5 The function level algebraic semantics of FP

Our object here is to give some examples of the kind of equations that might make up an algebraic description of the semantics of FP. The following equations do not constitute a complete description; the object of a complete description would be to give a practically useful set of equations, some of which might be redundant, as may be the case here. Some of the following equations will be used to derive propositions of the next section, others only illustrate the kind of equations that could be used.

2.6 Preliminaries

In what follows all variables will range over the space of FP functions [except in the form ~x, where x ranges over objects, or sel(n) or selr(n), where n ranges over integers, a subset of the objects] and are universally quantified. An equation may be qualified; thus we write

p ->-> f = g

to mean that, for all objects x, f:x is the same object as g:x provided that p:x = T. For example, if p:x = T and f:x = $\perp$, then the above asserts that g:x = $\perp$ ["=" denotes strong equality]. If p:x is anything but T, even undefined (=$\perp$), then the above makes no assertion.

We shall need the function "def" to qualify equations, where

def = ~T.

Since every function is strict, def•f yields T in exactly the domain in which f is defined, and $\perp$ elsewhere.

2.7 Semantics of primitive functions, examples of function level equations

selectors, append left, and tail

1) $al \cdot [x,[]] = [x]$

2) $al \cdot [x,[y_1,\ldots,y_n]] = [x,y_1,\ldots,y_n]$

3) $def \cdot [x_2,\ldots,x_n] \rightarrow\!\!> 1 \cdot [x_1,\ldots,x_n] = x_1$

4) $def \cdot x_1 \rightarrow\!\!> tl \cdot [x_1,\ldots,x_n] = [x_2,\ldots,x_n]$ if $n \geq 2$
$= []$ if $n=1$

5) $ar \cdot [[],x] = [x]$

6) $ar \cdot [[x_1,\ldots,x_n],y] = [x_1,\ldots,x_n,y]$

7) $def \cdot [x_1,\ldots,x_{n-1}] \rightarrow\!\!> lr \cdot [x_1,\ldots,x_n] = x_n$

8) $def \cdot x_n \rightarrow\!\!> tlr \cdot [x_1,\ldots,x_n] = [x_1,\ldots,x_{n-1}]$ if $n \geq 2$
$= []$ if $n=1$

9) $not \cdot null \rightarrow\!\!> al \cdot [1,tl] = id$

10) $not \cdot null \rightarrow\!\!> ar \cdot [tlr,lr] = id$

11) $f^n = f \cdot f^{n-1}$ for $n \geq 1$ $f^0 = id$

12) $sel(n) = 1 \cdot tl^{n-1}$, $selr(n) = lr \cdot tlr^{n-1}$ for $n \geq 1$

Identity

13) $id \cdot x = x \cdot id = x$

Constant functions

14) $def \cdot y \rightarrow\!\!> \tilde{}x \cdot y = \tilde{}x$

Atom

15) $atom \cdot \tilde{}a = \tilde{}T$ for each atom a (this represents an infinite set of equations, one for each atom)

16) $atom \cdot [] = \tilde{}F$

17) $def \cdot al \cdot [x,y] \rightarrow\!\!> atom \cdot al \cdot [x,y] = \tilde{}F$

null

18) $null \cdot [] = \tilde{}T$

19) $def \cdot al \cdot [x,y] \rightarrow\!\!> null \cdot (al \cdot [x,y]) = \tilde{}F$

length.

20) $len \cdot x = null \cdot x \rightarrow \tilde{}0; (\tilde{}1 + len \cdot tl \cdot x)$

2.8 Semantics of the combining forms, examples of function level equations

21) $f \cdot (g \cdot h) = (f \cdot g) \cdot h$

22) $f \cdot (p \rightarrow q;r) = p \rightarrow f \cdot q;\ f \cdot r$

23) $(p \rightarrow q;r) \cdot f = p \cdot f \rightarrow q \cdot f;\ r \cdot f$

24) $def \cdot g \rightarrow\!\!> [] \cdot g = []$

25) $[f_1,\ldots,f_n] \cdot g = [f_1 \cdot g,\ldots,f_n \cdot g]$

26) $[\ldots(p \rightarrow f;g)\ldots] = p \rightarrow [\ldots f \ldots];[\ldots g \ldots]$

27) /f•[x,y] = null•x -> y; f•[1•x, /f•[tl•x,y]]

28) \f•[x,y] = null•y -> x; f•[\f•[x,tlr•y], lr•y]

3. Some general function level results about programs

We summarize some general results from earlier papers [Backus 78, 81b] and derive a few others that we shall use in the examples to follow. The machinery outlined below for transforming and reasoning about programs has a somewhat mechanical character, involving a good deal of routine calculations using function level identities. However, we believe these techniques provide a uniform, accurate technology for building up a set of general results in a form that makes it easy to identify and apply them, either mechanically or by people.

3.1 Some simple general results

Here we state without proof a number of simple equations and results that can be derived from the algebraic semantics.

29) /f•[[x_1,...,x_n], f•[y,z]] = /f[[x_1,...,x_n,y], z]

30) \f•[f•[y,z], [x_1,...,x_n]] = \f•[y, [z,x_1,...,x_n]]

31) if h is associative and has unit u, then /h•[id,~u] = \h•[~u,id]

32) p -> q; p -> r; s = p -> q; s

3.2 The general solution of linear functional equations

The following linear expansion theorem and related facts about linearity are often generally useful for reasoning about recursively defined functions, as we shall see in later examples.

3.2.1 Definition of linear forms and equations. An equation of the form

f = p -> q; Hf

is linear if the expression or form Hf in the variable f is linear. A form Hf is linear if there exists a form H_tp such that the following two conditions hold:

L1) For all functions a, b, and c,

H(a -> b;c) = H_ta -> Hb; Hc

L2) For all objects x, if H~⊥:x ≠ ⊥, then, for all functions a,

H_ta:x = T .

H_t is the predicate transformer of H. Clearly Hf is linear with H_t if H and H_t satisfy (L1) and H~⊥ = ~⊥.

<u>3.2.2 Linear expansion theorem.</u> If Hf is a linear form, then the [least] solution of

f = p -> q; Hf

is

$f = p \rightarrow q; \ldots; H_t^n p \rightarrow H^n q; \ldots$

where the infinite condition on the right denotes the function that is the limit of the following increasingly defined functions:

$f_i = p \rightarrow q; \ldots; H_t^i p \rightarrow H^i q; \sim\bot$

and where $H^{n+1}f$ denotes $H(H^n f)$. Note that the distribution rules for condition apply to infinite conditions, in particular:

33) $(p_1 \rightarrow q_1; \ldots; p_i \rightarrow q_i; \ldots) \cdot g = (p_1 \cdot g \rightarrow q_1 \cdot g; \ldots; p_i \cdot g \rightarrow q_i \cdot g; \ldots)$

This extension of the rule (23) for finite conditions depends on the continuity of Hf = f•g, so that $H(\lim f_i) = \lim H(f_i)$; see [Williams 82].

<u>3.2.3 Composition of forms.</u> If H and G are two forms [function expressions each with one free variable], then by their <u>composition</u> HG we mean the form such that HGf = H(Gf) = the form obtained by replacing the free variable of H by the form Gf [G with its free variable replaced by the function or variable f]. Take care not to confuse HGf [form composition] with Hf•Gf [function composition].

<u>3.2.4 Theorem: composition of linear forms.</u> If H and G are linear, then HG [where HGf = H(Gf)] is linear with predicate transformer $(HG)_t = H_t G_t$.

<u>3.2.5 Theorem: basic linear forms.</u> In the following function expressions we use Roman letters for the free variable of a form and bold-faced letters for arbitrary fixed functions. The following forms Hf are linear in f with the given predicate transformer $H_t p$:

Hf = **r**	with $H_t p$ = ~T
Hf = **r**•f	with $H_t p$ = p = Ip
Hf = f•**r**	with $H_t p$ = p•**r** = Hp
Hf = [**g**,f]	with $H_t p$ = p = Ip
Hf = [...,f,...]	with $H_t p$ = p = Ip
Hf = **p** -> **q**;f	with $H_t r$ = **p** -> ~T;r
Hf = **p** -> f;**q**	with $H_t r$ = **p** -> r;~T
Hf = f -> **g**;**h**	with $H_t p$ = p = Ip

Applying the composition theorem to the above basic forms generates a great many linear forms and their predicate transformers. Note that <u>all</u> forms Hf with a single occurrence of f and built with composition, construction, and condition, are linear. Thus, for example, Hf = **h**•[**i**,f•**j**] is linear with $H_t p$ =p•**j**, since H = RST where

Rf = h•f, Sf = [i,f], Tf = f•j, and $H_t p = R_t S_t T_t p$ = II(p•j) = p•j. Many forms Hf with multiple occurrences of f are linear. For example,

Hf = s -> f•k; h•[i,f•j]

is linear with $H_t r$ = s -> r•k; r•j. Of course the usefulness of the linear expansion theorem to solve f = p -> q;Hf depends on being able to deal with the expressions $H_t^n p$ and $H^n q$; in this case these expressions become unmanageable unless the values of p, k, h, i, j allow simplifications, as may happen.

For some further results on linear forms see [Backus 81b] and for results on non-linear equations see [Williams 82].

3.3 Application of the linear expansion theorem: recursion removal

We use the linear expansion theorem to obtain a moderately general result that provides an iterative solution for a class of recursively defined functions. Similar techniques could be used to obtain similar results for other, more general recursive equations. Stronger results can be found in [Kieburtz and Shultis 81]; we present this one to illustrate a different method of proof with some useful intermediate facts. One of these is corollary 3.3.1, which provides useful data for reasoning about a larger class of functions.

3.3.1 Corollary of the linear expansion theorem.

Let f satisfy the following

34) f = p -> q; h•[i,f•j]

Then

35) $p\cdot j^n$ -> /h•[[i,i•j,...,$i\cdot j^{n-1}$],$q\cdot j^n$]

is the general term of the expansion for f.

3.3.2 Corollary of the linear expansion theorem.

Let f′ satisfy the following for all functions x and y:

36) f′•[x,y] = p•x -> h•[y,q•x]; f′•[j•x, h•[y,i•x]]

Then the general term of the expansion for f′ that is valid for pairs and $n \geq 1$ is

37) $p\cdot j^n\cdot 1$ -> \h•[2, [i,i•j,...,$i\cdot j^{n-1}$,$q\cdot j^n$]•1]

3.3.3 Recursion removal theorem.

Given

38) f = p -> q; h•[i, f•j]

39) f′•[x,y] = p•x -> h•[y,q•x]; f′•[j•x, h•[y, i•x]]

where h is associative with unit u, then

40) $f = f' \cdot [id, \sim u]$

Proof. From (37) we get the general term of the expansion for f′ on pairs and, using the law (33) to distribute [id,~u] over the expansion for f′ from the right, we get the general term of the expansion for f′·[id,~u′] and progressively transform it:

$p \cdot j^{n} \cdot 1 \cdot [id, \sim u] \rightarrow \backslash h \cdot [2, [i, i \cdot j, \ldots, i \cdot j^{n-1}, q \cdot j^{n}] \cdot 1\,] \cdot [id, \sim u]$

41) $p \cdot j^{n} \rightarrow \backslash h \cdot [\sim u, [i, i \cdot j, \ldots, i \cdot j^{n-1}, q \cdot j^{n}]]$ by (3) (12) (13) (25)

42) $p \cdot j^{n} \rightarrow /h \cdot [[i, i \cdot j, \ldots, i \cdot j^{n-1}, q \cdot j^{n}], \sim u]$ by (31)

since h associative

43) $p \cdot j^{n} \rightarrow /h \cdot [[i, i \cdot j, \ldots, i \cdot j^{n-1}], h \cdot [q \cdot j^{n}, \sim u]]$ by (29)

44) $p \cdot j^{n} \rightarrow /h \cdot [[i, i \cdot j, \ldots, i \cdot j^{n-1}], q \cdot j^{n}]$ since u is a unit of h

This last is the general term of the expansion of f. The initial term of f·[id,~u] is, by (39), p -> h·[~u,q], which reduces to p -> q . Therefore the expansions of f and of f′·[id,~u] are equal, therefore the functions are equal. Q.E.D.

4. The use of general theorems to obtain program transformations

Here we compare the use of general theorems in the approach of various authors. Let me first make a general comment about those that present their results in the form of object level equations. To take an example from [Wadler 81]: he gives the object level rule

```
(map f) (map g) xs
    => (map h) xs
        where   h x = f g x
```

This is exactly the FP law III.4 of [Backus 78]

```
@f·@g = @(f·g)
```

The point here is that the FP function level equation allows one to immediately substitute one side of the equation for the other in any function expression, whereas the object level statement of the principle is much less directly useful and requires the use of a lot of unnecessary indirection, as in the where clause that is required by failure to use the combining form composition. For this reason I submit that it is time we recognize that such principles are essentially function level ones and should uniformly be given as such.

4.1 The approach of [Kieburtz and Shultis 81] = K & S

This paper contains some good examples of the function level approach we are advocating and has a number of results that go beyond the simple recursion removal theorem presented here. It uses the FP function level technology to derive a general theorem [Prop 2] of which our recursion removal theorem is a corollary. We present our proof only because it represents a quite different approach to obtaining such results.

Their proof starts from the most basic function level identities and involves two inductive arguments, whereas ours relies on the linear expansion theorem, which was helpful in obtaining the formulation of the theorem by examining the general terms of the expansion for the given function f. Their proof is shorter and cleaner, ours involves some mechanical calculation that is tedious but easy.

They also prove several other interesting and generally useful theorems concerning recursion removal for the scheme f = p -> q; h•[i,f•j] of our theorem. They go on to prove several general theorems [props 5 & 6] for removing recursion from non-linear functions of the forms

f = p -> q; h•[f•r, f•s] and
f = p -> q; h•[r,h•[f•s, f•t]]

It is interesting to note that the basic scheme for the solutions of these last two questions came, according to the authors, from [Cohen 79]. However the function level presentation here of the result seems to make it much shorter and more accessible than in its more general presentation in terms of conventional programs in [Cohen 79].

This paper provides some of the most <u>directly useful</u> theorems for recursion removal that I am aware of. And they are presented as function level results that can be straightforwardly applied, rather than as principles and methods for obtaining results, or as lengthy, difficult to recognize, object level schemes.

4.2 The approach of [Burstall and Darlington 77] = B & D

It is interesting to examine the differences between the algebraic approach to program transformation and that of B & D, who originated the basic ideas about transforming recursively defined functions. Their approach is to provide a set of transformation rules to be applied to the transformation of an individual function or set of functions. This involves defining an auxiliary function in terms of the given one [the "eureka" step], then instantiating object variables, unfolding, simplification and folding; the object being to get a recursive expression for the defined function that does not depend on the given function, and then redefine the given one in terms of the new one. The goal is to define the new function so that the resulting redefinition of the given one is more efficient.

The approach we advocate is to develop general function level identities and theorems. The goal is to provide a basic uniform technology for transforming programs in a mechanical way that captures, once and for all, a lot of difficult details and reasoning, and to develop general theorems that can be easily applied to transform many different programs. Our methodology is just that of substitution in equations and needs no special rules. The above theorems and those of K & S and others are easier to apply [when they apply] than B & D's methods, and they preserve termination properties, whereas B & D's do not.

The most difficult programs to reason about are recursively defined ones. For linear recursive programs and for certain non-linear ones [Williams 82] the FP technology provides non-recursive expansions that often help one to understand and reason about such programs. Many programs that are normally defined recursively are easier to define in FP in a simple closed form in which one can reason about them directly using the FP technology. We have given above an example of a general theorem for transforming a class of linear recursive functions into iterative ones, K & S give several more plus two for classes of non-linear equations.

B & D's methods apply to many programs to which our theorems above do not. However, many of those programs may be handled by the results of K & S discussed above.

<u>4.2.1 Example 1. Factorial.</u> Given the definition of factorial

```
fact = eq0 -> ~1; *•[id, fact•subl]
```

we see that the recursion removal theorem applies since * is associative with unit 1. Thus in

```
f'•[x,y] = p•x -> h•[y,q•x]; f'•[j•x, h•[y, i•x]]
```

we need only replace p by eq0, h by *, q by ~1, i by id, and j by subl to give

```
f'•[x,y] = eq0•x -> *•[y,~1•x]; f'•[subl•x, *•[y, id•x]]
f'•[x,y] = eq0•x -> y; f'•[subl•x, *•[y,x]]
```

and fact = f'•[id,~1], which is essentially the same program derived by B & D. Their derivation of this example required some ingenuity and work, whereas our effort went into developing a general theorem which can be routinely applied to this example as well as many others.

In the FP programming style we would prefer to define

```
fact = /* • iota ,
```

a non-recursive definition using the iota function of APL, where

```
iota:n = <1,2,...,n> .
```

4.2.2 Example 2. Reverse. When reverse is defined using concatenate, as it is in B & D, it fits exactly the same scheme as factorial, since ++ is associative with unit []. Thus it can be treated exactly as above to obtain B & D's result. If we define reverse, using append left, as

    rev = null -> []; al•[1r, rev•tlr] ,

then Proposition 2 of [Kieburtz and Shultis 81] applies directly to provide an iterative solution that uses append right.

4.3 The approach of [Bird 84]

This is another paper that presents some general function level results. However it uses a somewhat idiosyncratic mixture of function and object level equations, curried functions, infix functions, "sections" and other special notation that makes for terse expressions but very difficult reading. One annoying feature of the use of curried functions is that it is difficult to determine the arity of a function in an expression unless the expression is object level, and even then it requires some analysis. I believe the use of uncurried functions promotes readability and eliminates the need for object level equations.

4.3.1 Example. The "promotion" scheme. We give the example first in Bird's terms [beginning with his paragraph 3, p491], and then give his derivation in an uncurried, function level FP version.

(D1) spec x = f(Hx)

(D2) H[] = ...

H(a;x) = h a (Hx)

(C1) f•(ha) = (h'a)•f

The equations (D1) and (D2) represent a specification in which we shall think of H as building an object from a sequence [a;x denotes appending a to the sequence x] by combining [with h] the head and H of the tail of its argument. For efficiency it would be preferable to compute spec by using the existence of an h' satisfying (C1). Here is the FP derivation of Bird's result, but using an uncurried version, uh, for his curried two-argument function h, and uh' for his h':

(D1')	spec = f•H	
(D2')	H = null -> r; uh•[1,H•tl]	r is supplied here for the [] case.
(C1')	f•uh•[x,y] = uh'•[x, f•y]	for all functions x, and y.

Thus we have the following

spec	= f•H	
	= f•(null -> r; uh•[1,H•tl])	def of H ["unfolding"]
	= null -> f•r; f•uh•[1,H•tl)	by (22)

$= null \rightarrow f \cdot r;\ uh' \cdot [1, f \cdot H \cdot tl]$ by (C1′)

$= null \rightarrow f \cdot r;\ uh' \cdot [1, spec \cdot tl]$ by (D1′)

which is Bird's result in FP form. Here it is obtained at the function level. The pattern-matching style of SASL equations, in which conditions are broken up into multiple equations using mutually exclusive arguments, requires that the arguments always tag along in Bird's object level derivation.

If one applies the corollary 3.3.1 to this last equation for spec, and then simplifies the general term in its expansion,

$$null \cdot tl^n \rightarrow /uh' \cdot [[1, 1 \cdot tl, \ldots, 1 \cdot tl^{n-1}], f \cdot r \cdot tl^n]$$
$$= \quad (len = n) \rightarrow /uh' \cdot [id, f \cdot r \cdot [\,]\,]$$

one then obtains the following closed form expression for spec

$$spec = /uh' \cdot [id, f \cdot r \cdot [\,]\,]$$

which is easier to reason about and to use.

5. Optimizing FP programs

The function level style of FP replaces the scattered form of "housekeeping" operations found in conventional programs with operations that rearrange data structures. This means that complex housekeeping operations can be defined and reused in FP, whereas such operations in conventional programs must be redeveloped for each program from scratch. Thus we believe that the FP style of housekeeping operations is the right way to think about and write such operations. But the price one pays in efficiency is enormous if programs with these operations are executed literally.

In addition to those due to FP housekeeping operations, there are many other potential sources of inefficiencies in FP programs. One of these is the creation of large and unnecessary intermediate results. This general problem for functional programs is discussed in [Wadler 84] where the author deals with the problem not only at the object level but by means of a "listless machine". Here we propose to deal with this issue at the function level in source language. Eliminating making copies of large objects is an important subcase of the problem of eliminating the creation of intermediate results.

We propose to optimize FP programs having these problems into programs in an FP language extended with additional constructs [which we laughingly call "Fortran constructs"]. These constructs denote pure functions but have obvious realizations as efficient Fortran or other conventional programs. We hope to optimize most FP programs that have potential efficient Fortran equivalents into those equivalents or something very close to them; that is, if an FP program is not essentially a "list processing" program, we hope to make it run as fast as a Fortran program. We hope to produce a Fortran-like program that addresses its data just as a Fortran program does, without using list structures or unnecessary levels of indirection in addressing.

The process of optimizing such FP programs into efficient extended FP programs will depend on the use of sophisticated strategies employing a collection of function level identities involving the Fortran constructs. The work on this approach is in an early stage, and it is not clear how far it can be carried nor how much of it can be automated. Much of its ultimate power and utility, if successful, will depend on:

a) Powerful strategies for employing functional identities or rewrite rules for transforming programs.

b) Systems for specifying programs and abstract data types within FP in terms of functional identities that are consistent with (a). These have yet to be completely developed, although an excellent basis already exists in [Guttag, Horning, Williams 81].

These goals are clearly difficult to achieve, but we believe they comprise an interesting effort to put together a practical, higher level functional style of programming. We have yet to study carefully the problems of optimizing programs that are recursively defined. We have some hope that some of the above machinery for linear programs will be helpful in this regard. For other programs some of the techniques of Burstall & Darlington should be helpful.

All we shall do here is to exhibit some examples and show how a given set of functional identities can be used to optimize them.

5.1 The basic optimizing strategy

Our basic strategy is the following. First, introduce some "Fortran constructs" into extended FP [which we will refer to as "constructs"]. These must have three properties:

a) any "Fortran-like" FP program can be represented as a construct.
b) any construct has an obvious Fortran-like program or similar counterpart that is directly efficiently realizable.
c) it is possible to write a construct that is the identity function for <u>all</u> the arguments of the function one wants to optimize.

Then, given such constructs as above, our strategy to optimize a program f is as follows:

1) Find a construct, C, that is the identity for all the valid arguments for f. This construct we call a "structure operator" for f [because it essentially describes the structure of f's arguments]. Thus f•C = f. We will work with f•C.

2) If f = G(g,C′) for some combining form G, then find an equation of the form G(g,C′) = C″; if no such equation exists, that is, if g is too complex, try to find H and K so that f = H(r,K(s,C′)) where s is simple enough that there is an equation K(s,C′) = C″ and then continue in this way until you get f = C‴. In our examples so far, this strategy involves just G = H = K = composition. In this case the strategy is just, e.g.,

   If f = g•h and f = f•C, and if

   h•C = C′ and g•C′ = C″ are instantiations of established equations then

   f = g•h•C => f = g•C′ => f = C″

   transforms f into the construct C″.

5.2 The **for** construct

Our most basic construct denotes a construction of a sequence of functions whose length may depend on the argument to which it is applied. The functions in the construction are generated by some expression E(i) that converts an integer i into a function E(i). We write the **for** construct using bold-faced square brackets as follows

45) **[**E(i) i=f,g**]**

where f and g are integer valued functions and i is a bound variable of the construct. The meaning of **[**E(i) i=f,g**]** is defined as follows for any object x: if $1 \leq f{:}x \leq g{:}x$, then

46) **[**E(i) i=f,g**]**:x = [E(f:x), E(f:x + 1),...,E(g:x)]:x

If f:x > g:x then

   [E(i) i=f,g**]**:x = <>

Thus, for example,

   [i i=~1,len**]**:<a,b,c> = [1,2,3]:<a,b,c> = <a,b,c>

Here E(i) = i (more properly we should write E(i) = sel(i), but here there is no doubt that the integer i is a selector function). So **[**i i=~1,len**]** is the identity function for every non-empty sequence.

Since a **for** construct denotes a construction, we will extend the definition of the **for** construct to include any construction whose elements are either functions or for constructs. Thus $[c_1,...,c_n]$ is a **for** construct whether or not some or all of the c_i's are for constructs.

5.2.1 The Fortran-like program corresponding to a for construct. Every construction $[f_1,\ldots,f_n]$ has a corresponding Fortran-like program that will compute $[f_1,\ldots,f_n]$:x -- given Fortran-like programs p(i,x) that will compute f_i:x -- (in the case of Fortran we must assume that f_i:x is a single word, in other languages we might not need this assumption). The program is as follows

```
for i = 1, n
  r[i] := p(i,x)
```

For a for construct [E(i) i=f,g], if p(E(i),x) is a program that will compute E(i):x for any integer i and any object x, then the program to compute [E(i) i=f,g]:x is

```
for i = f:x, g:x
  r[i] := p(E(i),x)
```

5.2.2 Nested for constructs. Consider the construct

[[E(i,j) i=f,g] j=r,s]

In the innermost construct the variable i is bound, so E′(j) = [E(i,j) i=f,g] is a legitimate expression in the free variable j that will produce a function for each integer j. For example

[[i•j i=~1,len•1] j=~1,len]

is the identity construct for every sequence of equal-length sequences [e.g., a rectangular matrix].

```
  [[i•j i=~1,len•1] j=~1,len]:<<a>,<b>>
= [[i•1 i=~1,len•1], [i•2 i=~1,len•1]]:<<a>,<b>>
= <[i•1 i=~1,len•1]:<<a>,<b>>, [i•2 i=~1,len•1]:<<a>,<b>>]
= <[1•1]:<<a>,<b>>, [1•2]:<<a>,<b>> >
= <<a>,<b>>
```

similarly

[[i•j i=~1,len•j] j=~1,len]

is the identity construct for every sequence of sequences.

5.3 The insertion construct

5.3.1 Simplified left and right insert for optimization. In the following and in the examples it is convenient to introduce a simplified notion for left and right insert, in which /f and \f are functions of one argument rather than two. We define these one argument versions in terms of the two argument ones used above as follows [these one argument inserts are those used in the original FP]. We use bold-faced /f and \f for the two argument versions and /f and \f for the one argument one.

47) **def** /f = /f•[id,~u] if f has a right unit u
    = /f•[tlr,lr] otherwise

48) **def** \f = \f•[~u,id] if f has a left unit u
    = \f•[1,tl] otherwise

5.3.2 The insertion construct. Our second construct is the insertion construct. If C is a for construct, then /f•C and \f•C are insertion constructs.

5.3.3 Fortran-like programs corresponding to insertion constructs. The program to compute \f•[E(i) i=g,h]:x is

```
r := u_f                       where u_f is a left unit of f
for i = g:x, h:x
   r := f:<r, p(E(i),x)>
```

5.4 Algebraic laws for constructs

The following identities have the form f•C = C′ where C and C′ are for constructs and f is a primitive FP function.

49) trans•[[E(i,j) i=f,g] j=r,s] = [[E(i,j) j=r,s] i=f,g]

50) trans•[[$E_1(i_1)$ i_1=f,g],...,[$E_n(i_n)$ i_n=f,g]] = [[E_1(i),...,E_n(i)] i=f,g]

51) distr•[[E(i) i=f,g], h] = [[E(i),h] i=f,g]

52) distl•[h, [E(i) i=f,g]] = [[h,E(i)] i=f,g]

53) @f•[E(i) i=g,h] = [f•E(i) i=g,h]

54) tl•[E(i) i=f,g] = [E(i) i=(f + ~1),g] |provided the left construct

55) tlr•[E(i) i=f,g] = [E(i) i=f,(g - ~1)] | does not produce <>.

These identities concern change of variables or bounds, distribution of a function from the right.

56) [E(i) i=f,g] = [E(j) j=f,g] [change bound variable]

57) [E(i) i=~n,f] = [E(i + ~(n-1)) i=~1, f - ~(n-1)]

58) [E(i) i=f,g] • h = [E(i)•h i=f•h,g•h]

There are many other possible identities, but these will suffice for our examples and indicate the possibilities. Note that they are easily derivable from the function level semantics of FP.

5.5 Examples of program optimization

<u>5.5.1 Inner Product.</u> The following FP program IP for inner product has the right form: it is concise and clear; it does not depend on the size of its arguments; it is non-recursive and therefore easy to reason about and transform. But its first operation is an inefficient housekeeping operation, transpose, which our optimization should eliminate.

```
def IP = \+ • @* • trans
```

Thus, for example

```
IP•[[a,b,c],[d,e,f]] = \+•@*•[[a,d],[b,e],[c,f]]
   = (a*d) + (b*e) + (c*f)       using infix function level notation
```

Now the structure operator for IP -- the identity construct for any proper argument of IP, a pair of equal-length sequences -- is the for construct

```
C = [[i•j i=~1,len•1] j=~1,~2]
```

We proceed to transform IP = IP•C by using the above rewrite rules

```
IP•C = \+•@*•trans•[[i•j i=~1,len•1] j=~1,~2]
     = \+•@*•[[i•j j=~1,~2] i=~1,len•1]                by (49)
     = \+•[ *•[i•j j=~1,~2] i=~1,len•1]                by (53)
     = \+•[ *•[i•1,i•2] i=~1,len•1]           by def of for construct
```

This last program corresponds directly to the following Fortran-like program for inner product for the argument <a,b>, where a and b are sequences of equal length.

```
r := 0              [the unit of +]
for i = 1, len:a
   r := r + (a[i] * b[i])
```

So our transformation has eliminated the need to perform transpose or to create the intermediate result @*•trans:<a,b> [provided we implement insertion constructs properly]. It has produced the standard conventional program. Using the same approach we could develop the general identity for IP•C:

```
59)  IP•[[E(i) i=~1,f], [F(j) j=~1,f]]  =  \+•[*•[E(i),F(i)] i=~1,f]
```

<u>5.5.2 Matrix multiplication.</u> Here is the standard FP matrix multiplication program, MM, that will multiply any pair of conformable matrices, <A,B>, where each matrix is represented as the sequence of its rows.

```
def MM = @(@IP)•@distl•distr•[1,trans•2]
```

Again, this is a good program that concisely gives the essence of the operation, but it is very space and time inefficient. The first three operations of this non-recursive program are data-rearranging "housekeeping" operations; all of them should be eliminated by our transformation [and they are]. Our goal is to transform MM into the following construct:

60) MM = [[\+•[*•[k•i•1, j•k•2] k=~1,len•2] j=~1,len•1•2]] i=~1,len•1]

The detailed transformation is given below. The construct (60) represents an efficient Fortran-like program. Let matrices A and B be given with

n_1 = len:A = no. of rows of A

n_2 = len:B = no. of rows of B = len•1:A = length of a row of A, and

n_3 = len•1:B = length of a row of B

Then the construct (60) corresponds exactly to the following Fortran-like program for the product of A and B, i.e., MM:<A,B>

```
for i = 1, n1
   for j = 1, n3
     r[i,j] := 0
     for  k = 1, n2
        r[i,j] := r[i,j] + (A[i,k] * B[k,j])
```

Thus our transformation has turned all the FP "housekeeping" operations [the first three operations of MM] into Fortran-like housekeeping operations. In a sense the transformation has turned the original program MM inside out; it has eliminated all the unnecessary intermediate results of the FP program and produced the standard matrix multiplication program.

<u>Details of the transformation of MM.</u> To use our strategy we wish to find a construct that is the identity for pairs of conformable matrices. Let

61) MI = [E(i) i=~1,len]   where

62) E(i) = [k•i k=~1,len•1]

Then MI is the identity construct for any matrix, and

63) [MI•1, MI•2]   is the identity for any pair of matrices

64) len•1•1 = len•2 is the conformability requirement that the length of a row of A (len•1•1) equals the length of a column [i.e., the no. of rows] of B (len•2)

So our strategy is to transform MM = MM•[MI•1,MI•2] by the above laws, where we can assume that (64) holds.

65) [1,trans•2]•[MI•1, MI•2] = [MI•1, trans•MI•2]   by (25) (3) (12)
                             = [MI•1, MI′•2]

where

66) MI′ = trans•MI = [E′(j) j=~1,len•1]   by (49) (56)
      where E′(j) = [j•k k=~1,len]

So we now have

```
67)  MM = @(@IP)•@distl•distr•[MI•1, MI′•2]
68)     = @(@IP)•@distl•distr•[[E(i)•1 i=~1,len•1], MI′•2]   by (61) (58) (14)
69)     = @(@IP)•@distl•[ [E(i)•1, MI′•2] i=~1,len•1]         by (51)
70)     = @(@IP)•[ distl•[E(i)•1, MI′•2] i=~1,len•1]          by (53)
71)     = @(@IP)•[ distl•[E(i)•1, [E′(j)•2 j=~1,len•1•2] ] i=~1,len•1]
                                                              by (66) (58) (14)
72)     = @(@IP)•[ [[E(i)•1, E′(j)•2] j=~1,len•1•2] ] i=~1,len•1]   by (52)
73)     = [ [IP•[E(i)•1, E′(j)•2] j=~1,len•1•2] i=~1,len•1]   by (53) twice
```

Now

```
IP•[E(i)•1, E′(j)•2] = IP•[[k•i•1 k=~1,len•1•1], [j•k•2 k=~1,len•2]]
                                                              by (62) (66)
```

Therefore we can use the identity for IP (59) to get

```
IP•[E(i)•1, E′(j)•2] = \+•[ *•[k•i•1, j•k•2] k=~1,len•2]
```

since by (64) len•1•1 = len•2. Substituting this last result in (73) gives (60), the desired construct.

5.5.3 Eliminating copying. Our final example demonstrates that our optimization technique can eliminate unnecessary copying of arguments. The preceding examples proceeded uniformly "downhill". That is, whenever a rule applies it eliminates a function from the composition that we are working on: g•C => C′. In this example we see a step g•C => g•C′ that requires a "pattern matching" step that merely alters the construct so that a reducing rule g•C′ => C′′ will then apply.

Consider the following program

```
f = @+ • trans • [tlr,tl]
```

This program, given $\langle x_1,\ldots,x_n\rangle$, computes $\langle x_1+x_2,\ldots,x_{n-1}+x_n\rangle$. The first operation constructs two almost-complete copies of the argument, whereas it would be simpler to compute the result from the original argument. Beginning with C = [i i=~1,len], the identity construct for sequences, the proper arguments for f, we transform f = f•C as follows

```
f = @+•trans•[tlr,tl]•[i i=~1,len]
  = @+•trans•[tlr•[i i=~1,len], tl•[i i=~1,len]]     by (25)
  = @+•trans•[[i i=~1,(len - ~1)], [i i=~2,len]]     by (54) (55)
```

Now the rule (50) would apply if the bounds on i were the same in both constructs, but they are not. However (57) applied to the second construct makes the bounds equal. This is the pattern-matching step.

```
  = @+•trans•[[i i=~1,(len - ~1)], [(i+~1) i=~1,(len - ~1)]]   by (57)
  = @+•[[i,i+~1] i=~1,(len - ~1)]                               by (50)
```

```
= [+•[i,i+~1] i=~1,(len - ~1)]                    by (53)
```

Thus again we have reduced our program to a construct, one that represents the following Fortran-like program to compute f:x

```
for i = 1, (len:x - 1)
    r[i] := x[i] + x[i+1]
```

This has eliminated all intermediate results and copying.

6. Conclusions

We have argued that programs with good mathematical properties must be built at the function level. We have exhibited a system whose semantics are [or can be] given by function level equations, equations whose operations are program-forming operations and whose constants and variables range over programs. From these equations theorems are derived whose conclusions are again equations, for example, one that asserts the equality of a recursive and an iterative function.

These equations, together with many others, constitute a fairly large body of general knowledge, knowledge that can be easily and uniformly applied to a fragment [i.e., subtree] of any function level program. We hope the examples show that this is a clear and preferable way to present general results about programs, so that they can be easily recognized and applied by users.

Although some functional languages are stunningly elegant for expressing certain programs, notably SASL and KRC [Turner 81], they often tie themselves to the object level in one or more ways: by using lambda abstraction, by using curried functions, and by separating conditional expressions into separate equations: When lambda abstraction is used to express an object-to-object function it requires the use of object variables. If h is an arbitrary curried function, then one cannot perceive its arity in (h a) unless one knows that, e.g., (h a x y) is an object. Thus for general expressions to be readable they must be object level [even then, considerable analysis may be needed to resolve the arity of h]. Finally, SASL-style pattern matching equations without conditions depend on the presence of mutually exclusive object level expressions as arguments, hence they must be object level.

I believe languages like SASL are excellent for <u>writing</u> programs, but for the reasons outlined above, they tend to be object level ones and hence lack many of the advantages of the function level technology for <u>general</u> reasoning about and transforming programs [activities that are <u>essentially</u> function level]. With the addition of higher level functions and infinite sequences as in FP_{84} [Halpern, Williams,

Wimmers, Winkler 85], it seems clear that any object level program can be transliterated into a function level equivalent, thereby making it subject to all the laws and theorems of that technology. And if it is a close transliteration, as I believe it can be in most cases, then that means that function level programs should be as clear, and as easy to read and to write as SASL programs.

Uncurried function level expressions [using construction as a combining form] help to make clear the structure of a function's argument in a sharper way than in curried expressions; they also reduce the number of function levels needed, reserving higher levels for truly higher level matters. Uncurried function level definitions that use construction and function variables correspond exactly to object level ones; they also serve as useful function level laws about the defined function.

The examples of FP program optimization give some hope that FP data-rearranging housekeeping operations [when they can be transformed into conventional ones], are practically viable, even though they represent extreme inefficiencies when done literally. If that turns out to be true, this should be a considerable advance for programming since conventional housekeeping operations represent one of the most central but disorderly and non-reusable aspects of programming today. The FP housekeeping operations, on the other hand, are compact and reusable, and have useful mathematical properties.

The work on optimization reported here is, I believe, just the beginning of a line of interesting and useful research. It shows that some programs [that one might call "fixed-data processing" in contrast to "list-processing", and that are much clearer and more orderly than Fortran programs] can be transformed into Fortran programs that run as fast as if they had been written in Fortran in the first place. It shows that the strategy for doing this, in some cases, is just the use of mechanical "downhill" rules: find a general equation $f \cdot C = C'$, where f is the rightmost operation in the function to be optimized, substitute the proper constants, replace $f \cdot C$ by the construct C', and continue.

At least for certain kinds of programs, the optimization technology suggested here has a potential for producing higher speed results than various combinator technologies [Hughes 82, Stoye, Clarke & Norman 84, Turner 79]. Our approach is source level, leading to programs of a fairly standard kind with known, easy to estimate efficiencies. Although these other methods employ some interesting source level transformations, their target language is a list-processing one and has several complexities [e.g., pointer manipulation, garbage collection] behind the language "curtain" that make estimations of running times complex. This makes it hard to evaluate alternative strategies for optimization, and in any case these target languages do not have the potential efficiency that Fortran-like languages do.

Many questions remain, of course. How to distinguish between the fixed-data and list-processing cases? Is it possible to develop mixed techniques for handling the two, in which some data will have relatively fixed allocations and be addressed by Fortran-like techniques, and some will be stored and treated by list-processing techniques? How can recursively-defined functions be optimized? What recursively defined functions can be transformed into closed form definitions that can then be optimized? Can other constructs be added that will enable us to use algebraic techniques in doing a lot of lower level optimizations also at the extended-source level? How complex will optimization strategies have to be to deal effectively with a large class of practical programs?

In spite of all the remaining questions, the work on optimization indicates that making full use of the mathematical properties of function level programs promises to have considerable payoffs. These will certainly be required to do the really extensive optimization that is needed to make easy-to-write function level programs run at the speed we have come to expect of Fortran programs.

Acknowledgments It is a pleasure to acknowledge the valuable help I received from John H. Williams on a number of the examples, and the many helpful discussions about various issues in the paper with him and with Edward L. Wimmers.

References

Backus, J. (1978) Can programming be liberated from the von Neumann style? A functional style and its algebra of programs. CACM **21**:8 (August)

Backus, J. (1981a) Is computer science based on the wrong fundamental concept of "program"? An extended concept. in Algorithmic Languages, de Bakker & van Vliet, eds., North-Holland, Amsterdam (1981)

Backus, J. (1981b) The algebra of functional programs: function level reasoning, linear equations, and extended definitions. Lecture Notes in Computer Science #107, Springer-Verlag (April)

Backus, J. (1981c) Function level programs as mathematical objects. Proc. Conf. on Functional Programming Languages and Computer Architecture. ACM (October)

Bird, R. S. (1984) The promotion and accumulation strategies in transformational programming. Trans. Prog. Langs. & Systs. **6**,4 ACM (October)

Burstall, R. M., and Darlington, J. (1977) A transformation system for developing recursive programs. JACM **24**,1 (January)

Cohen, N. H. (1979) Characterization and elimination of redundancy in recursive programs. Proc. 6th Symp. on Principles of Prog. Langs. ACM

Dijkstra, E. W. (1976) A Discipline of Programming. Prentice-Hall

Guttag, J., Horning, J., and Williams, J. (1981) FP with data abstraction and strong typing. Proc. Conf. on Functional Programming Languages and Computer Architecture. ACM (October)

Halpern, J. Y., Williams, J. H., Wimmers, E. L., and Winkler, T. C. (1985) Denotational semantics and rewrite rules for FP. Proc. 12th Symp. on Princ. Prog. Langs., ACM (January)

Kieburtz, R. B., and Shultis, J. (1981) Transformations of FP program schemes. Proc. Conf. on Functional Programming Languages and Computer Architecture. ACM (October)

Hoare, C. A. R. (1969) An axiomatic basis for computer programming. CACM 12,10 (October)

Hughes, R. J. M. (1982) Super combinators: a new implementation method for applicative languages. Proc. Symp. on LISP and Functional Programming ACM (August)

Manna, Z. (1974) Mathematical Theory of Computation. McGraw-Hill

Mills, H. D. (1975) The new math of computer programming. CACM 18,1 (January)

Stoye, W. R., Clarke, T. J. W. and Norman, A. C. (1984) Some practical methods for rapid combinator reduction. Proc. Symp. on LISP and Functional Programming ACM (August)

Turner, D. A. (1979) A new implementation technique for applicative languages, Software Practice & Experience, vol 9, pp31-49

Turner, D. A. (1981) The semantic elegance of applicative languages. Proc. Conf. on Functional Programming Languages and Computer Architecture. ACM (October)

Wadler, P. (1981) Applicative style programming, program transformation, and list operators. Proc. Conf. on Functional Programming Languages and Computer Architecture. ACM (October)

Wadler, P. (1984) Listlessness is better than laziness: lazy evaluation and garbage collection at compile time. Proc. Symp. on LISP and Functional Programming ACM (August)

Williams, J. H. (1982) On the development of the algebra of functional programs. Trans. Prog. Langs. & Systs. 4,4 ACM (October)

INDUCTIVELY DEFINED FUNCTIONS

(Extended Abstract)

R.M. Burstall
Dept. of Computer Science
University of Edinburgh
King's Buildings, Mayfield Road
Edinburgh EH9 3JZ
Scotland, U.K.

1. Introduction

A number of people have advocated the use of initial algebras to define data types in specification languages, see for example Burstall and Goguen (1981). Two aspects of this have worried me somewhat

- we do not have a really convenient way to define functions using the unique homomorphism property of the initial algebra
- we do not have any obvious way to prove inequations about the data elements.

I sketch here a proposal for defining functions by the unique homomorphism property, and I show how we can prove inequations using such functions. The function definition mechanism can also be seen as a programming language proposal for "inductive" case expressions and I formulate it in ML syntax (Milner 1984).

2. Definitions in ML

A new data type is introduced in ML by giving the alternative ways of constructing elements of that type. Thus for example to introduce (linear) lists of integers

*type rec intlist = nil | cons of (int * intlist)*

This defines the new type *intlist* together with the constructors

nil: *intlist*
cons: *int * intlist -> intlist*

The natural numbers could be defined by

type rec nat = zero | succ of nat

To define functions over such data types we resort to recursion. We use **val** to define values, just as **type** defines types. Thus

```
val rec length l = case l of
                      nil        . zero
                      cons(i, l1) . 1 + length l1
```

In each case a constructor on the left introduces a number of variables which are bound by matching, for example *i* and *l*1. Similarly

```
val rec plus(m, n) = case m of
                        zero    . n
                        succ m1 . succ(plus(m1, n))
```

We can easily make definitions by recursion which do not terminate. But "obvious" termination is rather common in practical programming, since many functions are defined by primitive recursion.

3. Defining functions inductively by cases

I would like to propose a variant of the ML case construction which makes the termination immediate from the syntax. We will write "ind case" for "inductive case". The syntax of an "ind case" expression is the same as that of a case expression.

Let us call the expression after the word **case** the argument of the case expression. Now the new feature for **ind case** is that if a variable *v* appearing on the left in the matching position inside a constructor has the same type as the argument then not only is *v* declared for use on the right but so is another special variable named **$*v***. This **$*v*** is bound to the value of the whole case expression in which the argument has been replaced by *v*. The original **ind case** then becomes simply **case**. Some examples will help.

```
val plus(m, n) = ind case m of
                    zero    . n
                    succ m1 . succ($m1)
```

Here the new variable **$*m1*** represents the value of the whole **ind case** expression replacing *m* by *m1*. Thus we could expand to

```
val plus(m, n) = case m of
                  zero    . n
                  succ m1 . succ( ind case m1 of z
                                   zero    . n
                                   succ m1 . succ $m1 )
```

Further such expansions will push the **ind case** expression arbitrarily deep in a nested case expression and enable us to calculate *plus*(*m*, *n*) for any finite *m*. By this informal argument we see that since all elements of ML data types are finitely deep **ind case** expressions always terminate. This is the advantage they have over explicit recursion.

We may also note that the **$*m1*** replaces a recursive call of *plus* in the previous definition. We could think of the **ind case** expression in general as standing for some anonymous recursive function applied to the argument expression; the **$** sign then corresponds to a recursive call of this function. This recursive call must, by our syntax, be applied to a component of the original argument. Hence the guarantee of termination.

The length example is similarly accomplished without recursion

```
val length l = ind case l of
                  nil         . zero
                  cons(i, l1) . succ $l1
```

Another familiar example

```
val fact n = ind case n of
               zero    . succ zero
               succ n1 . n * $n1
```

A tree example, summing the integers on the nodes

```
type rec tree = niltree | node of tree * int * tree

val sum t = ind case t of
              niltree          . 0
              node(t1, i, t2) . $t1 + i + $t2
```

Consider however an alternative definition of *plus*

```
val rec plus(m, n) = case m of
                       zero    . n
                       succ m1 . plus(m1, succ n))
```

Here we apply *plus* recursively to *m1*, but with the parameter *n* increased to *succ n*. There seems to be no way to express such definitions using **ind case**. We can however "curry" the definition of *plus*, and then translate it (noting that in ML **fun** means lambda)

```
plus : nat -> (nat->nat)

val rec plus m n = case m of
                        zero    . (fun n. n)
                        succ m1 . (fun n. plus m1 (succ n))
```

This becomes

```
val plus m n = ind case m of
                        zero    . (fun n. n)
                        succ m1 . (fun n. $m1 (succ n))
```

This is not particularly nice but the best I can do. Any better ideas?

The Fibonacci function which recurses on both n-1 and n-2 also presents a problem, but one can overcome this using the ML **as** construction which binds a variable to a subpattern.

Of course one can use second order functionals, like *maplist*, to capture primitive recursion, but they still need termination proofs and programs using them are not very readable.

4. Equational data types

The notation used in ML to introduce a recursive data type is just a cute way of defining a signature. The data type is the initial algebra on this signature. In specification languages we may be interested in defining the initial algebra on a signature subject to some equations. Finite strings, bags (alias multisets) and sets are all easily definable by adding equations for identity, associativity, commutativity and absorption. So for specification purposes let us extend the ML syntax slightly to allow equations, introducing a keyword **under**. Using __ as an infixed operator for appending, we define strings thus

```
type rec intstring = empty | unit int | intstring __ intstring
    under  empty __ s = s
    and    s __ empty = s
    and    s __ (t __ u) = (s __ t) __ u
```

We can define functions recursively on these equational data types, using cases.

For example

```
val slength s = ind case s of
                    empty . 0
                    unit i . 1
                    s1 __ s2 . $s1 + $s2
```

But in order for this definition to be deterministic we have to check some equations derived from the ones for strings

$$0 + n = n$$
$$n + 0 = 0$$
$$l + (m + n) = (l + m) + n$$

All these are elementary properties of +.

To derive the equations to be checked one may note that the right hand sides define operations corresponding to *empty*, *unit* and __ and we have to show that these operations obey the same equations as the constructors. (I am still a little fuzzy about a good way to say this precisely.) These new operations are the operations of the target algebra of the unique homomorphism which is being defined by the **ind case** expression.

5. Proving inequations

From the defining equations it is easy to prove other equations by using the usual properties of equality, substitution, transitivity, etc. But how can we prove inequations?

This is less obvious. Do we have to show somehow that a certain equation is <u>not</u> provable from the defining ones?

I want to show how inequations can be proved using another approach. First we note that if there are no equations terms are unequal just if they have different constructors, or (recursively) if they have the same constructor but some pair of components are unequal. This gives us some inequations to start off with e.g. true $\neq$ false, zero $\neq$ succ n.

But what if there are defining equations? We must use the basic property of the initial algebra, the existence of a unique homomorphism to any other algebra which satisfies the equations. Suppose this homomorphism is f. Then we can prove x $\neq$ y by observing that f(x) $\neq$ f(y). Now f(x) and f(y) may take their values in a data type where we already know some inequations. If not we must apply a similar trick until we get back to a type with no defining equations for which, as we have seen, the inequations are immediate.

The function f, acts as a discriminator, relating the type to another one which is already known. This is of course reminiscent of Guttag's idea of sufficient completeness.

Let us consider bags as an example. Suppose "++" has been declared syntactically to be an infixed operator. We define bags to be unordered sequences, with possible repetitions

type rec *bag* = *empty* | *int*++*bag*
  under *x*++*y*++*b* = *y*++*x*++*b*

It is convenient to write δ_{xy} for ***if*** $x = y$ ***then*** 1 ***else*** 0

val *count*(*x*,*b*) = ***ind case*** *b* ***of***
  empty . 0
  y++*c* . \$*c* + δ_{xy}

To ensure determinacy of this definition we check that $\lambda(y,\$c).\$c + \delta_{xy}$ satisfies the equation for ++, that is

$$(\$c + \delta_{xy'}) + \delta_{xz} = (\$c + \delta_{xz'}) + \delta_{xy}$$

We will write b_x for *count*(*x*,*b*), as an abbreviation.

Suppose we want to show that *empty* $\neq$ *x*++*empty*. Since type *nat* has no equations we know that *zero* $\neq$ *succ zero*. But $empty_x = zero$ and $(x{+}{+}empty)_x = succ\ zero$. So *empty* $\neq$ *x*++*empty*. Note how this depends on the deterministic property of count. Similarly we might show that *x*++*b* $\neq$ *b*. (My thanks are due to Horst Reichel for help with this example.)

But how do we know that *count* is sufficient to discriminate between <u>all</u> unequal bags? We need to show that different bags have a different count for some *x*. We wish to prove

Theorem $(\forall x. b_x = c_x) \Rightarrow b = c$

For the proof of this theorem we need an auxiliary definition. Assume that "—" has been declared as an infix.

val *b*—*y* = ***ind case*** *b* ***of***
  empty . *empty*
  x++*c* . ***if*** $x = y$ ***then*** *c* ***else*** *x*++\$*c*

Thus *b*—*y* deletes one occurrence of *y* from *b* if possible. We need three lemmas for the proof.

Lemma 1. $\forall x. b_x = 0 \Rightarrow b = empty$

Lemma 2. If $b_y > 0$ then $(b - y)_x = b_x - \delta_{xy}$

Lemma 3 If $b_x > 0$ then $x{+}{+}(b - x) = b$

Lemma 1 is immediate, the other two are proved by induction.

The proof of the theorem is then by induction on b.

For the data type *set* the function analogous to *count* would be membership.

Notice that the initial algebra gives rise to an induction principle and to use this we have to invent a suitable predicate to prove by induction. This comes from the 'no junk' property of the initial algebra. The 'no confusion' property gives rise to inequations, and here to do proofs we have to invent a suitable discriminant function. There is some pleasant feeling of duality here.

We have made our data definitions in equational logic, but drawn conclusions from them using inequalities and quantifiers. This is an example of the use of two different 'institutions' in one specification language, a trick called 'duplicity' in Burstall and Goguen (1981) and Goguen and Burstall (1984).

The deadline for this invited paper arrived before I had fully understood even these rather elementary matters. Please forgive the sketchy and tentative nature of this contribution.

Acknowledgements

I would like to thank Joe Goguen, Horst Reichel, Robin Milner and Andrzej Tarlecki (among others) for illuminating discussions. I am grateful to SERC and BP for support and to Eleanor Kerse for rapid scribing.

References

Burstall, R. and Goguen, J. An informal introduction to specification using Clear. In Boyer, R. and Moore, J (editor), The Correctness Problem in Computer Science, pages 185-213. Academic Press, 1981.

Goguen, J. and Burstall, R. Introducing institutions. In Logics of Programs. Springer LNCS No. 164, (eds. Clarke and Kozen), 1984.

Milner, R. The Standard ML core language. Computer Science Dept. Report, Univ. of Edinburgh, 1984.

THREE APPROACHES TO TYPE STRUCTURE†

John C. Reynolds

Syracuse University

Syracuse, New York 13210, U.S.A.

ABSTRACT We examine three disparate views of the type structure of programming languages: Milner's type deduction system and polymorphic let construct, the theory of subtypes and generic operators, and the polymorphic or second-order typed lambda calculus. These approaches are illustrated with a functional language including product, sum and list constructors. The syntactic behavio: of types is formalized with type inference rules, but their semantics is treated intuitively.

1. INTRODUCTION

The division of programming languages into two species, typed and untyped, has engendered a long and rather acrimonious debate. One side claims that untyped languages preclude compile-time error checking and are succinct to the point of unintelligibility, while the other side claims that typed languages preclude a variety of powerful programming techniques and are verbose to the point of unintelligibility.

From the theorist's point of view, both sides are right, and their arguments are the motivation for seeking type systems that are more flexible and succinct than those of existing typed languages. This goal has inspired a substantial volume of theoretical work in the last few years. In this paper I will attempt to survey some of this work at a level that I hope will reveal its implications for future languages.

The main difficulty that I face is that type theory has moved in several directions that, as far as we presently know, are incompatible with one another. This situation dictates the organization of this paper: Section 2 lays a groundwork that is common to all directions, while each later section is devoted to a particular direction. Thus the later sections are largely independent of one another.

Primarily because of my limited knowledge, this survey will be far from comprehensive. Neither algebraic data types [1-3] nor conjunctive types [4,5] will be covered. Nor will I describe any of several fascinating systems [6-9] in which types and values are so intertwined that types become full-blown program specifications and programs become constructive proofs that such specifications are satisfiable.

Moreover, I will consider only functional languages, since the essential character of type structure is revealed more clearly without the added complexity of imperative features. (I believe that the proper type structure for Algol-like languages is obtained from the subtype discipline of Section 4 by taking "types" to be "phrase

†Work supported by National Science Foundation Grant MCS-8017577.

types", built from primitive types such as <u>integer</u> <u>expression</u>, <u>real</u> <u>variable</u>, and <u>command</u> [10-12].)

To be accessible to readers who are untrained in mathematical semantics, our exposition will be formal but not rigorous. In particular, we will often discuss semantics at an intuitive level, speaking of the set denoted by a type or the function denoted by an expression when rigorously we should speak of the domain denoted by a type or the continuous function denoted by an expression. Regrettably, this level of discourse will obscure some profound controversies about the semantics of types.

Finally, I must apologize for omissions and errors due to either ignorance or haste. This is a preliminary report, and I would welcome suggestions from readers for corrections or extensions.

2. THE BASE LANGUAGE

Although the approaches to type structure that we are going to survey are incompatible with one another, they are all built upon a common view of what types are all about. In this section, we will formalize this view as a "base" language, in terms of which the various approaches can be described as extensions or modifications. First we will introduce the expressions of the base language as though it were a typeless language. Then we will introduce types and give rules for inferring the types of expressions. Finally we will show how expressions can be augmented to contain enough type information that the inference of their types becomes trivial.

Because it is intended for illustrative purposes, the base language is more complicated than a well-designed functional language should be. In several instances, it contains similar constructs (e.g. numbered and named alternatives) that are both included only because they exhibit significant differences in some extension of the language.

2a. Expressions

To define expressions we will give their abstract syntax, avoiding any formalization of precedence or implicit parenthesization. In this definition we write K for the set of positive integers, Z for the set of all integers, I for the (countably infinite) set of identifiers, and E for the set of expressions. The simplest kinds of expressions are identifiers (used as variables):

E ::= I (S1)

constants denoting integers:

E ::= Z (S2)

boolean values:

E ::= <u>true</u> | <u>false</u> (S3)

and primitive operations on integers and boolean values, such as

$$E ::= \underline{add} \mid \underline{equals} \tag{S4}$$

Less trivially, we have lambda expressions to denote functions, and a notation for function application:

$$E ::= \lambda I.\ E \mid E\ E \tag{S5}$$

Note that we do not require the operand of function application to be parenthesized, so that one can write f x instead of f(x). We will assume that application is left associative and that λ has a lower precedence than application, e.g. λx. λy. f y x means λx. (λy. ((f y) x)).

Informally, the meaning of functions is "explained" by the rule of beta-reduction

$$(\lambda i.\ e_1)\ e_2 = e_1\big|_{i \to e_2}\ , \tag{R5}$$

where the right side denotes the result of substituting e_2 for i in e_1 (with renaming of bound identifiers in e_1 that occur free in e_2). For example, (λx. f x y x)(g a b) has the same meaning as f (g a b) y (g a b).

The rule of beta-reduction implies that our language has "call-by-name" semantics. For example, if i does not occur in e_1 then $(\lambda i.\ e_1)\ e_2$ has the same meaning as e_1, even if e_2 is an expression whose evaluation never terminates. While much of what we are going to say is equally applicable to call by value, we prefer the greater elegance and generality of call by name (particularly since the development of lazy evaluation [13-14] has led to its efficient implementability).

To avoid any special notation for functions of several arguments, we will use the device of Currying, e.g. we will regard add as a function that accepts an integer and yields a function that accepts an integer and yields an integer, so that add 3 4 = 7. In general, where one might expect $\lambda(x_1, \ldots, x_n).\ e$ we will write $\lambda x_1. \ldots \lambda x_n.\ e$, and where one might expect $f(e_1, \ldots, e_n)$ we will write $(\ldots (f\ e_1) \ldots e_n)$ or, with implicit parenthesization, $f\ e_1 \ldots e_n$.

Next, we introduce notation for the construction and analysis of records. Here there are two possible approaches, depending upon whether fields are numbered or named. For records with numbered fields, we use the syntax

$$E ::= \langle E, \ldots, E\rangle \mid E.K \tag{S6}$$

For example, <x, y> denotes a two-field record whose first field is x and second field is y, and if z denotes a two-field record then z.1 and z.2 denote its fields. In general, the meaning of these constructions is determined by the reduction rule

$$\langle e_1, \ldots, e_n\rangle.k = e_k \qquad \text{when } 1 \le k \le n\ . \tag{R6}$$

Notice that this rule (as with beta reduction) implies a call-by-name semantics. For example, $\langle e_1, e_2\rangle.1 = \langle e_2, e_1\rangle.2 = e_1$, even when e_2 does not terminate.

For records with named fields, we use the syntax

$$E ::= \langle I\text{: } E, \dots , I\text{: } E\rangle \mid E.I \tag{S7}$$

with the restriction that the identifiers preceding the colons must be distinct. The reduction rule is

$$\langle i_1\text{: } e_1, \dots , i_n\text{: } e_n\rangle.i_k = e_k \quad \text{when } 1 \le k \le n\ . \tag{R7}$$

With named fields, the value of a record is independent of the order of its fields, e.g. $\langle \text{real: } x, \text{imag: } y\rangle = \langle \text{imag: } y, \text{real: } x\rangle$.

We also introduce notation for alternatives (often called a sum, disjoint union, or variant-record construct):

$$E ::= \underline{\text{inject}}\ K\ E \mid \underline{\text{choose}}(E, \dots , E) \tag{S8}$$

The value of inject k e is the value of e "tagged" with k. If $f_1, \dots , f_n$ are functions, then choose($f_1, \dots , f_n$) is a function that, when applied to the value x tagged with k, yields f_k x. Thus the appropriate reduction rule is

$$\underline{\text{choose}}(f_1, \dots , f_n)(\underline{\text{inject}}\ k\ e) = f_k\ e \quad \text{when } 1 \le k \le n\ . \tag{R8}$$

(Strictly speaking, "tagging" is pairing, so that inject k e is a pair like $\langle k, e\rangle$. But we consider these to be different kinds of pairs so that, for example, (inject k e).1 is meaningless.)

In some languages, the analysis of alternatives is performed by some form of case construction that can be defined in terms of choose, e.g.

$$\underline{\text{altcase}}\ i\text{: } e\ \underline{\text{of}}\ (e_1, \dots , e_n) \equiv \underline{\text{choose}}(\lambda i.e_1, \dots , \lambda i.e_n)\ e\ .$$

However, the choose construction is conceptually simpler since it does not involve identifier binding.

The tags of alternatives, like the fields of records, can be named instead of numbered. For named alternatives, we will use the syntax

$$E ::= \underline{\text{inject}}\ I\ E \mid \underline{\text{choose}}(I\text{: } E, \dots , I\text{: } E) \tag{S9}$$

with the restriction that the identifiers preceding the colons must be distinct. The reduction rule is

$$\underline{\text{choose}}(i_1\text{: } f_1, \dots , i_n\text{: } f_n)(\underline{\text{inject}}\ i_k\ e) = f_k\ e \quad \text{when } 1 \le k \le n\ , \tag{R9}$$

and the meaning of choose(i_1: $f_1, \dots , i_n$: f_n) is independent of the order of the components i_k: f_k.

For the construction of lists, we will use the primitives of LISP:

$$E ::= \underline{\text{nil}} \mid \underline{\text{cons}}\ E\ E \tag{S10a}$$

where nil denotes the empty list and cons x y denotes the list whose first element is x and whose remainder is y. For the analysis of lists, however, we will deviate substantially from LISP:

$$E ::= \underline{\text{lchoose}}\ E\ E \tag{S10b}$$

The value of lchoose e f is a function that, when applied to the empty list, yields e and, when applied to a list with first element x and remainder y, yields f x y. More formally, we have the reduction rules

$$(\underline{\text{lchoose}}\ e\ f)\ \underline{\text{nil}} = e\ ,\qquad \text{(R10)}$$
$$(\underline{\text{lchoose}}\ e\ f)\ (\underline{\text{cons}}\ x\ y) = f\ x\ y\ .$$

The lchoose operation can be defined in terms of the conventional LISP primitives:

$$\underline{\text{lchoose}}\ e\ f = \lambda\ell.\ \underline{\text{if}}\ \text{null}\ \ell\ \underline{\text{then}}\ e\ \underline{\text{else}}\ f\ (\text{car}\ \ell)\ (\text{cdr}\ \ell)\ ,$$

and the LISP primitives can be defined in terms of lchoose and an error operation:

$$\text{null} = \underline{\text{lchoose}}\ \underline{\text{true}}\ (\lambda x.\ \lambda y.\ \underline{\text{false}})\ ,$$
$$\text{car} = \underline{\text{lchoose}}\ \underline{\text{error}}\ (\lambda x.\ \lambda y.\ x)\ ,$$
$$\text{cdr} = \underline{\text{lchoose}}\ \underline{\text{error}}\ (\lambda x.\ \lambda y.\ y)\ .$$

However, in a typed language lchoose is preferable to the LISP primitives since it converts the common error of applying car or cdr to the empty list into a type error.

For the definition of identifiers we use Landin's let construction [15] (albeit with a call-by-name rather than a call-by-value semantics). The syntax is

$$E ::= \underline{\text{let}}\ I = E\ \underline{\text{in}}\ E \qquad \text{(S11)}$$

and the reduction rule is

$$\underline{\text{let}}\ i = e_2\ \underline{\text{in}}\ e_1 = e_1\big|_{i \to e_2}\ .$$

Of course, as noted by Landin, let can be defined in terms of λ and application:

$$\underline{\text{let}}\ i = e_2\ \underline{\text{in}}\ e_1 \equiv (\lambda i.\ e_1)\ e_2\ . \qquad \text{(R11)}$$

However, we will regard let $i = e_2$ in e_1 as an independent construction in its own right, since its typing behavior is significantly different than that of $(\lambda i.\ e_1)\ e_2$.

Finally, we introduce a conditional expression

$$E ::= \underline{\text{if}}\ E\ \underline{\text{then}}\ E\ \underline{\text{else}}\ E \qquad \text{(S12)}$$

with the reduction rules

$$\underline{\text{if}}\ \underline{\text{true}}\ \underline{\text{then}}\ e_1\ \underline{\text{else}}\ e_2 = e_1\ , \qquad \text{(R12)}$$
$$\underline{\text{if}}\ \underline{\text{false}}\ \underline{\text{then}}\ e_1\ \underline{\text{else}}\ e_2 = e_2\ ,$$

a case (branch-on-integer) expression

$$E ::= \underline{\text{case}}\ E\ \underline{\text{of}}\ (E, \ldots, E) \qquad \text{(S13)}$$

with the reduction rule

$$\underline{\text{case}}\ k\ \underline{\text{of}}\ (e_1, \ldots, e_n) = e_k \quad \text{when } 1 \le k \le n\ , \qquad \text{(R13)}$$

and a fixed-point expression

$$E ::= \underline{\text{fix}}\ E \qquad \text{(S14)}$$

with the reduction rule

$\underline{\text{fix}}\ e = e(\underline{\text{fix}}\ e)$. (R14)

The last is the key to recursive definition. For example, one can use it to define McCarthy's [16]

$\underline{\text{label}}\ i\text{: } e \equiv \underline{\text{fix}}(\lambda i.\ e)$,

or Landin's

$\underline{\text{letrec}}\ i = e_2\ \underline{\text{in}}\ e_1 \equiv \underline{\text{let}}\ i = \underline{\text{fix}}(\lambda i.\ e_2)\ \underline{\text{in}}\ e_1$.

However, just as with choose, fix is conceptually simpler since it does not involve identifier binding. (We are purposely neglecting the complications of the multiple letrec, which is needed to define simultaneously recursive functions, and whose definition in terms of fix is rather messy.)

It should be noticed that our choice of call-by-name semantics implies that fix can be used to define infinite lists. For example,

$\underline{\text{fix}}(\lambda \ell.\ \underline{\text{cons}}\ 0\ \ell)$

denotes the infinite list (0, 0, ...), and

$\underline{\text{fix}}(\lambda f.\ \lambda n.\ \underline{\text{cons}}\ n\ (f\ (\underline{\text{add}}\ 1\ n)))$

denotes the function mapping an integer n into the infinite list (n, n+1, ...).

2b. Types and their Inference Rules

We now introduce the types of our base language. Intuitively:

int denotes the set of integers.

bool denotes the set {true, false}.

$\omega \rightarrow \omega'$ denotes the set of functions that map values belonging to (the set denoted by) ω into values belonging to (the set denoted by) ω'.

$\underline{\text{prod}}(\omega_1, \ldots, \omega_n)$ denotes the set of n-field records in which, for $1 \leq k \leq n$, the kth field belongs to ω_k.

$\underline{\text{prod}}(i_1{:}\omega_1, \ldots, i_n{:}\omega_n)$ denotes the set of records with fields named $i_1, \ldots, i_n$, in which each i_k names a field belonging to ω_k.

$\underline{\text{sum}}(\omega_1, \ldots, \omega_n)$ denotes the set of tagged values such that the tag is an integer between 1 and n, and a value with tag k belongs to ω_k.

$\underline{\text{sum}}(i_1{:}\omega_1, \ldots, i_n{:}\omega_n)$ denotes the set of tagged values such that the tag belongs to $\{i_1, \ldots, i_n\}$, and a value with tag i_k belongs to ω_k.

list ω denotes the set of lists whose elements belong to ω.

More formally, the set Ω of types is defined by the abstract syntax

$$\begin{aligned}\Omega ::= \; & \underline{\text{int}} \mid \underline{\text{bool}} \\ & \mid \underline{\text{prod}}(\Omega, \ldots, \Omega) \mid \underline{\text{prod}}(I{:}\Omega, \ldots, I{:}\Omega) \\ & \mid \underline{\text{sum}}(\Omega, \ldots, \Omega) \mid \underline{\text{sum}}(I{:}\Omega, \ldots, I{:}\Omega) \\ & \mid \underline{\text{list}}\ \Omega\end{aligned}$$

with the proviso that the pairs $I{:}\Omega$ in $\underline{\text{prod}}(I{:}\Omega, \ldots, I{:}\Omega)$ or $\underline{\text{sum}}(I{:}\Omega, \ldots, I{:}\Omega)$ must begin with distinct identifiers, and that the ordering of these pairs is irrelevant.

Occasionally, we will need to speak of type expressions, which are defined by the same syntax with the added production

$$\Omega ::= T$$

where T is a countably infinite set of type variables.

We will assume that $\to$ is right associative and has a lower precedence than the other type operators. Thus for example, $\underline{\text{int}} \to \underline{\text{list}}\ \underline{\text{int}} \to \underline{\text{int}}$ stands for $\underline{\text{int}} \to ((\underline{\text{list}}\ \underline{\text{int}}) \to \underline{\text{int}})$.

Roughly speaking, an expression has a type if its value belongs to that type. But of course, just as the value of an expression depends upon the values of the identifiers occurring free within it, so the type of an expression will depend upon the types of the identifiers occurring free within it. To deal with this complication, we introduce the notion of a typing.

Let e be an expression, π (often called a type assignment) be a mapping of (at least) the identifiers occurring free in e into types, and ω be a type. Then

$$\pi \vdash e\colon \omega$$

is called a typing, and read "e has type ω under π". For example, the following are valid typings:

$$x\colon \underline{\text{int}},\ f\colon \underline{\text{int}} \to \underline{\text{int}} \vdash f(f\ x)\colon \underline{\text{int}}$$

$$f\colon \underline{\text{int}} \to \underline{\text{int}} \vdash \lambda x.\ f(f\ x)\colon \underline{\text{int}} \to \underline{\text{int}}$$

$$\vdash \lambda f.\ \lambda x.\ f(f\ x)\colon (\underline{\text{int}} \to \underline{\text{int}}) \to \underline{\text{int}} \to \underline{\text{int}}$$

$$\vdash \lambda f.\ \lambda x.\ f(f\ x)\colon (\underline{\text{bool}} \to \underline{\text{bool}}) \to \underline{\text{bool}} \to \underline{\text{bool}}\ .$$

Notice that, as illustrated by the last two lines, a typing of a closed expression can have an empty type assignment, and two typings can give different types to the same expression, even under the same type assignment.

We will now give rules for inferring valid typings of our base language. Each of these inference rules consists of zero or more premises separated by a long horizontal line from a conclusion, and contains various symbols called metavariables. An instance of a rule is obtained by replacing the metavariables by phrases of the appropriate kind as described by the following table (and occasionally subject to restrictions stated with the rule itself):

Metavariable	Kind of phrase
π	type assignments
$i\ i_1\ i_n\ i_k$	identifiers
$\omega\ \omega'\ \omega_1\ \omega_k\ \omega_n$	types
z	integers
$e\ e_1\ e_2\ e_n$	expressions
k	positive integers $(k > 0)$
n	nonnegative integers $(n \geq 0)$

Every instance of every rule has the property that, if the premises of the instance are all valid typings, then the conclusion of the instance is a valid typing. Thus a typing can be proved to be valid by giving a list of typings, ending with the typing in question, in which every typing is the conclusion of an instance of a rule whose premises all occur earlier in the list.

The following is a list of the inference rules, ordered to parallel the abstract syntax of the base language:

$$\frac{}{\pi \vdash i\colon \omega} \quad \text{when } i \text{ is in the domain of } \pi, \text{ and } \pi \text{ assigns } \omega \text{ to } i. \qquad \text{(I1)}$$

$$\frac{}{\pi \vdash z\colon \underline{\text{int}}} \quad \frac{}{\pi \vdash \underline{\text{true}}\colon \underline{\text{bool}}} \quad \frac{}{\pi \vdash \underline{\text{false}}\colon \underline{\text{bool}}} \qquad \text{(I2,I3)}$$

$$\frac{}{\pi \vdash \underline{\text{add}}\colon \underline{\text{int}} \to \underline{\text{int}} \to \underline{\text{int}}} \quad \frac{}{\pi \vdash \underline{\text{equals}}\colon \underline{\text{int}} \to \underline{\text{int}} \to \underline{\text{bool}}} \qquad \text{(I4)}$$

$$\frac{\pi, i\colon \omega \vdash e\colon \omega'}{\pi \vdash \lambda i.\ e\colon \omega \to \omega'} \quad \frac{\pi \vdash e_1\colon \omega \to \omega' \qquad \pi \vdash e_2\colon \omega}{\pi \vdash e_1\ e_2\colon \omega'} \qquad \text{(I5)}$$

$$\frac{\pi \vdash e_1\colon \omega_1 \quad \cdots \quad \pi \vdash e_n\colon \omega_n}{\pi \vdash \langle e_1, \ldots, e_n\rangle\colon \underline{\text{prod}}(\omega_1, \ldots, \omega_n)} \qquad \text{(I6a)}$$

$$\frac{\pi \vdash e\colon \underline{\text{prod}}(\omega_1, \ldots, \omega_n)}{\pi \vdash e.k\colon \omega_k} \quad \text{when } 1 \leq k \leq n \qquad \text{(I6b)}$$

$$\frac{\pi \vdash e_1\colon \omega_1 \quad \cdots \quad \pi \vdash e_n\colon \omega_n}{\pi \vdash \langle i_1\colon e_1, \ldots, i_n\colon e_n\rangle\colon \underline{\text{prod}}(i_1\colon \omega_1, \ldots, i_n\colon \omega_n)} \qquad \text{(I7a)}$$

$$\frac{\pi \vdash e\colon \underline{\text{prod}}(i_1{:}\omega_1, \ldots, i_n{:}\omega_n)}{\pi \vdash e.i_k\colon \omega_k} \quad \text{when } 1 \le k \le n \qquad \text{(I7b)}$$

$$\frac{\pi \vdash e\colon \omega_k}{\pi \vdash \underline{\text{inject}}\ k\ e\colon \underline{\text{sum}}(\omega_1, \ldots, \omega_n)} \quad \text{when } 1 \le k \le n \qquad \text{(I8a)}$$

$$\frac{\begin{array}{l}\pi \vdash e_1\colon \omega_1 \to \omega\\ \vdots\\ \pi \vdash e_n\colon \omega_n \to \omega\end{array}}{\pi \vdash \underline{\text{choose}}(e_1, \ldots, e_n)\colon \underline{\text{sum}}(\omega_1, \ldots, \omega_n) \to \omega} \qquad \text{(I8b)}$$

$$\frac{\pi \vdash e\colon \omega_k}{\pi \vdash \underline{\text{inject}}\ i_k\ e\colon \underline{\text{sum}}(i_1{:}\omega_1, \ldots, i_n{:}\omega_n)} \quad \text{when } 1 \le k \le n \qquad \text{(I9a)}$$

$$\frac{\begin{array}{l}\pi \vdash e_1\colon \omega_1 \to \omega\\ \vdots\\ \pi \vdash e_n\colon \omega_n \to \omega\end{array}}{\pi \vdash \underline{\text{choose}}(i_1{:}e_1, \ldots, i_n{:}e_n)\colon \underline{\text{sum}}(i_1{:}\omega_1, \ldots, i_n{:}\omega_n) \to \omega} \qquad \text{(I9b)}$$

$$\frac{}{\pi \vdash \underline{\text{nil}}\colon \underline{\text{list}}\ \omega} \qquad \frac{\begin{array}{l}\pi \vdash e_1\colon \omega\\ \pi \vdash e_2\colon \underline{\text{list}}\ \omega\end{array}}{\pi \vdash \underline{\text{cons}}\ e_1\ e_2\colon \underline{\text{list}}\ \omega} \qquad \text{(I10a,b)}$$

$$\frac{\begin{array}{l}\pi \vdash e_1\colon \omega'\\ \pi \vdash e_2\colon \omega \to \underline{\text{list}}\ \omega \to \omega'\end{array}}{\pi \vdash \underline{\text{lchoose}}\ e_1\ e_2\colon \underline{\text{list}}\ \omega \to \omega'} \qquad \text{(I10c)}$$

$$\frac{\begin{array}{l}\pi \vdash e_2\colon \omega\\ \pi, i\colon \omega \vdash e_1\colon \omega'\end{array}}{\pi \vdash \underline{\text{let}}\ i = e_2\ \underline{\text{in}}\ e_1\colon \omega'} \qquad \text{(I11)}$$

$$\frac{\begin{array}{l}\pi \vdash e\colon \underline{\text{bool}}\\ \pi \vdash e_1\colon \omega\\ \pi \vdash e_2\colon \omega\end{array}}{\pi \vdash \underline{\text{if}}\ e\ \underline{\text{then}}\ e_1\ \underline{\text{else}}\ e_2\colon \omega} \qquad \frac{\begin{array}{l}\pi \vdash e\colon \underline{\text{int}}\\ \pi \vdash e_1\colon \omega\\ \vdots\\ \pi \vdash e_n\colon \omega\end{array}}{\pi \vdash \underline{\text{case}}\ e\ \underline{\text{of}}\ (e_1, \ldots, e_n)\colon \omega} \qquad \text{(I12,I13)}$$

$$\frac{\pi \vdash e\colon \omega \rightarrow \omega}{\pi \vdash \underline{\text{fix}}\ e\colon \omega} \qquad \text{(I14)}$$

In rules I5a and I11, the notation π, i:ω denotes the type assignment that assigns ω to i and assigns to all other identifiers in the domain of π the same type that is assigned by π.

To illustrate the use of these rules we give a proof of the validity of a typing of the expression

$\underline{\text{fix}}$(λap. λx. λy. $\underline{\text{lchoose}}$ y (λn. λz. $\underline{\text{cons}}$ n (ap z y)) x) ,

which denotes a function for appending two lists. At the right of each typing in the proof we indicate the rule of which it is a conclusion. To save space, we write π_1 and π_2 to abbreviate the following type assignments:

π_1 = ap: list int → list int → list int, x: list int, y: list int

π_2 = π_1, n: int, z: list int

Then the proof is:

$\pi_2 \vdash$ ap: list int → list int → list int (I1)
$\pi_2 \vdash$ z: list int (I1)
$\pi_2 \vdash$ ap z: list int → list int (I5b)
$\pi_2 \vdash$ y: list int (I1)
$\pi_2 \vdash$ ap z y: list int (I5b)
$\pi_2 \vdash$ n: int (I1)
$\pi_2 \vdash$ cons n (ap z y): list int (I10b)
π_1, n: int $\vdash$ λz. cons n (ap z y): list int → list int (I5a)
$\pi_1 \vdash$ λn. λz. cons n (ap z y): int → list int → list int (I5a)
$\pi_1 \vdash$ y: list int (I1)
$\pi_1 \vdash$ lchoose y (λn. λz. cons n (ap z y)): list int → list int (I10c)
$\pi_1 \vdash$ x: list int (I1)
$\pi_1 \vdash$ lchoose y (λn. λz. cons n (ap z y)) x: list int (I5b)
ap: list int → list int → list int, x: list int (I5a)
    $\vdash$ λy. lchoose y (λn. λz. cons n (ap z y)) x: list int → list int
ap: list int → list int → list int (I5a)
    $\vdash$ λx. λy. lchoose y (λn. λz. cons n (ap z y)) x:
        list int → list int → list int
$\vdash$ λap. λx. λy. lchoose y (λn. λz. cons n (ap z y)) x: (I5a)
        (list int → list int → list int) → list int → list int → list int
$\vdash$ fix(λap. λx. λy. lchoose y (λn. λz. cons n (ap z y)) x): (I14)
        list int → list int → list int

As a second example, the reader might prove the typing

$\vdash$ <u>fix</u>(λred. λℓ. λf. λa. <u>lchoose</u> a (λn. λz. f n (red z f a)) ℓ):
<u>list</u> <u>int</u> → (<u>int</u> → <u>bool</u> → <u>bool</u>) → <u>bool</u> → <u>bool</u>

of the "reduce" function that, when applied to a list (x_1, ... , x_n), a function f, and a value a, gives f x_1 (f x_2 (... (f x_n a) ...)).

2c. Explicit Typing

So far we have taken the attitude that types are properties of expressions that can be inferred but do not actually appear within expressions. This view is subject to several criticisms:

The problem of generating typings of an expression is an instance of proof generation, for which - in general - there may be no efficient algorithm or even no algorithm at all. (Although we will see in the next section that an efficient typing algorithm is known for a slight variation of our base language, such algorithms are not known for some of the language extensions that will be discussed later.)

An expression will often have more than one valid typing. For example, the typings shown in the previous subsection remain valid if <u>int</u> and <u>bool</u> are replaced by arbitrary types. Thus, if one takes the view that different typings lead to different meanings, then our base language is ambiguous.

Presumably the competent programmer knows the types of the programs he writes. By preventing him from communicating this knowledge in his programs, we are precluding valuable error checking and degrading the intelligibility of programs.

These criticisms suggest modifying our base language so that expressions contain type information. We will call such a modified language <u>explicitly typed</u> if it satisfies two criteria:

(1) Every expression, under a particular type assignment, has at most one type. (Thus there is a partial function, called a <u>typing function</u>, that maps an expression e and a type assignment π into the type of e under π.)

(2) The type of any expression, under a particular type assignment, is a function of the types of its immediate subexpressions under particular (though perhaps different) type assignments.

In the specific case of our base language, to obtain explicit typing we must require type information to appear in four contexts: lambda expressions, <u>inject</u> operations for numbered and named alternatives, and <u>nil</u>. (In each of these contexts, we will write the type information as a subscript.) Thus explicit typing requires the modification of four pairs of abstract syntax and type-inference rules:

$$E ::= \lambda I_{\Omega}.\ E \qquad \frac{\pi,\ i{:}\ \omega \vdash e{:}\ \omega'}{\pi \vdash \lambda i_{\omega}.\ e{:}\ \omega \to \omega'} \tag{E5a}$$

$$E ::= \underline{\text{inject}}_{\Omega,\dots,\Omega}\ K\ E \qquad \frac{\pi \vdash e{:}\ \omega_k \qquad \text{when } 1 \le k \le n}{\pi \vdash \underline{\text{inject}}_{\omega_1,\dots,\omega_n}\ k\ e{:}\ \underline{\text{sum}}(\omega_1,\ \dots\ ,\ \omega_n)} \tag{E8a}$$

$$E ::= \underline{\text{inject}}_{I:\Omega,\dots,I:\Omega}\ I\ E \qquad \frac{\pi \vdash e{:}\ \omega_k \qquad \text{when } 1 \le k \le n}{\pi \vdash \underline{\text{inject}}_{i_1:\omega_1,\dots i_n:\omega_n}\ i_k\ e{:}\ \underline{\text{sum}}(i_1{:}\omega_1,\ \dots\ ,\ i_n{:}\omega_n)} \tag{E9a}$$

$$E ::= \underline{\text{nil}}_{\Omega} \qquad \frac{}{\pi \vdash \underline{\text{nil}}_{\omega} : \underline{\text{list}}\ \omega} \tag{E10a}$$

(Actually, one further restriction must be imposed on the base language to obtain explicit typing: we exclude the degenerate expressions $\underline{\text{case}}\ e\ \underline{\text{of}}\ ()$ and $\underline{\text{choose}}()$, which could have types ω and $\underline{\text{sum}}() \to \omega$ for arbitrary ω.)

Somewhat surprisingly, it is not necessary to modify the let construction, since the type of $\underline{\text{let}}\ i = e_2\ \underline{\text{in}}\ e_1$ under π must be the type of e_1 under $\pi,\ i{:}\omega$, where ω is the type of e_2 under π. (However, a similar argument does not work for the letrec construction.)

For example, the following are explicitly typed versions of the expressions whose typing was discussed in the previous subsection:

$\underline{\text{fix}}(\lambda ap_{\text{list int} \to \text{list int} \to \text{list int}}.\ \lambda x_{\text{list int}}.\ \lambda y_{\text{list int}}.$

$\quad \underline{\text{lchoose}}\ y\ (\lambda n_{\text{int}}.\ \lambda z_{\text{list int}}.\ \underline{\text{cons}}\ n\ (ap\ z\ y))\ x)$,

$\underline{\text{fix}}(\lambda red_{\text{list int} \to (\text{int} \to \text{bool} \to \text{bool}) \to \text{bool} \to \text{bool}}.$

$\quad \lambda \ell_{\text{list int}}.\ \lambda f_{\text{int} \to \text{bool} \to \text{bool}}.\ \lambda a_{\text{bool}}.$

$\quad \underline{\text{lchoose}}\ a\ (\lambda n_{\text{int}}.\ \lambda z_{\text{list int}}.\ f\ n\ (red\ z\ f\ a))\ \ell)$.

Although the rigorous semantics of types is beyond the scope of this paper, it should be mentioned that the pragmatic arguments about implicit versus explicit typing reflect profoundly different views of the meaning of types. Consider, for example, the untyped expression $\lambda x.\ x$, and the explicitly typed expressions $\lambda x_{\text{int}}.\ x$ and $\lambda x_{\text{int} \to \text{int}}.\ x$. Three views can be taken of the meaning of these expressions:

(1) All three expressions have the same meaning. The types in the explicitly typed expressions are merely assertions about how these expressions will be used [17].

(2) The expressions have different meanings, but the meanings of the typed expressions are functions of the meanings of the untyped expression and the meanings of the types int and int → int [18, 19].

(3) The typed expressions have meanings, but the untyped expression does not. It is not sensible to speak of a function that is applicable to all conceivable values [20].

3. TYPE DEDUCTION AND THE POLYMORPHIC let

3a. Type Deduction

An efficient algorithm has been discovered (independently) by J. R. Hindley and R. Milner [21,22] that is capable of deducing the valid typings of expressions in our base language (with the partial exception of inject expressions). It is based upon the concept of a principal typing of an expression.

A typing scheme is a typing containing type variables; more precisely, it is like a typing except that type expressions may occur in place of types (including within type assignments). A principal typing of an expression is a typing scheme such that the valid typings of the expression are exactly those that can be obtained from the principal typing by substituting arbitrary types for the type variables (and perhaps extending the type assignment to irrelevant identifiers).

For example, the following are principal typings:

$x{:}\alpha,\ f{:}\ \alpha \to \alpha \vdash f(f\ x){:}\ \alpha$,

$f{:}\ \alpha \to \alpha \vdash \lambda x.\ f(f\ x){:}\ \alpha \to \alpha$,

$\vdash \lambda f.\ \lambda x.\ f(f\ x){:}\ (\alpha \to \alpha) \to \alpha \to \alpha$,

$\vdash$ fix(λap. λx. λy. lchoose y (λn. λz. cons n (ap z y)) x):

list α → list α → list α ,

$\vdash$ fix(λred. λℓ. λf. λa. lchoose a (λn. λz. f n (red z f a)) ℓ):

list α → (α → β → β) → β → β .

Throughout this paper we will use lower case Greek letters as type variables. It is clear that the choice of variables in a principal typing is arbitrary.

Hindley and Milner showed that an expression has a principal typing if it has any typing, and that a principal typing (or nonexistence thereof) of an expression can be computed from principal typings of its subexpressions by using Robinson's unification algorithm [23]. Although we will not give a complete description of their algorithm, its essence can be seen by considering function applications.

When σ denotes a substitution, we write $\omega\sigma$ for the type expression obtained by applying σ to ω, and $\pi\sigma$ for the type assignment obtained by applying σ to each type expression assigned by π. Now suppose that e_1 and e_2 are expressions with principal typings $\pi_1 \vdash e_1{:}\ \omega_1$ and $\pi_2 \vdash e_2{:}\ \omega_2$. (For simplicity we assume that π_1 and π_2 assign to the same identifiers.) Then the set of conclusions of instances of inference rule (I5b) whose premises are valid typings of e_1 and e_2 is

$$\{\pi \vdash e_1\ e_2\colon \omega' \mid (\exists \sigma_1, \sigma_2, \omega)\ \pi_1\sigma_1 = \pi_2\sigma_2 = \pi \text{ and } \omega_1\sigma_1 = \omega \to \omega' \text{ and } \omega_2\sigma_2 = \omega\}\ .$$

Let α be a type variable not occurring in π_2 or ω_2. Since we can extend σ_2 to substitute any type expression for α, the above set of conclusions is

$$\{\pi \vdash e_1\ e_2\colon \omega' \mid (\exists \sigma_1, \sigma_2, \omega)\ \pi_1\sigma_1 = \pi_2\sigma_2 = \pi \text{ and } \omega_1\sigma_1 = (\omega_2 \to \alpha)\sigma_2 = \omega \to \omega'\}\ .$$

Now we can use unification to determine whether there are any σ_1 and σ_2 such that $\pi_1\sigma_1 = \pi_2\sigma_2$ and $\omega_1\sigma_1 = (\omega_2 \to \alpha)\sigma_2$ and, if so, to produce "most general" substitutions $\hat\sigma_1$ and $\hat\sigma_2$ such that all such σ_1 and σ_2 can be obtained by composing $\hat\sigma_1$ and $\hat\sigma_2$ with some further substitution σ. In this case, the set of conclusions is

$$\{\pi \vdash e_1\ e_2\colon \omega' \mid (\exists \sigma)\ \pi = (\pi_1\hat\sigma_1)\sigma \text{ and } \omega' = (\alpha\hat\sigma_2)\sigma\}$$

This is the set of typings that can be obtained by substitution from

$$(\pi_1\hat\sigma_1) \vdash e_1\ e_2\colon (\alpha\hat\sigma_2)\ ,$$

which is therefore a principal typing of $e_1\ e_2$.

To illustrate the Hindley-Milner algorithm, the following is a list of principal typings of the subexpressions of the append function, as they would be generated by a naive version of the algorithm. Note that a proof of a particular typing, such as is given in Section 2b, can be obtained from the list of principal typings by substitution. (Also note that the choice of type variables in each line is arbitrary.)

$\mathrm{ap}\colon \alpha \vdash \mathrm{ap}\colon \alpha$

$\mathrm{z}\colon \alpha \vdash \mathrm{z}\colon \alpha$

$\mathrm{ap}\colon \alpha \to \beta,\ \mathrm{z}\colon \alpha \vdash \mathrm{ap\ z}\colon \beta$

$\mathrm{y}\colon \alpha \vdash \mathrm{y}\colon \alpha$

$\mathrm{ap}\colon \alpha \to \beta \to \gamma,\ \mathrm{z}\colon \alpha,\ \mathrm{y}\colon \beta \vdash \mathrm{ap\ z\ y}\colon \gamma$

$\mathrm{n}\colon \alpha \vdash \mathrm{n}\colon \alpha$

$\mathrm{n}\colon \gamma,\ \mathrm{ap}\colon \alpha \to \beta \to \underline{\mathrm{list}}\ \gamma,\ \mathrm{z}\colon \alpha,\ \mathrm{y}\colon \beta \vdash \underline{\mathrm{cons}}\ \mathrm{n\ (ap\ z\ y)}\colon \underline{\mathrm{list}}\ \gamma$

$\mathrm{n}\colon \gamma,\ \mathrm{ap}\colon \alpha \to \beta \to \underline{\mathrm{list}}\ \gamma,\ \mathrm{y}\colon \beta \vdash \lambda \mathrm{z}.\ \underline{\mathrm{cons}}\ \mathrm{n\ (ap\ z\ y)}\colon \alpha \to \underline{\mathrm{list}}\ \gamma$

$\mathrm{ap}\colon \alpha \to \beta \to \underline{\mathrm{list}}\ \gamma,\ \mathrm{y}\colon \beta \vdash \lambda \mathrm{n}.\ \lambda \mathrm{z}.\ \underline{\mathrm{cons}}\ \mathrm{n\ (ap\ z\ y)}\colon \gamma \to \alpha \to \underline{\mathrm{list}}\ \gamma$

$\mathrm{y}\colon \alpha \vdash \mathrm{y}\colon \alpha$

$\mathrm{ap}\colon \underline{\mathrm{list}}\ \gamma \to \underline{\mathrm{list}}\ \gamma \to \underline{\mathrm{list}}\ \gamma,\ \mathrm{y}\colon \underline{\mathrm{list}}\ \gamma \vdash$
$\quad \underline{\mathrm{lchoose}}\ \mathrm{y}\ (\lambda \mathrm{n}.\ \lambda \mathrm{z}.\ \underline{\mathrm{cons}}\ \mathrm{n\ (ap\ z\ y)})\colon \underline{\mathrm{list}}\ \gamma \to \underline{\mathrm{list}}\ \gamma$

$\mathrm{x}\colon \alpha \vdash \mathrm{x}\colon \alpha$

$\mathrm{ap}\colon \underline{\mathrm{list}}\ \gamma \to \underline{\mathrm{list}}\ \gamma \to \underline{\mathrm{list}}\ \gamma,\ \mathrm{y}\colon \underline{\mathrm{list}}\ \gamma,\ \mathrm{x}\colon \underline{\mathrm{list}}\ \gamma \vdash$
$\quad \underline{\mathrm{lchoose}}\ \mathrm{y}\ (\lambda \mathrm{n}.\ \lambda \mathrm{z}.\ \underline{\mathrm{cons}}\ \mathrm{n\ (ap\ z\ y)})\ \mathrm{x}\colon \underline{\mathrm{list}}\ \gamma$

$\mathrm{ap}\colon \underline{\mathrm{list}}\ \gamma \to \underline{\mathrm{list}}\ \gamma \to \underline{\mathrm{list}}\ \gamma,\ \mathrm{x}\colon \underline{\mathrm{list}}\ \gamma \vdash$
$\quad \lambda \mathrm{y}.\ \underline{\mathrm{lchoose}}\ \mathrm{y}\ (\lambda \mathrm{n}.\ \lambda \mathrm{z}.\ \underline{\mathrm{cons}}\ \mathrm{n\ (ap\ z\ y)})\ \mathrm{x}\colon \underline{\mathrm{list}}\ \gamma \to \underline{\mathrm{list}}\ \gamma$

$\mathrm{ap}\colon \underline{\mathrm{list}}\ \gamma \to \underline{\mathrm{list}}\ \gamma \to \underline{\mathrm{list}}\ \gamma \vdash$
$\quad \lambda \mathrm{x}.\ \lambda \mathrm{y}.\ \underline{\mathrm{lchoose}}\ \mathrm{y}\ (\lambda \mathrm{n}.\ \lambda \mathrm{z}.\ \underline{\mathrm{cons}}\ \mathrm{n\ (ap\ z\ y)})\ \mathrm{x}\colon \underline{\mathrm{list}}\ \gamma \to \underline{\mathrm{list}}\ \gamma \to \underline{\mathrm{list}}\ \gamma$

$\vdash$ λap. λx. λy. lchoose y (λn. λz. cons n (ap z y)) x:

$(\text{list}\ \gamma \to \text{list}\ \gamma \to \text{list}\ \gamma) \to \text{list}\ \gamma \to \text{list}\ \gamma \to \text{list}\ \gamma$

$\vdash$ fix(λap. λx. λy. lchoose y (λn. λz. cons n (ap z y)) x):

$\text{list}\ \gamma \to \text{list}\ \gamma \to \text{list}\ \gamma$

The Hindley-Milner algorithm requires a certain amount of auxilliary information in the inject operation. If $\pi \vdash e: \omega$ is a principal typing of e, then the principal typing of inject k e should be

$$\pi \vdash \text{inject}\ k\ e: \text{sum}(\alpha_1, \ldots, \alpha_{k-1}, \omega, \alpha_{k+1}, \ldots, \alpha_n),$$

where $\alpha_1, \ldots, \alpha_{k-1}, \alpha_{k+1}, \ldots, \alpha_n$ are distinct type variables that do not occur in π or ω. However, inject k e does not contain any information that determines the number n of alternatives. Thus we must alter syntax rule (S8a) and inference rule (I8a) to provide this information explicitly:

$$E ::= \text{inject}_N\ K\ E \qquad \frac{\pi \vdash e: \omega_k}{\pi \vdash \text{inject}_n\ k\ e: \text{sum}(\omega_1, \ldots, \omega_n)} \quad \text{when } 1 \le k \le n. \quad \text{(I8a')}$$

Similarly (but less pleasantly), we must require the inject operation for named alternatives to contain a list of the identifiers used as names. Rules (S9a) and (I9a) become:

$$E ::= \text{inject}_{I,\ldots,I}\ I\ E$$

$$\frac{\pi \vdash e: \omega_k}{\pi \vdash \text{inject}_{i_1,\ldots,i_n}\ i_k\ e: \text{sum}(i_1{:}\omega_1, \ldots, i_n{:}\omega_n)} \quad \text{when } 1 \le k \le n. \quad \text{(I9a')}$$

3b. The Polymorphic let

Suppose we use the reduce function to define the following:

let red = fix(λred. λℓ. λf. λa.
    lchoose a (λn. λz. f n (red z f a)) ℓ)
  in λℓℓ. red ℓℓ (λℓ. λs. add(red ℓ add 0) s) 0 .

Intuitively, if ℓ is a list of integers, then red ℓ add 0 is its sum, so that λℓ. λs. add(red ℓ add 0) s is a function that accepts a list of integers and an integer and produces their sum. Thus the function defined above should sum a list of lists of integers.

But in fact this expression has no typing. From the principal typing of fix(λred. ...) and the inference rule (I11) for let, it is clear that the let expression can only have a typing if its body has a typing of the form

π, red: $\text{list}\ \alpha \to (\alpha \to \beta \to \beta) \to \beta \to \beta \vdash$

λℓℓ. red ℓℓ (λℓ. λs. add(red ℓ add 0) s) 0: ω

for some particular substitution of types for α, β, and ω. But the unique type int → int → int of add forces $\alpha = \beta =$ int for red ℓ add 0 to make sense, and then λℓ. λs. add(red ℓ add 0) s must have type list int → int → int, so that the outer occurrence of red requires $\alpha =$ list int, contradicting $\alpha =$ int. The difficulty is that the above definition only makes sense if red is understood as a polymorphic function, i.e. a function that is applicable to a variety of types.

In designing the type structure of the language ML [22], Milner realized that his type deduction algorithm permitted the treatment of this kind of polymorphism. If e_2 has the principal typing $\pi \mid\!- e_2 : \omega$, and if $\alpha_1, \ldots, \alpha_n$ are type variables occurring in ω but not in π, then in typing let i = e_2 in e_1, different occurrences of i in e_1 can be assigned different types that are instances of ω obtain by different substitutions for $\alpha_1, \ldots, \alpha_n$. For instance, in the above example, one can assign the inner and outer occurrences of red the different types

list int → (int → int → int) → int → int

and

list list int → (list int → int → int) → int → int

to obtain the type list list int → int for the entire example.

In this scheme, however, (in contrast to Section 5b) polymorphic functions can be bound by let but not by λ, so that one cannot define higher-order functions that accept polymorphic functions. For instance, if we convert let i = e_2 in e_1 to (λi. e_1) e_2 in the above example, we obtain

(λred. λℓℓ. red ℓℓ (λℓ. λs. add(red ℓ add 0) s) 0)

(fix(λred. ...)) ,

in which the first line has no typing.

The term "polymorphism" was coined by Christopher Strachey [24], who distinguished between "parametric" polymorphic functions, such as the reduce function, that treat all types in the same way, and "ad hoc" polymorphic functions that can behave differently for different types. In this paper, we shall reserve the word "polymorphism" for the parametric case, and call the ad hoc functions "generic". (Strachey's definition of "parametric" was intuitive; its precise semantic formulation is still controversial [20].)

3c. Infinite Types

Without explicitly mentioning it, we have assumed that types are finite phrases. However, nothing that we have done precludes infinite types. For example, the infinite type

prod(int, prod(int, prod(int, ...)))

is the type of infinite streams of integers, and list ω can be regarded as

sum(prod(), prod(ω, sum(prod(), prod(ω, sum(prod(), prod(ω, ...)))))).

(Note that prod() denotes a set with one element: the empty record <>.)

To make use of such infinite types, however, we must have finite type expressions to denote them, i.e. we must introduce the recursive definition of types. A simple approach is to introduce type expressions with the syntax

$\Omega ::= \underline{\text{rectype}}\ T\colon \Omega$

where T denotes the set of type variables, and the intuitive semantics that $\underline{\text{rectype}}\ \alpha\colon \omega$ denotes an infinite type obtained by the endless sequence of substitutions:

$$\alpha\big|_{\alpha \to \omega}\big|_{\alpha \to \omega}\big|_{\alpha \to \omega} \cdots .$$

For example, $\underline{\text{rectype}}\ \alpha\colon \underline{\text{prod}}(\underline{\text{int}}, \alpha)$ denotes the type of infinite streams, while $\underline{\text{rectype}}\ \alpha\colon \underline{\text{sum}}(\underline{\text{prod}}(), \underline{\text{prod}}(\omega, \alpha))$ denotes the type $\underline{\text{list}}\ \omega$.

It should be emphasized that the rectype construction is a type expression rather than a type. For example,

$\underline{\text{rectype}}\ \alpha\colon \underline{\text{prod}}(\underline{\text{int}}, \alpha)$
$\underline{\text{rectype}}\ \beta\colon \underline{\text{prod}}(\underline{\text{int}}, \beta)$
$\underline{\text{rectype}}\ \alpha\colon \underline{\text{prod}}(\underline{\text{int}}, \underline{\text{prod}}(\underline{\text{int}}, \alpha))$

all denote the same infinite type.

If $\underline{\text{list}}\ \omega$ is regarded as an abbreviation for $\underline{\text{rectype}}\ \alpha\colon \underline{\text{sum}}(\underline{\text{prod}}(), \underline{\text{prod}}(\omega, \alpha))$, then the various forms of expressions we have introduced for list manipulation can also be regarded as abbreviations:

$\text{nil} \equiv \underline{\text{inject}}_2\ 1\ \langle\rangle$
$\underline{\text{cons}}\ e_1\ e_2 \equiv \underline{\text{inject}}_2\ 2\ \langle e_1, e_2\rangle$
$\underline{\text{lchoose}}\ e_1\ e_2 \equiv \underline{\text{choose}}(\lambda x.\ e_1, \lambda x.\ e_2\ x.1\ x.2)$,

where x is an identifier not occurring in e_1 or e_2.

An obvious question is whether the Hindley-Milner algorithm can be extended to encompass infinite, recursively defined types. But the answer is obscure. It is known that the unification algorithm can be extended to expressions involving rectype (by treating such expressions as cyclic structures and omitting the "occurs" check) [25,26], and I have heard researchers say that this extension can be applied to the Hindley-Milner algorithm without difficulty. But I have been unable to find anything in the literature about the question.

4. SUBTYPES AND GENERIC OPERATORS

In many programming languages there is a subtype relation between types such that, if ω is a subtype of ω', then there is an implicit conversion from values of type ω to values of type ω', so that any expression of type ω can be used in any context allowing expressions of type ω'. Experience with languages such as PL/I and Algol 68 has shown that a rich subtype structure, particularly in conjunction with generic operators, can produce a language with quirkish and counterintuitive behavior. This experience has led to research on subtypes and generic operators, using category theory as a tool, that has established design criteria for avoiding such behavior [27,28].

To see the problem, suppose <u>int</u> is a subtype of <u>real</u>, and <u>add</u> is a generic operator mapping pairs of integers into integers and pairs of reals into reals. Then an expression such as <u>add</u> 5 6, occurring in a context calling for a real expression, can be interpreted as either the integer-to-real conversion of the integer sum of 5 and 6, or as the real sum of the integer-to-real conversions of 5 and 6.

In this case, the laws of mathematics insure that the two interpretations are equivalent(except for the roundoff behavior of machines with unfortunate arithmetic). On the other hand, suppose <u>digit</u> <u>string</u> is a subtype of <u>int</u> with an implicit conversion that interprets digit strings as decimal representations, and <u>equals</u> is a generic operator applicable to either pairs of digit strings or pairs of integers. Then an expression such as <u>equals</u> "1" "01" is ambiguous, since the implicit conversion maps unequal digit strings into equal integers.

4a. Subtypes

We will write $\omega \leq \omega'$ to indicate that ω is a subtype of ω'. Then the idea that any expression of type ω can be used as an expression of type ω' is formalized by the inference rule

$$\frac{\pi \vdash e: \omega}{\pi \vdash e: \omega'} \quad \text{when } \omega \leq \omega' . \tag{I15}$$

It is natural to assume that $\leq$ is a preorder, i.e. that it satisfies the laws

(a) (Reflexivity) $\omega \leq \omega$.

(b) (Transitivity) if $\omega \leq \omega'$ and $\omega' \leq \omega''$ then $\omega \leq \omega''$.

(This assumption can be justified by noting that the effect of (I15) remains unchanged if an arbitrary relation $\leq$ is replaced by its reflexive and transitive closure.)

Semantically, the laws of reflexivity and transitivity imply the existence of implicit conversions that must "fit together" correctly if the meaning of our language is to be unambiguous. For each ω, $\omega \leq \omega$ implies that there is an implicit conversion that can be applied to any value of type ω without changing its type; to avoid ambiguity this conversion must be an identity function. When $\omega \leq \omega'$ and $\omega' \leq \omega''$, one can convert from ω to ω'' either directly or through the intermediary of ω'; to avoid ambiguity the direct conversion from ω to ω'' must equal the functional composition of the conversions from ω to ω' and from ω' to ω''.

Mathematically, these restrictions on implicit conversions are tantamount to requiring that there be a functor from the preorder of types to the category of sets and functions (or some other appropriate category) that maps each type into the set that it denotes, and maps each pair ω, ω' such that $\omega \leq \omega'$ into the implicit conversion function from ω to ω'. This is the fundamental connection between subtypes and category theory.

If $\omega \leq \omega'$ and $\omega' \leq \omega$ then, under a given type assignment, an expression will have type ω if and only if it has type ω'. It is tempting to assume that such types are identical, i.e. to impose

(c) (Antisymmetry) if $\omega \leq \omega'$ and $\omega' \leq \omega$ then $\omega = \omega'$,

so that $\leq$ is a partial order. In fact, we will assume that $\leq$ is a partial order throughout this exposition, but it should be noted that most of the theory goes through in the more general case of preorders, and that one can conceive of applications that would use such generality (e.g., where ω and ω' denote abstractly equivalent types with different representations).

For each of the type constructors we have introduced, the existence of implicit conversions for the component types induces a natural conversion for the constructed type. For instance, suppose c is an implicit conversion from ω_2 to ω_1 and c' is an implicit conversion from ω_1' to ω_2'. Then it is natural to convert a function f of type $\omega_1 \rightarrow \omega_1'$ to the function $\lambda x_{\omega_2}.\ c'(f(c\ x))$ of type $\omega_2 \rightarrow \omega_2'$. If this conversion of functions is taken as an implicit conversion from $\omega_1 \rightarrow \omega_1'$ to $\omega_2 \rightarrow \omega_2'$, then the operator $\rightarrow$ will satisfy

If $\omega_2 \leq \omega_1$ and $\omega_1' \leq \omega_2'$ then $\omega_1 \rightarrow \omega_1' \leq \omega_2 \rightarrow \omega_2'$, (≤a)

i.e. $\rightarrow$ will be monotone in its second operand but antimonotone in its first operand.

If $c_1, \ldots, c_n$ are implicit conversions from ω_1 to $\omega_1', \ldots, \omega_n$ to ω_n', then it is natural to convert a record $\langle x_1, \ldots, x_n \rangle$ of type $\underline{\text{prod}}(\omega_1, \ldots, \omega_n)$ to the record $\langle c_1\ x_1, \ldots, c_n\ x_n \rangle$ of type $\underline{\text{prod}}(\omega_1', \ldots, \omega_n')$. Thus prod should be monotone in all of its operands:

If $\omega_1 \leq \omega_1', \ldots, \omega_n \leq \omega_n'$ then $\underline{\text{prod}}(\omega_1, \ldots, \omega_n) \leq \underline{\text{prod}}(\omega_1', \ldots, \omega_n')$. (≤b)

Similarly, for products with named fields, we have

If $\omega_1 \leq \omega_1', \ldots, \omega_n \leq \omega_n'$ then

$$\underline{\text{prod}}(i_1{:}\omega_1, \ldots, i_n{:}\omega_n) \leq \underline{\text{prod}}(i_1{:}\omega_1', \ldots, i_n{:}\omega_n') . \qquad (\leq c)$$

In this case, however, there is another kind of conversion that is both well-behaved and useful: if $<i_1:x_1, \ldots, i_n:x_n>$ is a record of type $\underline{prod}(i_1:\omega_1, \ldots, i_n:\omega_n)$, and if $\{i_1, \ldots, i_m\}$ is any subset of $\{i_1, \ldots, i_n\}$, then by forgetting fields it is natural to convert this record to $<i_1:x_1, \ldots, i_m:x_m>$. This leads to

$$\underline{prod}(i_1:\omega_1, \ldots, i_n:\omega_n) \leq \underline{prod}(i_1:\omega_1, \ldots, i_m:\omega_m) \quad \text{when } 0 \leq m \leq n\,. \qquad (\leq d)$$

(Theoretically, field-forgetting conversions are also applicable to records with numbered fields, but the constraint that the field numbers must form a consecutive sequence renders these conversions much more limited.)

If $c_1, \ldots, c_n$ are implicit conversions from ω_1 to $\omega_1', \ldots, \omega_n$ to ω_n', then it is natural to convert $\underline{inject}\ k\ x$ of type $\underline{sum}(\omega_1, \ldots, \omega_n)$ to $\underline{inject}\ k\ (c_k\ x)$ of type $\underline{sum}(\omega_1', \ldots, \omega_n')$. Thus $\underline{sum}$ is also monotone:

$$\text{If } \omega_1 \leq \omega_1', \ldots, \omega_n \leq \omega_n' \text{ then } \underline{sum}(\omega_1, \ldots, \omega_n) \leq \underline{sum}(\omega_1', \ldots, \omega_n')\,. \qquad (\leq e)$$

Similarly, for named alternatives we have

$$\text{If } \omega_1 \leq \omega_1', \ldots, \omega_n \leq \omega_n' \text{ then} \quad \underline{sum}(i_1:\omega_1, \ldots, i_n:\omega_n) \leq \underline{sum}(i_1:\omega_1', \ldots, i_n:\omega_n')\,. \qquad (\leq f)$$

Just as with products, however, names give a richer subtype structure than numbers. Whenever $\{i_1, \ldots, i_m\}$ is a subset of $\{i_1, \ldots, i_n\}$, $\underline{sum}(i_1:\omega_1, \ldots, i_m:\omega_m)$ is a subset of $\underline{sum}(i_1:\omega_1, \ldots, i_n:\omega_n)$, and the identity injection is a natural implicit conversion. Thus

$$\underline{sum}(i_1:\omega_1, \ldots, i_m:\omega_m) \leq \underline{sum}(i_1:\omega_1, \ldots, i_n:\omega_n) \quad \text{when } 0 \leq m \leq n\,. \qquad (\leq g)$$

The implicit conversions of forgetting named fields in products and adding named alternatives in sums have been investigated by L. Cardelli [29], who shows that they generalize the subclass concept of SIMULA 67 and also provide a suitable type structure for object-oriented languages such as SMALLTALK.

Finally, if c is an implicit conversion from ω to ω', then it is natural to convert $(x_1, \ldots, x_n)$ of type $\underline{list}\ \omega$ to $(c\ x_1, \ldots, c\ x_n)$ of type $\underline{list}\ \omega'$, so that $\underline{list}$ is also monotone:

$$\text{If } \omega \leq \omega' \text{ then } \underline{list}\ \omega \leq \underline{list}\ \omega'\,. \qquad (\leq h)$$

4b. Explicit Minimal Typing

With the introduction of subtypes, it is no longer possible to achieve explicit typing in the sense of Section 2c, since an expression that has type ω under some type assignment will also have type ω' whenever $\omega \leq \omega'$. However, we can still hope to arrange things so that, if an expression (under a given type assignment) has any type, then its set of types will have a least member (which must be unique since $\leq$ is a partial order).

We write

$\pi \vdash_m e: \omega$,

called a <u>minimal typing</u>, to indicate that ω is the least type of e under π. In other words $\pi \vdash_m e: \omega$ means that $\pi \vdash e: \omega'$ holds for just those ω' such that $\omega \leq \omega'$.

If the expressions of our base language are to have minimal typings then the partial ordering of types must satify two dual properties:

(LUB) If ω_1 and ω_2 have an upper bound then ω_1 and ω_2 have a least upper bound.

(GLB) If ω_1 and ω_2 have a lower bound then ω_1 and ω_2 have a greatest lower bound.

Fortunately these properties are consistent with the ordering laws given in the previous subsection. It can be shown that the least partial ordering satisfying ($\leq$a) to ($\leq$h) and including some partial ordering of primitive types will satisfy (LUB) and (GLB) if the partial ordering of primitive types satisfies (LUB) and (GLB). (Moreover, this situation will remain true as additional ordering laws are introduced in Sections 4d and 4e.)

To see why these properties are needed for minimal typing, consider the conditional expression. Suppose $\pi \vdash_m e$: <u>bool</u>, $\pi \vdash_m e_1: \omega_1$, and $\pi \vdash_m e_2: \omega_2$. Then the types ω such that $\pi \vdash$ <u>if</u> e <u>then</u> e_1 <u>else</u> $e_2: \omega$ will be the upper bounds of ω_1 and ω_2, so that $\pi \vdash_m$ <u>if</u> e <u>then</u> e_1 <u>else</u> $e_2: \omega$ will only hold if ω is a least upper bound of ω_1 and ω_2.

(By a similar argument, the minimal typing of <u>case</u> e <u>of</u> $(e_1, \ldots, e_n)$ requires that, if the finite set $\{\omega_1, \ldots, \omega_n\}$ has an upper bound, then it must have a least upper bound. Fortunately, for n > 0 this property is implied by (LUB).)

Assuming that (LUB) and (GLB) hold, we can give inference rules for the explicit minimal typing of our base language that effectively define a partial function mapping each e and π into the least type of e under π. As with the explicit typing in the absence of subtypes described in Section 2c, we must require type information to appear in lambda expressions, <u>inject</u> expressions for numbered alternatives, and the expression <u>nil</u>, and we must exclude the vacuous expressions <u>choose</u>() and <u>case</u> e <u>of</u> (). But now, type information is no longer needed in <u>inject</u> expressions for named alternatives, because of the implicit conversions that add alternatives. If ω is the least type of e then <u>sum</u>(i:ω) is the least type of <u>inject</u> i e.

(It is curious that numbered alternatives require less type information under the Hindley-Milner approach of Section 3a, while named alternatives require less type information under the present approach.)

The following is a list of the inference rules for minimal typing. In the provisos of some of the rules, we write lub for least upper bound and glb for greatest lower bound.

$$\frac{}{\pi \vdash_m i\colon \omega}$$ when i is in the domain of π, and π assigns ω to i. (M1)

$$\frac{}{\pi \vdash_m z\colon \underline{\text{int}}} \qquad \frac{}{\pi \vdash_m \underline{\text{true}}\colon \underline{\text{bool}}} \qquad \frac{}{\pi \vdash_m \underline{\text{false}}\colon \underline{\text{bool}}}$$ (M2,M3)

$$\frac{}{\pi \vdash_m \underline{\text{add}}\colon \underline{\text{int}} \to \underline{\text{int}} \to \underline{\text{int}}} \qquad \frac{}{\pi \vdash_m \underline{\text{equals}}\colon \underline{\text{int}} \to \underline{\text{int}} \to \underline{\text{bool}}}$$ (M4)

$$\frac{\pi, i\colon \omega \vdash_m e\colon \omega'}{\pi \vdash_m \lambda i_\omega .\ e\colon \omega \to \omega'} \qquad \frac{\pi \vdash_m e_1\colon \omega_1 \to \omega' \quad \pi \vdash_m e_2\colon \omega_2}{\pi \vdash_m e_1\ e_2\colon \omega'}$$ when $\omega_2 \preceq \omega_1$ (M5)

$$\frac{\pi \vdash_m e_1\colon \omega_1 \quad \cdots \quad \pi \vdash_m e_n\colon \omega_n}{\pi \vdash_m \langle e_1, \ldots , e_n\rangle\colon \underline{\text{prod}}(\omega_1, \ldots , \omega_n)}$$ (M6a)

$$\frac{\pi \vdash_m e\colon \underline{\text{prod}}(\omega_1, \ldots , \omega_n)}{\pi \vdash_m e.k\colon \omega_k}$$ when $1 \le k \le n$ (M6b)

$$\frac{\pi \vdash_m e_1\colon \omega_1 \quad \cdots \quad \pi \vdash_m e_n\colon \omega_n}{\pi \vdash_m \langle i_1{:}e_1, \ldots , i_n{:}e_n\rangle\colon \underline{\text{prod}}(i_1{:}\omega_1, \ldots , i_n{:}\omega_n)}$$ (M7a)

$$\frac{\pi \vdash_m e\colon \underline{\text{prod}}(i_1{:}\omega_1, \ldots , i_n{:}\omega_n)}{\pi \vdash_m e.i_k\colon \omega_k}$$ when $1 \le k \le n$ (M7b)

$$\frac{\pi \vdash_m e\colon \omega_k}{\pi \vdash_m \underline{\text{inject}}_{\omega_1,\ldots,\omega_n}\ k\ e\colon \underline{\text{sum}}(\omega_1, \ldots , \omega_n)}$$ when $1 \le k \le n$ (M8a)

$$\frac{\pi \vdash_m e_1\colon \omega_1 \to \omega_1' \quad \cdots \quad \pi \vdash_m e_n\colon \omega_n \to \omega_n'}{\pi \vdash_m \underline{\text{choose}}(e_1, \ldots , e_n)\colon \underline{\text{sum}}(\omega_1, \ldots , \omega_n) \to \omega'}$$ when ω' is the lub of $\{\omega_1', \ldots , \omega_n'\}$ (M8b)

$$\frac{\pi \vdash_m e\colon \omega}{\pi \vdash_m \underline{\text{inject}}\ i\ e\colon \underline{\text{sum}}(i{:}\omega)}$$ (M9a)

$$\frac{\begin{array}{l}\pi \vdash_m e_1\colon \omega_1 \to \omega_1' \\ \vdots \\ \pi \vdash_m e_n\colon \omega_n \to \omega_n'\end{array}}{\pi \vdash_m \underline{\text{choose}}(i_1{:}e_1, \ldots , i_n{:}e_n)\colon \underline{\text{sum}}(i_1{:}\omega_1, \ldots , i_n{:}\omega_n) \to \omega'} \quad \begin{array}{l}\text{when } \omega' \text{ is the lub of} \\ \{\omega_1', \ldots , \omega_n'\}\end{array} \qquad \text{(M9b)}$$

$$\frac{}{\pi \vdash_m \underline{\text{nil}}_\omega\colon \underline{\text{list}}\ \omega} \qquad \frac{\begin{array}{l}\pi \vdash_m e_1\colon \omega_1 \\ \pi \vdash_m e_2\colon \underline{\text{list}}\ \omega_2\end{array}}{\pi \vdash_m \underline{\text{cons}}\ e_1\ e_2\colon \underline{\text{list}}\ \omega} \quad \begin{array}{l}\text{when } \omega \text{ is the lub} \\ \text{of } \omega_1 \text{ and } \omega_2\end{array} \qquad \text{(M10a,b)}$$

$$\frac{\begin{array}{l}\pi \vdash_m e_1\colon \omega_1' \\ \pi \vdash_m e_2\colon \omega_1 \to \underline{\text{list}}\ \omega_2 \to \omega_2'\end{array}}{\pi \vdash_m \underline{\text{lchoose}}\ e_1\ e_2\colon \underline{\text{list}}\ \omega \to \omega'} \quad \begin{array}{l}\text{when } \omega' \text{ is the lub of } \omega_1' \text{ and } \omega_2' \\ \text{and } \omega \text{ is the glb of } \omega_1 \text{ and } \omega_2\end{array} \qquad \text{(M10c)}$$

$$\frac{\begin{array}{l}\pi \vdash_m e_2\colon \omega \\ \pi, i\colon \omega \vdash_m e_1\colon \omega'\end{array}}{\pi \vdash_m \underline{\text{let}}\ i = e_2\ \underline{\text{in}}\ e_1\colon \omega'} \qquad \text{(M11)}$$

$$\frac{\begin{array}{l}\pi \vdash_m e\colon \omega_0 \\ \pi \vdash_m e_1\colon \omega_1 \\ \pi \vdash_m e_2\colon \omega_2\end{array}}{\pi \vdash_m \underline{\text{if}}\ e\ \underline{\text{then}}\ e_1\ \underline{\text{else}}\ e_2\colon \omega} \quad \begin{array}{l}\text{when } \omega_0 \le \underline{\text{bool}} \text{ and } \omega \text{ is} \\ \text{the lub of } \omega_1 \text{ and } \omega_2\end{array} \qquad \text{(M12)}$$

$$\frac{\begin{array}{l}\pi \vdash_m e\colon \omega_0 \\ \pi \vdash_m e_1\colon \omega_1 \\ \vdots \\ \pi \vdash_m e_n\colon \omega_n\end{array}}{\pi \vdash_m \underline{\text{case}}\ e\ \underline{\text{of}}\ (e_1, \ldots , e_n)\colon \omega} \quad \begin{array}{l}\text{when } \omega_0 \le \underline{\text{int}} \text{ and } \omega \text{ is the} \\ \text{lub of } \{\omega_1, \ldots , \omega_n\}\end{array} \qquad \text{(M13)}$$

$$\frac{\pi \vdash_m e\colon \omega_1 \to \omega_2}{\pi \vdash_m \underline{\text{fix}}\ e\colon \omega_2} \quad \text{when } \omega_2 \le \omega_1 \qquad \text{(M14)}$$

4c. Generic Operators

We now consider extending our base language to include primitive operators that act upon values of several types, performing possibly different operations for different types. For example, if the partial ordering of primitive types is

$$\begin{array}{ccc}\underline{\text{real}} & & \\ | & & \underline{\text{bool}} \\ \underline{\text{int}} & & \end{array}$$

we might want add to denote integer addition, real addition, and boolean disjunction, and equals to denote equality tests for integers, reals, and truth values. In this case we would replace inference rules (I4) by

$$\frac{\pi \vdash e_1\colon \underline{int} \quad \pi \vdash e_2\colon \underline{int}}{\pi \vdash \underline{add}\ e_1\ e_2\colon \underline{int}} \qquad \frac{\pi \vdash e_1\colon \underline{real} \quad \pi \vdash e_2\colon \underline{real}}{\pi \vdash \underline{add}\ e_1\ e_2\colon \underline{real}} \qquad \frac{\pi \vdash e_1\colon \underline{bool} \quad \pi \vdash e_2\colon \underline{bool}}{\pi \vdash \underline{add}\ e_1\ e_2\colon \underline{bool}}$$

$$\frac{\pi \vdash e_1\colon \underline{int} \quad \pi \vdash e_2\colon \underline{int}}{\pi \vdash \underline{equals}\ e_1\ e_2\colon \underline{bool}} \qquad \frac{\pi \vdash e_1\colon \underline{real} \quad \pi \vdash e_2\colon \underline{real}}{\pi \vdash \underline{equals}\ e_1\ e_2\colon \underline{bool}} \qquad \frac{\pi \vdash e_1\colon \underline{bool} \quad \pi \vdash e_2\colon \underline{bool}}{\pi \vdash \underline{equals}\ e_1\ e_2\colon \underline{bool}}$$

The general situation, for an n-ary operator op, can be described as follows: there will be an index set Γ and functions $\alpha_1, \ldots, \alpha_n, \rho$ from Γ to Ω such that the inference rules for op will be the instances of

$$\frac{\pi \vdash e_1\colon \alpha_1\gamma \quad \vdots \quad \pi \vdash e_n\colon \alpha_n\gamma}{\pi \vdash \underline{op}\ e_1 \ldots e_n\colon \rho\gamma} \tag{I4'}$$

for each index γ in Γ.

For example, our generic add operator would be described by $\Gamma = \{\underline{int}, \underline{real}, \underline{bool}\}$, with α_1, α_2, and ρ all being the identity injection from Γ to Ω. The equals operator would be described similarly, except that ρ would be a constant function giving bool.

An obvious question is under what conditions this kind of inference rule scheme provides minimal typing. A sufficient condition is that Γ must possess a partial ordering such that

The function $\alpha_1, \ldots, \alpha_n$, and ρ are all monotone and, for all $\omega_1, \ldots, \omega_n$, if the set

$$\{\gamma \mid \omega_1 \le \alpha_1\gamma, \ldots, \omega_n \le \alpha_n\gamma\} \tag{*}$$

is nonempty, then this set possesses a least member.

If this condition is satisfied (as is the case for add and equals) then minimal typing is given by the rule

$$\frac{\pi \vdash_m e_1\colon \omega_1 \quad \vdots \quad \pi \vdash_m e_n\colon \omega_n}{\pi \vdash_m \underline{op}\ e_1 \ldots e_n\colon \rho\gamma_0} \quad \text{when } \gamma_0 \text{ is the least member of } \{\gamma \mid \omega_1 \le \alpha_1\gamma, \ldots, \omega_n \le \alpha_n\gamma\} \tag{M4'}$$

Semantically, however, there is a further issue. If the meaning of our language is to be unambiguous, then the meanings of op for various indices must satisfy the following relationship with the implicit conversion functions:

For each index γ, let $\overline{op}_\gamma$ be the function of type $\alpha_1\gamma \to \ldots \to \alpha_n\gamma \to \rho\gamma$ that is the meaning of <u>op</u> for γ. For types ω and ω' such that $\omega \leq \omega'$, let $c_{\omega\leq\omega'}$ be the conversion function from ω to ω'. Then for all indices γ and γ' such that $\gamma \leq \gamma'$, and all values $x_1, \ldots, x_n$ of types $\alpha_1\gamma, \ldots, \alpha_n\gamma$, $\overline{op}_\gamma$ and $\overline{op}_{\gamma'}$ must satisfy

$$c_{\rho\gamma\leq\rho\gamma'}(\overline{op}_\gamma\ x_1\ \ldots\ x_n) = \overline{op}_{\gamma'}(c_{\alpha_1\gamma\leq\alpha_1\gamma'}\ x_1)\ \ldots\ (c_{\alpha_n\gamma\leq\alpha_n\gamma'}\ x_n).$$

In conjunction with (*), this relationship (which can be expressed category-theoretically by saying $\overline{op}$ is a certain kind of natural transformation) is sufficient to preclude ambiguous meanings.

For our examples of <u>add</u> and <u>equals</u>, the relationship becomes

$$c_{\underline{int}\leq\underline{real}}(\overline{add}_{\underline{int}}\ x_1\ x_2) = \overline{add}_{\underline{real}}(c_{\underline{int}\leq\underline{real}}\ x_1)(c_{\underline{int}\leq\underline{real}}\ x_2)$$

and

$$\overline{equals}_{\underline{int}}\ x_1\ x_2 = \overline{equals}_{\underline{real}}(c_{\underline{int}\leq\underline{real}}\ x_1)(c_{\underline{int}\leq\underline{real}}\ x_2)\ .$$

The condition for <u>add</u> is the classical homomorphic relation between integer and real addition, while the condition for <u>equals</u> (assuming the meanings of $\overline{equals}$ are equality predicates) is equivalent to requiring $c_{\underline{int}\leq\underline{real}}$ to map distinct integers into distinct reals. Note that there is no constraint on $\overline{add}_{\underline{bool}}$ or $\overline{equals}_{\underline{bool}}$.

4d. <u>The Universal Type</u>

Suppose we expand the partially ordered set of types by introducing a new type that is a subtype of every type:

For all ω, $\underline{univ} \leq \omega$. (<u><</u>i)

Several pleasantries occur. We can give minimal typings for the vacuous <u>case</u> and <u>choose</u> operations:

$$\frac{\pi \vdash_m e{:}\omega_0}{\pi \vdash_m \underline{case}\ e\ \underline{of}\ ()\colon \underline{univ}} \quad \text{when } \omega_0 \leq \underline{int} \qquad \text{(M13')}$$

$$\frac{}{\pi \vdash_m \underline{choose}()\colon \underline{sum}() \to \underline{univ}} \qquad \text{(M8b')}$$

for an <u>inject</u> operation for numbered alternatives with less explicit type information:

$$\frac{\pi \vdash_m e\colon \omega_k}{\pi \vdash_m \underline{inject}_n\ k\ e\colon \underline{sum}(\underline{univ}, \ldots, \underline{univ}, \omega_k, \underline{univ}, \ldots, \underline{univ})} \quad \text{when } 1 \leq k \leq n \qquad \text{(M8a')}$$

and for a <u>nil</u> expression without type information:

$$\frac{}{\pi \vdash_m \underline{nil}\colon \underline{list}\ \underline{univ}} \qquad \text{(M10a')}$$

Less happily, however, the introduction of univ complicates the nature of generic operators by making (*) in the previous subsection harder to satisfy. For instance, taking $\omega_1 = \dots = \omega_n = \underline{\text{univ}}$, (*) implies that Γ, if nonempty, must have a least element.

For our example of add and equals, the introduction of univ forces us to add a least index (which we will also call univ) to Γ, with $\alpha_1\ \underline{\text{univ}} = \alpha_2\ \underline{\text{univ}} = \rho\ \underline{\text{univ}} = \underline{\text{univ}}$ (except that we could take $\rho\ \underline{\text{univ}} = \underline{\text{bool}}$ for equals). In effect, we must introduce vacuous versions of add and equals corresponding to the inference rules

$$\frac{\pi \vdash e_1\colon \underline{\text{univ}} \qquad \pi \vdash e_2\colon \underline{\text{univ}}}{\pi \vdash \underline{\text{add}}\ e_1\ e_2\colon \underline{\text{univ}}} \qquad \frac{\pi \vdash e_1\colon \underline{\text{univ}} \qquad \pi \vdash e_2\colon \underline{\text{univ}}}{\pi \vdash \underline{\text{equals}}\ e_1\ e_2\colon \underline{\text{univ}}\ \ (\text{or } \underline{\text{bool}})}$$

Semantically, one can interpret univ as denoting a domain with a single element $\perp_{\text{univ}}$, with an implicit conversion function to any other domain that maps $\perp_{\text{univ}}$ into the least element of that domain. A more intuitive understanding can be obtained from the following expressions of type univ:

$\underline{\text{case}}\ 1\ \underline{\text{of}}\ ()$

$\underline{\text{fix}}(\lambda x_{\text{univ}}.\ x)$

If evaluated, the first expression gives an error stop and the second never terminates. These are both meanings that make sense for any type.

4e. The Nonsense Type

We have developed a system in which, for a given type assignment, every expression with a type has a least type, but there are still nonsensical expressions with no type at all. Thus the function that maps e and π into the least type of e under π is only partial.

To avoid the mathematical complications of partial functions, we can introduce a new type of which every type is a subtype,

$$\text{For all } \omega,\ \omega \le \underline{\text{ns}} \qquad (\le j)$$

and make ns a type of every expression by adding the inference rule

$$\frac{}{\pi \vdash e\colon \underline{\text{ns}}} \qquad (\text{I16})$$

Now (under a given type assignment), since even nonsensical expressions have the type ns, every expression has at least one type, and therefore a least type. The inference rules for minimal typing remain correct if one adds a "metarule" that $\pi \vdash_m e\colon \underline{\text{ns}}$ holds whenever $\pi \vdash_m e\colon \omega$ cannot be inferred for any other ω.

This idea was introduced by the author in [27]. Today, I remain bemused by its elegance, but much less sanguine about its practical utility. The difficulty is that it permits nonsensical expressions to occur within sensible ones. For example,

$(\lambda x_{ns}.\ 3)$(add 1 nil)

<3, add 1 nil>.1

are both integer expressions containing the nonsensical subexpression add 1 nil. Abstractly, perhaps, they have the value 3, but pragmatically they should cause compile-time errors.

4f. Open Questions

As far as I know, no one has dealt successfully with the following questions: Can the above treatment of generic operators be extended to the definition of such operators by some kind of let construction? Can the Hindley-Milner algorithm be extended to deal with subtypes? Can the theory of subtypes be extended to encompass infinite, recursively defined types?

5. TYPE DEFINITIONS AND EXPLICIT POLYMORPHISM

5a. Type Definitions

In this section we will extend the explicitly typed language of Section 2c to permit the definition of types. We begin by adding Ω ::= T to the definition of Ω, so that Ω becomes the set of type expressions built from the type variables in T. Then we introduce the new expressions

$$E ::= \underline{\text{lettype}}\ T = \Omega\ \underline{\text{in}}\ E \mid \underline{\text{lettran}}\ T = \Omega\ \underline{\text{in}}\ E \qquad \text{(S17,S18)}$$

which satisfy similar reduction rules

$$\underline{\text{lettype}}\ \alpha = \omega\ \underline{\text{in}}\ e = e\big|_{\alpha \to \omega} \qquad \underline{\text{lettran}}\ \alpha = \omega\ \underline{\text{in}}\ e = e\big|_{\alpha \to \omega}\,, \qquad \text{(R17,R18)}$$

where the right sides denote the result of substituting ω for α in the type expressions embedded within the explicitly typed expression e. (Both lettype and lettran bind the occurrences of α in the type expressions embedded in e.)

The difference between lettype and lettran lies in their inference rules:

$$\frac{\pi \vdash e\colon \omega'}{\pi \vdash \underline{\text{lettype}}\ \alpha = \omega\ \underline{\text{in}}\ e\colon (\omega'\big|_{\alpha \to \omega})} \quad \begin{array}{l}\text{when } \alpha \text{ does not occur (free)}\\ \text{in any type expression}\\ \text{assigned by } \pi\end{array} \qquad \text{(E17)}$$

$$\frac{\pi \vdash (e\big|_{\alpha \to \omega})\colon \omega'}{\pi \vdash \underline{\text{lettran}}\ \alpha = \omega\ \underline{\text{in}}\ e\colon \omega'} \qquad \text{(E18)}$$

The second rule shows that lettran $\alpha = \omega$ in e is a transparent type definition that simply permits α to be used as a synonym for ω within e. It makes sense whenever $e\big|_{\alpha \to \omega}$ makes sense (and has the same meaning).

On the other hand, lettype $\alpha = \omega$ in e is an opaque or abstract type definition, which only makes sense if e makes sense when α is regarded as an arbitrary type, independent of ω.

The proviso that α must not occur free in any type expression assigned by π is necessary since α has independent meanings inside and outside the scope of <u>lettype</u>. For example, without this proviso we could infer

$$f\colon \alpha \to \alpha \;\vdash\; \lambda x_\alpha .\ f\ x\colon \alpha \to \alpha$$

$$f\colon \alpha \to \alpha \;\vdash\; \underline{\text{lettype}}\ \alpha = \underline{\text{int}}\ \underline{\text{in}}\ \lambda x_\alpha .\ f\ x\colon \underline{\text{int}} \to \underline{\text{int}}$$

$$\vdash \lambda f_{\alpha \to \alpha} .\ \underline{\text{lettype}}\ \alpha = \underline{\text{int}}\ \underline{\text{in}}\ \lambda x_\alpha .\ f\ x\colon (\alpha \to \alpha) \to \underline{\text{int}} \to \underline{\text{int}}$$

$$\vdash \underline{\text{lettype}}\ \alpha = \underline{\text{bool}}\ \underline{\text{in}}\ \lambda f_{\alpha \to \alpha} .\ \underline{\text{lettype}}\ \alpha = \underline{\text{int}}\ \underline{\text{in}}\ \lambda x_\alpha .\ f\ x\colon (\underline{\text{bool}} \to \underline{\text{bool}}) \to \underline{\text{int}} \to \underline{\text{int}}$$

But two applications of (R17) reduce the expression in the last line to

$$\lambda f_{\text{bool} \to \text{bool}} .\ \lambda x_{\text{int}}\ f\ x\ ,$$

which has no type.

Our main interest is in abstract definitions, in which the "abstract type" α is defined by a "representation" ω. However, for such a definition to be useful, one must be able to include definitions of primitive functions (or constants) on the abstract type in terms of its representation. This seems to require a more complex definitional expression such as

$$E ::= \underline{\text{lettype}}\ T = \Omega\ \underline{\text{with}}\ I_\Omega = E,\ \dots\ ,\ I_\Omega = E\ \underline{\text{in}}\ E \qquad \text{(S19)}$$

where each triplet $I_\Omega = E$ specifies an identifier denoting a primitive function, its type in terms of the abstract type, and its definition in terms of the representation. For example,

$$\begin{aligned}
&\underline{\text{lettype}}\ \text{complex} = \underline{\text{prod}}(\underline{\text{real}}, \underline{\text{real}})\\
&\underline{\text{with}}\ i_{\text{complex}} = \langle 0, 1\rangle,\\
&\qquad \text{addc}_{\text{complex} \to \text{complex} \to \text{complex}} =\\
&\qquad \lambda x_{\text{prod(real, real)}} .\ \lambda y_{\text{prod(real, real)}} .\ \langle \underline{\text{add}}\ x.1\ y.1,\ \underline{\text{add}}\ x.2\ y.2\rangle\\
&\underline{\text{in}}\ \dots
\end{aligned}$$

However, because our language provides higher-order functions, this more complex definitional form can be defined as an abbreviation in terms of the original <u>lettype</u> construct:

$$\underline{\text{lettype}}\ \alpha = \omega\ \underline{\text{with}}\ i_{1\omega_1} = e_1,\ \dots\ ,\ i_{n\omega_n} = e_n\ \underline{\text{in}}\ e$$

$$\equiv (\underline{\text{lettype}}\ \alpha = \omega\ \underline{\text{in}}\ \lambda i_{1\omega_1} .\ \dots\ \lambda i_{n\omega_n} .\ e)\ e_1 \dots e_n\ .$$

This definition reduces to

$$e\big|_{\alpha \to \omega}\big|_{i_1 \to e_1,\ \dots\ ,\ i_n \to e_n}$$

(where because of the effect of bound-identifier renaming, the substitution of the e_k's for i_k's is simultaneous), which embodies the right semantics. More critically, one can derive the inference rule

$$\frac{\begin{array}{l}\pi,\ i_1{:}\omega_1,\ \ldots\ ,\ i_n{:}\omega_n \vdash e{:}\ \omega_0 \\ \pi \vdash e_1{:}\ \omega_1\big|_{\alpha \to \omega} \\ \vdots \\ \pi \vdash e_n{:}\ \omega_n\big|_{\alpha \to \omega}\end{array}}{\pi \vdash \underline{\text{lettype}}\ \alpha = \omega\ \underline{\text{with}}\ i_{1\omega_1} = e_1,\ \ldots\ ,\ i_{n\omega_n} = e_n\ \underline{\text{in}}\ e{:}\ \omega_0\big|_{\alpha \to \omega}} \quad \begin{array}{l}\text{when } \alpha \text{ does not occur (free)} \\ \text{in any type expression} \\ \text{assigned by } \pi\end{array} \qquad \text{(E19)}$$

which captures the notion of abstraction since ω does not occur in the typing of e. (Some authors [30] have suggested that the principle of abstraction necessitates a restriction on this rule that α should not occur (free) in ω_0, so that $\omega_0\big|_{\alpha \to \omega} = \omega_0$ and the type of the lettype ... with ... expression is the same as the type of its body. This restriction is also justified by the alternative definition of the lettype ... with ... expression that will be given in Section 5e.

For full generality, we should further extend lettype to permit the simultaneous definition of several abstract types with primitive functions, e.g.

lettype point = ... , line = ...
with $\text{intersect}_{\text{line} \to \text{line} \to \text{point}}$ = ... ,
    $\text{connect}_{\text{point} \to \text{point} \to \text{line}}$ = ...
in ... .

Such a generalization can still be defined in terms of our simple lettype, but the details are messy.

It is an open question how this kind of type definition can be combined with the subtype discipline of Section 4. There are clearly problems if the programmer can define his own implicit conversions. But even if he can only select a subset of the conversions that are induced by existing conversions between representations, it is not clear how to preserve the conditions (LUB) and (GLB) of Section 4b.

5b. Explicit Polymorphism

In [31], I defined a language that has come to be known as the polymorphic, or second-order typed lambda calculus, in which polymorphic functions can be defined by abstraction on type variables, and such functions can be applied to type expressions. (Somewhat to my chagrin, it was only much later that I learned that a similar, somewhat more general language had been invented earlier by J.-Y. Girard [32].)

This facility can be added to our explicitly typed base language (with type variables permitted in type expressions) by introducing the new expressions

$$E ::= \Lambda T.\ E \mid E[\Omega] \qquad \text{(S20)}$$

in which Λ binds type variables just as λ binds identifiers, with the rule of type beta reduction:

$$(\Lambda\alpha.\ e)[\omega] = e\big|_{\alpha \to \omega} \ . \qquad \text{(R20)}$$

For example,

$$\Lambda\alpha.\ \lambda f_{\alpha \to \alpha}.\ \lambda x_{\alpha}.\ f(f\ x)$$

is a polymorphic "doubling" function that can be applied to any type to obtain a doubling function for that type, e.g.

$$(\Lambda\alpha.\ \lambda f_{\alpha \to \alpha}.\ \lambda x_{\alpha}.\ f(f\ x))[\underline{int}] = \lambda f_{int \to int}.\ \lambda x_{int}.\ f(f\ x)$$

$$(\Lambda\alpha.\ \lambda f_{\alpha \to \alpha}.\ \lambda x_{\alpha}.\ f(f\ x))[\underline{real} \to \underline{bool}] =$$
$$\lambda f_{(real \to bool) \to (real \to bool)}.\ \lambda x_{real \to bool}.\ f(f\ x)\ .$$

Less trivially,

$$\Lambda\alpha.\ \underline{fix}(\lambda ap_{list\ \alpha \to list\ \alpha \to list\ \alpha}.\ \lambda x_{list\ \alpha}.\ \lambda y_{list\ \alpha}.$$
$$\underline{lchoose}\ y\ (\lambda n_{\alpha}.\ \lambda z_{list\ \alpha}.\ \underline{cons}\ n\ (ap\ z\ y))\ x)$$

is a polymorphic function that, when applied to a type α, yields a function for appending lists of type <u>list</u> α. Similarly,

$$\Lambda\alpha.\ \Lambda\beta.\ \underline{fix}(\lambda red_{list\ \alpha \to (\alpha \to \beta \to \beta) \to \beta \to \beta}.$$
$$\lambda \ell_{list\ \alpha}.\ \lambda f_{\alpha \to \beta \to \beta}.\ \lambda a_{\beta}.$$
$$\underline{lchoose}\ a\ (\lambda n_{\alpha}.\ \lambda z_{list\ \alpha}.\ f\ n\ (red\ z\ f\ a))\ \ell)$$

is a polymorphic function that, when applied to types α and β, yields a function for reducing lists of type <u>list</u> α to values of type β.

Even in 1974, the idea of passing types as arguments to functions was fairly widespread. The novelty was to extend the set of type expressions to provide types for polymorphic functions that were sufficiently refined to permit explicit typing:

$$\Omega ::= \Delta T.\ \Omega$$

Here Δ is an operator that binds type variables within type expressions. (Note that this implies that bound type variables can be renamed. We will regard renamed type expressions such as $\Delta\alpha.\ \alpha \to \alpha$ and $\Delta\beta.\ \beta \to \beta$ to be identical.) The idea is that if e has type ω' then Λα. e has type Δα.ω', and if e has type Δα. ω' then e[ω] has type $\omega'|_{\alpha \to \omega}$. Thus the polymorphic functions of type Δα. ω' can be thought of as functions that, when applied to a type α, give a value of type ω'.

For example, the types of the polymorphic doubling, append, and reduce functions given above (under the empty type assignment) are:

$$\Delta\alpha.\ (\alpha \to \alpha) \to \alpha \to \alpha\ \ ,$$
$$\Delta\alpha.\ \underline{list}\ \alpha \to \underline{list}\ \alpha \to \underline{list}\ \alpha\ \ ,$$
$$\Delta\alpha.\ \Delta\beta.\ \underline{list}\ \alpha \to (\alpha \to \beta \to \beta) \to \beta \to \beta\ \ .$$

More precisely, explicit typing is provided by the inference rules

$$\frac{\pi \vdash e:\ \omega'}{\pi \vdash \Lambda\alpha.\ e:\ \Delta\alpha.\ \omega'} \quad \text{when } \alpha \text{ does not occur free in any type expression assigned by } \pi \qquad \text{(E20a)}$$

and

$$\frac{\pi \mid\!\!- e\!: \Delta\alpha.\ \omega'}{\pi \mid\!\!- e[\omega]\!:(\omega'\big|_{\alpha \to \omega})} \tag{E20b}$$

Once type abstraction and application have been introduced, the <u>lettype</u> expression of the previous subsection can be defined as an abbreviation (just as the ordinary <u>let</u> can be defined in terms of ordinary abstraction and application):

$$\underline{\text{lettype}}\ \alpha = \omega\ \underline{\text{in}}\ e \ \equiv\ (\Lambda\alpha.\ e)[\omega] \quad .$$

From this definition, one can use reduction rule (R20) to derive (R17), and inference rules (E20a) and (E20b) to derive (E17).

More important, we have a notion of explicit polymorphism that includes that of Section 3b, but goes further to permit higher-order functions that accept polymorphic functions, albeit at the expense of a good deal of explicit type information. For instance, we can mirror the example of a polymorphic <u>let</u> in Section 3b by

<u>let</u> red = "polymorphic reduce function" <u>in</u>

$\lambda\ell\ell_{\text{list list int}}.\ \text{red}[\underline{\text{list}}\ \underline{\text{int}}, \underline{\text{int}}]\ \ell\ell$

$(\lambda\ell_{\text{list int}}.\ \lambda s_{\text{int}}.\ \underline{\text{add}}(\text{red}[\underline{\text{int}}, \underline{\text{int}}]\ \ell\ \underline{\text{add}}\ 0)\ s)\ 0\ .$

But now we can go further and rewrite this <u>let</u> as the application of a higher-order function to the polymorphic reduce function:

$(\lambda\text{red}_{\Delta\alpha.\ \Delta\beta.\ \text{list}\ \alpha \to (\alpha \to \beta \to \beta) \to \beta \to \beta}.$

$\lambda\ell\ell_{\text{list list int}}.\ \text{red}[\underline{\text{list}}\ \underline{\text{int}}, \underline{\text{int}}]\ \ell\ell$

$(\lambda\ell_{\text{list int}}.\ \lambda s_{\text{int}}.\ \underline{\text{add}}(\text{red}[\underline{\text{int}}, \underline{\text{int}}]\ \ell\ \underline{\text{add}}\ 0)\ s)\ 0)$

("polymorphic reduce function") .

5c. Higher-Order Polymorphic Programming

At first sight, functions that accept polymorphic functions seem exotic beasts of dubious utility. But the work of a number of researchers sugests that such functions may be the key to a novel programming style. They have studied an austere subset of the language we are considering in which the fixpoint operator <u>fix</u> is excluded, and have shown that this restricted language has extraordinary properties [32,33]. On the one hand, all expressions have normal forms, i.e. their evaluation always terminates (though the proof of this fact requires "second-order arithmetic" and cannot be obtained from Peano's axioms). On the other hand, the variety of functions that can be expressed goes far beyond the class of primitive recursive functions. (Indeed, one can express any program whose termination can be proved in second-order arithmetic.) Beyond this, they have shown that certain types of polymorphic functions provide rich structures akin to data algebras. [20, 34, 35]. (In particular, every many-sorted, anarchic, free algebra is "mimicked" by some polymorphic type.)

Our purpose here is not to give the details of this theoretical work, but to illustrated the unusual programming style that underlies it. The starting point is a variant of the way in which early investigators of the (untyped) lambda calculus encoded truth values and natural numbers.

Suppose we regard bool as an abbreviation for a polymorphic type, and true and false as abbreviations for certain functions of this type, as follows:

$$\underline{\text{bool}} \equiv \Delta\alpha.\ \alpha \to (\alpha \to \alpha) \quad ,$$
$$\underline{\text{true}} \equiv \Lambda\alpha.\ \lambda x_\alpha.\ \lambda y_\alpha.\ x \quad ,$$
$$\underline{\text{false}} \equiv \Lambda\alpha.\ \lambda x_\alpha.\ \lambda y_\alpha.\ y \quad .$$

Then, when e_1 and e_2 have type ω, we can define the conditional expression by

$$\underline{\text{if}}\ b\ \underline{\text{then}}\ e_1\ \underline{\text{else}}\ e_2 \equiv b[\omega]\ e_1\ e_2 \quad .$$

Moreover, we can define such functions as

$$\text{not} \equiv \lambda b_{\text{bool}}.\ \Lambda\alpha.\ \lambda x_\alpha.\ \lambda y_\alpha.\ b[\alpha]\ y\ x \quad ,$$
$$\text{and} \equiv \lambda b_{\text{bool}}.\ \lambda c_{\text{bool}}.\ \Lambda\alpha.\ \lambda x_\alpha.\ \lambda y_\alpha.\ b[\alpha](c[\alpha]\ x\ y)\ y \quad .$$

Similarly (ignoring negative numbers), we can define int and the natural numbers by

$$\underline{\text{int}} \equiv \Delta\alpha.\ (\alpha \to \alpha) \to (\alpha \to \alpha) \quad ,$$
$$0 \equiv \Lambda\alpha.\ \lambda f_{\alpha \to \alpha}.\ \lambda x_\alpha.\ x \quad ,$$
$$1 \equiv \Lambda\alpha.\ \lambda f_{\alpha \to \alpha}.\ \lambda x_\alpha.\ f\ x \quad ,$$
$$n \equiv \Lambda\alpha.\ \lambda f_{\alpha \to \alpha}.\ \lambda x_\alpha.\ f^n\ x \quad ,$$

where $f^n\ x$ denotes

$$\underbrace{f(\ \dots\ (f\ x)\ \dots\ )}_{n\ \text{times}} \quad ,$$

so that the number n becomes a polymorphic function accepting a function and giving its n-fold composition. (For example, 2 is the polymorphic doubling function.) Then we can define the function

$$\text{succ} \equiv \lambda n_{\text{int}}.\ \Lambda\alpha.\ \lambda f_{\alpha \to \alpha}.\ \lambda x_\alpha.\ f(n[\alpha]\ f\ x)$$

that increases its argument by one.

Now suppose g of type $\underline{\text{int}} \to \omega$, c of type ω, and h of type $\omega \to \omega$ satisfy the equations

$$g\ 0 = c \quad ,$$
$$g\ (n + 1) = h\ (g\ n) \quad .$$

Then $g\ n = h^n\ c$, so that

$$g = \lambda n_{\text{int}}.\ n[\omega]\ h\ c \quad .$$

For example (taking $\omega = \underline{\text{int}}$),

$$\text{add m } 0 = \text{m}, \qquad \text{add m } (\text{n} + 1) = \text{succ(add m n)},$$

so that

$$\text{add m} = \lambda \text{n}_{\text{int}}.\ \text{n}[\underline{\text{int}}]\ \text{succ m},$$

or more abstractly,

$$\text{add} \equiv \lambda \text{m}_{\text{int}}.\ \lambda \text{n}_{\text{int}}.\ \text{n}[\underline{\text{int}}]\,\text{succ m}.$$

Similarly, we can define

$$\text{mult} \equiv \lambda \text{m}_{\text{int}}.\ \lambda \text{n}_{\text{int}}.\ \text{n}[\underline{\text{int}}]\ (\text{add m})\ 0,$$

$$\text{exp} \equiv \lambda \text{m}_{\text{int}}.\ \lambda \text{n}_{\text{int}}.\ \text{n}[\underline{\text{int}}]\ (\text{mult m})\ 1.$$

More generally, suppose f of type $\underline{\text{int}} \to \omega$, c of type ω, and h of type $\underline{\text{int}} \to \omega \to \omega$ satisfy the equations

$$\text{f } 0 = \text{c}, \qquad \text{f } (\text{n} + 1) = \text{h n (f n)}.$$

Let g of type $\underline{\text{int}} \to \underline{\text{int}} \times \omega$ be such that $\text{g n} = \langle \text{n}, \text{f n}\rangle$, (where $\omega \times \omega'$ is $\underline{\text{prod}}(\omega,\omega')$).

$$\text{g } 0 = \langle 0, \text{c}\rangle,$$

$$\begin{aligned}\text{g } (\text{n} + 1) &= \langle \text{n} + 1, \text{f } (\text{n} + 1)\rangle = \langle \text{succ n}, \text{h n (f n)}\rangle \\ &= (\lambda \text{z}_{\text{int} \times \omega}.\ \langle \text{succ z.1}, \text{h z.1 z.2}\rangle)\ \langle \text{n}, \text{f n}\rangle \\ &= (\lambda \text{z}_{\text{int} \times \omega}.\ \langle \text{succ z.1}, \text{h z.1 z.2}\rangle)\ (\text{g n}).\end{aligned}$$

Thus by the argument given above,

$$\text{g} = \lambda \text{n}_{\text{int}}.\ \text{n}[\underline{\text{int}} \times \omega]\ (\lambda \text{z}_{\text{int} \times \omega}.\ \langle \text{succ z.1}, \text{h z.1 z.2}\rangle)\ \langle 0, \text{c}\rangle,$$

so that

$$\text{f} = \lambda \text{n}_{\text{int}}.\ (\text{n}[\underline{\text{int}} \times \omega]\ (\lambda \text{z}_{\text{int} \times \omega}.\ \langle \text{succ z.1}, \text{h z.1 z.2}\rangle)\ \langle 0, \text{c}\rangle).2.$$

Thus if we define primrec of type $\Delta\alpha.\ (\underline{\text{int}} \to \alpha \to \alpha) \to \alpha \to \underline{\text{int}} \to \alpha$ by

$$\begin{aligned}\text{primrec} \equiv \Lambda\alpha.\ &\lambda \text{h}_{\text{int} \to \alpha \to \alpha}.\ \lambda \text{c}_{\alpha}. \\ &\lambda \text{n}_{\text{int}}.\ (\text{n}[\underline{\text{int}} \times \alpha]\ (\lambda \text{z}_{\text{int} \times \alpha}.\ \langle \text{succ z.1}, \text{h z.1 z.2}\rangle)\ \langle 0, \text{c}\rangle).2,\end{aligned}$$

then $\text{f} = \text{primrec}[\omega]\ \text{h c}$.

For example, since the predecessor function satisfies

$$\text{pred } 0 = 0, \qquad \text{pred}(\text{n} + 1) = \text{n} = (\lambda \text{n}_{\text{int}}.\ \lambda \text{m}_{\text{int}}.\ \text{n})\ \text{n (pred n)},$$

we can define

$$\text{pred} \equiv \text{primrec}[\underline{\text{int}}](\lambda \text{n}_{\text{int}}.\ \lambda \text{m}_{\text{int}}.\ \text{n})\ 0,$$

and since the factorial function satisfies

$$\text{fact } 0 = 1, \qquad \text{fact}(\text{n} + 1) = \text{mult n (fact n)},$$

we can define

$$\text{fact} \equiv \text{primrec}[\underline{\text{int}}]\ \text{mult } 1.$$

Moreover, we can define functions from integers to other types. For example, primrec[list int] cons (nil_{int}) is the function from int to list int that maps n into (n − 1, ... , 0).

However, our ability to define numerical functions is not limited to the scheme of primitive recursion. For example, the exponentiation laws $f^m \cdot f^n = f^{m+n}$ and $(f^m)^n = f^{m \times n}$ lead to the definitions

$$\text{add} \equiv \lambda m_{int}.\ \lambda n_{int}.\ \Lambda\alpha.\ \lambda f_{\alpha \to \alpha}.\ \lambda x_{\alpha}.\ n[\alpha]\ f\ (m[\alpha]\ f\ x) \quad ,$$

$$\text{mult} \equiv \lambda m_{int}.\ \lambda n_{int}.\ \Lambda\alpha.\ \lambda f_{\alpha \to \alpha}.\ n[\alpha](m[\alpha]\ f) \quad ,$$

and the law $(\lambda f.\ f^m)^n = \lambda f.\ f^{(m^n)}$ (which can be proved by induction on n) leads to

$$\text{exp} \equiv \lambda m_{int}.\ \lambda n_{int}.\ \Lambda\alpha.\ n[\alpha \to \alpha](m[\alpha]) \quad .$$

More spectacularly, suppose we define

$$\text{aug} \equiv \lambda f_{int \to int}.\ \lambda n_{int}.\ \text{succ}\ n\ [\underline{int}]\ f\ 1 \quad ,$$

of type (int → int) → int → int. Then

$$\text{aug}\ f\ 0 = f\ 1 \ , \qquad \text{aug}\ f\ (n + 1) = f(\text{aug}\ f\ n) \ .$$

Thus, if we define

$$\text{ack} \equiv \lambda m_{int}.\ m[\underline{int} \to \underline{int}]\ \text{aug succ} \quad ,$$

of type int → int → int, we get

$$\text{ack}\ 0 = \text{succ} \ , \qquad \text{ack}\ (m + 1) = \text{aug}\ (\text{ack}\ m) \quad ,$$

so that

$$\begin{aligned} &\text{ack}\ 0\ n = n + 1 \quad , \\ &\text{ack}\ (m + 1)\ 0 = \text{aug}\ (\text{ack}\ m)\ 0 = \text{ack}\ m\ 1 \quad , \\ &\text{ack}\ (m + 1)\ (n + 1) = \text{aug}\ (\text{ack}\ m)\ (n + 1) = \text{ack}\ m\ (\text{aug}\ (\text{ack}\ m)\ n) \\ &\qquad = \text{ack}\ m\ (\text{ack}\ (m + 1)\ n) \quad , \end{aligned}$$

i.e., $\lambda n_{int}.$ ack n n is Ackermann's function.

In addition to the primitive types int and bool, various compound types can be defined in terms of polymorphic functions. For numbered products we can define

$$\underline{prod}(\omega_1, \ldots , \omega_n) \equiv \Delta\alpha.\ (\omega_1 \to \ldots \to \omega_n \to \alpha) \to \alpha$$

and, when $e_1, \ldots , e_n$ have types $\omega_1, \ldots , \omega_n$ and r has type $\underline{prod}(\omega_1, \ldots , \omega_n)$,

$$\langle e_1, \ldots , e_n \rangle \equiv \Lambda\alpha.\ \lambda f_{\omega_1 \to \ldots \to \omega_n \to \alpha}.\ f\ e_1 \ldots e_n \quad ,$$

$$r.k \equiv r[\omega_k]\ (\lambda x_{1\omega_1}.\ \ldots\ \lambda x_{n\omega_n}.\ x_k) \quad ,$$

since we have the reduction

$$\begin{aligned} &\langle e_1, \ldots , e_n \rangle.k \\ &\quad = (\Lambda\alpha.\ \lambda f_{\omega_1 \to \ldots \to \omega_n \to \alpha}.\ f\ e_1 \ldots e_n)\ [\omega_k](\lambda x_{1\omega_1}.\ \ldots\ \lambda x_{n\omega_n}.\ x_k) \\ &\quad = (\lambda f_{\omega_1 \to \ldots \to \omega_n \to \omega_k}.\ f\ e_1 \ldots e_n)(\lambda x_{1\omega_1}.\ \ldots\ \lambda x_{n\omega_n}.\ x_k) \end{aligned}$$

$$= (\lambda x_{1\omega_1}. \ldots \lambda x_{n\omega_n}. x_k)\, e_1 \ldots e_n = e_k .$$

Similarly, for numbered sums, we can define

$$\underline{\text{sum}}(\omega_1, \ldots, \omega_n) \equiv \Delta\alpha.\ (\omega_1 \to \alpha) \to \ldots \to (\omega_n \to \alpha) \to \alpha$$

and, when e has type ω_k and $f_1, \ldots, f_n$ have types $\omega_1 \to \omega, \ldots, \omega_n \to \omega$,

$$\underline{\text{inject}}_{\omega_1,\ldots,\omega_n}\ k\ e \equiv \Lambda\alpha.\ \lambda h_{1\omega_1 \to \alpha}. \ldots \lambda h_{n\omega_n \to \alpha}.\ h_k\ e ,$$

$$\underline{\text{choose}}(f_1, \ldots, f_n) \equiv \lambda s_{\text{sum}(\omega_1,\ldots,\omega_n)}.\ s[\omega]\ f_1 \ldots f_n ,$$

since we have the reduction

$$\underline{\text{choose}}(f_1, \ldots, f_n)(\underline{\text{inject}}_{\omega_1,\ldots,\omega_n}\ k\ e)$$
$$= (\lambda s_{\text{sum}(\omega_1,\ldots,\omega_n)}.\ s\,[\omega]\ f_1 \ldots f_n)(\Lambda\alpha.\ \lambda h_{1\omega_1 \to \alpha}. \ldots \lambda h_{n\omega_n \to \alpha}.\ h_k\ e)$$
$$= (\Lambda\alpha.\ \lambda h_{1\omega_1 \to \alpha}. \ldots \lambda h_{n\omega_n \to \alpha}.\ h_k\ e)\ [\omega]\ f_1 \ldots f_n$$
$$= (\lambda h_{1\omega_1 \to \omega}. \ldots \lambda h_{n\omega_n \to \omega}.\ h_k\ e)\ f_1 \ldots f_n = f_k\ e .$$

Finally, we consider lists, drawing upon a discovery by C. Böhm [35]. We define

$$\underline{\text{list}}\ \omega \equiv \Delta\beta.\ (\omega \to \beta \to \beta) \to \beta \to \beta ,$$

with the idea that, if $\ell = (x_1, \ldots, x_n)$, f has type $\omega \to \omega' \to \omega'$, and a has type ω',

$$\ell[\omega']\ f\ a = f\ x_1\ (f\ x_2 \ldots (f\ x_n\ a) \ \ldots) .$$

In other words, we regard a list as its own reduce function. Then we define

$$\underline{\text{nil}}_\omega \equiv \Lambda\beta.\ \lambda f_{\omega \to \beta \to \beta}.\ \lambda a_\beta.\ a$$

and, when e_1 has type ω and e_2 has type $\underline{\text{list}}\ \omega$,

$$\underline{\text{cons}}\ e_1\ e_2 \equiv \Lambda\beta.\ \lambda f_{\omega \to \beta \to \beta}.\ \lambda a_\beta.\ f\ e_1\ (e_2\ f\ a) .$$

Now we can carry out a development analogous to that for natural numbers. If g of type $\underline{\text{list}}\ \omega \to \omega'$, c of type ω', and h of type $\omega \to \omega' \to \omega'$ satisfy

$$g\ \underline{\text{nil}}_\omega = c , \qquad g(\underline{\text{cons}}\ x\ \ell) = h\ x\ (g\ \ell) ,$$

then

$$g = \lambda\ell_{\text{list}\ \omega}.\ \ell[\omega']\ h\ c .$$

For example, consider the function rappend, of type $\underline{\text{list}}\ \omega \to \underline{\text{list}}\ \omega \to \underline{\text{list}}\ \omega$, that is similar to the append function but with interchanged arguments. Since this function satisfies

$$\text{rappend}\ m\ \underline{\text{nil}}_\omega = m , \qquad \text{rappend}\ m\ (\underline{\text{cons}}\ x\ \ell) = \underline{\text{cons}}\ x\ (\text{rappend}\ m\ \ell) ,$$

we have

$$\text{rappend}\ m = \lambda\ell_{\text{list}\ \omega}.\ \ell[\underline{\text{list}}\ \omega]\ (\lambda x_\omega.\ \lambda y_{\text{list}\ \omega}.\ \underline{\text{cons}}\ x\ y)\ m ,$$

and thus

$$\text{append} \equiv \lambda\ell_{\text{list}\ \omega}.\ \lambda m_{\text{list}\ \omega}.\ \ell[\underline{\text{list}}\ \omega]\ (\lambda x_\omega.\ \lambda y_{\text{list}\ \omega}.\ \underline{\text{cons}}\ x\ y)\ m .$$

Next, consider the function flatten, of type $\underline{\text{list}}\ \underline{\text{list}}\ \omega \to \underline{\text{list}}\ \omega$, that appends together the elements of its arguments. Since this function satisfies

$$\text{flatten}\ \underline{\text{nil}}_{\text{list}\ \omega} = \underline{\text{nil}}_{\omega}\ , \qquad \text{flatten}\ (\underline{\text{cons}}\ x\ \ell) = \text{append}\ x\ (\text{flatten}\ \ell)\ ,$$

we have

$$\text{flatten} \equiv \lambda\ell_{\text{list list}\ \omega}.\ \ell[\underline{\text{list}}\ \omega]\ \text{append}\ \underline{\text{nil}}_{\omega}\ .$$

By similar arguments,

$$\text{length} \equiv \lambda\ell_{\text{list}\ \omega}.\ \ell[\underline{\text{int}}]\ (\lambda x_{\omega}.\ \lambda y_{\text{int}}.\ \text{succ}\ y)\ 0\ ,$$

$$\text{sumlist} \equiv \lambda\ell_{\text{list int}}.\ \ell[\underline{\text{int}}]\ \underline{\text{add}}\ 0\ ,$$

$$\text{mapcar} \equiv \lambda\ell_{\text{list}\ \omega}.\ \lambda f_{\omega \to \omega'}.\ \ell[\underline{\text{list}}\ \omega']\ (\lambda x_{\omega}.\ \lambda y_{\text{list}\ \omega'}.\ \underline{\text{cons}}\ (f\ x)\ y)\ \underline{\text{nil}}_{\omega'}\ ,$$

$$\text{null} \equiv \lambda\ell_{\text{list}\ \omega}.\ \ell[\underline{\text{bool}}]\ (\lambda x_{\omega}.\ \lambda y_{\text{bool}}.\ \underline{\text{false}})\ \underline{\text{true}}\ ,$$

$$\text{car} \equiv \lambda\ell_{\text{list}\ \omega}.\ \ell[\omega]\ (\lambda x_{\omega}.\ \lambda y_{\omega}.\ x)\ \underline{\text{error}}\ .$$

More generally, suppose f of type $\underline{\text{list}}\ \omega \to \omega'$, c of type ω' and h of type $\omega \to \underline{\text{list}}\ \omega \to \omega' \to \omega'$ satisfy

$$f\ \underline{\text{nil}}_{\omega} = c\ , \qquad f(\underline{\text{cons}}\ x\ \ell) = h\ x\ \ell\ (f\ \ell)\ .$$

Let g, of type $\underline{\text{list}}\ \omega \to (\underline{\text{list}}\ \omega) \times \omega'$, be such that $g\ \ell = \langle \ell, f\ \ell \rangle$. Then

$$g\ \underline{\text{nil}}_{\omega} = \langle \underline{\text{nil}}_{\omega}, c \rangle\ ,$$

$$g\ (\underline{\text{cons}}\ x\ \ell) = \langle \underline{\text{cons}}\ x\ \ell, f\ (\underline{\text{cons}}\ x\ \ell)\rangle = \langle \underline{\text{cons}}\ x\ \ell, h\ x\ \ell\ (f\ \ell)\rangle$$

$$= (\lambda x_{\omega}.\ \lambda z_{(\text{list}\ \omega) \times \omega'}.\ \langle \underline{\text{cons}}\ x\ z.1, h\ x\ z.1\ z.2\rangle)\ x\ (g\ \ell)\ ,$$

so that

$$g = \lambda\ell_{\text{list}\ \omega}.\ \ell[(\underline{\text{list}}\ \omega) \times \omega']$$
$$(\lambda x_{\omega}.\ \lambda z_{(\text{list}\ \omega) \times \omega'}.\ \langle \underline{\text{cons}}\ x\ z.1, h\ x\ z.1\ z.2\rangle)\ \langle \underline{\text{nil}}_{\omega}, c\rangle\ ,$$

and $f\ \ell = (g\ \ell).2$. Thus, if we define

$$\text{listrec} \equiv \Lambda\alpha.\ \Lambda\beta.\ \lambda h_{\alpha \to \text{list}\ \alpha \to \beta \to \beta}.\ \lambda c_{\beta}.\ \lambda\ell_{\text{list}\ \alpha}.$$
$$(\ell[(\underline{\text{list}}\ \alpha) \times \beta](\lambda x_{\alpha}.\ \lambda z_{(\text{list}\ \alpha) \times \beta}.\ \langle \underline{\text{cons}}\ x\ z.1, h\ x\ z.1\ z.2\rangle)\langle \underline{\text{nil}}_{\omega}, c\rangle).2$$

we have $f = \text{listrec}[\omega][\omega']\ h\ c$.

For example, since cdr of type $\underline{\text{list}}\ \omega \to \underline{\text{list}}\ \omega$ satisfies

$$\text{cdr}\ \underline{\text{nil}}_{\omega} = \underline{\text{error}}\ , \qquad \text{cdr}(\underline{\text{cons}}\ x\ \ell) = \ell\ ,$$

we can define

$$\text{cdr} \equiv \text{listrec}[\omega][\underline{\text{list}}\ \omega](\lambda x_{\omega}.\ \lambda y_{\text{list}\ \omega}.\ \lambda z_{\text{list}\ \omega}.\ y)\ \underline{\text{error}}\ ,$$

and since, when a has type ω' and f has type $\omega \to \underline{\text{list}}\ \omega \to \omega'$,

$$\underline{\text{lchoose}}\ a\ f\ \underline{\text{nil}}_{\omega} = a\ , \qquad \underline{\text{lchoose}}\ a\ f\ (\underline{\text{cons}}\ x\ \ell) = f\ x\ \ell\ ,$$

we can define

$$\underline{\text{lchoose}}\ a\ f \equiv \text{listrec}[\omega][\omega'](\lambda x_{\omega}.\ \lambda y_{\text{list}\ \omega}.\ \lambda z_{\omega'}.\ f\ x\ y)\ a\ .$$

Less trivially, consider the function insertandapply, of type $\underline{\text{int}} \rightarrow (\underline{\text{list int}} \rightarrow \underline{\text{list int}}) \rightarrow \underline{\text{list int}} \rightarrow \underline{\text{list int}}$, such that when ℓ is an ordered list of integers, insertandapply n f ℓ inserts n into the proper position of ℓ and applies f to the portion of ℓ following this position. This function satisfies

$$\text{insertandapply n f } \underline{\text{nil}}_{\text{int}} = \underline{\text{cons}} \text{ n } (\text{f } \underline{\text{nil}}_{\text{int}}) ,$$

$$\text{insertandapply n f } (\underline{\text{cons}} \text{ x } \ell) =$$
$$\underline{\text{if}} \text{ n} \le \text{x } \underline{\text{then}} \text{ } \underline{\text{cons}} \text{ n } (\text{f } (\underline{\text{cons}} \text{ x } \ell)) \text{ } \underline{\text{else}} \text{ } \underline{\text{cons}} \text{ x } (\text{insertandapply n f } \ell) ,$$

so that we can define

$$\begin{aligned}&\text{insertandapply} \equiv \lambda n_{\text{int}} .\ \lambda f_{\text{list int} \rightarrow \text{list int}} .\ \text{listrec}[\underline{\text{int}}][\underline{\text{list int}}]\\ &\quad (\lambda x_{\text{int}} .\ \lambda \ell_{\text{list int}} .\ \lambda m_{\text{list int}} .\\ &\quad\quad \underline{\text{if}} \text{ n} \le \text{x } \underline{\text{then}} \text{ } \underline{\text{cons}} \text{ n } (\text{f } (\underline{\text{cons}} \text{ x } \ell)) \text{ } \underline{\text{else}} \text{ } \underline{\text{cons}} \text{ x m})\\ &\quad (\underline{\text{cons}} \text{ n } (\text{f } \underline{\text{nil}}_{\text{int}}))\ .\end{aligned}$$

Then let merge, of type $\underline{\text{list int}} \rightarrow \underline{\text{list int}} \rightarrow \underline{\text{list int}}$, be the function that merges two ordered lists. This function satisfies

$$\text{merge } \underline{\text{nil}}_{\text{int}} = \lambda m_{\text{list int}} .\ m ,$$

$$\text{merge } (\underline{\text{cons}} \text{ x } \ell) = \text{insertandapply x } (\text{merge } \ell) .$$

Thus we can define

$$\text{merge} \equiv \lambda \ell_{\text{list int}} .\ \ell[\underline{\text{list int}} \rightarrow \underline{\text{list int}}] \text{ insertandapply } (\lambda m_{\text{list int}} .\ m) .$$

Semantically, the introduction of type abstraction and application raises thorny problems. Since, in the absence of $\underline{\text{fix}}$, all programs terminate, one might expect to obtain a semantics in which types denote ordinary sets and $\omega \rightarrow \omega'$ denotes the set of all functions from ω to ω'. But it can be shown that no such set is possible [36]. (Specifically, one can show that if the polymorphic type

$$\Delta\alpha.\ (((\alpha \rightarrow \underline{\text{bool}}) \rightarrow \underline{\text{bool}}) \rightarrow \alpha) \rightarrow \alpha$$

denoted some set, then from this set one could construct a set P that is isomorphic to $(P \rightarrow \underline{\text{bool}}) \rightarrow \underline{\text{bool}}$, which is impossible.)

The only known models [18,19] are domain-theoretic ones that give a semantics to the language with $\underline{\text{fix}}$ as well as without. Moreover, polymorphic types in these models denote huge domains that include generic functions that are not parametric in the sense of Strachey [24].

It should be mentioned that [18] and [19] provide models for a more general language than is described here, in which one can define functions from types to types and (in a somewhat different sense than in Section 3c, recursively defined types.

5d. Type Deduction Revisited

As the examples in the last subsection have made all too clear, explicit typing requires a dismaying amount of type information. This raises the problem of type deduction.

Consider taking an explicitly typed expression and erasing all type information by carrying out the replacements

$$\lambda i_{\omega}.\ e \Rightarrow \lambda i.\ e \qquad \Lambda\alpha.\ e \Rightarrow e \qquad e[\omega] \Rightarrow e$$

(For simplicity, we are only considering the pure polymorphic typed lambda calculus here, as defined by syntax equations (S1), (S5a,b) and (S20a,b).) This erasure maps our language back into the original (lambda-calculus subset of the) base language. But now the base language has a richer type structure than in Section 2b, since $\Delta\alpha.\ \omega$ remains a type expression and inference rules (E20a,b) become

$$\frac{\pi \vdash e\colon \omega'}{\pi \vdash e\colon \Delta\alpha.\ \omega'} \quad \text{when } \alpha \text{ does not occur free in any type expression assigned by } \pi \tag{I20a}$$

$$\frac{\pi \vdash e\colon \Delta\alpha.\ \omega'}{\pi \vdash e\colon (\omega'|_{\alpha \to \omega})} \tag{I20b}$$

An obvious question is whether one can give an algorithm for type deduction (in the sense of Section 3a) with these additional rules, i.e. (roughly speaking) an algorithm for recovering the erased type information. The answer, however, is far from obvious; despite strenuous efforts the question remains open. However, it has proved possible to devise useful but more limited algorithms that allow the programmer to omit some, though not all, type information [37,38].

5e. Existential Types

In [32], Girard introduced, in addition to Δ, another binding operator $\exists$ that produces types called existential types, with the syntax

$$\Omega ::= \exists T.\ \Omega$$

Recently, Mitchell and Plotkin [30] suggested that an existential type can be thought of as the signature of an abstract data type (in roughly the sense of algebraic data types), and that values of the existential type can be thought of as representations of the abstract data type.

For example, in a type definition of the form

$\underline{\text{lettype}}\ \alpha = \omega\ \underline{\text{with}}\ i_{1\omega_1} = e_1, \ldots, i_{n\omega_n} = e_n\ \underline{\text{in}}\ e$,

the type ω and the expression $\langle e_1, \ldots, e_n\rangle$ together constitute a data-type representation that is a value of the existential type $\exists\alpha.\ \underline{\text{prod}}(\omega_1, \ldots, \omega_n)$. Note that this existential type provides exactly the information to ascertain that the

e_k's bear the right relationship to ω, i.e. that each must have the type $\omega_k|_{\alpha \to \omega}$.

To "package" such representations, we introduce the operator rep, with the syntax

$$E ::= \underline{\text{rep}}_{\exists T.\Omega}\ \Omega\ E \qquad \text{(S21a)}$$

For example, the above representation would be the value of

$$\underline{\text{rep}}_{\exists\alpha.\text{prod}(\omega_1,\ldots,\omega_n)}\ \omega\ \langle e_1, \ldots, e_n\rangle\ .$$

To unpackage representations, we introduce the operator abstype, with the syntax

$$E ::= \underline{\text{abstype}}\ T\ \underline{\text{with}}\ I = E\ \underline{\text{in}}\ E\ . \qquad \text{(S21b)}$$

These operators are related by the reduction rule

$$\underline{\text{abstype}}\ \alpha\ \underline{\text{with}}\ i = (\underline{\text{rep}}_{\exists\alpha.\omega'}\ \omega\ e_2)\ \underline{\text{in}}\ e_1 = e_1|_{\alpha \to \omega}|_{i \to e_2}\ . \qquad \text{(R21)}$$

(Here we can assume the two occurrence of α are the same type variable since the occurrence in the subscript is a bound variable that can be renamed.)

Using these new constructs, the lettype expression given above can be defined as an abbreviation for

$$\underline{\text{abstype}}\ \alpha\ \underline{\text{with}}\ i = (\underline{\text{rep}}_{\exists\alpha.\text{prod}(\omega_1,\ldots,\omega_n)}\ \omega\ \langle e_1, \ldots, e_n\rangle)$$
$$\underline{\text{in}}\ \underline{\text{let}}\ i_1 = i.1\ \underline{\text{in}}\ \ldots\ \underline{\text{let}}\ i_n = i.n\ \underline{\text{in}}\ e\ ,$$

where i is an identifier not occurring free in e. Notice that this definition reduces to

$$(\underline{\text{let}}\ i_1 = i.1\ \underline{\text{in}}\ \ldots\ \underline{\text{let}}\ i_n = i.n\ \underline{\text{in}}\ e)|_{\alpha \to \omega}|_{i \to \langle e_1, \ldots, e_n\rangle}\ ,$$

and then to

$$e|_{\alpha \to \omega}|_{i_1 \to e_1, \ldots, i_n \to e_n}\ ,$$

which coincides with the reduction of the alternative definition of lettype ... with given in Section 5a.

The new operations satisfy the following inference rules:

$$\frac{\pi \vdash e\colon (\omega'|_{\alpha \to \omega})}{\pi \vdash \underline{\text{rep}}_{\exists\alpha.\omega'}\ \omega\ e\colon \exists\alpha.\ \omega'} \qquad \text{(E21a)}$$

$$\frac{\pi,\ i\colon \omega' \vdash e_1\colon \omega_0 \qquad \pi \vdash e_2\colon \exists\alpha.\ \omega'}{\pi \vdash \underline{\text{abstype}}\ \alpha\ \underline{\text{with}}\ i = e_2\ \underline{\text{in}}\ e_1\colon \omega_0} \quad \text{when } \alpha \text{ does not occur free in } \omega_0 \text{ or in any type expression assigned by } \pi \qquad \text{(E21b)}$$

(Note that the presence of the subscript $\exists\alpha.\ \omega'$ is necessary to make the first rule explicit.) From these rules and the definition of lettype ... with ... in terms of abstype and rep, one can derive an inference rule that is similar to (E19) except for the addition of a proviso that α must not occur free in ω_0 (which Mitchell and Plotkin believe is necessary to make lettype ... with ... truly abstract.)

Of course, existential types would be uninteresting if their only purpose was to provide an alternative definition of <u>lettype</u> ... <u>with</u> ... . The real point of the matter is that they permit abstract data type representations to be "first class" values that can be bound to identifiers and given as arguments to functions. For instance, the example in Section 5a can now be rewritten as

$(\lambda r_{\exists \text{complex. prod(complex, complex} \rightarrow \text{complex} \rightarrow \text{complex)}}\cdot$

  <u>abstype</u> complex <u>with</u> prims = r <u>in</u>

    <u>let</u> i = prims.1 <u>in</u> <u>let</u> addc = prim.2 <u>in</u> ...)

$(\underline{\text{rep}}_{\exists \text{complex. prod(complex, complex} \rightarrow \text{complex} \rightarrow \text{complex)}}$

  <u>prod</u>(<u>real</u>, <u>real</u>)

  $\langle\langle 0, 1\rangle, \lambda x_{\text{prod(real,real)}}\cdot\ \lambda y_{\text{prod(real,real)}}\cdot$

    $\langle \underline{\text{add}}\ x.1\ y.1,\ \underline{\text{add}}\ x.2\ y.2\rangle\rangle)$.

Here the first three lines denote a function that could be defined separately and applied to a variety of representations of type

$\exists$ complex. <u>prod</u>(complex, complex → complex → complex) ,

such as representations of complex numbers using different coordinate systems.

The analogy with algebraic data types is that $\exists\alpha.\ \omega$ is a signature and that values of the type are algebras of the signature. However, the analogy breaks down in two ways. On the one hand, $\exists\alpha.\ \omega$ cannot contain equations constraining the algebra, i.e. one can only mirror anarchic algebras this way. On the other hand, existential types go beyond algebra in permitting the primitive operations to be higher-order functions.

ACKNOWLEDGEMENT

This survey has been inspired by conversations with numerous researchers in the area of type theory. I am particularly indebted to Gordon Plotkin, Nancy McCracken, and Lockwood Morris.

REFERENCES

1. Guttag, J.V., and Horning, J.J., "The Algebraic Specification of Abstract Data Types", Acta Informatica 10 (1), 1978, pp. 27-52.
2. Goguen, J.A., Thatcher, J.W., and Wagner, E.G., "Initial Algebra Approach to the Specification, Correctness, and Implementation of Abstract Data Types", Current Trends in Programming Methodology, 4, Data Structuring, ed. R.T. Yeh, Prentice-Hall, Englewood Cliffs, 1978.
3. Kapur, D., "Towards a Theory for Abstract Data Types", Tech. Rep. TR-237, Laboratory for Computer Science, M.I.T., May 1980.
4. Coppo, M., and Dezani-Ciancaglini, M., "A New Type Assignment for λ-terms", Archive f. math. Logik u. Grundlagenforschung 19 (1979) 139-156.
5. Barendregt, H., Coppo, M., and Dezani-Ciancaglini, M., "A Filter Lambda Model and the Completeness of Type Assignment", to appear in the Journal of Symbolic Logic.
6. Martin-Lof, P., "Constructive Mathematics and Computer Programming", Proceedings of the Sixth (1979) International Congress for Logic, Methodology and Philosophy of Science, North-Holland, Amsterdam, 1979.
7. Constable, Robert L. and Zlatin, D.R., "The Type Theory of PL/CV3", ACM Transactions on Programming Languages and Systems, 6 (1984), 94-117.
8. Coquand, T., and Huet, G., "A Theory of Constructions", unpublished.
9. Burstall, R., and Lampson, B., "A Kernel Language for Abstract Data Types and Modules", Semantics of Data Types, eds. G. Kahn, D.B. MacQueen, and G. Plotkin, Lecture Notes in Computer Science 173, Springer-Verlag, Berlin (1984), pp. 1-50.
10. Reynolds, J.C., "The Essence of Algol", Algorithmic Languages, eds. J.W. de Bakker and J.C. van Vliet, North-Holland, 1981, pp. 345-372.
11. Oles, F.J., "A Category-Theoretic Approach to the Semantics of Programming Languages", Ph. D. dissertation, Syracuse University, August 1982.
12. Oles, F.J., "Type Algebras, Functor Categories, and Block Structure", Algebraic Methods in Semantics, eds. M. Nivat and J.C. Reynolds, Cambridge University Press (1985).
13. Henderson, P., and Morris, J.H., "A Lazy Evaluator". Proc. 3rd annual ACM SIGACT-SIGPLAN Symposium on Principles of Programming Languages, Atlanta, 1976, 95-103.
14. Friedman, D.P., and Wise, D.S., "CONS Should not Evaluate its Arguments". In Automata, Languages and Programing, eds. Michaelson and Milner, Edinburgh University Press, 1976, 257-284.
15. Landin, P.J., "A Correspondence Between Algol 60 and Church's Lambda-Notation", Comm. ACM 8, (February-March 1965), pp. 89-101 and 158-165.
16. McCarthy, J., "Recursive Functions of Symbolic Expressions and Their Computation by Machine, Part I", Comm. ACM 3 (April 1960), pp. 184-195.
17. MacQueen, D.B., and Sethi, R., "A Semantic Model of Types for Applicative Languages", ACM Symposium on LISP and Functional Programming, 1982, pp. 243-252.
18. McCracken, N., "An Investigation of a Programming Language with a Polymorphic Type Structure", Ph. D. dissertation, Syracuse University, June 1979.
19. McCracken, N.J., "A Finitary Retract Model for the Polymorphic Lambda-Calculus", submitted to Information and Control.

20. Reynolds, J.C., "Types, Abstraction and Parametric Polymorphism", Information Processing 83, ed. R.E.A. Mason, Elsevier Science Publishers B.V. (North-Holland) 1983, pp. 513-523.

21. Hindley, R., "The Principal Type-scheme of an Object in Combinatory Logic", Trans. Amer. Math. Society 146 (1969) 29-60.

22. Milner, R., "A Theory of Type Polymorphism in Programming", Journal of Computer and System Sciences 17 (1978), 348-375.

23. Robinson, J.A., "A Machine-oriented Logic Based on the Resolution Principle", JACM 12, 1 (1965), 23-41.

24. Strachey, C., "Fundamental Concepts in Programming Languages", Lecture Notes, International Summer School in Computer Programming, Copenhagen, August 1967.

25. Huet, G., "Resolution d'Équations dans des Langages d'Ordre 1,2,...,ω", doctoral thesis, University of Paris VII (September 1976).

26. Morris, F.L., "On List Structures and Their Use in the Programming of Unification", School of Computer and Information Science, Report 4-78, Syracuse University, 1978.

27. Reynolds, J.C., "Using Category Theory to Design Implicit Conversions and Generic Operators", Semantics-Directed Compiler Generation, Proceedings of a Workshop, Aarhus, Denmark, January 14-18, 1980, ed. N.D. Jones, Lecture Notes in Computer Science 94, Springer-Verlag, New York, pp. 211-258.

28. Goguen, J.A., "Order Sorted Algebras: Exceptions and Error Sorts, Coercions and Overloaded Operators", Semantics and Theory of Computation Report 14, Computer Science Department, UCLA (December 1978). To appear in Journal of Computer and Systems Science.

29. Cardelli, L., "A Semantics of Multiple Inheritance", Semantics of Data Types, eds. G. Kahn, D.B. MacQueen and G. Plotkin, Lecture Notes in Computer Science 173, Springer-Verlag, Berlin (1984), pp. 51-67.

30. Mitchell, J.C., and Plotkin, G.D., "Abstract Types Have Existential Type", Proc. 12th Annual ACM Symposium on Principles of Programming Languages, New Orleans, 1985.

31. Reynolds, J.C., "Towards a Theory of Type Structure, Proc. Colloque sur la Programmation, Lecture Notes in Computer Science 19, Springer-Verlag, New York, 1974, pp. 408-425.

32. Girard, J.-Y., "Interprétation Fonctionelle et Elimination des Coupures dans l'Arithmétique d'Ordre Supérieur", Thèse de Doctorat d'État, Paris, 1972.

33. Fortune, S., Leivant, D., and O'Donnell, M., "The Expressiveness of Simple and Second-Order Type Structures", Journal of the ACM 30, 1 (January 1983), pp. 151-185.

34. Leivant, D., "Reasoning About Functional Programs and Complexity Classes Associated with Type Disciplines", Twenty-fourth Annual Symposium on Foundation of Computer Science (1983) 460-469.

35. Böhm, C., and Berarducci, A., "Automatic Synthesis of Typed Λ-Programs on Term Algebras", submitted for publication.

36. Reynolds, J.C., "Polymorphism is not Set-Theoretic", Semantics of Data Types, eds. G. Kahn, D.B. MacQueen, and G. Plotkin, Lecture Notes in Computer Science 173, Springer-Verlag, Berlin (1984), pp. 145-156.

37. Leivant, D., "Polymorphic Type Inference", Proc. 10th Annual ACM Symposium on Principles of Programming Languages, Austin, 1983.

38. McCracken, N.J., "The Typechecking of Programs with Implicit Type Structure", Semantics of Data Types, eds. G. Kahn, D.B. MacQueen, and G. Plotkin, Lecture Notes in Computer Science 173, Springer-Verlag, Berlin (1984), pp. 301-315.

ON THE MAXIMUM SIZE OF RANDOM TREES

M. Protasi*, M. Talamo**

* Dipartimento di Matematica, Università dell'Aquila, Via Roma 33, 67100 L'Aquila. Incaricato di ricerca presso lo IASI-CNR.

** Istituto di Analisi dei Sistemi ed Informatica del C.N.R., Viale Manzoni 30, 00185 Roma.

ABSTRACT

In this paper we prove a conjecture of Erdös and Palka on the maximum size of random trees. Furthermore, while, generally speaking, in the probabilistic analysis the results are proved only when the size of the graphs tends to infinity, in this case, with extremely small probability of error, the results also hold for graphs of small size.

1. INTRODUCTION

The problem of finding the size of the largest induced tree in a graph is NP-complete (see, for instance, [4]) and because of its computational intractability has been studied in the last years using alternative approaches to the classical worst case analysis. In particular, some research effort has been spent analyzing this problem on random graphs, with the aim of evaluating the optimal solution and finding very good approximate solutions for almost all graphs. The achieved results can be summarized as follows. If we assume the constant density model, after some first results in [5], in [1] and in [6], independently, it has been shown that the size of the largest tree is about $\frac{2 \log n}{\text{constant}}$ for almost all graphs of size n. Furthermore in [6], it has been proved that a simple greedy algorithm achieves a solution whose value is one half of the value of the optimal solution. As above written, these results are stated in the constant density model, that is we assume a constant probability p of existence of an edge between two nodes. The situation is different if we are in a more general setting.

In [1], Erdös and Palka posed the following open problem. Let now p be a function on n, i.e. p = p(n) with p tending to zero as n tends to infinity. Find such a value of the edge probability p for which a random graph has the largest induced tree. They conjectured that for a suitable p(n) a random graph contains a tree of size b(c)·n, where b(c) depends only on the constant c. In this paper we prove the conjecture evaluating a lower and upper bound to the size of the tree. In order to achieve this result we will exploit a general probabilistic model introduced in [8] and [9] for studying the max clique and the max independent set problems on random graphs. This general model allows to study graphs of different density and in particular sparse graphs as in the case of the open problem posed by Erdös.

The other important point of the paper regards the asymptoticity of the analysis. In fact with a probability of error tending to zero very fast, we are able to state our results not only, as generally happens, when the number of the nodes of the graph is sufficiently large, but also for graphs of small size. This characteristics, of course, has interesting practical consequences because in the applications we are interested in graphs having a reasonable number of nodes. The possibility of performing such a study has due substantially to the properties of the model we use and in [9] other non asymptotic results have been found.

2. THE PROBLEM

First of all, we recall the problem we want to study. Given a graph, we want to find the largest induced tree, that is, the induced tree having the maximum number of nodes. As we said in the introduction, the problem has been already solved from a probabilistic point of view if we assume a constant probability p $(0 < p < 1)$ of existence of an edge between two nodes independently from the presence or absence of any other edges. With this model the number of nodes of the largest induced tree is $2 \log n/[\log(1/(1-p)]+o(\log n)$ for almost all graphs of size n. However, assuming a constant density model, we are able to deal only with dense graphs.

The probabilistic analysis of our problem becomes much more interesting and complete if we also include the case of the sparse graphs. It is natural to expect that in this case we are able to find trees of larger cardinality than for dense graphs. This situation has lead Erdös and Palka [1] to formulate the following open problem:
Let now p be a function on n, i.e. $p = p(n)$ with $p(n)$ tending to zero as n tends to infinity. Find such a value of the edge probability p for which a random graph has the largest induced tree.

Erdös and Palka conjectured that for a suitable p(n) a random graph contains a tree of size b(c)·n, where b(c) depends only on c and c is a constant.

In order to prove, in the following paragraph, that the conjecture is true we will use a general probabilistic model already used in [8] and [9].

DEFINITION 1. Let V be a set of n nodes. Every pair (i,j) with $i,j \in V$ is an edge with probability $p(n) = 1-c^{-c/n}$ independently from the presence or absence of any other edges.

The type of graphs we study depends, of course, on the value of c we choose. Therefore if we allow that c can assume values belonging to a large range, we are able to perform a very general analysis. In particular, for suitable values of c, we are able to deal, at the same time, with dense and sparse graphs. To prove the conjecture we will have to find particular values of c such that the corresponding values of p(n) make the random graph have induced trees as large as possible.

3. MAIN RESULTS

Before giving the main results, we need to state some preliminary lemmas.

Let z_n be the value of the maximum induced tree of a graph of n nodes.

In order to bound some combinatorial inequalities more easily, in the following instead of z_n, we will often use the formula θn.

LEMMA 1.

$$1-c^{-\frac{c}{n}} = \frac{c\ln c}{n} \quad \text{as } n \to \infty$$

PROOF. It is trivial applying some elementary analytic steps.

LEMMA 2. For any fixed $\varepsilon > 0$ and $c \geq e$ constant

$$z_n > \left(\frac{1}{c} - \varepsilon\right)\frac{\log\log n}{\log n} \cdot n$$

for a suitable probability p(n).

PROOF. See [7].

LEMMA 3.

$$\binom{n}{\theta n} \leq \left(\theta^{\theta}(1-\theta)^{(1-\theta)}\right)^{-n}$$

PROOF. See [3] or [9].

Let us now prove the conjecture

THEOREM 1. $0{,}42n \leq z_n \leq 0{,}91n$ almost surely.

PROOF. Let $k = \theta n$. Given a tree T, let $|T|$ denote the number of nodes of T. The proof is divided in two parts. First of all we will find an upper bound to z_n.

1) $\text{Prob}(\exists T/|T| \geq k) \leq \binom{n}{k} \cdot k^{k-2} \cdot p^{k-1} \cdot q^{\binom{k}{2}-(k-1)}$ (where $q = 1-p$)

(Since $k = \theta n$ and $p = 1-c^{-\frac{c}{n}}$)

$$\leq \binom{n}{\theta n} \cdot k^{k-2} (1-c^{-\frac{c}{n}})^{k-1} \cdot (c^{-\frac{c}{n}})^{\frac{(k-1)(k-2)}{2}}$$

(Exploiting Lemmas 1 and 3)

$$\leq (\theta^{\theta}(1-\theta)^{(1-\theta)})^{-n} \cdot (\theta n)^{\theta n-2} (\frac{c \ln c}{n})^{\theta n-1} \cdot (c^{-\frac{c}{n}})^{\frac{(\theta n-1)(\theta n-2)}{2}}$$

$$\leq (\theta(1-\theta)^{\frac{1}{\theta}-1})^{-\theta n} \cdot \theta^{\theta n-2} \cdot n (c \ln c)^{\theta n-1} \cdot c^{-\frac{\theta c}{2}(\theta n-1)}$$

(Exploiting Lemma 2)

$$\leq ((\theta(1-\theta)^{\frac{1}{\theta}-1})^{-1} \cdot \theta c \ln c \cdot c^{-\frac{\theta c}{2}})^{\theta n-1} = \left(\frac{c^{(1-\frac{\theta c}{2})} \ln c}{(1-\theta)^{(\frac{1}{\theta}-1)}} \right)^{\theta n-1}$$

This quantity tends to zero if $\frac{c^{(1-\theta)} \cdot \ln c}{(1-\theta)^{(\frac{1}{\theta}-1)}} < 1$. If we succeed, among the possible values of c that verify the last inequality, in finding the value that maximes θn, we will be immediately able to evaluate an upper bound to z_n. We study the threshold

$$\frac{c^{(1-\frac{\theta c}{2})} \ln c}{(1-\theta)^{(\frac{1}{\theta}-1)}} = 1$$

Applying numerical methods we obtain that the maximum value of θ such that θn is the value of the largest induced tree is less than 0,91.

2) In order to find the lower bound to z_n and complete the proof of the proof of the conjecture, we consider the following inequality. Let y_k be the random variable that denotes the number of trees of size $< k$

$$\mathrm{Prob}(y_k = 0) \leq \frac{\sigma^2(n,k)}{E^2(n,k)}$$

where σ^2 is the variance and E is the expectation of y_k. Now we bound the ratio $\frac{\sigma^2(n,k)}{E^2(n,k)}$

$$\frac{\sigma^2(n,k)}{E^2(n,k)} \leq \frac{\sum_{l=1}^{k} k^{k-2}(k-l)^{k-l-2} k^{k-l-1} (k-l+1)^{l} \binom{k}{l}\binom{n}{k}\cdot\binom{n-k}{k-l}\left(\frac{c\ln c}{n}\right)^{2k-4-l} c^{-\frac{c}{n}\left(2\binom{k}{2}-2k-\binom{l}{2}\right)}}{\binom{n}{k}^2 k^{2k-4}\left(\frac{c\ln c}{n}\right)^{2k-4} c^{-\frac{c}{n}2\binom{k}{2}-2k+4}}$$

(By doing some combinatorial and algebraic steps and remembering that $k = \theta n$)

$$\leq \frac{\left(\frac{\theta}{1-\theta}\right)^{\theta n}\left(\frac{n}{c\ln c}\right)^{\theta n} c^{\frac{c\theta^2 n}{2}}}{(\theta n)^{\theta n}} \cdot n^2$$

$$\leq \left(\frac{\left(\frac{\theta}{1-\theta}\right) c^{\frac{c\theta}{2}}}{\theta c\ln c}\right)^{\theta n}(1+\varepsilon) \qquad (\varepsilon > 0)$$

This quantity tends to zero if $\theta \leq 0{,}42$.

QED

After proving the conjecture, we will give another theorem that shows that the result also hold for graphs pf small size, so strengthening remarkably the achieved result.

Let p_e be the probability of error that z_n does not belong to the interval $[0{,}42n, 0{,}91n]$.

THEOREM 2. $p_e < d^n$ for every n, where d is a constant < 1.

Therefore Theorem 2 assures that, with a very small probability of error, that is, with a probability of error tending to zero very fast, the result of Theorem 1 holds for graphs of every size.

4. REFERENCES

[1] P. Erdös, Z. Palka: *Trees in random graphs*, Discr. Math., Vol. 46 (1983).

[2] P. Erdös, A. Renyi: *On the evolution of random graphs*, Publ. Math. Inst. Hung. Acad. Sci., Vol. 5A (1960).

[3] J. Friedman: *Constructing* 0(n log n) *size monotone formulae for the* k-th *elementary symmetric polynomial of* n *boolean variables*, Proc. 25th Symp. on Foundations of Computer Science (1984).

[4] M.R. Garey, D.S. Johnson: *Computers and intractability. A guide to the theory of NP completeness*, Freeman, San Francisco (1978).

[5] M. Karonski, Z. Palka: *On the size of a maximal induced tree in a random graph*, Math. Slovaca, Vol. 30 (1980).

[6] A. Marchetti Spaccamela, M. Protasi: *The largest tree in a random graph*, Theor. Comp. Sci., Vol. 23 (1983).

[7] Z. Palka, A. Rucinski: *On the order of the largest induced tree in a random graph*, (to appear).

8] M. Protasi, M. Talamo: *A new probabilistic model for the study of algorithmic properties of random graph problems*, Proc. Conference on Foundations of Computation Theory, Borgholm, Lect. Notes in Comp. Sci., Springer Verlag (1983).

9] M. Protasi, M. Talamo: *A general analysis of the max independent set and related problems on random graphs*, Tech. Rep. n. 3/84, Dipartimento di Matematica, Università del-l'Aquila (1984).

FAST SEARCHING IN A REAL ALGEBRAIC MANIFOLD WITH APPLICATIONS TO GEOMETRIC COMPLEXITY

Bernard Chazelle
Department of Computer Science
Brown University
Providence, RI 02912, USA

Abstract

This paper generalizes the multidimensional searching scheme of Dobkin and Lipton [SIAM J. Comput. 5(2), pp. 181–186, 1976] for the case of arbitrary (as opposed to linear) real algebraic varieties. Let d, r be two positive constants and let $P_1, \ldots, P_n$ be n rational r-variate polynomials of degree $\leq d$. Our main result is an $O(n^{2^{r+6}})$ data structure for computing the predicate $[\exists i\ (1 \leq i \leq n) \mid P_i(x) = 0]$ in $O(\log n)$ time, for any $x \in E^r$. The method is intimately based on a decomposition technique due to Collins [Proc. 2nd GI Conf. on Automata Theory and Formal Languages, pp. 134–183, 1975]. The algorithm can be used to solve problems in computational geometry via a *locus* approach. We illustrate this point by deriving an $o(n^2)$ algorithm for computing the time at which the convex hull of n (algebraically) moving points in E^2 reaches a steady state.

1. Introduction

Let F be a family of n hyperplanes in E^r, with r taken as a constant. Dobkin and Lipton [DL] have shown how to represent F, using a polynomial amount of storage, so that whether a given point lies in any of the hyperplanes of F can be checked in $O(\log n)$ time. Our goal is to generalize the technique used to achieve this result for the case where F is a family of algebraic varieties. Put more

---

This research was supported in part by NSF grants MCS 83–03925 and the Office of Naval Research and the Defense Advanced Research Projects Agency under contract N00014–83–K–0146 and ARPA Order No. 4786.

formally, let d, r be two positive constants and let $F = \{P_1, \ldots, P_n\}$ be a set of n polynomials of degree $\leq d$ in r real variables with rational coefficients. We consider the problem of preprocessing F so that, for any $x \in E^r$, the predicate $[\exists i \ (1 \leq i \leq n) \mid P_i(x) = 0]$ can be evaluated efficiently. If the predicate is true, any one of the indices i for which $P_i(x) = 0$ should be reported — note that requiring the report of all such indices might by itself rule out a fast response. If the predicate is false, the point x lies in an open, maximally connected region, over which the value of each P_i keeps a constant sign. Assuming that these regions have been labelled in preprocessing, retrieving the label corresponding to the region containing x will also be required. In this form, the problem is a direct generalization of the well-known *planar point location* problem. Previous work on point location with non-linear boundaries has been limited to the case $r = 2$, culminating in the optimal algorithms of Edelsbrunner, Guibas and Stolfi [EGS], and Cole [Co].

Our main result is a data structure for answering any such query in $O(\log n)$ time. The space and time necessary to construct the data structure are in $O(n^{2^{r+6}})$, i.e. polynomial in n. The main consequence of our result is to open the door to *locus*-based methods for solving previously untouchable problems of computational geometry: the locus approach for retrieval problems involves considering each query as a higher-dimensional point and partitioning the underlying space into regions providing the same answer — see [O] for a discussion of this approach. We illustrate this notion on a specific example by considering a problem posed by Atallah in [A]. Given n points in E^2, each moving as a fixed degree polynomial function of time, is it possible to compute in $o(n^2)$ time the first instant at which the convex hull of the points will enter its final (steady) configuration? We use our generalized point location algorithm to solve this problem in the affirmative.

In the next section (Section 2), we review the necessary algebraic tools and describe the point location algorithm in Section 3. As in Dobkin and Lipton's method, the search proceeds by iterated projections on canonical hyperplanes. The preprocessing is inspired by Collins' cylindrical algebraic decomposition [C], adding some refinements due to Schwartz and Sharir [SS]. Ultimately, the spirit of the method goes back to Tarski's fundamental work on the decidability of elementary algebra [T]. In the remainder of the paper (Sections 4–5), we present our solution to Atallah's problem and conclude with directions for further research.

2. The Algebraic Machinery

Most of the algebraic notions involved in this work can be found exposed in great detail in two milestone papers by Collins [C] and Schwartz and Sharir [SS]. We have tried to adhere to the terminology used in these papers as much as possible. The fundamental algebraic concepts can be found in van der Waerden's classic text [W]; for the specialized treatment of resultants and subresultants used in the paper the reader should turn to Brown and Traub [BT].

I) *Collins' Decidability Theorem*

In 1948, Tarski proved that every statement in elementary algebra (i.e. in the elementary theory of real-closed fields) is decidable. The non-elementary procedure given by Tarski was subsequently (computationally) improved in a number of different ways by several researchers (e.g. Seidenberg, Cohen, Collins, Monk/Solovay, Ben-Or/Kozen/Reif). For the purpose of the present work, we shall use Collins' decision procedure as a guiding framework. Let a ***standard prenex formula*** be any logical sentence of the form $(Q_k x_k)(Q_{k+1} x_{k+1}) \ldots (Q_r x_r)\phi(x_1, \ldots, x_r)$, where Q_i is the universal or existential quantifier and $\phi(x_1, \ldots, x_r)$ is a quantifier-free formula made of boolean connectives, standard comparators, and polynomials with rational coefficients in the real variables $x_1, \ldots, x_r$. A logical sentence is called an ***atomic formula*** if it is free of quantifiers and logical connectives.

Theorem 1. (Collins [C]) Let Φ be an arbitrary standard prenex formula with r variables, c atomic formulas, m polynomials of degree $\leq d$, and no integral coefficient of length $\geq n$. Whether Φ is true or false can be decided in $cn^3(2d)^{2^{2r+8}} m^{2^{r+6}}$ operations.

II) *The Cylindrical Algebraic Decomposition*

A ***Collins decomposition*** of E^r is a scheme for partitioning E^r in order to discriminate among the connected regions of E^r induced by a real algebraic variety. Before reviewing the main components of Collins' technique, we need define the fundamental notion of ***cylindrical algebraic decomposition*** (c.a.d., for short). A c.a.d. of E^r is a partitioning of E^r defined recursively as follows. For $r = 1$, a c.a.d. is a finite set of disjoint open intervals, along with the algebraic numbers bounding them, whose union form E^1. For $r > 1$, a c.a.d. K is defined in terms of a c.a.d. K' of E^{r-1} and an r-variate polynomial $P(x_1, \ldots, x_{r-1}, y)$ with rational coefficients. Let $K' = \{c_1, \ldots, c_\mu\}$; for each $c_i \in K'$, there exists ν_i such that for each $x = (x_1, \ldots, x_{r-1}) \in c_i$, $P(x, y)$, regarded as a polynomial in y, has ν_i real roots $f_{i,1}(x) < \ldots < f_{i,\nu_i}(x)$, each of which is a continuous function in x over c_i. If $\nu_i = 0$, set $c_{i,1} = c_i \times E^1$. If $\nu_i > 0$, set $c_{i,2j} = \{(x, f_{i,j}(x)) \mid x \in c_i\}$ for $1 \leq j \leq \nu_i$, set

$c_{i,2j+1} = \{(x,y) \mid x \in c_i \ \&\ f_{i,j}(x) < y < f_{i,j+1}(x)\}$ for $1 \le j < \nu_i$. Also, set $c_{i,1} = \{(x,y) \mid x \in c_i \ \&\ y < f_{i,1}(x)\}$ and $c_{i,2\nu_i+1} = \{(x,y) \mid x \in c_i \ \&\ f_{i,\nu_i}(x) < y\}$. Finally, K is defined as the set of cells $\{c_{1,1}, \ldots, c_{1,2\nu_1+1}, \ldots, c_{\mu,1}, \ldots, c_{\mu,2\nu_\mu+1}\}$. Informally, the cells can be formed by considering the cylinders based at each $c \in K'$ and chopping them off with the real hypersurface $P(x_1, \ldots, x_{r-1}, y) = 0$. P (resp. K') is called the base polynomial (resp. cylindrical algebraic decomposition) of K. Since a c.a.d. is defined in terms of a unique base c.a.d. of lesser dimension, by induction, K defines an *induced* c.a.d for each E^k $(1 \le k < r)$. Incidentally, one should note that each cell of K is "well-behaved," in the sense that it is topologically equivalent to an open cell of dimension $\le r$.

To be of interest here, a cylindrical algebraic decomposition must provide a framework for discriminating among several algebraic varieties. So, let F be a family of n functions of r variables. We say that a c.a.d. K is *F-invariant* if for each $c \in K$ and each $f \in F$, we have: $f(x) = 0$ for each $x \in c$, $f(x) < 0$ for each $x \in c$, or $f(x) > 0$ for each $x \in c$. To prevent topological anomalies and thus facilitate the computation, we may enforce a cylindrical algebraic decomposition to be *well-based*, using a probabilistic procedure [SS]. K is said to be well-based if its base polynomial $P(x, y)$, regarded as a polynomial in y, is not identically zero for any given value of x in E^{r-1}. With these conditions, Schwartz and Sharir [SS] have shown that each root function $f_{i,j}$ (defined over $c_i \in K'$) can be extended continuously over the closure of c_i. This implies that the closure of every cell in K is a union of cells. Informally, this fairly intuitive result prevents K from displaying any pathological configuration. In particular, this means that every line $(x_1, \ldots, x_{r-1}) \times E^1$ intersects the algebraic variety $P(x, y) = 0$ a finite number of times. These intersections will form the basis of the binary search underlying the point location algorithm to be presented in the next section.

Following Collins' terminology, an *algebraic sample* of K is a set of points with algebraic coordinates, one in each cell of K (recall that a number is algebraic if it is a root of a polynomial). An algebraic sample is *cylindrical* (c.a.s) if 1) $r = 1$, or 2) the set of $r - 1$ first coordinates of each point form a c.a.s of K', the base c.a.d. of K. If $\{c_{i,1}, \ldots, c_{i,2\nu_i+1}\}$ is the set of cells of K associated with the cell c_i of K', the sample points in each $c_{i,j}$ $(1 \le j \le 2\nu_i + 1)$ all share the same $r - 1$ first coordinates. We conclude this string of definitions with a word on the representation of a c.a.d. The *standard definition* of a c.a.d. $K = \{d_1, \ldots, d_\nu\}$ is a sequence of quantifier-free formulas $\{\phi_1(x), \ldots, \phi_\nu(x)\}$ $(x = (x_1, \ldots, x_r))$, where $\phi_i(x)$ has only $x_1, \ldots, x_r$ as free variables and $d_i = \{x \in E^r \mid \phi_i(x) \text{ is true }\}$. $\phi_i(x)$ is called the *defining* formula of d_i.

III) *The Collins Construction*

Let d, r be two positive constants and let $F = \{P_1, \ldots, P_n\}$ be a set of n polynomials of degree $\leq d$ in r real variables with rational coefficients. A Collins decomposition, K, is an F-invariant cylindrical algebraic decomposition. The construction of K proceeds recursively. Sharir and Schwartz solve the problem by considering the product $P = \Pi_{1 \leq i \leq n} P_i$ and using the well-known fact that P (as a polynomial in x_r) has exactly $d_1 - d_2$ distinct real roots, where d_1 is the degree of P and d_2 is the degree of R, the gcd of P and its x_r-derivative. It then suffices to find a c.a.d. of E^{r-1} such that on each cell the degrees of both P and R remain constant. We follow the same strategy; however, because of its prohibitively high degree, we must avoid considering the product P in the actual construction of the cylindrical algebraic decomposition. Instead, although we will keep P as the base polynomial of the c.a.d., we will use the fact that P is a product of smaller-degree polynomials when carrying out the construction. Intuitively, it appears that taking all pair-products $P_i \times P_j$ should be sufficient, for the number of distinct roots change either when two roots of P_i "merge" into one or when a root of P_i "merges" with a root of P_j. In the present case, degeneracies among the coefficients (which are, we should recall, polynomials in $r-1$ variables) necessitates a more careful treatment.

To begin with, we recall basic facts from elimination theory. Let Q be an r-variate polynomial with rational coefficients. We regard $Q(x_1, \ldots, x_r) = \sum_{0 \leq i \leq p} Q_i(x_1, \ldots, x_{r-1}) x_r^i$ as a polynomial in x_r with coefficients in the ring of polynomials in $r-1$ variables. Let $\deg(Q) = p$ be the degree of Q and $\mathrm{ldcf}(Q) = Q_p(x_1, \ldots, x_{r-1})$ be the leading coefficient of Q. Following Tarski and Collins, we define the *reductum* of Q, $\mathrm{red}(Q) = \sum_{0 \leq i \leq p-1} Q_i(x_1, \ldots, x_{r-1}) x_r^i$. We also introduce $\mathrm{red}^0(Q) = Q$ and for each $k \geq 0$, $\mathrm{red}^{k+1}(Q) = \mathrm{red}(\mathrm{red}^k(Q))$. Finally, let $\mathrm{der}(Q)$ denote the x_r-derivative of Q. Let $A(x)$ and $B(x)$ be two polynomials in the real variable x with $\deg(A) = a$, $\deg(B) = b$. The *Sylvester matrix* of A and B is the $(a+b) \times (a+b)$ matrix M obtained by placing the coefficients of the polynomials $x^{b-1}A(x), \ldots, xA(x), A(x), x^{a-1}B(x), \ldots, xB(x), B(x)$ in consecutive rows of M, with the coefficients of x^i appearing in column $a+b-i$. For $0 \leq i \leq j \leq \min(a,b)$, M_j is the matrix obtained by deleting the last j rows of A coefficients, the last j rows of B coefficients, and all the last $2j$ columns. We can then define $\mathrm{psc}^j(A,B)$ (the jth principal subresultant coefficient of A and B) as the determinant of M_j (see [C] for details). The unique factorization theorem for polynomials implies that A and B have exactly j common roots (i.e. j is the degree of $\gcd(A,B)$) if and only if j is the least index for which $\mathrm{psc}^j(A,B) \neq 0$. This result is at the basis of the recursive construction of an F-invariant c.a.d.

Next, we define what arguments should be passed to the decomposition algorithm at the second recursive call. To do so, we define the *projection* of F, denoted G, as the union of G_2, G_3, G_4, with $G_1 = \{\mathrm{red}^k(P) \mid P \in F \;\&\; k \geq 0 \;\&\; \deg(\mathrm{red}^k(P)) \geq 1\}$, $G_2 = \{\mathrm{ldcf}(P) \mid P \in G_1\}$, $G_3 = \{\mathrm{psc}^k(P, \mathrm{der}(P)) \mid P \in G_1 \;\&\; 0 \leq k < \deg(\mathrm{der}(P))\}$, $G_4 = \{\mathrm{psc}^k(P,Q) \mid P,Q \in G_1 \;\&\; 0 \leq k < \min(\deg(P), \deg(Q))\}$. The notion of projection generalizes Dobkin and Lipton's idea of pairing up hyperplanes: the pairing takes place in G_4, while G_2 and G_3 account for the occasional losses of rank in each variety. The following result is proven in [C]: let K' be a G-invariant c.a.d. of E^{r-1} and let c_i be any cell of K'; the total number of distinct roots of $P_1(x_1, \ldots, x_r), \ldots, P_n(x_1, \ldots, x_r)$, as polynomials in x_r, remains constant as $(x_1, \ldots, x_{r-1})$ varies in c_i. These roots form a well-ordered set of continuous functions over c_i: $f_{i,1}(x_1, \ldots, x_{r-1}), \ldots, f_{i,\nu_i}(x_1, \ldots, x_{r-1})$. As a result, for each $c_i \in K'$, the partition of $c_i \times E^1$ induced by the hypersurfaces $f_{i,1}(x) = 0, \ldots, f_{i,\nu_i}(x) = 0$ $(x \in E^{r-1})$ defines an F-invariant c.a.d. of E^r.

This provides a recursive scheme for computing an F-invariant c.a.d. K of E^r. The algorithm takes F as input and recurses by calling itself with G, the projection of F, as argument. The output of the algorithm will be a c.a.s. of F, $\{\beta_1, \ldots, \beta_\nu\}$, where for each $i = 1, \ldots, \nu$ each coordinate of $\beta_i \in E^r$ is represented by a quantifier-free defining formula. It turns out that Collins' sophisticated method for computing a standard definition of K is not really necessary (although doable within the same asymptotic running time). Instead, we need an additional piece of information: a correspondence between sample points and their defining polynomials in F. If $r > 1$, K has a base c.a.d. $K' = \{c_1, \ldots, c_\mu\}$ (which is G-invariant). Let $\{\beta'_1, \ldots, \beta'_\mu\}$ be the c.a.s. of K', computed recursively. For each $i = 1, \ldots, \mu$, let $\{\beta_{i,1}, \ldots, \beta_{i,2\nu_i+1}\}$ be the points of the c.a.s. of K, ordered in ascending x_r-order, whose first $r-1$ coordinates form the point β'_i. Each point $\beta_{i,2j}$ $(1 \leq j \leq \nu_i)$ lies on at least one algebraic variety of the form $P_l(x) = 0$. Let $l_{i,j}$ be any such value of l and let $\beta_{i,2j} = (a_1, \ldots, a_r)$; we define $m_{i,j}$ as the number of distinct roots of $Q(y)$ that are strictly smaller than a_r, where $Q(y) = P_{l_{i,j}}(a_1, \ldots, a_{r-1}, y)$ is regarded as a polynomial in y. As part of the output, we require the sequence $\{(l_{i,1}, m_{i,1}), \ldots, (l_{i,\nu_i}, m_{i,\nu_i})\}$ for each $i = 1, \ldots, \mu$. This sequence will be necessary later on in order to carry out the binary searches underlying the point location algorithm.

The next step is to show how to derive these sequences from $\{\beta_{i,1}, \ldots, \beta_{i,2\nu_i+1}\}$ $(1 \leq i \leq \mu)$. Recall that the latter sequences are provided directly by the Collins construction. Let $\phi(x)$ be the quantifier-free defining formula for $\beta_{i,2j}$ $(1 \leq j \leq \nu_i)$. Trivially, we can test the predicate $[\exists x \in E^r \mid \phi(x) \;\&\; (P_l(x) = 0)]$ for each $l = 1, \ldots, n$, and pick as $l_{i,j}$, say, the first value of l found to satisfy the predicate. To obtain $m_{i,j}$, it suffices to express with a prenex formula the fact, F_k, that z is a root of

Q and $Q(y) = P_{l_{i,j}}(a_1, \ldots, a_{r-1}, y)$ has exactly k distinct roots strictly smaller than z. We can express F_k with the formula $R^{l_{i,j}}_{a_1,\ldots,a_{r-1},k}(z) = [(\exists y_1, \ldots, y_k)(\forall x) \mid ((Q(z))^2 + (Q(y_1))^2 + \ldots + (Q(y_k))^2 = 0) \ \& \ (y_1 < \ldots < y_k < z) \ \& \ (Q(x) \neq 0 \text{ or } z \leq x \text{ or } (y_1 - x)^2 + \ldots + (y_k - x)^2 = 0)]$. The value of $m_{i,j}$ is then given by the unique index k for which $R^{l_{i,j}}_{a_1,\ldots,a_{r-1},k}(z)$ is true, with $\beta_{i,2j} = (\beta'_i, z)$. In analyzing the complexity of the algorithm, we assume that only rational symbolic calculations are used during the course of the computation. We omit the proofs of the following complexity results, which can be found in [C]. Let d be the maximum degree of any polynomial in F in any variable and let l be an upper bound on the *norm-length* of any polynomial of F. The norm-length of a polynomial is the number of bits needed to represent the sum of the absolute values of its coefficients. As usual, we assume that r, d, and l are independent of n. The F-invariant c.a.d. produced by the Collins construction consists of $O\left((2d)^{3^{r+1}} n^{2^r}\right) = O(n^{2^r})$ cells. The total number of polynomials defined in the various projections introduced in the decomposition is bounded above by $O\left((2d)^{3^r} n^{2^{r-1}}\right) = O(n^{2^{r-1}})$ and the degree of each polynomial is at most $\frac{1}{2}(2d)^{2^{r-1}} = O(1)$. The norm-length of each polynomial is $\leq (2d)^{2^r} l = O(1)$. Consider now the c.a.s. of the decomposition. Each algebraic point is represented by its coordinates. We can represent an algebraic number in two ways, depending on the interpretation we give to it. Either it is a real root α of a polynomial $A(x)$ with algebraic coefficients. We isolate the root by specifying an interval I with rational endpoints, so α is represented by the pair (A, I). The coefficients of A (if non-rational) are represented recursively. In some cases, an algebraic number β will appear as an element of the algebraic number field $Q(\alpha)$ (recall that $Q(\alpha)$ is the intersection of all the extension fields of Q which contain α, or equivalently, the smallest subfield of $\Re$ which contains Q and α). In this case, we represent β as a rational polynomial $B(\alpha)$. The degree of any polynomial used in the definition of all the c.a.s.'s is dominated by $(2d)^{2^{2r-1}} = O(1)$ and more interestingly, the norm-length of each polynomial is $\leq l(2d)^{2^{2r+3}} n^{2^{r+1}} = O(n^{2^{r+1}})$. Implementing the Collins construction proper requires $O(l^3(2d)^{2^{2r+8}} n^{2^{r+6}}) = O(n^{2^{r+6}})$ operations. Using Theorem 1 and the previous upper bounds, it is easy to see that this running time asymptotically dominates the overhead of computing the sequences of the form $\{(l_{i,1}, m_{i,1}), \ldots, (l_{i,\nu_i}, m_{i,\nu_i})\}$.

3. The Generalized Point Location Algorithm

Most of the ingredients entering the composition of the algorithm have already been introduced. The data structure $DS(F)$ is defined recursively as follows: it includes 1) $DS(G)$, where G is the projection of F; 2) a c.a.s. of K; 3) a set of ν one-word memory cells $C^f_1, \ldots, C^f_\nu$ (which we conveniently associate with the cells of K). Let $C_1^g, \ldots, C_\mu^g$ be the memory cells associated with $DS(G)$ (in one-to-one correspondence with the cells of $K' = \{c_1, \ldots, c_\mu\}$). Each cell C_i^g ($1 \leq i \leq \mu$) stores a pointer to the sequence $\{l_{i,1}, \ldots, l_{i,\nu_i}\}$ previously defined. Recall that the cell C_i^g is associated with $2\nu_i + 1$ cells of K (each projecting exactly on c_i). Let $W_i = \{C_{i,1}^f, \ldots, C_{i,2\nu_i+1}^f\}$ be the corresponding memory cells, in ascending x_r-order. Consider the sequence $S_i = \{l_{i,1}, \ldots, l_{i,\nu_i}\}$ as an ordered set of *keys*. The possible outcomes of a binary search in this set is a sequence of $2\nu_i + 1$ keys and open intervals, which we put in one-to-one correspondence with W_i. The data structure is now complete, so we can describe the algorithm.

The input is a family of polynomials F, assumed to be preprocessed as previously described. The *generalized point location* problem to be solved can be stated as follows: "given a query point $q = (q_1, \ldots, q_r) \in E^r$, compute the index i such that C_i^f corresponds to the unique cell of K that contains q." Note in passing the practical importance of having a c.a.s. of K. In most cases, indeed, the point location problem arises when one wishes to compute a function f from E^r to some range (e.g. set of integers or reals). The c.a.d. K partitions E^r into regions on which f is invariant. The availability of a c.a.s. allows us to precompute the unique value of f over each such region, thus reducing the original problem of computing $f(q)$ for arbitrary $q \in E^r$ to a generalized point location problem. If $r = 1$, the algorithm is trivial, so assume that $r > 1$. Recursively, we assume that we have available the index of the cell C_i^g that contains $(q_1, \ldots, q_{r-1})$. Perform a binary search in S_i with respect to x_r, and report the element of W_i corresponding to the result of the search. Without concern for efficiency, we implement the generic comparison against $l_{i,j}$ as the two-fold question:

1. Does $P_{l_{i,j}}(q) = 0$?
2. Is q_r strictly larger or smaller than the $(m_{i,j} + 1)$st real root of $P_{l_{i,j}}(q_1, \ldots, q_{r-1}, y)$, regarded here as a polynomial in y?

The latter question is answered by testing the predicate $[(\forall y) \mid (y > q_r) \text{ or } \neg(R^{l_{i,j}}_{q_1, \ldots, q_{r-1}, m_{i,j}}(y))]$. For K to be well-based ensures consistency in the search process, i.e. the intersection of the line $(q_1, \ldots, q_{r-1}) \times E^1$ and the variety $P_{l_{i,j}}$ always consists of a discrete set of points. To analyze the complexity of the algorithm, we use the fact that each polynomial occurring in any projection has degree $O(1)$, so from Theorem 1 it easily follows that any comparison can be decided in constant time. Since the total number of these polynomials is in $O(n^{2^{r-1}})$, so is the length of any sequence over which a binary search is performed. As a result, each binary search requires $O(2^r \log n)$ time. As mentioned earlier, the preprocessing costs are in $O(n^{2^{r+6}})$, both in time and space. We conclude with the main

result of this section.

Theorem 2. **Let $F = \{P_1, \ldots, P_n\}$ be a family of n fixed-degree r-variate polynomials with rational coefficients and let $S = \{x \in E^r \mid \Pi_{1 \le i \le n} P_i(x) \ne 0\}$. In $O(n^{2^{r+6}})$ time and space, it is possible to compute a set of algebraic points, one in each connected region of S, as well as set up a data structure for computing the predicate $[\exists i\ (1 \le i \le n) \mid P_i(q) = 0]$, for any $q \in E^r$. In $O(2^r \log n)$ time, the algorithm will return an index i such that $P_i(q) = 0$ if such an index is to be found, otherwise it will return the algebraic point associated with the unique region of S that contains q.**

4. Applications to Computational Geometry

In [A], Atallah poses the following problem. Consider n points moving in the plane as a polynomial function of the time. What is the first time their convex hull enters a steady-state, i.e. a combinatorially invariant configuration? We will assume that the real roots of any univariate polynomial of fixed degree with real coefficients independent of n can be computed in constant time with any desired precision — note that this assumption does not in any way follow from Collins' theorem. The naive algorithm consists of computing the steady convex hull in $O(n \log n)$ time [A], and then retrieving the first time each point achieves its steady positioning with respect to each edge on the hull. The maximum of all these times provides the desired value. Can this quadratic algorithm be improved? We will show that it can — at least theoretically. More precisely, we will use the generalized point location algorithm of the previous section to produce an $O(n^{2-\epsilon})$ time algorithm, for a very small positive constant ϵ. Let $V = \{p_1, \ldots, p_n\}$ be a set of $n > 2$ points in the Euclidean plane, subject to algebraic motion. This assumes the existence of $2n$ univariate polynomials $p_1^x, p_1^y, \ldots, p_n^x, p_n^y$ of degree d with real coefficients, such that for each i $(1 \le i \le n)$, $p_i^x(t)$ and $p_i^y(t)$ are respectively the x and y coordinates of p_i at time $t \ge 0$. We assume that d as well as all the polynomials' coefficients are independent of n. Let $p_{i_1}, \ldots, p_{i_k}$ be the points on the boundary of the convex hull of V at time t, given in clockwise order with $i_1 < \min(i_2, \ldots, i_k)$ (if $p_{i_j}, \ldots, p_{i_{j'}}$ coincide, their indices appear in the order $i_j < \ldots < i_{j'}$). Let $H(t)$ be the (uniquely defined) sequence $\{i_1, \ldots, i_k\}$. It is trivial to show that $H(t)$ converges as t grows to infinity. We define the *threshold* of $H(+\infty)$ as the smallest value of $t \ge 0$ such that $[(\forall t' \ge t) \mid H(t) = H(t')]$.

Let $p_i^x(t) = \sum_{0 \le j \le d} a_{i,j} t^j$ and $p_i^y(t) = \sum_{0 \le j \le d} b_{i,j} t^j$ for $i = 1, \ldots, n$. Wlog, assume that $H(+\infty) = \{1, \ldots, m\}$ $(m \le n)$ and that all n points $(a_{i,0}, \ldots, a_{i,d}, b_{i,0}, \ldots, b_{i,d})$ of $\Re^{2d+2}$ are pairwise

distinct. In $O(n \log n)$ time, compute $H(+\infty)$ [A] and check all pairs (p_i, p_{i+1}) $(1 \le i \le m)$ in order to determine the largest $t_0 \ge 0$ such that, for some i, we have $p_i^x(t_0) = p_{i+1}^x(t_0)$ and $p_i^y(t_0) = p_{i+1}^y(t_0)$; note that we may have $t_0 = -\infty$. Here (as in the following), index arithmetic is taken mod m. Similarly, we ensure the convexity of the polygon $\{p_1, \ldots, p_m\}$ by considering the function $f_i(q) = (p_i^y(t) - p_{i+1}^y(t))q_x + (p_{i+1}^x(t) - p_i^x(t))q_y + p_i^x(t)p_{i+1}^y(t) - p_i^y(t)p_{i+1}^x(t)$. The point $q = (q_x, q_y) \in E^2$ lies to the right (resp. on, to the left) of the oriented line $(p_i, \overrightarrow{p_i p_{i+1}})$ iff $f_i(q) < 0$ (resp. $f_i(q) = 0$, $f_i(q) > 0$). For each p_i $(1 \le i \le m)$ compute the largest real root of $f_{i+1}(p_i)$ as a polynomial in t; discard every case where the polynomial is identically zero. Let t_1 be the largest value thus obtained (or $-\infty$ if there is none), and let $t_2 = \max(0, t_0, t_1)$. Trivially, t_2 can be computed in $O(n)$ time, once $H(+\infty)$ is available. All that remains to be done is to compute the first instants at which each p_j $(m < j \le n)$ lies inside $H(+\infty)$ for good. To do so, we allow ourselves some preprocessing. Let $q(t) = (q_x(t), q_y(t))$ be a point in E^2, with $q_x(t) = \sum_{0 \le j \le d} q_j^x t^j$ and $q_y(t) = \sum_{0 \le j \le d} q_j^y t^j$. The point $\chi = (q_0^x, \ldots, q_d^x, q_0^y, \ldots, q_d^y)$ belongs to E^{2d+2} and is independent of n. Let $\text{sign}(A) = -1$ (resp. $= 0, 1$) if $A < 0$ (resp. $A = 0$, > 0). We define $t(\chi) = [\min t \in \Re \mid t \ge t_2 \;\&\; (\forall i; 1 \le i \le m)(\forall t' > t)\; \text{sign}(f_i(q(t))) = \text{sign}(f_i(q(t')))]$. We next describe a fast algorithm for computing $t(\chi)$ based on the generalized point location algorithm of the preceding section.

For each i $(1 \le i \le m)$, let $F = \{\phi_1(\chi, t), \ldots, \phi_m(\chi, t)\}$, where $\phi_i(\chi, t)$ denotes the $(2d+3)$-variate polynomial of degree $2d+1$, $f_i(\sum_{0 \le j \le d} q_j^x t^j, \sum_{0 \le j \le d} q_j^y t^j)$. Let K be the F-invariant c.a.d. of E^{2d+3} provided by the procedure described in Section 2, and let $K' = \{c_1, \ldots, c_\mu\}$ be its base c.a.d. (i.e. the induced c.a.d. of E^{2d+2}). Recall that for each $c_i \in K'$ the procedure provides us with a sequence of indices (possibly empty) $S_i = \{l_{i,1}, \ldots, l_{i,\nu_i}\}$ with the following meaning: for any given $\chi \in c_i$ the line $\chi \times E^1$ provides an increasing sequence of real roots for the univariate polynomials $\phi_{l_{i,1}}(\chi, t), \ldots, \phi_{l_{i,\nu_i}}(\chi, t)$. The interpretation of this sequence is trivial: it gives the indices of the lines passing through $p_i p_{i+1}$ that are intersected by the trajectory of χ in chronological order (from $t = -\infty$ to $t = +\infty$). If the sequence is empty, χ never intersects such a line. Once K' has been preprocessed for point location, computing $t(\chi)$ is straightforward. Locate the cell c_i that contains χ and check whether the sequence S_i is empty. If yes, set $t(\chi) = t_2$. If the sequence is not empty, the trajectory of χ intersects the line passing through $p_{l_{i,\nu_i}} p_{l_{i,\nu_i}+1}$ at some time t and does not intersect any other such line subsequently. We obtain t by computing the largest real root of $\phi_{l_{i,\nu_i}}(\chi, t)$ as a polynomial in t (which must exist). Finally we set $t(\chi) = \max(t_2, t)$. From Theorem 2, we immediately conclude.

Lemma 1. In $O(m^{2^{2d+9}})$ time and space, it is possible to construct a data structure so that the function $t(\chi)$ can be evaluated at any point $\chi \in E^{2d+2}$ in $O(4^d \log m)$ time.

We are now ready to attack Atallah's problem. We use a batching strategy inspired by Yao's work on higher-dimensional MST [Y]. Partition the boundary of $H(+\infty)$ into p polygonal lines $H_1, \ldots, H_p$, each consisting of α edges; we set $m = \alpha(p-1)+r$, so one of the polygonal lines will have $r < \alpha$ edges. Each H_i can be regarded as an unbounded convex polygon by stretching its end-edges to infinity. This allows us to apply Lemma 1 with respect to each point $p_{m+1}, \ldots, p_n$ and each polygon $H_1, \ldots, H_p$. The maximum of the set formed by t_2 and the $(n-m)p$ values thus obtained is exactly the threshold of $H(+\infty)$. We easily see that the time complexity of the algorithm is in $O\big(p(n-m)4^d \log \alpha + p\alpha^{2^{2d+9}}\big)$. Setting $\alpha = (n \log n)^{1/2^{2d+9}}$, we conclude

Theorem 3. In $O\big(n^{2-1/2^{2d+9}} (\log n)^{1-1/2^{2d+9}}\big) = O(n^{2-1/4^{d+5}})$ time, it is possible to compute the threshold of n points moving according to a polynomial function of time of degree d.

5. Conclusions

The main contribution of this work has been to show that Dobkin and Lipton's method [DL] for searching among hyperplanes can be generalized to handle arbitrary algebraic varieties. Our method is an adaptation of a quantifier-elimination procedure due to Collins [C]. This feature gives even more generality to our algorithm than mentioned earlier. Indeed, we do not have to limit ourselves to real algebraic varieties but may consider the more general problem of discriminating among *semi-algebraic sets*. Recall that a set $S \subseteq E^r$ is semi-algebraic if there exists a first-order sentence $\phi(x_1, \ldots, x_r)$ in the theory of real numbers, with $x_1, \ldots, x_r$ as the only free variables of ϕ, such that $S = \{x \in E^r \mid \phi(x)$ is true $\}$. The preprocessing involves eliminating each quantifier by means of Collins projections (one projection per quantifier), and the point location takes place in the induced cylindrical algebraic decomposition of E^r. A similar technique was implicitly used in eliminating the time variable in the preprocessing of Section 4.

Further work includes the (difficult) problem of drastically reducing the high space-complexity of the generalized point location algorithm. Even the case of hyperplanes is still open. Regarding the problem of computing thresholds in steady-state computations, one will observe that our technique is general enough to be applied to other problems (e.g. closest/farthest pairs). An interesting question is to determine whether ad hoc treatment of these problems leads to more efficient solutions and thus our technique is in a sense too general, or if Theorem 3 is essentially all we can hope for. For example, a continuity argument easily shows that ensuring the local coherence of the steady-state Voronoi diagram is sufficient to compute its threshold (e.g. checking the non-zero length of its edges). It is then fairly simple to devise an $O(n \log n)$ algorithm for computing the steady-state Voronoi diagram of n moving points as well as its threshold. Note that the same argument can be made for convex hulls if all the points are guaranteed to lie on it. One essential feature of these easy cases is that the output *involves* all the input. Is this in general a necessary condition of efficiency?

Acknowledgments: Thanks to Janet Incerpi and Chee Yap for useful comments about this manuscript.

REFERENCES

[A] Atallah, M.J. *Dynamic computational geometry*, Proc. 24th Annual FOCS Symp., pp. 92–99, Nov. 1983.

[BT] Brown, W., Traub, J.F. *On Euclid's algorithm and the theory of subresultants*, J. ACM 18, pp. 505–514, 1971.

[Co] Cole, R. *Searching and storing similar lists*, Tech. Rep. No. 88, New York University, Oct. 1983.

[C] Collins, G.E., *Quantifier elimination for real closed fields by cylindric algebraic decomposition*, Proc. 2nd GI Conf. on Automata Theory and Formal Languages, Springer-Verlag, LNCS 35, Berlin, pp. 134–183, 1975

[DL] Dobkin, D.P., Lipton, R.J. *Multidimensional searching problems*, SIAM J. Comput. 5(2), pp. 181–186, 1976.

[EGS] Edelsbrunner, H., Guibas, L., Stolfi, J. *Optimal point location in a monotone subdivision*, to appear.

[O] Overmars, M.H. *The locus approach*, Tech. Rept. RUU–CS–83–12, Univ. Utrecht, July 1983.

[SS] Schwartz, J.T., Sharir, M. *On the "piano movers" problem. II: General techniques for computing topological properties of real algebraic manifolds*, Adv. in Appl. Math 4, pp. 298–351, 1983.

[T] Tarski, A. *A decision method for elementary algebra and geometry*, Univ. of Calif. Press, 1948, 2nd edition, 1951.

[W] van der Waerden, B.L. *Modern Algebra*, Ungar Co., New York, 1953.

[Y] Yao, A.C. *On constructing minimum spanning tree in k-dimensional space and related problems*, SIAM J. Comput. 11(4), pp. 721–736, 1982.

TYPED CATEGORICAL COMBINATORY LOGIC

P-L. Curien

CNRS-Université Paris VII, LITP, Tour 55-56 1er étage,
3 Place Jussieu, 75221 PARIS CEDEX 05

ABSTRACT

The subject of the paper is the connection between the typed λ-calculus and the cartesian closed categories, pointed out by several authors. Three languages and their theories, defined by equations, are shown to be equivalent: the typed λc-calculus (i.e. the λ-calculus with explicit products and projections) λc_K, the free cartesian closed category CCC_K, and a third intermediary language, the typed categorical combinatory logic CCL_K, introduced by the author. In contrast to CCC_K, CCL_K has the same types as λc_K, and roughly the terminal object in CCC_K is replaced by the application and couple operators in CCL_K. In CCL_K β-reductions as well as evaluations w.r.t. environments (the basis of most practical implementations of λ-calculus based languages) may be simulated in the well-known framework of a same term rewriting system. Finally the introduction of CCL_K allowed the author to understand the untyped underlying calculus, investigated in a companion paper. Another companion paper describes a general setting for equivalences between equational theories and their induced semantic equivalences, the equivalence between CCL_K and CCC_K is an instance of which.

1. Introducing categorical combinators

Categories and λ-calculus are alternative theories of functionality, based on composition of functions (more abstractly arrows), substitution of actual parameters to formal ones respectively. A part from the interest in itself to be able to connect two different formalisms, and to let benefit one from the other (see the end of section 2 for an example), there is an operational significance: roughly λ-calculus is well-suited for programming, and combinators (of Curry, or those introduced here) allow for implementations getting rid of some difficulties in the scope of variables. Indeed we intend to develop implementations of functional programming languages based on categorical combinators, which we introduce now, letting them arise from the known principle that a formal semantic description yields a compilation.

Suppose that x has value $\underline{3}$ (3 is underlined to stress that $\underline{3}$ is the representation of 3), and that we want to express the function associating $y(\underline{3})$ (also written simply $y\underline{3}$) with every function y. The λ-calculus provides the elegant notation $\lambda y.yx$ (M in the sequel) for that function.

Hence $N = \lambda f.f(fx)$ will designate the function associating with f the result of applying f to $f(3)$. One may like to relate those two so-called λ-expressions and to point out some modularity by showing that the second expression may be built from the first one and a third expression.

Indeed two constructions are involved in the informal definition of N: first we associate with $f = x \mapsto f(x)$ the function $f \circ f = x \mapsto f(f(x))$, and then we apply the function described by M. This can be summarized by

$$N' = M \circ (\lambda f.f \circ f)$$

where we have mixed the λ-notation with the notation for composing functions, the basic concept of the theory of categories. We may turn N' into a pure λ-expression (i.e. code the composition in terms of λ-notation): we obtain

$$P = \lambda f.(\lambda y.yx)((\lambda yz.y(yz))f)$$

It is an interesting exercise for readers not familiar with λ-calculus to experiment with the β-reduction, which is the formal application of a function to its argument, involving substitution of the formal argument at each occurrence of the formal parameter. Repeated application of this rule yields N back from P:

$$P \rightarrow \lambda f.(\lambda y.yx)(\lambda z.f(fz)) \rightarrow \lambda f.(\lambda z.f(fz))x \rightarrow \lambda f.f(fx)$$

It will be clear from the rest of the section that we could have made the other choice, i.e. express N' in a pure categorical notation.

Now we turn back to the initial goal: the formal description of the meaning of say $\lambda y.yx$ which we shall give in terms of the meanings of the sub-expressions x, y, yx. Those expressions clearly depend on the value of at least x. The values of the variables are kept in an environment (a pile if one thinks of an implementation). We represent the environment as follows: one considers one more variable z, D_x, D_y, D_z are the sets of possible values of x,y and z, and ".." is the rest of the environment which needs not be detailed for the present description, and × denotes the usual cartesian product of set theory.

$$Env = ((..\times D_z)\times D_x)\times D_y$$

(Represented as a tree, *Env* looks like a comb.)

The meaning of an expression depends on *Env*. For instance the meaning of x (y) is obtained by having access to the D_x (D_y) part of the environment, which may be done through the first and second projections, denoted by *Fst* and *Snd*. Whence the meanings of x and y, denoted by $[\![x]\!]$ and $[\![y]\!]$ (more generally $[\![M]\!]$ denotes the semantics of M):

$$[\![x]\!] = Snd \circ Fst$$
$$[\![y]\!] = Snd$$

The behaviour of $\circ$,*Fst* and *Snd* may be described by equations:

$$\begin{cases} (ass)\ (x \circ y)z = x(yz) \\ (fst)\ Fst(x,y) = x \\ (snd)\ Snd(x,y) = y \end{cases}$$

(applying a function to its argument is denoted by simple juxtaposition; of course x, y, z are fresh variables and have nothing to do with those in $\lambda y.yx$)

Now we define $[\![yx]\!]$ from $[\![y]\!]$ and $[\![x]\!]$, which are functions from *Env* to D_y, from *Env* to D_x respectively. First we may form the pair $<[\![y]\!],[\![x]\!]>$, which is a function from *Env* to $D_y \times D_x$. To fix ideas we set $D_x = \mathbb{N}$ and $D_y = \mathbb{N} \Rightarrow \mathbb{N}$, the set of functions from $\mathbb{N}$ to $\mathbb{N}$ (which is coherent with the above value of x). Remarking that the semantics of yx has to be a function from *Env* to $\mathbb{N}$ we obtain

$$[\![yx]\!] = App \circ <[\![y]\!],[\![x]\!]>$$

where *App* is the application function from $(\mathbb{N} \Rightarrow \mathbb{N}) \times \mathbb{N}$ to $\mathbb{N}$. The following equations describe the behaviour of *App*, $< >$:

$$\begin{cases} (app)\ App(x,y) = xy \\ (dpair)\ <x,y>z = (xz, yz) \end{cases}$$

The meaning of $\lambda y.yx$ depends on *Env* for x, but not for y which is bound (see below). It is a function from *Env* to $D_y \Rightarrow \mathbb{N}$ whereas $[\![yx]\!]$ is a function from $Env' \times D_y$ to $\mathbb{N}$, where $Env' = (..\times D_z) \times D_x$. We are tempted to use the currying which transforms a function f with two arguments a and b into a function $\Lambda(f)$ of a having as result a function of b such that

$$(\Lambda(f)(a))(b) = f(a,b)$$

or in equational form

$$\begin{cases} (d\Lambda)\ (\Lambda(x)y)z = x(y,z) \end{cases}$$

So we would like to write

$$[\![\lambda y.yx]\!] = \Lambda([\![yx]\!])$$

relying on the intuition that $[\![yx]\!]$ is the function associating yx with x and y, and that $[\![\lambda y.yx]\!]$ is the function associating with x the function associating yx with y. But then we loose symmetry since $\Lambda([\![yx]\!])$ is not a function from *Env* to $\mathbb{N}$, but from *Env'* to $\mathbb{N}$. So we have to take some care to ensure that the semantics takes always its argument in *Env*. We define $[\![\lambda y.yx]\!]$ as the currying of a function in $Env \times D_y \Rightarrow \mathbb{N}$ which itself is the composition of $[\![yx]\!]$ and a function $Subst_y$ from $Env \times D_y$ to *Env* which associates with a couple (ρ, a) a modified environment $\rho[y \leftarrow a]$, where only the component y has been changed (to a). We leave the reader check that in the present case

$$Subst_y = <Fst \circ Fst, Snd>$$

yielding

$$\llbracket \lambda y.yx \rrbracket = \Lambda(\llbracket yx \rrbracket \circ Subst_y) = \Lambda((App \circ \langle Snd, Snd \circ Fst\rangle) \circ \langle Fst \circ Fst, Snd\rangle)$$

Obviously the last expression may be simplified using the following equations (which the reader may "check" by applying both members to a same formal argument):

$$(Ass)\quad (x \circ y) \circ z = x \circ (y \circ z)$$

$$\begin{cases} (DPair)\ \langle x,y\rangle \circ z = \langle x \circ z, y \circ z\rangle \\ (Snd)\ Snd \circ \langle x,y\rangle = y \\ (Fst)\ Fst \circ \langle x,y\rangle = x \end{cases}$$

Using these rules we obtain

$$\llbracket \lambda y.yx \rrbracket = \Lambda(App \circ \langle Snd, Snd \circ (Fst \circ Fst)\rangle$$

We have introduced all the categorical combinators but the identity constant, which will arise below.

Now we present another way of associating a categorical term, i.e. of the kind built above, with any λ-expression, i.e. a term built from variables by application (MN) and abstraction $(\lambda x.M)$. We shall use the following notation.

$Snd \circ Fst^n = n!$

(By *Ass* this is unambiguous.)

$App \circ \langle A,B\rangle = S(A,B)$

So we have

(1) $\llbracket \lambda y.yx \rrbracket = \Lambda(S(0!,2!))$

Now we manipulate a more involved term:

$$Q = (\lambda x.(\lambda z.zx)y)((\lambda t.t)z)$$

Q has a disguised form of M as a subterm, namely $\lambda z.zx$ (exercise: define $\llbracket \lambda z.zx \rrbracket$ as above and check $\llbracket \lambda z.zx \rrbracket = \llbracket \lambda y.yx \rrbracket$ using the above equations).

This observation that the name of bound variables is indifferent is the basis of a variable name free notation due do N. De Bruijn, which we describe now.

N. De Bruijn's idea is to replace bound variable names by a number recording where they are bound in an expression, which is the only important information about them. Free variables are included in this treatment by considering in our example

$R = \lambda zxy.Q$ ($\lambda zxy.Q$ is an abbreviation for $\lambda z.(\lambda x.(\lambda y.Q)))$)

where the order z,x,y is consistent with the discussion above. The number associated with any occurrence u of a variable, a leaf in the tree representation of R, is the number of nodes labelled λv, with $v \neq u$, which are met in the path from that leaf to the root until a node λu is encountered. The result of that transformation is

$R' = (\lambda.(\lambda.01)1)((\lambda.0)2)$

Now we make only a textual tranformation and replace λ by Λ, "." by S, n by n! and obtain what we shall call the De Bruijn translation of Q and denote by $Q_{DB(z,x,y)}$:

$Q_{DB(z,x,y)} = S(\Lambda(S(\Lambda(S(0!,1!)),1!)),S(\Lambda(0!),2!))$

The reader may check that (1) above indeed defines $(\lambda y.yx)_{DB(x,y)} = (\lambda y.yx)_{DB(z,x,y)}$.

We end this introduction by suggesting that one may compute with the categorical expressions (that they may be called so will result from the precise connection with cartesian closed categories established in the next section).

Q reduces by β-reduction to yz, which is 5 if y is *succ* (the successor function) and z is 4. But first the outermost β-reduction yields

$Q' = (\lambda u.u((\lambda t.t)z))y$

(The bound variable z was renamed to avoid the free occurrence of z becoming bound after substitution.)

We show that $Q_{DB(z,x,y)}$ reduces to $Q'_{DB(z,x,y)}$; we shall need some more equations.

We decompose $Q_{DB(z,x,y)}$ as follows

$Q_{DB(z,x,y)} = App \circ <\Lambda(A),B>$ where

$B = App \circ <\Lambda(Snd),Snd \circ Fst \circ Fst>$

$A = App \circ <\Lambda(C),Snd \circ Fst>$ where

$C = App \circ <Snd,Snd \circ Fst>$

First we use the following rule, which may be "checked" as above

(*Beta*) $App \circ <\Lambda(x),y> = x \circ <Id,y>$

We get

$Q_{DB(z,x,y)} = A \circ E$ where

$E = <Id,B>$

Now we lift E down to the leaves of the tree representation of A. Combining *Ass* and *DPair* allows to distribute E along an S node:

$A \circ E = App \circ <\Lambda(C) \circ E,(Snd \circ Fst) \circ E>$

The leaf corresponding to the free occurrence of y in Q has already been reached; we may use *Ass*, *DPair*, *Fst* and the right identity equation

(*IdR*) $x \circ Id = x$

We obtain

$App \circ <\Lambda(C) \circ E,(Snd \circ Fst) \circ E> = App \circ <\Lambda(C) \circ E,Snd>$

Now we need an equation allowing to distribute E inside $\Lambda(C)$:

$$\{(D\Lambda)\ \ \Lambda(x)\circ y = \Lambda(x\circ <y\circ Fst, Snd>)$$

We get

$App\circ <\Lambda(C)\circ E, Snd>$

$= App\circ <\Lambda(App\circ <Snd\circ <E\circ Fst, Snd>, (Snd\circ Fst)\circ <E\circ Fst, Snd>>), Snd>$

After some dressing

$$App\circ <\Lambda(C)\circ E, Snd> = App\circ <\Lambda(App\circ <Snd, Snd\circ (E\circ Fst)>), Snd>$$
$$= App\circ <\Lambda(App\circ <Snd, B\circ Fst>), Snd>$$

remembering $E = <Id, B>$. Now we compute $B\circ Fst$ by distributing Fst in the same way:

$$B\circ Fst = App\circ <\Lambda(Snd\circ <Fst\circ Fst, Snd>), Snd\circ Fst\circ Fst\circ Fst>$$
$$= App\circ \Lambda(Snd, Snd\circ Fst^3) = S(\Lambda(0!), 3!)$$

Finally

$Q_{DB(z,x,y)} = S(\Lambda(S(0!, S(\Lambda(0!), 3!))), 0!) = Q'_{DB(z,x,y)}$

We have simulated a β-reduction by categorical rewritings. These rewritings have been able to recompute the number associated with the free occurrence of y in Q which is 1 in $Q_{DB(z,x,y)}$ and 0 in $P_{DB(z,x,y)}$ because the node λx has disappeared; they also recompose the fact that the free occurrence of z becomes 2 in $Q_{DB(z,x,y)}$ and 3 in $Q'_{DB(z,x,y)}$ because a node λz is inserted in the sequence of nodes λv up to the root.

Now we compute Q completely in the environment suggested above, using the rules with only lower case letters (*ass* rather than *Ass*, etc..). This looks very much like usual implementations of applicative languages. We start from

$S(\Lambda(A), B)\rho$ where

$\rho = (((\rho', \underline{4}), x), succ)$

We get by *ass*, *dpair* and *app*

$S(\Lambda(A), B)\rho = (\Lambda(A)\rho)(B\rho)$

We use $d\Lambda$ and set

$\rho' = (\rho, B\rho)$

We get

$(\Lambda(A)\rho)(B\rho) = A\rho'$

We manipulate A similarly and get

$A\rho' = C\rho''$ where

$\rho'' = (\rho', 1!\rho')$

Then

$C\rho'' = (0!\rho'')(1!\rho'')$

Making the leftmost reductions, and remembering the definitions of of ρ'', ρ', we get

$(0!\rho'')(1!\rho'') = (1!\rho')(1!\rho'') = (0!\rho)(1!\rho'') = succ\,(1!\rho'')$

Now we reduce the argument of *succ*.

$succ\,(1!\rho'') = succ\,(0!\rho') = succ\,(B\rho) = succ\,(0!(\rho,2!\rho)) = succ\,(2!\rho) = succ\,(\underline{4}) = \underline{5}$

Summarizing, we have introduced categorical combinators and we have suggested that their world was full of computations corresponding to those known in the λ-calculus world (β-reduction, abstract interpretation machines based on environment manipulations). Moreover all these computations are described in the unified framework of a first order rewriting system, whereas the formalisms of β-conversion and P. Landin's SECD machine [Lan] are quite different. The rest of the paper describes the typed categorical combinators formally.

2. Typed categorical combinators

First we define λc_K and CCL_K formally.

2.1. Definition

The **K-typed λc-calculus λc_K** and the **K-typed categorical combinatory logic** CCL_K are defined as follows:

K is a set of **basic types**; each term has a type, which is a term of $T_{\times,\Rightarrow}(K)$, and if M has the type σ, we write

M^σ or $M: \sigma$.

We agree that × has precedence over ⇒, and we write

$\sigma_1\times\sigma_2..\times\sigma_n = (..(\sigma_1\times\sigma_2)..\times\sigma_n)$

The structure of terms is as follows:

For λc_K:

- If x is a variable and σ is a type, then $x: \sigma$ is a term
- if $M: \sigma\Rightarrow\tau$ and $N: \sigma$, then $MN: \tau$
- if $x: \sigma$ and $M: \tau$, then $\lambda x.M: \sigma\Rightarrow\tau$
- if $M: \sigma$ and $N: \tau$, then $(M,N): \sigma\times\tau$
- if $M: \sigma\times\tau$, then $fst(M): \sigma$
- if $M: \sigma\times\tau$, then $snd(M): \tau$

For CCL_K

-	If x is a variable and σ is a type, then $x: \sigma$ is a term
-	if $A: \sigma_2 \Rightarrow \sigma_3$ and $B: \sigma_1 \Rightarrow \sigma_2$, then $A \circ B: \sigma_1 \Rightarrow \sigma_3$
-	$Id: \sigma \Rightarrow \sigma$
-	if $A: \sigma \Rightarrow \tau_1$ and $B: \sigma \Rightarrow \tau_2$, then $<A,B>: \sigma \Rightarrow \tau_1 \times \tau_2$
-	$Fst: \sigma \times \tau \Rightarrow \sigma$ (we shall often write $Fst^{\sigma,\tau}$)
-	$Snd: \sigma \times \tau \Rightarrow \tau$ (we shall often write $Snd^{\sigma,\tau}$)
-	if $A: \sigma_1 \times \sigma_2 \Rightarrow \sigma_3$, then $\Lambda(A): \sigma_1 \Rightarrow (\sigma_2 \Rightarrow \sigma_3)$
-	$App: (\sigma \Rightarrow \tau) \times \sigma \Rightarrow \tau$ (we shall often write $App^{\sigma,\tau}$)
-	if $A: \sigma \Rightarrow \tau$ and $B: \sigma$, then $AB: \tau$
-	if $A: \sigma$ and $B: \tau$, then $(A,B): \sigma \times \tau$

Hence CCL_K is an algebra of first order terms. The theories are:

$\beta\eta SP_K$

(*beta*) $(\lambda x^{\sigma}.M^{\tau})N^{\sigma} = M[x \leftarrow N]$
(η) $\lambda x^{\sigma}.M^{\sigma \Rightarrow \tau}x = M$ if $x \notin FV(M)$
(*fst*) $fst(M^{\sigma}, N^{\tau}) = M$
(*snd*) $snd(M^{\sigma}, N^{\tau}) = N$
(*SP*) $(fst(M^{\sigma \times \tau}), snd(M)) = M$

AA_K:

$$(Ass)\ (x^{\sigma_3\Rightarrow\sigma_4}\circ y^{\sigma_2\Rightarrow\sigma_3})\circ z^{\sigma_1\Rightarrow\sigma_2} = x\circ(y\circ z)$$
$$(IdL)\ Id^{\tau\Rightarrow\tau}\circ x^{\sigma\Rightarrow\tau} = x^{\sigma\Rightarrow\tau}$$
$$(IdR)\ x^{\sigma\Rightarrow\tau}\circ Id^{\sigma\Rightarrow\sigma} = x$$
$$(Fst)\ Fst^{\tau_1,\tau_2}\circ\langle x^{\sigma\Rightarrow\tau_1}, y^{\sigma\Rightarrow\tau_2}\rangle = x$$
$$(Snd)\ Snd^{\tau_1,\tau_2}\circ\langle x^{\sigma\Rightarrow\tau_1}, y^{\sigma\Rightarrow\tau_2}\rangle = y$$
$$(DPair)\ \langle x^{\sigma_1\Rightarrow\tau_1}, y^{\sigma_1\Rightarrow\tau_2}\rangle\circ z^{\sigma\Rightarrow\sigma_1} = \langle x\circ y, x\circ z\rangle$$
$$(Beta)\ App^{\sigma_2,\sigma_3}\circ\langle\Lambda(x^{\sigma_1\times\sigma_2\Rightarrow\sigma_3}), y^{\sigma_1\Rightarrow\sigma_2}\rangle = x\circ\langle Id^{\sigma_1\Rightarrow\sigma_1}, y\rangle$$
$$(D\Lambda)\ \Lambda(x^{\sigma_1\times\sigma_2\Rightarrow\sigma_3})\circ y^{\sigma\Rightarrow\sigma_1} = \Lambda(x\circ\langle y\circ Fst^{\sigma,\sigma_2}, Snd^{\sigma,\sigma_2}\rangle)$$
$$(AI)\ \Lambda(App^{\sigma,\tau}) = Id^{(\sigma\Rightarrow\tau)\Rightarrow(\sigma\Rightarrow\tau)}$$
$$(FSI)\ \langle Fst^{\sigma,\tau}, Snd^{\sigma,\tau}\rangle = Id^{\sigma\times\tau\Rightarrow\sigma\times\tau}$$
$$(ass)\ (x^{\sigma_1\Rightarrow\sigma_2}\circ y^{\sigma\Rightarrow\sigma_1})z^{\sigma} = x(yz)$$
$$(fst)\ Fst^{\sigma_1,\sigma_2}(x^{\sigma_1}, y^{\sigma_2}) = x$$
$$(snd)\ Snd^{\sigma_1,\sigma_2}(x^{\sigma_1}, y^{\sigma_2}) = y$$
$$(dpair)\ \langle x^{\sigma\Rightarrow\tau_1}, y^{\sigma\Rightarrow\tau_2}\rangle z^{\sigma} = (xz, yz)$$
$$(app)\ App^{\sigma,\tau}(x^{\sigma\Rightarrow\tau}, y^{\sigma}) = xy$$
$$(Quote\,1)\ \Lambda(Fst^{\sigma,\sigma_2})x^{\sigma}\circ y^{\sigma_1\Rightarrow\sigma_2} = \Lambda(Fst^{\sigma,\sigma_1})x$$
$$(Quote\,2)\ App^{\sigma_2,\sigma_3}\circ\langle x^{\sigma\Rightarrow(\sigma_2\Rightarrow\sigma_3)}\circ\Lambda(Fst^{\sigma,\sigma_1})y^{\sigma}, z^{\sigma_1\Rightarrow\sigma_2}\rangle = xy\circ z$$

Some of these equations must be applied with caution. For instance we only can replace Id by $\Lambda(App)$, $\langle Fst, Snd\rangle$ if Id is of type $(\sigma\Rightarrow\tau)\Rightarrow(\sigma\Rightarrow\tau)$, $\sigma\times\tau\Rightarrow\sigma\times\tau$ respectively.

The following lemma states some equational consequences of AA_K.

2.2. Lemma

The following equations are consequences of AA_K:

$$(Quote\,3)\ \Lambda(x^{\sigma\times\sigma_1\Rightarrow\sigma_2})y^{\sigma} = x\circ\langle\Lambda(Fst^{\sigma,\sigma_1})y, Id^{\sigma_1}\rangle$$
$$(id)\ Id^{\sigma\Rightarrow\sigma}x^{\sigma} = x$$
$$(d\Lambda)\ \Lambda(x^{\sigma_1\times\sigma_2\Rightarrow\sigma_3})y^{\sigma_1}z^{\sigma_2} = x(y,z)$$

The system AA_K is equivalent to the system obtained by replacing *Quote* 2 by the two following equations:

$$(Quote\,2a)\ \Lambda(Fst^{\sigma_1\Rightarrow\sigma_2,\sigma})x^{\sigma_1\Rightarrow\sigma_2} = \Lambda(x\circ Snd^{\sigma,\sigma_1})$$
$$(Quote\,2b)\ \Lambda(Fst^{\sigma_2,\sigma})(x^{\sigma_1\Rightarrow\sigma_2}y^{\sigma_1}) = x\circ\Lambda(Fst^{\sigma_1,\sigma})y$$

Proof: We only prove *Quote* 2b from *Quote* 2.

$$\Lambda(Fst)(xy) =_{ass} (\Lambda(Fst)\circ x)y =_{D\Lambda,Fst} \Lambda(x\circ Fst)y$$
$$=_{Quote\,3} x\circ Fst\circ\langle\Lambda(Fst)y, Id\rangle =_{Fst} x\circ\Lambda(Fst)y \quad ■$$

Now we define formally last section's De Bruijn's translation as well as the translations between λc_K and CCL_K.

2.3. Definition

Let $M:\sigma \in \lambda c_K$, and $x_0:\sigma_0,\ldots, x_n:\sigma_n$ be s.t. $FV(M) \subseteq \{x_0,\ldots,x_n\}$. We define $M_{DB_K(x_0,\ldots,x_n)}$ as follows:

$$x^{\sigma}_{DB_K(x_0^{\sigma_0},\ldots,x_n^{\sigma_n})} = Snd^{\sigma\times\sigma_n \cdots \times\sigma_{i+1},\sigma_i} \circ Fst^{\sigma\times\sigma_n \cdots \times\sigma_i,\sigma_{i-1}} \circ \cdots \circ Fst^{\sigma\times\sigma_n \cdots \times\sigma_1,\sigma_0}$$

where i is minimum s.t. $x=x_i$

$$(\lambda x.M)^{\sigma\Rightarrow\tau}_{DB_K(x_0,\ldots,x_n)} = \Lambda(M_{DB_K(x,x_0,\ldots,x_n)})$$

$$(M^{\sigma\Rightarrow\tau}N^{\sigma})_{DB_K} = App^{\sigma,\tau} \circ <M_{DB_K}, N_{DB_K}>$$

$$(M,N)^{\sigma\times\tau}_{DB_K} = <M_{DB_K}, N_{DB_K}>$$

$$fst\,(M^{\sigma\times\tau})_{DB_K} = Fst^{\sigma,\tau} \circ M_{DB_K}$$

$$snd\,(M^{\sigma\times\tau})_{DB_K} = Snd^{\sigma,\tau} \circ M_{DB_K}$$

One has

$$M^{\tau}_{DB_K(x_0^{\sigma_0},\ldots,x_n^{\sigma_n})} : \sigma\times\sigma_n \cdots \times\sigma_0 \Rightarrow \tau$$

($\sigma_0,\ldots,\sigma_n,\tau$ are determined by $M, x_0,\ldots,x_n$ while σ is any type).

We define

$$M_{CCL_K} = M^{\tau}_{DB_K(x_0^{\sigma_0},\ldots,x_n^{\sigma_n})}(y^{\sigma}, x_n^{\sigma_n},\ldots,x_0^{\sigma_0})$$

where y is different from all x_i and has the type σ in M_{DB_K} (we apply the De Bruijn's translation to the environment formally).

We define the translation in the reverse direction by

$$x^{\sigma}_{\lambda c_K} = x^{\sigma}$$

$$Id^{\sigma\Rightarrow\sigma}_{\lambda c_K} = \lambda x^{\sigma}.x$$

$$Fst^{\sigma,\tau}_{\lambda c_K} = \lambda x^{\sigma\times\tau}.fst(x)$$

$$Snd^{\sigma,\tau}_{\lambda c_K} = \lambda x^{\sigma\times\tau}.snd(x)$$

$$App^{\sigma,\tau}_{\lambda c_K} = \lambda x^{(\sigma\Rightarrow\tau)\times\sigma}.fst(x)snd(x)$$

$$(A^{\sigma_2\Rightarrow\sigma_3} \circ B^{\sigma_1\Rightarrow\sigma_2})_{\lambda c_K} = \lambda x^{\sigma_1}.A_{\lambda c_K}(B_{\lambda c_K}x)$$

$$(A^{\sigma\Rightarrow\tau}B^{\sigma})_{\lambda c_K} = A_{\lambda c_K}B_{\lambda c_K}$$

$$<A^{\sigma\Rightarrow\tau_1}, B^{\sigma\Rightarrow\tau_2}>_{\lambda c_K} = \lambda x^{\sigma}.(A_{\lambda c_K}x, B_{\lambda c_K}x)$$

$$(A^{\sigma}, B^{\tau})_{\lambda c_K} = (A_{\lambda c_K}, B_{\lambda c_K})$$

$$\Lambda(A^{\sigma_1\times\sigma_2\Rightarrow\sigma_3})_{\lambda c_K} = \lambda x^{\sigma_1}y^{\sigma_2}.A_{\lambda c_K}(x,y)$$

Clearly

$M^{\tau}_{CCL_K}:\tau$ and $A^{\tau}_{\lambda c_K}:\tau$

We suppose that x,y do not appear in A,B .

In [CuCCL,CuTh] the untyped version of these calculi and translations is defined. The main difference is that in the untyped case the application and couple operators are defined and not primitive. There we proved a First equivalence theorem of which the Second theorem below is just a typed copy (we refer to [CuCCL] for a sketchy proof and to [CuTh] for a full proof).

2.4. Second equivalence theorem

For any terms $M,N \in \lambda c_K$, $A,B \in CCL_K$, the following holds:

(1) $M_{CCL_K \lambda c_K} =_{\beta\eta SP_K} M$

(2) $A_{\lambda c_K CCL_K} =_{AA_K} A$

(3) $A =_{AA_K} B \Rightarrow A_{\lambda c_K} =_{\beta\eta SP_K} B_{\lambda c_K}$

(4) $M =_{\beta\eta SP_K} N \Rightarrow M_{CCL_K} =_{AA_K} N_{CCL_K}$.

Actually neither η nor SP are needed in (1) (see [CuTh]).

Now we introduce CCC_K.

2.5. Definition

Let K be a set of **basic objects.** The types are now couples written $\sigma \to \tau$ of terms σ,τ of $T_{\times,\Rightarrow}(K \cup \{\varepsilon\})$ where ε, called terminal object, is different from all the elements of K. The elements of $T_{\times,\Rightarrow}(K \cup \{\varepsilon\})$ are the **objects.**

The **free cartesian closed category** CCC_K is defined as follows:

- if x is a variable and σ,τ are objects, then $x: \sigma \to \tau$ is a term
- if $f: \sigma_2 \to \sigma_3$ and $g: \sigma_1 \to \sigma_2$ are terms, then $f \circ g: \sigma_1 \to \sigma_3$ is a term
- $Id: \sigma \to \sigma$ is a term
- if $f: \sigma \to \tau_1$ and $g: \sigma \to \tau_2$ are terms, then $<f,g>: \sigma \to \tau_1 \times \tau_2$ is a term
- $Fst: \sigma \times \tau \to \sigma$ is a term
- $Snd: \sigma \times \tau \to \tau$ is a term
- $1: \sigma \to \varepsilon$ is a term
- if $f: \sigma_1 \times \sigma_2 \to \sigma_3$ is a term, then $\Lambda(f): \sigma_1 \to (\sigma_2 \Rightarrow \sigma_3)$ is a term
- $App: (\sigma \Rightarrow \tau) \times \sigma \to \tau$ is a term

We use as above the notation $Fst^{\sigma,\tau}$, $Snd^{\sigma,\tau}$ and $App^{\sigma,\tau}$, and we also write Id^{σ} for $Id: \sigma \to \sigma$ and 1^{σ} for $1: \sigma \to \varepsilon$.

CCC_K is the set of equations $CCL\beta\eta SP + Ter$ where

$(Ter)\ 1^{\sigma \to \varepsilon} = x^{\sigma \to \varepsilon}$.

$CCL\beta\eta SP$ consists of the equations from Ass until FSI included in AA_K above (in the types some $\Rightarrow$ have to be replaced by $\to$) (more on $CCL\beta\eta SP$ in [CuCCL]).

Here typing is critical since *Ter* without types would reduce to: "everything equals 0". The difference to the definition 2.1 is the absence of application and couple operators, and the presence of a family of constants 1, the unique arrows to the terminal object.

Now we establish the equivalence of CCL_K, AA_K and CCC_K, CCC_K. First we have to connect the types of both theories. We shall use the well-known isomorphism between $A \to B$ and $1 \to (A \Rightarrow B)$ in a cartesian closed category (A, B are any objects, 1 is the terminal object), which is as follows in our setting:

$$\left[\begin{array}{l}(x^{\sigma \to \tau})^{+} = \Lambda(x \circ Snd^{\varepsilon,\sigma}) \\ (x^{\varepsilon \to \sigma \Rightarrow \tau})^{-} = App^{\sigma,\tau} \circ \langle x \circ 1^{\sigma \to \varepsilon}, Id^{\sigma} \rangle\end{array}\right.$$

One proves easily the following equations:

$$\left[\,((x^{\sigma \to \tau})^{+})^{-} =_{CCC_K} x \text{ and } ((x^{\varepsilon \to \sigma \Rightarrow \tau})^{-})^{+} =_{CCC_K} x\right.$$

2.6. Definition

With every object σ we associate

$$\left\{\sigma^{*} \in T_{\times,\Rightarrow}(K) \cup \{\varepsilon\}\ ,\ \sigma^{-}: \sigma \to \sigma^{*} \in CCC_K\ ,\ \sigma^{+}: \sigma^{*} \to \sigma \in CCC_K\right.$$

defined as follows:

- $\sigma^{*} = \sigma$, $\sigma^{+} = \sigma^{-} = Id^{\sigma}$ if $\sigma \in K \cup \{\varepsilon\}$

For the product we proceed by cases:

- $\sigma_1^{*}, \sigma_2^{*} \neq \varepsilon$:

  $(\sigma_1 \times \sigma_2)^{*} = \sigma_1^{*} \times \sigma_2^{*}$, $(\sigma_1 \times \sigma_2)^{+} = \langle \sigma_1^{+} \circ Fst, \sigma_2^{+} \circ Snd \rangle$, $(\sigma_1 \times \sigma_2)^{-} = \langle \sigma_1^{-} \circ Fst, \sigma_2^{-} \circ Snd \rangle$
- $\sigma_1^{*} \neq \varepsilon$, $\sigma_2^{*} = \varepsilon$

  $(\sigma_1 \times \sigma_2)^{*} = \sigma_1^{*}$, $(\sigma_1 \times \sigma_2)^{+} = \langle Id, \sigma_2^{+} \circ 1^{\sigma_1} \rangle \circ \sigma_1^{+}$, $(\sigma_1 \times \sigma_2)^{-} = \sigma_1^{-} \circ Fst$
- $\sigma_1^{*} = \varepsilon$, $\sigma_2^{*} \neq \varepsilon$: symmetric
- $\sigma_1^{*}, \sigma_2^{*} = \varepsilon$

  $(\sigma_1 \times \sigma_2)^{*} = \varepsilon$, $(\sigma_1 \times \sigma_2)^{+} = \langle \sigma_1^{+}, \sigma_2^{+} \rangle$, $(\sigma_1 \times \sigma_2)^{-} = 1$

Now the exponentiation:

- $\sigma_1^{*}, \sigma_2^{*} \neq \varepsilon$

  $(\sigma_1 \Rightarrow \sigma_2)^{*} = \sigma_1^{*} \Rightarrow \sigma_2^{*}$

  $(\sigma_1 \Rightarrow \sigma_2)^{+} = \Lambda(\sigma_2^{+} \circ App \circ \langle Fst, \sigma_1^{-} \circ Snd \rangle)$, $(\sigma_1 \Rightarrow \sigma_2)^{-} = \Lambda(\sigma_2^{-} \circ App \circ \langle Fst, \sigma_1^{+} \circ Snd \rangle)$
- $\sigma_1^{*} = \varepsilon$, $\sigma_2^{*} \neq \varepsilon$

  $(\sigma_1 \Rightarrow \sigma_2)^{*} = \sigma_2^{*}$

$(\sigma_1\Rightarrow\sigma_2)^+ = \Lambda(Fst)\circ\sigma_2^+$, $(\sigma_1\Rightarrow\sigma_2)^- = \sigma_2^-\circ App\circ <Id,\sigma_1^+\circ 1^{\sigma_1\Rightarrow\sigma_2}>$

$\sigma_2^\bullet=\varepsilon$

$(\sigma_1\Rightarrow\sigma_2)^\bullet = \varepsilon$, $(\sigma_1\Rightarrow\sigma_2)^+ = \Lambda(\sigma_2^+\circ 1^{\varepsilon\times\sigma_1})$, $(\sigma_1\Rightarrow\sigma_2)^- = 1$ ■

We omitted many types, and shall do so in the sequel. $\sigma^\bullet$ can be viewed as a canonical representent for σ when identifying $\sigma\times\varepsilon$, $\varepsilon\times\sigma$, $\varepsilon\Rightarrow\sigma$ with σ, and $\sigma\Rightarrow\varepsilon$ with ε. This is justified by the following lemma:

2.7. Lemma

For any $\sigma\in T_{\times,\Rightarrow}(K\cup\{\varepsilon\})$ the following holds:

$$\sigma^+\circ\sigma^- =_{CCC_K} Id^\sigma \text{ and } \sigma^-\circ\sigma^+ =_{CCC_K} Id^{\sigma^\bullet}$$

Proof: We only check one case.

$$<Id,\sigma_2^+\circ 1^{\sigma_1}>\circ\sigma_1^+\circ\sigma_1^-\circ Fst^{\sigma_1\times\sigma_2} =_{rec,Ter} <Fst,\sigma_2^+\circ 1^{\sigma_1\times\sigma_2}>$$
$$=_{Ter} <Fst,\sigma_2^+\circ 1^{\sigma_2}\circ Snd^{\sigma_1\times\sigma_2}>$$
$$=_{rec} <Fst,Snd> = Id$$ ■

Now we define the translations between CCL_K and CCC_K

2.8. Definition

With any term A: σ of CCL_K we associate a term A_{CCC_K}: $\varepsilon\to\sigma$ of CCC_K defined as follows:

$$x^\sigma_{CCC_K} = x^{\varepsilon\to\sigma}$$
$$A_{CCC_K} = A^+, \text{ if } A = Id, Fst, Snd, App$$
$$(A\circ B)_{CCC_K} = (A^-_{CCC_K}\circ B^-_{CCC_K})^+$$
$$<A,B>_{CCC_K} = <A^-_{CCC_K}, B^-_{CCC_K}>^+$$
$$\Lambda(A)_{CCC_K} = \Lambda(A^-_{CCC_K})^+$$
$$(AB)_{CCC_K} = A^-_{CCC_K}\circ B_{CCC_K}$$
$$(A,B)_{CCC_K} = <A_{CCC_K}, B_{CCC_K}>$$

Conversely with any term $f:\sigma\to\tau$ of CCC_K s.t. $(\sigma\Rightarrow\tau)^\bullet\neq\varepsilon$ (i.e. $\tau^\bullet\neq\varepsilon$), we associate a term f_{CCL_K}: $(\sigma\Rightarrow\tau)^\bullet$ of CCL_K defined by

$$x^{\sigma\rightarrow\tau}_{CCL_K} = x^{(\sigma\Rightarrow\tau)^*}$$

$$Id^{\sigma}_{CCL_K} = Id^{\sigma^*\Rightarrow\sigma^*}$$

$$Fst^{\sigma,\tau}_{CCL_K} = Fst^{\sigma^*,\tau^*} \quad \text{if } \tau^*\neq\varepsilon \qquad = Id^{\sigma^*\Rightarrow\sigma^*} \quad \text{if } \tau^*=\varepsilon$$

Symmetrically for *Snd*

$$App^{\sigma,\tau}_{CCL_K} = App^{\sigma^*,\tau^*} \quad \text{if } \sigma^*\neq\varepsilon \qquad = Id^{\tau^*\Rightarrow\tau^*} \quad \text{if } \sigma^*=\varepsilon$$

$$(f^{\sigma_2\rightarrow\sigma_3} \circ g^{\sigma_1\rightarrow\sigma_2})_{CCL_K} = f_{CCL_K} \circ g_{CCL_K} \quad \text{if } \sigma_1^*,\sigma_2^*\neq\varepsilon$$
$$= f_{CCL_K} g_{CCL_K} \quad \text{if } \sigma_1^*=\varepsilon\ ,\ \sigma_2^*\neq\varepsilon$$
$$= \Lambda(Fst^{\sigma_3^*,\sigma_1^*}) f_{CCL_K} \quad \text{if } \sigma_1^*\neq\varepsilon\ ,\ \sigma_2^*=\varepsilon$$
$$= f_{CCL_K} \quad \text{if } \sigma_1^*,\sigma_2^*=\varepsilon$$

$$<f^{\sigma\rightarrow\tau_1}, g^{\sigma\rightarrow\tau_2}>_{CCL_K} = <f_{CCL_K}, g_{CCL_K}> \quad \text{if } \sigma^*,\tau_1^*,\tau_2^*\neq\varepsilon$$
$$= (f_{CCL_K}, g_{CCL_K}) \quad \text{if } \sigma^*=\varepsilon\ ,\ \tau_1^*,\tau_2^*\neq\varepsilon$$
$$= f_{CCL_K} \quad \text{if } \tau_1^*\neq\varepsilon\ ,\ \tau_2^*=\varepsilon$$
$$= g_{CCL_K} \quad \text{if } \tau_1^*=\varepsilon\ ,\ \tau_2^*\neq\varepsilon$$

$$\Lambda(f^{\sigma_1\times\sigma_2\rightarrow\sigma_3})_{CCL_K} = \Lambda(f_{CCL_K}) \quad \text{if } \sigma_1^*,\sigma_2^*\neq\varepsilon$$
$$= f_{CCL_K} \quad \text{if } \sigma_1^*=\varepsilon \text{ or } \sigma_2^*=\varepsilon \ .$$

Now we may state the Third equivalence theorem.

2.9. Third equivalence theorem

For all terms A,B of CCL_K and f,g of CCC_K of appropriate types, the following holds:

(1) $A =_{AA_K} B \;\Rightarrow\; A_{CCC_K} =_{CCC_K} B_{CCC_K}$

(2) $f^{\sigma\rightarrow\tau} =_{CCC_K} g^{\sigma\rightarrow\tau} \;\Rightarrow\; f_{CCL_K} =_{AA_K} g_{CCL_K}$ if $\tau^*\neq\varepsilon$

(3) $A_{CCC_K,CCL_K} =_{CCC_K} A$

(4) $f^{\sigma\rightarrow\tau}_{CCL_K,CCC_K} =_{AA_K} \overline{\sigma\Rightarrow\tau}(f\,[x_0\leftarrow\underline{\sigma\Rightarrow\tau}(x_0),\ldots,x_n\leftarrow\underline{\sigma\Rightarrow\tau}(x_n)])$

where $V(A) = \{x_0,\ldots,x_n\}$ and $\overline{\sigma\Rightarrow\tau}$, $\underline{\sigma\Rightarrow\tau}$ are defined by

$\overline{\sigma\Rightarrow\tau}(f^{\sigma\rightarrow\tau}) = (\tau^-\circ f\circ\sigma^+)^+$ if $\sigma^*\neq\varepsilon$ $\qquad = \tau^-\circ f\circ\sigma^+$ if $\sigma^*=\varepsilon$

$\underline{\sigma\Rightarrow\tau}\,(g^{\varepsilon\rightarrow(\sigma\Rightarrow\tau)^*}) = \tau^+\circ f^-\circ\sigma^-$ if $\sigma^*\neq\varepsilon$

$\qquad = \tau^+\circ f\circ\sigma^-$ if $\sigma^*=\varepsilon$

Proof: Tedious but easy. We only check (4) for *App*.

$App^{\sigma,\tau}_{CCL_K,CCC_K} = A$

$(\sigma^*\neq\varepsilon)$ $\quad (App^{\sigma^*,\tau^*})^+$

We have to check

$$\vdash \tau^{-} \circ App^{\sigma,\tau} \circ ((\sigma \Rightarrow \tau) \times \sigma)^{+} = App^{\sigma^*,\tau^*}$$

Let

$$(\sigma \Rightarrow \tau)^{+} = \Lambda(B)$$

$$\begin{aligned}\tau^{-} \circ App^{\sigma,\tau} \circ ((\sigma \Rightarrow \tau) \times \sigma)^{+} &= \tau^{-} \circ App \circ <\Lambda(B) \circ Fst, \sigma^{+} \circ Snd> \\ &= \tau^{-} \circ \tau^{+} \circ App \circ <Fst, \sigma^{-} \circ Snd> \circ <Fst, \sigma^{+} \circ Snd> \\ &= App \circ <Fst, \sigma^{-} \circ \sigma^{+} \circ Snd> = App\end{aligned}$$

$$(\sigma^* = \varepsilon) \quad A = (Id^{\tau^*})^{+}$$

We have to check

$$\vdash \tau^{-} \circ App^{\sigma,\tau} \circ ((\sigma \Rightarrow \tau) \times \sigma)^{+} = Id^{\tau^*}$$

$$\begin{aligned}\tau^{-} \circ App^{\sigma,\tau} \circ ((\sigma \Rightarrow \tau) \times \sigma)^{+} &= \tau^{-} \circ App \circ <Id, \sigma^{+} \circ 1> \circ \Lambda(Fst) \circ \tau^{+} \\ &= \tau^{-} \circ App \circ <\Lambda(Fst), ..> \circ \tau^{+} \\ &= \tau^{-} \circ Fst \circ <Id, ..> \circ \tau^{+} = Id^{\tau^*} \blacksquare\end{aligned}$$

We end the section by pointing out that the two equivalence theorems of the section may be used to decide the equational equality in CCC_K (and also in CCL_K). Indeed the rewriting system obtained by orienting the rules of $\beta\eta SP_K$ from left to right is confluent (cf. [Pot]) and noetherian. We refer to [LamSco] for a proof of that property, which was actually established by J. Lambek and P.J. Scott for the same purpose. For concluding on decidability, we just have to remark

$$\vdash f^{\sigma \to \tau} =_{CCC_K} g^{\sigma \to \tau} \text{ iff } f_{CCL_K} =_{AA_K} g_{CCL_K} \text{ iff } f_{CCL_K, \lambda c_K} =_{\beta\eta SP_K} g_{CCL_K, \lambda c_K}$$

using

$$\overline{\sigma \to \tau}(\underline{\sigma \to \tau}(x)) =_{CCC_K} x \quad \text{et} \quad \underline{\sigma \to \tau}(\overline{\sigma \to \tau}(x)) =_{CCC_K} x$$

3. Conclusion

We have exhibited the connection between λ-calculus and cartesian closed categories, which goes back to [Lam,Sco] and quite independently to [BeSy,CuTh3], in a very syntactical and computational fashion. We refer to [CuTh,CuEq] for the semantic equivalences induced by the theorems in this paper.

It is very tempting to implement evaluators of categorical combinators. A result in [CuTh,CuTh] states that the evaluator last informally described in section 1, working by leftmost-outermost reductions, is complete with respect to the models of the underlying theory (namely CCL_K enriched with arithmetic combinators). Moreover the author devised a categorical abstract machine transforming categorical combinators into actual machine instructions. This machine will be described in a forthcoming paper with G. Cousineau, who significantly improved the original proposal.

Related (and independent) work appears in [PaGho,Poi,Dyb,LamSco]. [Poi],[LamSco] explicitely state an equivalence in the kind of this paper between (quite) λc_K and CCC_K, in a syntactic, a more semantic setting respectively. The differences of the present paper to these references are mainly the introduction of CCL_K, and the connection with De Bruijn's ideas, both contributing to an operational setting.

4. References

[BeSy] G. Berry, Some Syntactic and Categorical Constructions of Lambda-calculus models, Rapport INRIA 80 (1981).

[Bru] N.G. De Bruijn, Lambda-calculus Notation without Nameless Dummies, a Tool for Automatic Formula Manipulation, Indag Math. 34, 381-392 (1972).

[CuTh3] P-L. Curien, Algorithmes Séquentiels sur Structures de Données Concrètes, Thèse de Troisième Cycle, Université Paris VII (Mars 1979).

[CuTh] P-L. Curien, Combinateurs Catégoriques, Algorithmes Séquentiels et Programmation Applicative, Thèse d'Etat, Université Paris VII (Décembre 83), to be published in english as a monograph.

[CuCCL] P-L. Curien, Categorical Combinatory Logic, submitted to ICALP 85.

[CuEq] Syntactic Equivalences Inducing Semantic Equivalences, submitted to EUROCAL 1985.

[Dyb] P. Dybjer, Category-Theoretic Logics and Algebras of Programs, PhD Thesis, Chalmers University of Technology, Goteborg (1983).

[Lam] J. Lambek, From Lambda-calculus to Cartesian Closed Categories, in To H.B. Curry: Essays on Combinatory Logic, Lambda-calculus and Formalism, ed. J.P. Seldin and J.R. Hindley, Academic Press (1980).

[LamSco] J. Lambek and P.J. Scott, Introduction to Higher Order Categorical Logic, to be published by Cambridge University Press (1984).

[Lan] P.J. Landin, The Mechanical Evaluation of Expressions, Computer Journal 6, 308-320 (1964).

[PaGhoTh] K. Parsaye-Ghomi, Higher Order Abstract Algebras, PhD Thesis, UCLA (1981).

[Poi] A. Poigné, Higher Order Data Structures, Cartesian Closure Versus λ-calculus, STACS 84, Lect. Notes in Comput. Sci.

[Pot] G. Pottinger, The Church-Rosser Theorem for the Typed λ-calculus with Extensional Pairing, preprint, Carnegie-Mellon University, Pittsburgh (March 1979).

[Sco4] D. Scott, Relating Theories of the Lambda-calculus, cf. [Lam].

A PATH ORDERING FOR PROVING TERMINATION OF TERM REWRITING SYSTEMS

D. Kapur, P. Narendran
Computer Science Branch
Corporate Research and Development
General Electric Company
Schenectady, NY

G. Sivakumar†
Dept. of Computer Science
University of Illinois,
Urbana-Champaign, IL

ABSTRACT

A new partial ordering scheme for proving uniform termination of term rewriting systems is presented. The basic idea is that two terms are compared by comparing the paths through them. It is shown that the ordering is a well-founded simplification ordering and also a strict extension of the recursive path ordering scheme of Dershowitz. Terms can be compared under this path ordering in polynomial time.

1. INTRODUCTION

Term rewriting systems have been found to be widely applicable in many areas of computer science and mathematics, including word problems, unification problems, decision procedures for equational theories, theorem proving, program transformation and synthesis, polynomial simplification, analysis and design of specifications, proving properties by induction, etc. In most of these applications, the Knuth-Bendix completion procedure [11] and its extensions discussed in [1,8,9,13,15] play a crucial role. However, the successful use of the completion procedure crucially depends upon the ability to prove the termination of term rewriting systems that are generated during the course of the completion procedure to obtain a canonical set of rewrite rules. In this connection, many termination orderings have been proposed in the literature including the original Knuth-Bendix ordering based on weights [11], paths of subterm ordering [16], polynomial ordering [12], recursive path ordering (RPO) [2] and its extension based on lexicographic status (LRPO) [7], and recursive decomposition ordering (RDO) [6] and its extension based on lexicographic status (RDOS) [14].

In RRL, a rewrite rule laboratory under development at the Computer Science Branch at General Electric Corporate Research and Development Center, we have implemented RPO as a

† This work was done when Sivakumar was a graduate student in the Dept. of Mathematical Sciences, Rensselaer Polytechnic Institute, Troy, NY. Kapur and Sivakumar were partially supported for this research by the NSF grant MCS-8211621.

way to establish the termination of term rewriting systems [10]. By and large, our experience has been positive with LRPO in terms of its performance and its applicability to a wide range of term rewriting systems; however in some examples, we have found LRPO to be weak to handle terms which are intuitively simple to handle [10]. In this paper, we develop a new termination ordering based on paths in terms which is intuitively simple to understand and which is an extension of RPO.

In the next section, we study examples illustrating what is lacking in RPO. We analyze the definition of RPO pointing out weaknesses in different aspects of the definition of RPO. Section 3 introduces the ordering based on paths. In Section 4, it is shown that the path ordering is a simplification ordering and has the substitution property, which implies that it can be used for proving termination of term rewriting systems [2]. We also show that the new ordering is a strict extension of recursive path ordering. In Section 5, we prove that comparison of two terms under the path ordering can be done in polynomial time; we show an upper-bound of $O(|s|^5 * |t|^5)$, where $|s|$ and $|t|$ are the size of terms s and t being compared. Section 6 is a brief discussion of how the proposed ordering can be extended to incorporate lexicographic status of function symbols [7]. In Section 7, we outline an extension to the path ordering which further generalizes it. In Section 8, we discuss how the path ordering relates to RDO and RDOS.

2. A DISCUSSION OF RECURSIVE PATH ORDERING (RPO)

Consider the equation $a(b(x)) = c(d(x))$ with the precedence $a > d$ and $b > c$. Under RPO the terms $a(b(x))$ and $c(d(x))$ are incomparable, since $b(x) \not\gtrsim c(d(x))$ and $d(x) \not\gtrsim a(b(x))$. But, in our mind, it is clear to us that this equation should be oriented as $a(b(x)) \rightarrow c(d(x))$, since a 'takes care of' d and b 'takes care of' c. In the next section we make this notion of 'taking-care-of' more precise and thereby define the new ordering.

At this point it will be helpful to take a look at 'recursive path ordering' (RPO) and discuss its weaknesses.

2.1 DEFINITION OF RPO

Let $\gtrsim$ denote a quasi-ordering of a set E (i.e, a reflexive and transitive relation on E) and $\sim$ be the equivalence relation $\gtrsim \cap \lesssim$. Let $>$ denote the partial ordering $\gtrsim - \lesssim$; in other words, $a > b$ if and only if $a \gtrsim b$ and $\neg (b \gtrsim a)$. A partial ordering $>$ is *well-founded* if there exists no infinite descending sequence $e_1 > e_2 > \ldots$ of elements of E. A well-founded partial order $>$ on a set E can be extended to the set of *multisets* on E as follows:
$M_1 \gg M_2$ if and only if for each $x \in M_2 - M_1$, there is a $y \in M_1 - M_2$ such that $y > x$.
A finite multiset M is written by enumerating its elements and enclosing them within braces.

Let $\gtrsim$ be a quasi-ordering of a set of function symbols F. Then terms $s = f(s_1, \cdots, s_m)$ and $t = g(t_1, \cdots, t_n)$ are equivalent (denoted by $s \sim t$) if and only if $f \sim g$, $m = n$ and there is a permutation π of the set {1, ... ,n} such that $s_i \sim t_{\pi(i)}$ for all $1 \leq i \leq n$.

Given a quasi-ordering $\gtrsim$ of a set of function symbols F, the ***recursive path ordering*** ($\underset{rpo}{>}$) on terms generated by F can be defined as follows:

(a) $s = f(s_1, \cdots, s_m) \underset{rpo}{>} t = x$ if and only if $x \in Var(s)$, the set of variables in s.

(b) $s = f(s_1, \cdots, s_m) \underset{rpo}{>} t = g(t_1, \cdots, t_n)$ if and only if

(1) $f > g$ and $s \underset{rpo}{>} t_i$ for all i, $1 \leq i \leq n$, or,

(2) $f \sim g$ and $\{s_1, \ldots, s_m\} \underset{rpo}{\gg} \{t_1, \ldots, t_n\}$, or,

(3) $s_i \underset{rpo}{\gtrsim} t$ for some i, $1 \leq i \leq m$.

2.2 WHERE RPO DOES NOT WORK

We give below a few examples of terms which can definitely be ordered by the intuitive concept of paths but which RPO cannot order.

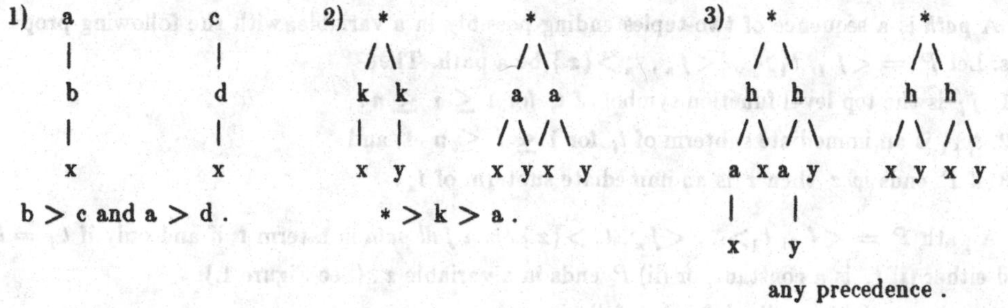

Let $t = f(t_1, \cdots, t_n)$, $s = g(s_1, \cdots, s_m)$, $M_1 = \{t_1, \cdots, t_n\}$ and $M_2 = \{s_1, \cdots, s_m\}$. Now when RPO tries to order these terms it first compares root-symbols. There are 3 cases:

1) $f > g$. Then RPO needs $t \underset{rpo}{>} s_i$. This is most powerful and needs no change.

2) $f \sim g$. Then RPO needs $M_1 \underset{rpo}{\gg} M_2$.

This is not really the best case for this demands that for every s_i in $M_2 - M_1$ there is a single term t_j in $M_1 - M_2$ such that $t_j \underset{rpo}{>} s_i$. This may not be suited when for each path in s_i there are paths in different t_j's which are "bigger". (See examples 2, 3). So we need to treat the terms collectively.

3) $f \not\gtrsim g$. Here again RPO is not ideal for it needs a single term t_j such that $t_j \underset{rpo}{\gtrsim} s$. We

also do not remember the root (topmost) operator f of t any more even though it may be needed to make a path "bigger".(See example 1).

3. PATH ORDERING

Let F be a finite set of function symbols of fixed arity and V be a denumerable set of variables. By $T(F,V)$ we denote the set of all possible terms that can be constructed using F and V. For a term t, $Var(t)$ denotes the set of all variables that occur in t. For example, $Var(f(x,y,g(y))) = \{x,y\}$. The *size* of a term s is the total number of function and variable symbol occurrences in s and is denoted by $|s|$. By $f(\ldots t \ldots)$ we denote a term that contains the term t as an immediate (top-level) subterm.

A *term rewriting system* P over a set of terms $T(F,V)$ is a finite set of rewrite rules of the form $l_i \to r_i$ where $l_i, r_i \in T(F,V)$. We write $t \Rightarrow t'$ to indicate that the term t' can be derived from the term t by a single application of a rule P to one of its subterms. P is said to be *uniformly terminating* if there exist no infinite sequences of terms $t_i \in T(F,V)$ such that $t_1 \Rightarrow t_2 \Rightarrow \ldots$.

3.1 PATHS

A *path* is a sequence of two-tuples ending possibly in a variable, with the following properties: Let $P = <f_1, t_1> \ldots <f_n, t_n>\{x\}$ be a path. Then

1. f_i is the top level function symbol of t_i for $1 \le i \le n$.
2. t_{i+1} is an immediate subterm of t_i for $1 \le i \le n-1$, and
3. if P ends in x then x is an immediate subterm of t_n.

A path $P = <f_1, t_1> \ldots <f_n, t_n>\{x\}$ is a *full path* in a term t if and only if $t_1 = t$ and either (i) t_n is a constant, or (ii) P ends in a variable x. (See Figure 1.)
Full paths can be formally defined as follows:

(a) If $t = x$, a variable, then x is the only full path in t.
(b) If $t = b$, a constant, then $<b, b>$ is the only full path in t.
(c) If $t = f(t_1, \cdots, t_m)$ then a full path in t is $<f, t> . p$, where p is a full path in some t_i.

Examples:
(1) $t = f(x,y)$. $<f, f(x,y)>x$ and $<f, f(x,y)>y$ are the full paths.
(2) $t = f(g(x),h(b,y))$. The full paths are

i. $<f, f(g(x),h(b,y))><g, g(x)>x$,

ii. $<f, f(g(x),h(b,y))><h, h(b,y)><b,b>$,

iii. $<f, f(g(x),h(b,y))><h, h(b,y)>y$.

3.2 PATH COMPARISON

i. Paths ending in different variables are incomparable.

ii. A variable is incomparable with any two-tuple. (This ensures that a path ending in a constant is never greater than a path ending in a variable.)

iii. To compare paths ending with the same variable we drop the variable from the end of both paths and compare the remaining sequences of two-tuples.

Let $P = \langle f_1, t_1\rangle\langle f_2, t_2\rangle \ldots \langle f_m, t_m\rangle\{x\}$ be a path. With every two-tuple $\langle f_i, t_i\rangle$ we can associate a *left-context* (LC) and a *right-context* (RC) defined as:

$$RC(\langle f_i, t_i\rangle, P) = \begin{cases} \langle f_{i+1}, t_{i+1}\rangle \ldots \langle f_m, t_m\rangle\{x\} & \text{where } i < m \text{, and} \\ \{x\} & \text{if } i = m. \end{cases}$$

$$LC(\langle f_i, t_i\rangle, P) = \begin{cases} \langle f_1, t_1\rangle \ldots \langle f_{i-1}, t_{i-1}\rangle, & \text{where } i > 1 \text{, and} \\ \lambda & \text{if } i = 1. \end{cases}$$

(See Figures 1 and 2. One can visualise the right-context of a tuple as *below* and the left-context as *above* the corresponding node in the tree representation of a term.)

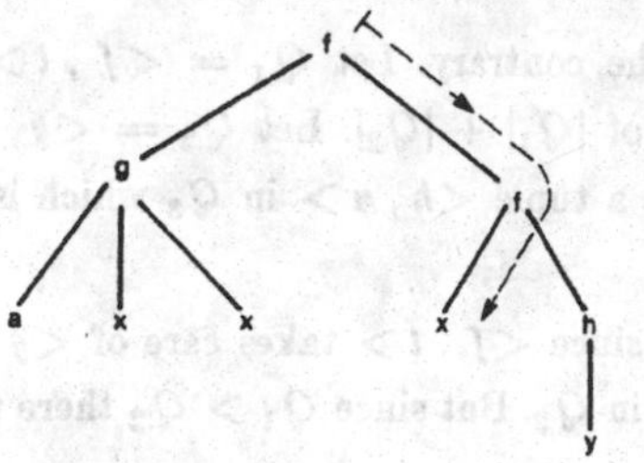

The path shown here is

⟨f, f(g(a,x,x), f(x,h(y)))⟩ ⟨f, f(x,h(y))⟩ x

Figure 1.

P = ________________ <f, t> ________________

<- - - LC - - ->   <- - - RC - - ->

Figure 2.

Let $P_1 = \langle f_1, t_1\rangle \ldots \langle f_m, t_m\rangle$ and $P_2 = \langle g_1, s_1\rangle \ldots \langle g_n, s_n\rangle$ be two sequences of two-tuples. $P_1 \underset{P}{>} P_2$ if and only if for all $\langle g_j, s_j\rangle$ in P_2 there exists $\langle f_i, t_i\rangle$ in P_1 such that

a. $f_i > g_j$ or

b. $f_i \sim g_j$ and

1. $RC(\langle f_i, t_i\rangle, P_1) \underset{P}{>} RC(\langle g_j, s_j\rangle, P_2)$ or

2. $RC(\langle f_i, t_i\rangle, P_1) \sim RC(\langle g_j, s_j\rangle, P_2)$ and $t_i \underset{T}{>} s_j$ or

3. $RC(\langle f_i, t_i\rangle, P_1) \sim RC(\langle g_j, s_j\rangle, P_2)$, $t_i \sim s_j$ and

$LC(\langle f_i, t_i\rangle, P_1) \underset{P}{>} LC(\langle g_j, s_j\rangle, P_2)$

where $P_1 \sim P_2$ (the paths are *equivalent*) if $m = n$ and $f_i \sim g_i$ and $t_i \sim s_i$ for all $1 \leq i \leq n$, and term comparison ($\underset{T}{>}$) is defined in the next section. (The subscript P in $\underset{P}{>}$ will be omitted whenever it is obvious from the context.)

We often express this as '$<f_i, t_i>$ *takes care of* $<g_j, s_j>$'. Clearly, if $<f_i, t_i>$ takes care of $<g_j, s_j>$, then $f_i \gtrsim g_j$. Checking whether a two-tuple in one path takes care of a two-tuple in another is done in the following sequence: First compare the two function symbols, then the right contexts, then the terms in the two-tuples and finally the left-contexts.

Lemma 1: Let P_1, P_2, P_3 be paths such that $P_3 = <f, t> P_1$ for some term t and $P_1 \gtrsim P_2$. Then $P_3 > P_2$.

Lemma 2: Let P_1, P_2, P_3 be paths such that $P_1 = <f, t> P_1'$, $P_3 = <g, s> P_2$, $P_1 > P_2$ and $f > g$. Then $P_1 > P_3$.

Sketch of the Proof: Assume the contrary. Let $Q_1 = <f, t> Q_1'$ and Q_2 constitute a shortest counterexample in terms of $|Q_1| + |Q_2|$. Let $Q_3 = <g, s> Q_2$ with $f > g$ such that $Q_1 \not> Q_3$. Then there exists a tuple $<h, u>$ in Q_3 which is not taken care of by any tuple in Q_1.

Clearly $<h, u> \neq <g, s>$ since $<f, t>$ takes care of $<g, s>$. Thus there exists j such that $<h, u> = <g_j, s_j>$ in Q_2. But since $Q_1 > Q_2$ there must exist a tuple $<f_i, t_i>$ in Q_1 such that either

(a) $f_i > g_j$, or
(b) $f_i \sim g_j$ and $\mathrm{RC}(<f_i, t_i>, Q_1) > \mathrm{RC}(<g_j, s_j>, Q_2)$, or
(c) $f_i \sim g_j$, $\mathrm{RC}(<f_i, t_i>, Q_1) \sim \mathrm{RC}(<g_j, s_j>, Q_2)$ and $t_i > s_j$, or
(d) $f_i \sim g_j$, $\mathrm{RC}(<f_i, t_i>, Q_1) \sim \mathrm{RC}(<g_j, s_j>, Q_2)$, $t_i \sim s_j$ and $\mathrm{LC}(<f_i, t_i>, Q_1) > \mathrm{LC}(<g_j, s_j>, Q_2)$.

It is easy to see that (d) is the only possibility. But then $\mathrm{LC}(<g_j, s_j>, Q_3) = <g, s> \mathrm{LC}(<g_j, s_j>, Q_2)$, and $|\mathrm{LC}(<f_i, t_i>, Q_1)| + |\mathrm{LC}(<g_j, s_j>, Q_2)| < |Q_1| + |Q_2|$ and the minimality of Q_1 and Q_2 leads us to a contradiction. ☐

Lemma 3: Let P_1, P_2, P_3, P_4 be paths such that $P_3 = <f, t> P_1$, $P_4 = <f, s> P_2$ for some s, t. Then $P_1 \gtrsim P_2$ and $t \underset{T}{>} s$ implies $P_3 > P_4$.

Lemma 4: Let P_1, P_2, Q_1 and Q_2 be paths such that $Q_1 = <f, t> P_1$, $Q_2 = <f', s> P_2$ and $f \sim f'$. Then $Q_1 > Q_2$ and $t \not\gtrsim s$ implies $P_1 > P_2$.

Lemma 5: Let P_1, P_2, P_3, P_4, P_5 be paths such that $P_4 = P_1 P_3$ and $P_5 = P_2 P_3$. Then $P_1 > P_2$ implies $P_4 > P_5$.
(Intuitively, attaching equivalent paths to the right does not alter the ordering relation.)

3.3 TERM COMPARISON

Let $M = \{ s_1, s_2, \ldots, s_m \}$ be a multiset of terms. By $MP(M)$ we denote the union of the multisets of all full paths in each s_i.

Example: Let $M = \{ f(a,b), f(a,b) \}$.

$$MP(M) = \{ <f, f(a,b)><a,a>, <f, f(a,b)><b,b>, <f, f(a,b)><a,a>, <f, f(a,b)><b,b> \}.$$

We deliberately abuse the notation when M is a singleton set and write $MP(t)$ instead of $MP(\{t\})$. Note that $MP(s) \sim MP(t)$ if and only if $s \sim t$.

Definition: (i) If s is a non-variable term and t is a variable then $s \underset{T}{>} t$ if and only if s contains t. (ii) $s = f(s_1, \ldots, s_m)$ and $t = g(t_1, \ldots, t_n)$. Let $M_1 = \{ s_1, \ldots, s_m \}$, $M_2 = \{ t_1, \ldots, t_n \}$. Then $s \underset{T}{>} t$ if and only if

a. $f > g$ and $s \underset{T}{>} t_i$ for all i, $1 \le i \le n$, or

b. $f \sim g$ and $MP(M_1) \underset{P}{\gg} MP(M_2)$, or

c. $f \not\gtrsim g$ and $MP(s) \underset{P}{\gg} MP(t)$.

Note that the path-comparisons that are done during term-comparison may necessitate further comparisons of terms. Let u_i and v_j be two terms that have to be compared while comparing s and t. It can be seen that either u_i must be a proper subterm of s or v_j must be a proper subterm of t. The only case where this may not be obvious is (ii)c, where $s = f(s_1, \ldots, s_m)$, $t = g(t_1, \ldots, t_n)$ and $f \not\gtrsim g$. But, even though every full path of s starts with the tuple $<f, s>$ and every full path of t starts with the tuple $<g, t>$, we have no occasion to compare s and t again, since $<f, s>$ cannot ever take care of $<g, t>$. Thus the algorithm for term-comparison terminates, or, in other words, the ordering scheme is well-defined.

Examples:

```
1) s = f            >        f = t
      /|\                   /|\
     / | \                 / | \
    /  |  \               x  g  c        with precedence b > c.
   a   g   b                /\  \
  /   /\    \              x  y  y
 x   x  y    y
```

Since the top-level function symbols are the same, we have to compare the multisets of all full paths in the immediate subterms, $MP(\{ a(x), g(x,y), b(y) \})$ and $MP(\{ x, g(x,y), c(y) \})$.

$MP(\{a(x), b(y)\}) = \{ <a, a(x)>x, <b, b(y)>y \}$. $MP(\{x, c(y)\}) = \{ x, <c, c(y)>y \}$.

Clearly $<a, a(x)>x \underset{P}{>} x$. $<b, b(y)>y \underset{P}{>} <c, c(y)>y$ since $<b, b(y)>$ takes care of $<c, c(y)>$.

(2)
```
s = f       >      g = t
   / \            / \
  /   \          /   \
 a     b        x     y        with a > g and b > g.
 |     |
 x     y
```

Since f and g are incomparable, we have to compare $MP(s)$ and $MP(t)$, where $MP(s) = \{<f,s><a,a(x)>x,\ <f,s><b,b(y)>y\}$ and $MP(t) = \{<g,t>x,\ <g,t>y\}$.

$<f,s><a,a(x)> \underset{P}{>} <g,t>$ since $<a,a(x)>$ takes care of $<g,t>$. Similarly

$<f, s><b, b(y)> \underset{P}{>} <g,t>$. (Note also that s and t are incomparable under RPO.)

(3)
```
s = a       >      p = t
    |              |
    b              a        with a > p and b ~ q.
    |              |
    c              q
    |              |
    x              x
```

Since $a > p$, we have to compare $a(b(c(x)))$ and $a(q(x))$ and then again since the top-level symbols are both a's, we end up comparing $<b,b(c(x))><c,c(x)>$ and $<q,q(x)>$. It can be seen that $<b,b(c(x))>$ takes care of $<q,q(x)>$ since the former has a bigger right context.

It should be noted here that in the case when all function symbols are monadic, we can treat paths merely as *strings* formed by the function symbols. Consider the two-tuples $<f,s>$ in path P and $<g,t>$ in path Q. Now it is not hard to observe that if $f \sim g$ and the right-contexts are equivalent, then $s \sim t$ also. Hence it is quite unnecessary to keep the terms around.

(4)
```
s = f       >      a = t
    |              |
    b              b
    |              |
    c              c        with f > a.
    |              |
    x              x
```

We compare $f\ b\ c$ and $a\ b\ c$. a is taken care of by f. The b in $a\ b\ c$ is taken care of by the b in $f\ b\ c$ since the latter has a bigger left-context. Similarly, $f\ b > a\ b$ and thus the c in $f\ b\ c$ takes care of the c in $a\ b\ c$.

Lemma 6: $s \underset{T}{>} t$ if and only if $MP(s) \underset{P}{\gg} MP(t)$.

$MP(s) \underset{P}{\gg} MP(t)$ implies that $MP(s) \cap MP(t) = \emptyset$.

Corollary 6.1: $s \underset{T}{>} t$ implies $f(\dots s \dots) \underset{T}{>} t$.

Theorem 1: (a) $\underset{T}{>}$ is a partial ordering on terms.

(b) $\underset{P}{>}$ is a partial ordering on paths.

4. PROPERTIES OF THE ORDERING

A partial ordering $>$ on terms is a *simplification ordering* if it satisfies the following properties (See [2]):

(1) $s > t$ implies $f(\dots s \dots) > f(\dots t \dots)$, (*replacement*)
(2) $f(\dots t \dots) > t$, (*subterm*)
for any terms $f(\dots s \dots)$ and $f(\dots t \dots)$.

Theorem 2: $\underset{T}{>}$ satisfies the subterm and replacement properties.

Proof: Follows from Lemma 6. □

Theorem 3: $\underset{T}{>}$ is closed under substitutions.

Basic Idea: A major share of the effort is in proving the following two propositions:

(1) For all substitutions σ, $s_1 \underset{T}{>} t_1$ implies $\sigma(s_1) \underset{T}{>} \sigma(t_1)$.

(2) Let $P_1 = \langle f_1, s_1 \rangle \dots \langle f_m, s_m \rangle$ and $P_2 = \langle g_1, t_1 \rangle \dots \langle g_n, t_n \rangle$ be two variable-free paths. Then for all substitutions σ,

$$\sigma(P_1) = \langle f_1, \sigma(s_1) \rangle \dots \langle f_m, \sigma(s_m) \rangle \underset{P}{>} \langle g_1, \sigma(t_1) \rangle \dots \langle g_m, \sigma(t_n) \rangle = \sigma(P_2).$$

We prove this by simultaneous induction on $|s_1| + |t_1|$. □

Theorem 4: $\underset{T}{>}$ is an extension of the recursive path ordering (RPO).

Sketch of the Proof: It can be easily observed that $s \underset{rpo}{\sim} t$ implies $s \underset{T}{\sim} t$. We prove that $s \underset{rpo}{>} t$ implies $s \underset{T}{>} t$ by induction on $|s| + |t|$.

The case when t is a variable and s is a term containing t is trivial.
Let $s = f(s_1, \dots, s_m)$, $t = g(t_1, \dots, t_n)$, $M_1 = \{ s_1, \dots, s_m \}$ and $M_2 = \{ t_1, \dots, t_n \}$. The following cases have to be considered:

1. $f > g$. Then, by the definition of RPO, $s \underset{rpo}{>} t_i$ for all i, $1 \le i \le n$. Since $|s| + |t_i| < |s| + |t|$ for every i, we get $s \underset{T}{>} t_i$ for $1 \le i \le n$. Thus $s \underset{T}{>} t$ by definition.

2. $f \sim g$. RPO requires that for every t_j in $M_2 - M_1$ there exist s_i in $M_1 - M_2$ such that $s_i \underset{rpo}{>} t_j$. Again, by induction hypothesis, $s_i \underset{T}{>} t_j$ implying $\mathrm{MP}(s_i) \underset{P}{\gg} \mathrm{MP}(t_j)$ by Lemma 6. Thus $\mathrm{MP}(M_1 - M_2) \underset{P}{\gg} \mathrm{MP}(M_2 - M_1)$ and the result follows.

3. $f \not\gtrsim g$. Therefore there exists i such that $s_i \underset{rpo}{\gtrsim} t$. If $s_i \underset{rpo}{\sim} t$ then $s_i \underset{T}{\sim} t$, $\mathrm{MP}(s_i) \underset{P}{\sim} MP(t)$ and by Lemma 1, $MP(s) \underset{P}{\gg} MP(t)$. Hence the result.

   If $s_i \underset{rpo}{>} t$, then by induction hypothesis $s_i \underset{T}{>} t$ and the rest of the proof can be carried out in a similar fashion. □

The path ordering also has the incrementality property desired of a termination ordering that is built on the fly to ensure the termination of rewriting systems generated by the Knuth-Bendix completion procedure in the process of computing a canonical system. The following theorem states that as the new precedence relations between function symbols are added, they do not upset the old ordering relation among terms; only terms which were not comparable earlier can be compared.

Theorem 5: Let $\overset{a}{>}$ and $\overset{b}{>}$ be two partial orderings of the set of function symbols F such that $\overset{b}{>}$ properly contains $\overset{a}{>}$. Then for all terms s, t in $T(F,V)$ $s \underset{T}{\overset{a}{>}} t$ implies $s \underset{T}{\overset{b}{>}} t$.

5. COMPLEXITY OF COMPARING TERMS USING PATH ORDERING

Given two terms s and t and a quasi-ordering $\gtrsim$ of the set of function symbols F, it can be determined whether $s \underset{T}{>} t$ or $s \underset{T}{\sim} t$ in time $0(|s|^5 * |t|^5)$. We show this upper-bound by doing the comparison by a method similar to dynamic programming.

Assume that all subterms of s have already been compared to all proper subterms of t and all subterms of t have already been compared to all proper subterms of s. (In other words, we have done everything short of comparing s and t themselves.) Assume also that the results are stored in a 2-dimensional array A that can be accessed easily. For instance, if s_i is a subterm of s at position p and t_j is a subterm of t at position q ($p \neq \lambda$ or $q \neq \lambda$), then $A(p,q)$ tells you whether $s_i \underset{T}{>} t_j$ or $s_i \sim t_j$. Our aim is to determine $TCOMP(s,t)$, the additional time required to compare s and t.

Note that the worst case is the one in which *every* full path in s has to be compared with *every* full path in t. Hence we have to analyze the time complexity of path comparison. This seems at first to be difficult, since path comparison involves further term comparisons. But note that in those comparisons at least one of the terms must be a *proper* subterm of s or of t. Thus this involves only looking up in the array A. Let P and Q be full paths in s and t respectively. It can be shown, under the above assumption (namely that every subterm of s

has been compared with every subterm of t, with the exception of s and t themselves), that P and Q can be compared in time $O(|P|^3 * |Q|^3)$. We prove this later using, again, a dynamic programming-like method.

An upper bound for $TCOMP(s,t)$ can now be derived in terms of $|s|$ and $|t|$. The number of full paths in any term u has an upper bound of $|u|$. Hence the number of full-path-comparisons required to compare s and t has an upper bound of $|s| * |t|$. Similarly, the maximum length of a full path in a term u is also bounded above by $|u|$. Thus $TCOMP(s,t)$ takes $O(|s|^4 * |t|^4)$ time.

All that we now have to do is to sum up all possible $TCOMP(s_i, t_j)$s, where s_i is a subterm of s and t_j is a subterm of t. The number of all possible subterms of a term u is quite clearly $|u|$, since every position in the term corresponds to a subterm and vice versa. Each $TCOMP(s_i, t_j)$ is bounded above by $TCOMP(s,t)$ and therefore takes $O(|s|^4 * |t|^4)$ time. Thus

$$\sum TCOMP(s_i,t_j) = O(|s|^5 * |t|^5)$$

and the claim is proved.

It remains to be shown that two paths P and Q can be compared in time $O(|P|^3 * |Q|^3)$. Let $|P| = m$ and $|Q| = n$. For all i, j such that $1 \leq i \leq j \leq m$, let P_{ij} denote the subpath of P from the i-th tuple through the j-th. Similarly, let Q_{kl} denote the subpath of Q from the k-th tuple through the l-th.

Assume, as we did for comparing terms, that all proper subpaths of P have been compared with all proper subpaths of Q. Let B be a 4-dimensional array in which the results are stored: $B(i,j,k,l)$ gives you the result of comparing P_{ij} with Q_{kl}. Denote by $PCOMP(P,Q)$ the (additional) time required to compare P and Q under these assumptions.

For any tuple $<g, s_i>$ in P, we can find out whether there exists a tuple in Q that takes care of it in $O(n)$ time by a straightforward two-pass algorithm: in the first path, it is checked whether there is any tuple $<f, t_j>$ in Q such that $f > g$; if none, then in the second pass, we check whether $f \sim g$ in which case the arrays B and A are used to compare right-contexts, subterms and left-contexts. Thus $PCOMP(P,Q) = O(mn) = O(|P| * |Q|)$. Note that there are $O(m^2)$ subpaths of P and similarly $O(n^2)$ subpaths of Q. Therefore, $O(m^2 * n^2)$ path-comparisons are necessary in the worst case and each of these is bounded above by $PCOMP(P,Q)$. The overall time, therefore, is

$$O(m^3 * n^3) = O(|P|^3 * |Q|^3).$$

This completes the proof. □

6. INCORPORATING STATUS IN PATH ORDERING

While comparing terms we have to sometimes introduce the concept of ***status*** of an operator to make the terms comparable. The most common example is the associativity law of certain operators like + and *: $(x * y) * z = x * (y * z)$.

In order to orient such equations as rules we have to assign left-to-right (l-r) or right-to-left (r-l) status to one or more operators. (e.g. the $*$ operator in the above equation.) Equivalent operators should have the same status. If the top-level operators of two terms are the same we compare the immediate subterms *not as multisets* but in a *lexicographical* way, either left-to-right or right-to-left.

Kamin and Levy [7] extend RPO to LRPO (lexicographic RPO) in the following way:

$s = f(s_1, \ldots, s_m) \underset{lrpo}{>} t = g(t_1, \ldots, t_n)$ if and only if one of the following three conditions hold:

(1) $f > g$ and $s \underset{lrpo}{>} t_i$ for all i, $1 \leq i \leq n$ (same as for RPO).

(2) $f \sim g$ and (a) if f and g have l-r status then there exists j such that $s_1 \sim t_1$, $s_{j-1} \sim t_{j-1}$, $s_j \underset{lrpo}{>} t_j$ and $s \underset{lrpo}{>} t_i$ for $j+1 \leq i \leq n$, (Similarly for r-l status.) whereas

(b) if f and g have no status then it must be that $\{ s_1, \ldots, s_m \} \underset{lrpo}{\gg} \{ t_1, \ldots, t_n \}$.

(3) $s_i \underset{lrpo}{\geq} t$ for some i, $1 \leq i \leq n$ (same as for RPO).

To incorporate status in the path ordering we have to modify path comparison and term comparison as follows:

Path Comparison (LP):

Let $P_1 = \langle f_1, t_1\rangle\langle f_2, t_2\rangle \ldots \langle f_m, t_m\rangle$ and

$P_2 = \langle g_1, s_1\rangle\langle g_2, s_2\rangle \ldots \langle g_n, s_n\rangle$ be two sequences of two-tuples. $P_1 \underset{LP}{>} P_2$ if and only if for all $\langle g_j, s_j\rangle$ in P_2 there exists $\langle f_i, t_i\rangle$ in P_1 such that

a. $f_i > g_j$ or

b. $f_i \sim g_j$ and if they have no status then

1. $RC(\langle f_i, t_i\rangle, P_1) \underset{LP}{>} RC(\langle g_j, s_j\rangle, P_2)$ or
2. $RC(\langle f_i, t_i\rangle, P_1) \sim RC(\langle g_j, s_j\rangle, P_2)$ and $t_i \underset{LT}{>} s_j$ or
3. $RC(\langle f_i, t_i\rangle, P_1) \sim RC(\langle g_j, s_j\rangle, P_2)$, $t_i \sim s_j$ and

   $LC(\langle f_i, t_i\rangle, P_1) \underset{LP}{>} LC(\langle g_j, s_j\rangle, P_2)$.

If they have l-r status, say, then

1. $t_i \underset{LT}{>} s_j$ or
2. $t_i \sim s_j$ and $LC(\langle f_i, t_i\rangle, P_1) \underset{LP}{>} LC(\langle g_j, s_j\rangle, P_2)$

Term Comparison (LT):

Let $s = f(s_1, \dots, s_m)$, $t = g(t_1, \dots, t_n)$, $M_1 = \{ s_1, \dots, s_m \}$ and $M_2 = \{ t_1, \dots, t_n \}$.

Then $s \underset{LT}{>} t$ if and only if one of the following three conditions hold:

a. $f > g$ and $s \underset{LT}{>} t_i$ for all i, $1 \leq i \leq n$.

b. $f \sim g$ and if f and g have l-r status, say, then there exists j such that $s_1 \sim t_1, \dots s_{j-1} \sim t_{j-1}$, $s_j \underset{LT}{>} t_j$ and $s \underset{LT}{>} t_i$ for $j+1 \leq i \leq n$. (Similarly for r-l status.)

If f and g have no status, then MP(M_1 - M_2) $\underset{LP}{\gg}$ MP(M_2 - M_1).

c. $f \not\gtrsim g$ and $MP(s) \underset{LP}{\gg} MP(t)$.

7. A FURTHER EXTENSION OF PATH ORDERING

A possible way of extending the path ordering even further is by allowing tuples from *different* paths to take care of tuples along a path. Thus we could remove the restriction that a path must be taken care of by another path and instead let tuples from a collection of paths take care of the tuples from another path. An example where such an idea would work is the following:

Example: $s = f(a(x), b(x))$, $t = g(h(x))$ with $a > g$ and $b > h$.

The tuple $<a, a(x)>$ takes care of $<g, g(h(x))>$ and $<b, b(x)>$ takes care of $<h, h(x)>$. Note that neither the path ordering nor RDO can compare s and t.

It must be noted of course that this new scheme is not clearly defined yet. But this looks like an interesting problem for future research and we are working on the details.

8. RELATIONSHIP TO RECURSIVE DECOMPOSITION ORDERING

Jean-Pierre Jouannaud [personal communication] has pointed out to us the similarity between our ordering scheme and the Recursive Decomposition Ordering Scheme (RDO) [6,14]. The way our ordering is formulated is similar to the "entire choice" scheme of RDO, since essentially *all* the paths in a term and *all* the tuples in a path are taken into account. (The reader is referred to [6,14] for the formal definition of RDO and further details.) However, the path ordering is *not* quite the same as RDO, since there are terms that the path ordering can compare but RDO cannot. An example is given below.

Example: $s = h(a(z), g(a(a(x)), x), g(b(b(y)), y))$ and
$t = h(a(z), g(a(x), b(y)), g(a(x), b(y)))$ with no precedence among the function symbols.

While comparing s with t, we eventually end up comparing the paths through the multisets $M_1 = \{ g(a(a(x)), x), g(b(b(y)), y) \}$ and $M_2 = \{ g(a(x), b(y)), g(a(x), b(y)) \}$. It can be easily seen that MP(M_1) $\underset{P}{\gg}$ MP(M_2).

When RDO compares terms s and t, it has to compare decompositions along the paths that end in z. (They are the 'leftmost' paths, expressed as '1.1' in positional notation.) Clearly the decomposition corresponding to h in s must take care of the decomposition corresponding to h in t. In other words, the decomposition

$$< h, a(z), \{ g(a(a(x)), x), g(b(b(y)), y) \}, \Box > = < h, a(z), M_1, \Box >$$

must be greater than

$$< h, a(z), \{ g(a(x), b(y)), g(a(x), b(y)) \}, \Box > = < h, a(z), M_2, \Box >.$$

But this is impossible since M_1 and M_2 are not comparable as multisets of *terms*.

Jouannaud has suggested a way to overcome this in RDO by introducing a varyadic function symbol, the details of which still need to be worked out. It will be interesting to examine whether the proposed extension of RDO is equivalent to the path ordering.

Jouannaud has also remarked that that the path ordering may not compare well with RDO in efficiency. There is no way to do this comparison at present; we are not aware of any complexity analysis of RDO; there do not exist implementations of path ordering and RDO on the same machine and in the same language either to compare their experimental performance.

9. CONCLUSION

We have found that recursive path ordering (RPO), though simple and elegant, cannot order certain equations which we intuitively 'feel' we can order. The example (given in Section 2.2) $a(b(x)) = c(d(x))$ with $b > c$ and $a > d$ is one such. We have devised a new ordering which compares terms using the paths in them. This ordering properly contains RPO and eliminates many of RPO's drawbacks. We also feel it is easy to understand.

ACKNOWLEDGEMENTS: We thank Nachum Dershowitz and Jean-Pierre Jouannaud for their comments on earlier drafts of this paper.

10. REFERENCES

[1] Buchberger, B., "A Theoretical Basis for the Reduction of Polynomials to Canonical Forms," *ACM-SIGSAM Bulletin,* 39, August 1976, pp. 19-29.

[2] Dershowitz, N., "Orderings for Term Rewriting Systems," *Theoretical Computer Science* 17 (1982) 279-301.

[3] Guttag, J.V., Kapur, D., and Musser, D.R., "On Proving Uniform Termination and Restricted Termination of Rewriting Systems," *SIAM Journal on Computing,* Vol. 12, No. 1, February, 1983, pp. 189-214.

[4] Huet, G., "Confluent Reductions: Abstract Properties and Applications to Term Rewriting Systems," *JACM,* Vol. 27, No. 4, Oct., 1980, pp. 797-821.

[5] Huet, G., and Lankford, D.S., "On the Uniform Halting Problem for Term Rewriting Systems," Rapport Laboria 283, INRIA, Paris, March 1978.

[6] Jouannaud, J.-P., Lescanne, P., Reinig, F., "Recursive Decomposition Ordering," *Conf. on Formal Description of Programming Concepts*, Garmisch, 1982.

[7] Kamin, S., and Levy, J-J., "Attempts for Generalizing the Recursive Path Ordering," Unpublished Manuscript, Feb. 1980.

[8] Kandri-Rody, A. and Kapur, D., "Algorithms for Computing the Grobner Bases of Polynomial Ideals over Various Euclidean Rings," Proceedings of *EUROSAM* '84, Cambridge, England, July, 1984, Springer Verlag Lecture Notes in Computer Science LNCS 174, (ed. J. Fitch), pp. 195-206.

[9] Kapur, D., and Narendran, P., "The Knuth-Bendix Completion Procedure and Thue Systems," *Third Conference on Foundation of Computer Science and Software Engg.*, Bangalore, India, December 1983, pp. 363-385.

[10] Kapur, D., and Sivakumar, G., "Architecture of and Experiments with RRL, a Rewrite Rule Laboratory," *Proceedings of the NSF Workshop on Rewrite Rule Laboratory*, Rensselaerville, NY, September 4-6, 1983.

[11] Knuth, D.E. and Bendix, P.B., "Simple Word Problems in Universal Algebras," in *Computational Problems in Abstract Algebras* (ed. J. Leech), Pergamon Press, 1970, pp. 263-297.

[12] Lankford, D.S., "On Proving Term Rewriting Systems are Noetherian," Memo MTP-3, Mathematics Department, Louisiana Technical University, Ruston, LA (1979).

[13] Lankford, D.S., and Ballantyne A.M., "Decision Procedures for Simple Equational Theories with Commutative-Associative Axioms: Complete Sets of Commutative-Associative Reductions," Memo ATP-39, Dept. of Mathematics and Computer Sciences, Univ. of Texas, Austin, TX (1979).

[14] Lescanne, P., "How to Prove Termination? An Approach to the Implementation of a New Recursive Decomposition Ordering," Proceedings of an *NSF Workshop on the Rewrite Rule Laboratory* Sept. 6-9 Sept. 1983, (eds. Guttag, Kapur, Musser), General Electric Research and Development Center Report 84GEN008, April, 1984, pp. 109-121.

[15] Peterson, G.E., and Stickel, M.E., "Complete Sets of Reductions for Some Equational Theories," *JACM*, Vol. 28, No. 2, pp. 233-264.

[16] Plaisted, D.A., "A Recursively Defined Ordering for Proving Termination of Term Rewriting Systems," Report R78-943, Dept. of Computer Science, Univ. of Illinois, Urbana, IL.

A Rewrite Rule Based Approach for Synthesizing Abstract Data Types

Deepak Kapur
Computer Science Branch
General Electric R & D Center, KWC264,
Schenectady, NY 12345, U.S.A.

Mandayam Srivas[1]
Department of Computer Science
State University of New York at Stony Brook
Stony Brook, NY 11794, U.S.A.

Abstract

An approach for synthesizing data type implementations based on the theory of term rewriting systems is presented. A specification is assumed to be given as a system of equations; an implementation is derived from the specification as another system of equations. The proof based approach used for the synthesis consists of reversing the process of proving theorems (i.e. searching for appropriate theorems rather than proving the given ones). New tools and concepts to embody this reverse process are developed. In particular, the concept of expansion, which is a reverse of rewriting (or reduction), is defined and analyzed. The proposed system consists of a collection of inference rules - instantiation, simplification, expansion and hypothesis tesing, and two strategies for searching for theorems depending upon whether the theorem being looked for is in the equational theory or in the inductive theory of the specification.

1. Introduction

In this paper we develop a formal system for automatically synthesizing implementations of abstract data types from their algebraic specifications. In our approach, the *implemented* data type (i.e., the data type which is being synthesized) and the *representing* data types (i.e., the data types used to represent the implemented type) are specified as algebraic axioms. In addition, a mapping, called the *abstraction function*, that relates the values of representing data types to the values of the implemented data type is also specified. The output of the synthesis procedure consists of implementations for the operations of the implemented data type in terms of the operations of the representing data types. Thus, the operations of the representing types are used as primitive functions in the implementation being synthesized. This approach to synthesis can be applied hierarchically to as many levels of abstraction as necessary until we obtain an implementation in terms of the operations of data types, such as arrays, that are directly supported by a programming language system.

Our approach is based on the theory of term rewriting systems developed recently in the context of reasoning and proving theorem automatically about algebraic structures and data types from their specifications [KnB70], [HuH80], [Mus80b], [Hsi82]. Systems built using rewrite rule based approach for abstract data types such as AFFIRM [Mus80a], OBJ [GoT79],

[1]Research supported in part by the National Science Foundation under grants DCR-8401624 and DCR-8319966

and other systems for manipulating general term rewriting systems such as REVE [Les83] and FORMEL [HuH80], have provided encouraging signs for using the approach for theorem proving applications. In our work on generating implementations from specifications, we use many of the same tools and concepts used in theorem proving. Since we view theory-based synthesis as reversing the process of proving theorems (i.e., searching for appropriate theorems rather than proving known ones), we develop new tools and concepts to embody this reverse process. Our system consists of a set of *inference rules*, and strategies for using the inference rules to synthesize implementations. Our experience in working out many examples by hand have suggested that program transformation, optimization and synthesis based on rewrite-rule-based theorem proving is highly promising.

A major advantage of using term rewriting theory is that we can address all issues related to data type synthesis within a uniform framework. We are able to provide formal justification of the soundness of the rules of inference of our system since the consistency of data type specifications can be characterized using term rewriting concepts. We characterize the conditions under which a strategy would successfully synthesize implementations. We present two strategies - one that synthesizes implementations in the *equational theory*, and another that operates in the *inductive theory* of the specification. Both the strategies require that the data type specifications be organized as *canonical* term rewriting systems. This requirement in turn means that there exists a well-founded ordering on terms (as in [Der82]) that can be used to guarantee the *uniform termination* property of the specification term rewriting system. Although the strategies work for any such ordering it is assumed that one such ordering is available. The use of such an ordering assures the termination of the programs synthesized.

The equational strategy is completely automatic in the sense that it does not need any human intervention for it to succesfully synthesize an implementation. It is guaranteed to synthesize an implementation provided there exists an implementation (in the equational theory) the termination of which can be demonstrated using the termination ordering being employed by the strategy. The equational strategy may not terminate unless such an implementation exists. The inductive strategy is only semi-automatic since it needs prompts from the user at strategic points. Furthermore, the strategy is not complete for the inductive theory in the sense it is not guaranteed to produce an implementation even if one exists in the inductive theory.

The rest of the introduction contains an overview of related works in the area. In the next section, we illustrate our approach on a small example. Section 3 gives the reader a background in the use of algebraic techniques for specifying and implementing data types. Section 4 describes our synthesis system: Section 4.1 presents the rules of inference of our system, their justification along with their theoretical basis. A detailed discussion of the strategies for putting various rules of inferences together is given in Section 4.2. This is followed by a section that illustrates the approach in detail on a couple of examples. (For lack of space we present only simple examples in the paper. For more detailed examples see [KaS84].)

1.1. Related Work

Burstall and Darlington [BuD77] and Manna and Waldinger [MaW80] proposed a set of general purpose transformations for refining programs from their high level specifications. Feather [Fea82] has extended Burstall and Darlington's approach in his ZAP system by providing the ability for a user to specify metaprograms to direct the transformation process. Manna and Waldinger have adopted strategies based on theorem proving methods in the first order predicate calculus in their DEADLAUS [MaW80] system. Our method is related to that of Burstall and Darlington. The algebraic specification language is similar to the recursive equational language used by Burstall and Darlington for specifying the behavior of functions in their program transformation work.

Our system offers several advantages over the one proposed by Darlington (([BuD77], [Dar82]) all of which arise as result of the use of the term rewriting framework. The inference mechanisms of our system subsume all the machinery developed by Darlington and Burstall; in this sense, we not only provide a theoretical basis for their approach but also extend it. (Kott [Kot82] has also independently provided theoretical justification for their method by giving conditions under which *fold/unfold* can be applied.) Our system operates in a richer theory (the *inductive theory*) of the specification. This enables us to derive a richer class of implementations. Our approach also appears more promising in developing intelligent strategies for synthesis. This is primarily because our framework is conducive to adapting theorem proving strategies (such as the inference mechanism based on the Knuth-Bendix completion procedure [KnB70], [Mus80b], [HuH80], [Lan81]) for synthesis.

2. Illustration of the Approach

Before getting into the details of our approach, we will illustrate some of our ideas informally on an example. A function **union** is defined on the data type **multiset**, which is constructed using two constructors - a constant function **nullset** that creates an empty multiset and a binary function **insert**: **int X multiset → multiset** that inserts an element into a multiset. The function **union**, which returns the union of two multisets, has the following primitive recursive definition expressed as a set of *rewrite rules*:

(T1) union(nullset, s2) → s2
(T2) union(insert(e1, s1), s2) → insert(e1, union(s1, s2))

The data type **multiset** is to be implemented using another data type, **sequence**, whose values are constructed using the constant function [ ], and a binary constructor **+ : int X sequence → sequence**. The *abstraction function h* specifying how sequences represent multisets is also given as rewrite rules:

(H1) $h([\,]) \rightarrow$ **nullset**
(H2) $h(\mathbf{e} + \mathbf{v}) \rightarrow$ **insert(e,** h**(v))**

This example is interesting because (1) a **multiset** can be constructed in more than one way using its constructors, and (2) several different sequences may represent the same multiset. Using the abstraction function h, the implementation of **union** denoted as **UNION**, can be derived as follows. First, we introduce the rewrite rule that completely characterizes **UNION** in terms of **union** and h. This rewrite rule is called the *specification rule* of **UNION**. (Note that for **UNION** to be correctly implementing **union** the latter has to be a homomorphic image of the former.)

(S1) h**(UNION(v1, v2)) → union(**h**(v1),** h**(v2))**

The objective is to derive an implementation for **UNION** independent of **union** and h. This is done by deriving enough rules of the form h**(UNION(t1, t2)) →** h**(t3)** so that we can derive the implementation of **UNION** (as a total function on *sequence*) by dropping h from both sides. Rules (H1) and (H2) defining h suggest that we can instantiate the right hand side of rule (S1) - by first instantiating **v1** to be **[ ]**, and then to be **insert(e1 + v1)**. (Our strategy will discover these instantiations automatically such that they completely cover the domain of the function being implemented.)

h**(UNION([ ], v2)) → union(**h**([ ]),** h**(v2))** *Instantiate (S1)*
h**(UNION([ ], v2)) →** h**(v2)** *Simplify using (H1), (T1)*

By dropping h on both sides, we get

(I1) UNION([ ], v2) → v2

Similarly, by instantiating **v1** to be **e1 + v1** in rule (S1) as suggested by rule (H2) of h, we have

h**(UNION(e1 + v1, v2))**
→ union(h**(e1 + v1),** h**(v2))** *Instantiation of (S1)*
→ insert(e1, union(h**(v1),** h**(v2)))** *Simplify using (H2), (T2)*

Now, we want to bring symbol h on the right hand side to its outer most level. This can be done by applying some of our rewrite rules in the reverse direction (called *expansion* later in the paper, which is similar to but, more general than, Burstall and Darlington's *fold* mechanism). In order to use rule (H2) for h in the reverse direction, we must first use rule (S1) in the reverse direction. These two expansion steps will result in the following rewrite rules.

h**(UNION(e1 + v1, v2)) → insert(e1,** h**(UNION(v1, v2)))**
→ h**(e1 + UNION(v1, v2))**

Dropping h on both sides, we have

(I2) UNION(e1 + v1, v2) → e1 + UNION(v1, v2)

(I1) and (I2) constitute an implementation for **UNION** because the two rewrite rules define **UNION** as a total function on **sequence**. Note that in this example the correspondence between **union** and **UNION** is very obvious because of the close correspondence between the constructors (**[ ]**, **+**) of *sequence* and the constructors (*nullset*, *insert*) of *multiset* (except for the relation on the constructors of *multiset*). The examples in Section 6 will demonstrate that the method can synthesize more interesting implementations.

The mechanisms used above are all embodied in our system as *inference rules* that act on a set of rewrite rules to produce a new rewrite rule. The theoretical justification (given later) for each of the inference rules is that it always produces rewrite rules consistent with the original set of rewrite rules. Specifically, the h dropping mechanism can be justified by the *hypothesis checking* inference rule which uses the Knuth-Bendix completion procedure; we need to have a criterion for h dropping especially when h is a many-to-one function. According to this rule of inference, if the new rule being hypothesized does not result in any

contradiction along with the existing rewrite rules, then the hypothesized rule can be added to the system.

We organize the rewrite rules in our specification into several groups based on the role they play during the synthesis of an implementation. The specification rules (such as S1) that specify the new function(s) to be implemented are grouped into the *specification set*. The rules that are used for expansion, such as (S1), (H1), and (H2), are grouped as the *expansion set*. The rules that are used for simplifying terms form the *reduction set*. The criteria that determine membership of a rewrite rule in these groups will be given later.

Although the transformation process at least for this simple example is similar to that of Burstall and Darlington [BuD77], there are several fundamental differences. We deduce the instantiations of the rules automatically from the left hand sides of the rules defining *h*. The inference rule in our system that does the instantiation uses the notion of *derived pairs* [GKM82]. Note that there are several possible sets of instantiations that completely span the domain. Our synthesis strategy enumerates all possible sets of instantiations systematically attempting to find implementation for each of them until it finds one. For instance, in the above example another set of instantiations possible for the arguments of UNION is {(v1, [ ]), (v1, e + v2)}. In this case, our system may derive the following pair of rewrite rules:

UNION(v1, [ ]) → UNION(v1, UNION([ ], [ ]))
UNION(v1, e + v2) → UNION(v1, INSERT(e, v2))

Our strategy would discard this implementation because the first rewrite rule cannot be ordered under any well-founded ordering.

There are four inference mechanisms being used in our approach:

(1) *Instantiation* of variables in the rewrite rules specifying the function to be implemented. (The instantiation is not arbitrary, but is directed by the definitions of other functions.)

(2) *Simplification* of a term to its irreducible form (a term is said to be irreducible if it cannot be further simplified.)

(3) *Expansion* to introduce recursion or other helping functions in the implementation.

(4) The *Knuth-Bendix completion* procedure for checking whether new hypotheses being made are indeed consistent with the existing definitions.

3. Abstract Data Types

3.1. Specification

Abstract data types are specified using the algebraic technique developed by Guttag [GuH78] and the ADJ group [GTW8.]. Our presentation of the specification is patterned after Guttag et. al. Specifically, abstract data types are defined one by one in a hierarchical way assuming other data types to be specified elsewhere. We use the initial algebra semantics [GTW8.] for our data type specifications. The data type sequence is specified in the figure below. (The data type item is assumed to be specified elsewhere.)

The operations of a data type are grouped into two classes: (i) *generators*, which generate all the values of the data type, and (ii) *defined functions*, which are defined on the values of the data type constructed by the generators. These classes are explicitly

identified in the specification along with the domain and range of every operation.

The construction of values of a data type in terms of its generators is not necessarily unique. The generators - **nullset** and **insert** - of the data type **multiset** we saw in section 2 can be used in more than one way to construct the same multiset. For instance, **insert(insert(nullset,1),2)** and **insert(insert(nullset,2),1)** both construct the same multiset **{1, 2}**. This equivalence relation is characterized by the equations relating the generators. We refer to the terms like **insert(insert(nullset, 2), 1)** constructed solely using the generators as *generator terms*. Generator terms that do not contain any variables are referred to as *generator constants*.

We require that the specification of every data type involved in the synthesis be complete and consistent. By completeness ([GuH78], [Kap80].), we mean that the equations in the specification are such that every defined function is defined for every generator constant of the data type. That is, every term that applies a defined function to generator constants, such as **union(insert(nullset, 1), nullset)**, can be shown to be equivalent to a generator constant (or the distinguished element **error**) of the range type of the function. Consistency ensures that every function of the data type forms a well-defined function. One way to guarantee the completeness and consistency of a specification is to ensure that it can be organized as a term rewriting system that is *well-spanned* [Sri82] and *canonical*[2]

Data Type Sequence

Generators

[ ] : → **sequence**
+ : **item X sequence → sequence**

Defined Functions

first : **sequence → item ∪ { Error }**
rest : **sequence → sequence ∪ {Error}**
rotate : **sequence → sequence**

Axioms

(1) **first([ ]) ≡ Error**
(2) **first(e + v) ≡ e**

(3) **rest([ ]) ≡ Error**
(4) **rest(e + v) ≡ v**

(7) **rotate([ ]) ≡ [ ]**
(8) **rotate(e + [ ]) ≡ e + [ ]**
(9) **rotate(e1 + (e2 + v)) ≡ e2 + rotate(e1 + v)**

Figure 1. Algebraic Specification of Data Type Sequence

[2] A term rewriting system is canonical if every sequence of rewrites emanating form a term α terminates with the same term β; β is said to be the normal form of α.

[HuH80]. We briefy describe the two properties below.

A term rewriting system is ***well-spanned*** if it can be structured such that every term of the form **F(g)**, where **F** is a defined function and **g** is a generator constant, is an instance of the left hand side of an equation in the specification. A canonical term rewriting system for a given set of equational axioms can be derived using the Knuth-Bendix completion procedure [HuH80]. It is assumed that the equational axioms specifying the data types can be oriented and made into rewrite rules using some termination ordering [Der82]. For instance, all the axioms in the specifications of **sequence** and **queue** can be oriented as rewrite rules from left to right using a ***recursive path ordering*** [Der82] in which the defined operation symbols are assigned more weight than the generators. In case any of the operations have the associative and/or commutative property or any other property which needs special handling, then appropriate termination ordering [DHJ83] incorporating such properties must be used. The Knuth-Bendix completion procedure or its generalization developed by Peterson and Stickel [PeS81] or Lankford and Ballantyne [LaB77] to deal with special properties such as associativity, commutativity is then applied to rules to obtain a canonical term rewriting system which gives the decision procedures for the equational theories of these data types. For instance, the specifications of **sequence** and **queue** are canonical because the Knuth-Bendix completion procedure when run on them does not generate any new rewrite rules.

The ***equational theory*** of a data type specified by equational axioms are all equational formulas that can be derived using the axioms of equality - reflexivity, symmetry, transitivity, substitution property and replacement. When a set of equations E can be organized into a canonical term rewriting system $\mathbf{R}$, the equational theory of E contains formulas $\alpha = \beta$ such that α and β have the same normal form. The ***inductive theory***, which contains the equational theory, is the set of all equational formulas that can be derived using the rules of equality and the following principle of induction. In the following, a generator-constant substitution is a substitution in which variables are substituted by generator constants.

$$\frac{\forall\ generator\text{-}constant\ substitution\ \sigma,\ \sigma(a) = \sigma(b) \in equational\ theory}{a = b \in inductive\ theory}$$

For example, consider the function **f** defined on natural numbers (with generators **0** and **S**) by the rewrite rules: $\mathbf{f(0)} \rightarrow 0$ and $\mathbf{f(S(x))} \rightarrow \mathbf{f(x)}$. The equation $\mathbf{f(x) = 0}$ is not in the equational theory of natural numbers with **f** but is in the inductive theory. For the data types **sequence** and **queue** for example, the equational formula expressing the associativity of append is in the inductive theory. It should be noted that this inference rule is not effective. Generally weaker forms of this inference rule are used which are practically powerful enough.

The equations in the inductive theory are not theorems in the logical sense because they do not hold good in all the models of the specification. They do hold good in the initial algebra model of the specification. They are useful for our purpose because we use the initial algebra semantics for data types. The inductive theory is the basis for our method since the programs synthesized by our inference rules lie in the inductive theory of the specification.

Henceforth, we will assume that the specifications of all data types being used in the synthesis procedure are well-spanned and each specification has a canonical term rewriting system associated with it (modulo an equivalence relation specified by special properties such as associativity, commutativity, etc.).

3.2. Specifying the Desired Implementation

Implementing a data type consists of choosing a representation for the data type, and implementing every operation of the data type in terms of the operations of the representation type. The implementation for the operations can, in general, be expressed in an arbitrary language. In the present work we express the implementation in a language identical to the one used to express the specification. That is, an implementation for an operation is expressed as a set of well-formed rewrite rules that defines the operation as a function on the chosen representation type.

In order to synthesize interesting implementations for a data type **D**, we require the user to furnish information about how the values of the representing type(s) **R** are used to represent the values of **D**. This information is specified by the user as an abstraction function *h* from **R** to **D** again as a set of rewrite rules. In general, not all the values of **R** may be used to represent the values of **D**; the subset of values of **R** that are used is specified by an *invariant predicate.* In this paper we will assume that the set of values of **R** used for representing **D** is identical to the domain of the abstraction function *h*. Some of the issues concerning data type synthesis in the presence of nontrivial invariants are discussed in [Sri82].

The abstraction mapping can be complex and may use additional auxiliary functions on **D** and/or **R** which are specified using the operations of **D** and **R** again as rewrite rules such that they are completely defined. Further, we will assume that these rewrite rules for *h* and auxiliary functions also satisfy the completeness and consistency conditions stated earlier. Our experience suggests that the more complex the abstraction funcion is, the more difficult it is to generate implementations for the operations of **D**. The rewrite rules (H1) through (H2) of the example in section 2 specify the abstraction function for an implementation of **multiset**. Specified below is another abstraction function for representing queues in terms of sequences. The empty queue is represented by the empty sequence. A nonempty queue is represented by a sequence whose elements are identical to the ones in the queue, but are arranged in the reverse order. The motivation for such a scheme is that the reading and deletion of elements from a queue can be performed efficiently. The specification of *h* uses an auxiliary function **add_at_head** on queues. This function adds an element at the front end of a queue.

(H1) h([ ]) → **nullq**
(H2) h(e + v) → **add_at_head**(h(e), h(v))

(H3) **add_at_head(e, nullq) → enqueue(e, nullq)**
(H4) **add_at_head(e1, enqueue(e2, q)) → enqueue(e2, add_at_head(e1, q))**

An implementation **F** for an operation **f** of **D** is then completely characterized by the following homomorphism property:

(*) h(**F(x1, ..., xn)**) → **f**(h(**x1**), ..., h(**xn**))

The mapping h is assumed to behave like an identity function on values of data types other than **R**. This is natural because we generate implementations for data types one at a time and in a hierarchical way. For each **f**, the above rewrite rule that specifies the implementing function **F** is said to be the *specification rule* of **F**.

4. Proof Based Approach for Synthesis

Our approach to synthesis is proof based because (1) we employ concepts used in theorem proving based on term rewriting systems, and (2) an implementation is derived as a theorem of the specification. The goal of our synthesis task, however, is fundamentally different from that of theorem proving. In theorem proving the goal is to establish that a given property is a theorem of a set of axioms (specification). In synthesis we have to search for a rewrite rule of an appropriate form that is known to be a theorem of the specification. The rewrite rules we are looking for are to constitute an implementation. Note that our characterization of the synthesis task is different from that of Waldinger [MaW80]. In [MaW80] a program is derived as a proof of a theorem which is an input/output specification of the program. Our approach is better suited for taking advantage of theorem proving ideas based on term rewriting systems.

In our approach, synthesis is performed with help of a system of *inference rules*. Every inference rule acts on a set of rewrite rules **R** and produces a new rewrite rule that is guaranteed to be consistent (i.e., a theorem in the inductive theory) with **R**. The set **R** initially consists of

(1) the specification of all data types,

(2) the specification of the abstraction function h, and

(3) the specification rule for the implementing functions.

Note that (1) and (2) form a well-formed system of rewrite rules. When (3) is added the set **R** remains canonical but is no longer well-spanned. This is because the functions (F) implementing the operations (**f**) are not yet defined on all values of the representation type. The synthesis process consists of repeatedly invoking appropriate inference rules on **R** so as to make it well-spanned. An implementation for an operation **f** of a data type is synthesized by deriving a well-spanned set of rewrite rules that implements **f** as a function on the representation type. The implementation is guaranteed to be correct since every rewrite rule, being derived by one or more application of an inference rule, is consistent with the specification rule of **f**. We first present the inference rules of the system, and then describe the strategies for invoking the inference rules.

4.1. Inference Rules

In the following we state the inference rules of our system. Every inference rule has the general form $\dfrac{c_1, c_2, \ldots, c_n}{\alpha \rightarrow \beta}$, where c_1, ..., c_n are a set of conditions, and $\alpha \rightarrow \beta$ is a new rewrite rule. The inference rule is to be read as "if the conditions c_1, ..., c_n hold good for a rewriting system S, then the rewrite rule $\alpha \rightarrow \beta$ may be added to S." The soundness of the inference rules is guaranteed by ensuring that the new rule $\alpha \rightarrow \beta$ is in the inductive theory of data types and the functions under consideration. In every inference rule it is also assumed that the new rewrite rule $\alpha \rightarrow \beta$ is added to S only if it preserves the uniform termination property of S. This can be ensured by checking that under the termination ordering $>$ being used $\alpha > \beta$. (In all our examples given later we use the recursive path ordering defined by Dershowitz [Der82].)

Instantiation

$$\frac{\alpha_1 \rightarrow \beta_1,\ \alpha_2 \rightarrow \beta_2 \in S,\quad <\gamma, \delta> \textit{ is a derived pair of the first rewrite rule on the second}}{\gamma \rightarrow \delta}$$

A ***derived pair*** [GKM82] $<\gamma, \delta>$ of $\alpha_1 \rightarrow \beta_1$ on $\alpha_2 \rightarrow \beta_2$ obtained by ***superposing*** [KnB70] β_1 on α_2 is defined as follows: consider a nonvariable subterm t of β_1 which unifies with α_2; let σ be the most general unifier of t and α_2. Then $\gamma = \sigma(\alpha_1)$, and $\delta = \sigma(\beta_{11})$, where β_{11} is obtained by replacing t in β_1 by β_2. For instance, in the case of the **union** example of section 2, the derived pair of rewrite rule (S1) on rewrite rule (H1) produces the rewrite rule *h* (UNION([ ], v2)) → union(*h* ([ ]), *h* (v2)).

Derived pairs are a generalization of reduction applied on rules except that derived pairs are constructed using unification instead of matching. Derived pair generation is similar to narrowing [LaB79]. Clearly, the new rewrite rule derived by this inference rule is in the equational theory of S; An advantage of using the derived pair mechanism rather than arbitrary instantiation (like done in [BuD77] [Fea82]) to instantiate rewrite rules is that it is possible to generate a well-spanned set of instantiations automatically. This is done (as will be explained more clearly in section 4.2) by computing derived pairs between a specification rule and every other possible rewrite rule. The fact that each of the functions in the specification are completely specified ensures that the function being implemented will also be defined completely when all the derived pairs are computed.

Simplification

Let $\beta \rightarrow^* \gamma$ stand for β simplifies to γ (i.e., γ is the normal form of β) using rewrite rules in S.

$$\frac{\alpha \rightarrow \beta \in S,\ \beta \rightarrow^* \gamma}{\alpha \rightarrow \gamma}$$

The justification of this rule of inference is obvious from the definition of reduction; the new rule in this case is also in the equational theory and preserves the termination ordering. The new rewrite rule obtained using this inference rule is put in to the same set of rewrite rules from where $\alpha \rightarrow \beta$ comes.

Expansion

We say γ ***expands to*** δ ***in*** S (written $\gamma \Leftarrow \delta$) using a rewrite rule $\alpha \rightarrow \beta \in S$ if the following conditions hold:

(1) There exists a subterm t of γ such that t is unifiable[3] with β.

[3] If any of the functions satisfy special properties such as associativity and commutativity then it is necessary to use unification algorithms under equational theories [Sti81].

(2) $\delta = \sigma(\gamma_1)$, where γ_1 is obtained by replacing t in γ by α.

The rule is

$$\frac{\alpha \rightarrow \beta \in S,\ \beta \Leftarrow \delta \textit{ with } \sigma \textit{ being the unifier used for expansion}}{\sigma(\alpha) \rightarrow \delta}$$

Expanding a term γ using a rewrite rule $\alpha \rightarrow \beta$ is roughly equivalent to reducing γ using the rule $\beta \rightarrow \alpha$. The difference lies in the fact that expansion uses unification (γ with β) whereas reduction uses matching (γ with β). Note that whenever $\beta \Leftarrow \delta$, δ necessarily reduces to $\sigma(\beta)$ for some substitution σ. The new rewrite rule obtained from the expansion rule of inference is also in the equational theory as the following diagram illustrates.

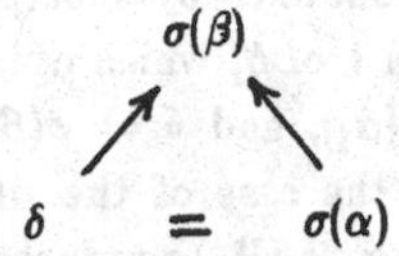

The difference between expansion and folding [BuD77] is that the former uses unification while the latter uses matching. To see the advantage of expansion over folding it would be instructive to consider the purpose an expansion/folding inference is serving in the synthesis process: To obtain an arbitrary term δ that is reducible to a given term β using the rewrite rules in the S. Folding (used repeatedly) is adequate for the purpose only if every rewrite rule *lhs* $\rightarrow$ *rhs* in S is *variable preserving* (i.e., is such that every variable in *lhs* also appears in *rhs*). However, if S has non-variable-preserving rewrite rules then folding alone is not sufficient, and we need expansion as illustrated by the following example. Let us suppose we wish to obtain from **cons(x, nil)** the term **Reml(Insertall(cons(x, nil)))** using the following set of rewrite rules. The following sequence of expansion steps achieves the desired result while no sequence folds does. Specifically, the last step in the sequence (in which rewrite rule (4) is used) cannot be performed if folding were being used.

(1) Reml(cons(x, nil)) → nil
(2) Reml(cons(x, cons(y, L))) → cons(x, Reml(cons(y, L)))
(3) Insertall(nil) → nil
(4) Insertall(cons(x, L)) → cons(x, cons(1, Insertall(L)))

cons(x, nil) $\Leftarrow^*$ **Reml(cons(x, cons(x*, Insertall(nil))))**
$\Leftarrow$ **Reml(Insertall(cons(x, nil))**

Note that every expansion step can in general be replaced by an arbitrary substitution for the variables followed by a folding. Mixing substitutions and folding, however, complicates the strategy for invocation of the inference rules since there are potenially infinite substitutions possible. The use of unification in expansion determines the productive substitutions automatically.

Further, while expanding a term β (with the hope of determining a term δ that is reducible to β) it is necessary to consider for unification only those variables that are newly introduced during expansion, but not the ones in β. New variables are introduced whenever a term is expanded using a rewrite rule that is not variable-preserving. We refer to such variables as *free* variables. For instance, the asterisked variables in the expansion sequence shown above are free variables. It can be shown that every term δ reducible to β is an instance of some term δ^* (for some substitution of the free variables in δ^*) that is obtained after performing a finite number of expansions on β. Intutively, the free variables are placeholders for an abitrary term. In our synthesis strategy the binding of the free variables is delayed until a decision is either automatically made by the unification performed during an expansion step (as in the above example), or expanding the term any further makes it bigger than a term even if the free variable is replaced by a least term in the ordering on the terms being used.

Hypothesis Testing

$$\frac{\{\alpha \rightarrow \beta\} \cup S \ \textit{is-KB-completable}}{\alpha \rightarrow \beta}$$

where is-KB-completable is a predicate that acts on a rewriting system **S**.

The above predicate, which is a partial function, characterizes the outcome of running the Indictive Knuth-Bendix completion proecedure ([HuH80]) on **S** which is a semidecision procedure for checking the *confluence* (i.e., consistency [KaM82]) of **S**. The predicate *is-KB-completable* returns true if the inductive Knuth-Bendix completion procedure terminates successfully; otherwise, it is undefined if the completion procedure does not terminate; in the other two cases, the preducate is false. This rule is powerful as it provides a way to check whether a hypothesis made based on derivations, or by generalizing a definition of a function on a class of examples (this technique is further discussed in the next section, and illustrated by the examples in section 5.) is indeed consistent with the rest of the specification. The new rewrite rule derived using this inference rule is in the inductive theory of the data types being considered.

The above inference rule is more powerful and provides a more effective way of introducing new definitions into the system than the *redefinition* mechanism [BuD77] of Burstall and Darlington. In the redifinition mechanism, to hypothesize a new definition for a function one adds it to the system, and one tries to generate the rewrite rules constituting the existing definition for the function using fold/unfold transformations. This, however, is only a sufficient condition for the new definition to be consistent. For instance, one might have to introduce definitions besides the one being hypothesized in order to obtain the original definition. The inductive KB-completion procedure does this automatically in a significant number of cases. Also, inductive KB-completion procedure uses only reductions (not expansions), and hence is more effective.

The h-dropping Rule

This inference rule can be used to obtain a new rewrite rule $\alpha \rightarrow \beta$ from a rewrite rule of the form $h(\alpha) \rightarrow h(\beta)$. It is not always sound to drop h, since h may be many-to-one and dropping it may cause inconsistencies. It would be sound to drop the symbol provided $\alpha \rightarrow \beta$ does not introduce any inconsistency, i.e., does not introduce any relationships among values that are distinct in the system. This condition is usually satisfied when α involves a function which is unimplemented in the system.

$$\frac{h(\alpha) \rightarrow h(\beta) \in S \ , \ \alpha \rightarrow \beta \cup S \ \textit{is-KB-completable}}{\alpha \rightarrow \beta}$$

The new rule derived this way is also in the inductive theory of data types being considered.

4.2. Strategies for Synthesis

A common mode in which the inference rules of our system is used to synthesize an implementation is to set up a *goal* that specifies the approximate form a program to be synthesized is supposed to take, and then make a judicious selection of the inference rules that achieves the goal. A *synthesis strategy* is a procedure which determines a sequence of invocations of the inference rules that will generate a new rewrite rule satisfying the desired goal. For instance, the goal in synthesizing an implementation for an operation f of a data type is to derive a well-spanned set of rewrite rules of the form $F(g_i) \rightarrow t_i$, where g is a generator term and t is an arbitrary term that does not involve any operations of the type being implemented. Each of these rewrite rules is derived by deriving theorems of the form $h(F(g_i)) \rightarrow h(t_i)$, and then dropping the symbol h on either side using the h-dropping inference rule.

In the following we present two general strategies for using the inference rules of our system to derive theorems of the form $h(F(g_i)) \rightarrow h(t_i)$. The two strategies differ in the theory to which the new rewrite rules being derived belong. The first one derives rewrite rules in the equational theory, while the second can also derive rewrite rules in the inductive theory.

Different rewrite rules in the specification play different roles during the synthesis process. Based on their role we have categorized the rewrite rules into the following groups. This categorization facilitates our synthesis strategies greatly.

(1) Rules specifying the functions to be synthesized form the *specification* set.

(2) Rules used for expanding terms form the *expansion*-set.

(3) Rules that are used only for simplification and/or computing derived pairs form the *reduction*-set.

In the case of data type synthesis, the specification set initially consists of the rewrite rules expressing the homomorphism property between the abstract operations and their implementing function. For example, in the informal derivation shown in section 2, the specification set initially consists of only the rewrite rule (S1). The rewrite rules that go into the expansion set will, in general, depend on the desired form of the new rewrite rule to be derived. In addition to the specification rule and the rules specifying the abstraction function and its auxiliary functions, the expansion set includes the rules determined as follows. Suppose F is the set of function symbols that are permitted to appear on the right hand side of the rewrite rule to be derived. (This information has to be furnished by the user, in general.) Let us suppose that we have a *uses* relationship defined on the function symbols that holds if the definition of a function uses another function. Let F∗ be the reflexive, transitive closure of *uses* applied to F. The expansion set will include the rewrite rules that define the functions in F∗. The reduction set will normally include the entire system.

4.2.1. Equational Strategy

The equational strategy is based on the property that a rewrite rule $lhs \rightarrow rhs$ in an implementation is in the equational theory of **R** if lhs and rhs have the same normal form in **R** . Thus, if *lhs* of the desired rewrite rule can be determined somehow, then *rhs* has to be a term that has the same normal form as *lhs* . The *lhs*'s of the rewrite rules are fixed based on the requirement that they have to be of the form $h(\mathbf{F}(\mathbf{g}_i))$ such that $\{\mathbf{g}_i\}$ forms a well-spanned set of generator terms.

This strategy synthesizes an implementation by repeatedly performing the following steps in sequence: The *Instantiation Step*, the *Simplification Step*, and the *Expansion Step*. Every iteration of the loop generates at most one rewrite rule $lhs_i \rightarrow rhs_i$ which is inserted into the set I (initially empty). The loop is terminated when a well-spanned set of rewrite rules are collected in I. The instantiation step consists of setting up lhs_i. The instantiation step is perfomed in such a way that every possible well-spanned set of lhs_i is generated after a finite number of iterations. The simplification step consists of simplifying the right hand side of the rewrite rule obtained in the first step to its normal form. The expansion step consists of repeatedly expanding the right hand side of the rewrite rule obtained in the simplification step. The rewrite rule returned by the expansion step is inserted into I. The expansion step, which is guaranteed to terminate (see below), may not yield an appropriate rhs_i for the lhs_i set up in the instantiation step. In such a case nothing is inserted into I during that iteration.

The instantiation step essentially consists of invoking the *Instantiation*-inference rule between a rewrite rule in the specification set and a rewrite rule outside the specification set. Although the exact rewrite rules which participate in the instantiation step are left unspecified, we assume that all the rewrite rules in the specifictaion set are treated *fairly*. This essentially means no rewrite rule in the specification set is ignored infinitely often. In other words the instantiation step has to ensure that the *Instantiation*-inference rule is invoked on every rewrite rule in the specification set and every other possible rewrite rules in the system after a finite number of iterations. This ensures every possible well-spanned set is generated after finite number of execution of the instaniation step. To see this is why, note that the specification set initially consists of the rewrite rule $h(F(x)) \rightarrow f(h(x))$. Computing derived pairs between this rewrite rule and every other possible rule generates the first well-spanned set. Each of the resulting rewrite rule (after simplification) is inserted into the specification set. Computation of derived pairs using each of these rules will generate the next well-spanned set, and so on.

repeat

(1) Apply *Instantiation*-inference between a rewrite rule in the specification set, and any other rewrite rule in the program.

(2) Repeatedly apply *Simplification*-inference to the new rewrite rule generated in step 1 until it is no longer applicable. Let the resulting rewrite rule be $\alpha \rightarrow \beta$. Add the simplifed rewrite rule to the specification set.

(3) Carry out the *Expansion step* (described below) on $\alpha \rightarrow \beta$ to obtain the rewrite rule $\alpha \rightarrow \gamma$.

(4) Replace any free variable in γ by an appropriate least element, and insert into the output set I.

until a well-spanned set of rewrite rules is obtained in I

The Expansion Step

The expansion step takes a rewrite rule $\alpha \rightarrow \beta$, and produces another rewrite rule $\alpha \rightarrow \gamma$ using the *Expansion*-inference rule repeatedly. The rewrite rules used for the expansion of β are picked from the expansion set. At each step there can be several different rewrite rules may be used for expanding. We assume that the algorithm uses a "dove tailing" technique in which all possible expansions are considered one step at a time. Also while checking if $\alpha < \beta$ below we assume that a free variable in β are treated as a least element in the ordering. It is important to note that the expansion step is guaranteed to terminate assuming there is a termination ordering on the terms. This is because the right hand side of a rewrite rule is expanded only as long as it is less than its left hand side.

while $\beta < \alpha$ and β is not of the desired form **do**
(1) Apply *Expansion*-inference rule between $\alpha \rightarrow \beta$ and a rule from the expansion set to obtain a new rewrite rule $\alpha \rightarrow \gamma$.

(2) Replace β by γ.

endwhile

The equational strategy will find an implementation as long there is an implementation $\{lhs_i \rightarrow rhs_i\}$ in the equational theory such that $lhs_i > rhs_i$. This is because the instantiation step is guranteed to generate the desired lhs_i. For such an lhs_i repeated expansion is guaranteed to find an rhs_i as long there is one that is less than (in the ordering $>$) lhs_i. If there is no such implementation then the strategy may not terminate. In such a case the user would have to interrupt the strategy himself.

4.2.2. Strategies for the Inductive Theory

When the equational strategy is unsuccessful, we end up in a partial implementation that defines the function being implemented on a subset of the domain values. Even when successful one might wish to derive a better implementation that is not in the equational theory. Under such circumstances we switch over to an inductive strategy. The KB-Completion inference rule is the one that gives our system the ability to derive rewrite rules in the inductive theory. Inference by KB-completion, unlike the rest of the inferences, does not derive a new rewrite rule by directly modifying an existing rewrite rule in the program. The KB-completion rule only gives a condition under which a candidate rewrite rule hypothesized to be consistent with the program can be added to the program. Thus, developing inductive strategies involves finding ways to systematically hypothesize candidate rewrite rules. The h-dropping inference rule provides one way of generating a candidate rewrite rule: $\alpha \rightarrow \beta$ is obtained from an existing rule of the form $h(\alpha) \rightarrow h(\beta)$.

Another inductive strategy that is more generally applicable uses the technique of *generalization*. This strategy is related to the technique of synthesis by example [Sum75]. This strategy is based on the following fact: If $\alpha \rightarrow \beta$ is a rewrite rule in the inductive theory then every ground instance of it is in the equational theory. We pick an instance of one of the rewrite rules that belongs to the partial implementation derived by the equational strategy. The rewrite rule so chosen is then "massaged" by applying a few expansion steps until the rewrite rule obtained appears to be an instance of the desired rewrite rule. The "massaged" rewrite rule is then *generalized* by replacing selected constants on either side of the rewrite rule by appropriate variables. The generalized rule is used as the candidate rule. Steps (3) and (4) need some human intervention in this strategy. In step (3) the user has to check if the rewrite rule derived has the suitable form; if not, the expansion step has to be continued further. In step (4) the user has to help the synthesis process in deciding which terms to generalize.

(1) Pick a rewrite rule $\alpha \rightarrow \beta$ (from a partial implementation derived by other means) that defines the function being synthesized on a subset of the domain values.

(2) Simplify β to its irreducible form (say γ).

(3) Apply the expansion step (described above) starting with the rewrite rule $\alpha \rightarrow \gamma$. Let the outcome of this step be $\alpha \rightarrow \delta$.

(4) *Generalize*: The candidate rewrite rule is $\alpha_1 \rightarrow \delta_1$ such that $\sigma(\alpha_1) = \alpha$ and $\sigma(\delta_1) = \delta$, where σ is an appropriate substitution.

5. Examples

In the following we present two examples. The first is a data type synthesis example; the second one is a short example presented mainly to illustrate the advantage of expansion over folding. To keep the presentation simple, we have not shown all the steps of the strategy, but only the interesting ones.

Example 1: Synthesis of Queue in terms of Sequence

We synthesize an implementation for **queue** (specified in section 2) using **sequence** as the representation type. The representation scheme used is the same as the one described in section 4. Here we show a complete derivation of two different implementations for the operation **enqueue** only. Derivation of the first implementation uses the equational strategy while the second employs the inductive strategy. Implementations for the remaining operations of **queue** can also be derived similarly.

For the derivation of the first implementation of **ENQUEUE**, we categorize the rewrite rules of the various specifications as follows. Note that the goal here is to derive rewrite rules of the form h **(ENQUEUE(g))** $\rightarrow$ h **(t)**, where **g** is a generator term of type **sequence**, and **t** is an abitrary term. The specification set will initially consists of the homomorphism rewrite rule specifying **ENQUEUE**. We first wish to synthesize a recursive implementation that does not use any defined function symbols besides **ENQUEUE**. Hence the expansion set will consist of only the rewrite rules defining the abstraction function and the functions it is dependent on. The reduction set will include all the rewrite rules in the specification.

Derivation of a Recursive Implementation

Expansion Set

(H1) $h([\,]) \rightarrow$ **nullq**
(H2) $h(e + v) \rightarrow$ **add_at_head**$(h(e), h(v))$

(H3) **add_at_head(e, nullq) → enqueue(e, nullq)**
(H4) **add_at_head(e1, enqueue(e2, q)) → enqueue(e2, add_at_head(e1, q))**

(S1) h**(ENQUEUE(e, v)) → enqueue(**h**(e),** h**(v))**

Specification Set

(S1) h**(ENQUEUE(e, v)) → enqueue(**h**(e),** h**(v))**

Reduction Set = Expansion Set ∪ Specification Set

h**(ENQUEUE(e, [ ]))**

→ **enqueue(**h**(e),** h**([ ]))**	*Derived pair of (S1) on (H1)*
→ **enqueue(**h**(e), nullq)**	*Simplify using (H1)*
→ **add_at_head(**h**(e), nullq)**	*Expansion using (H3)*
→ **add_at_head(**h**(e),** h**([ ]))**	*Expansion using (H1)*
→ h**(e + [ ])**	*Expansion using (H2)*

(I1) **ENQUEUE(e, [ ]) → e + [ ]** *h-dropping*

h**(ENQUEUE(e, e1 + v1))**

→ **enqueue(**h**(e),** h**(e1 + v1))**	*Derived pair of (S1) on (H2)*
→ **enqueue(**h**(e), add_at_head(**h**(e1),** h**(v1)))**	*Simplify using (H2)*
→ **add_at_head(**h**(e1), enqueue(**h**(e),** h**(v1)))**	*Expansion using (H4)*
→ **add_at_head(**h**(e1),** h**(ENQUEUE(e, v1)))**	*Expansion using (S1)*
→ h**(e1 + ENQUEUE(e, v1))**	*Expansion using (H2)*

(I2) **ENQUEUE(e, e1 + v1) → e1 + ENQUEUE(e, v1)** *h-dropping*

Rewrite rules (I1) and (I2) form a well-formed implementation for **ENQUEUE**.

Derivation of a Nonrecursive Implementation

The second implementation of **ENQUEUE** is intended to be a nonrecursive implementation in terms of only the operations of **sequence**. Hence, the expansion set in this case will only include the rewrite rules in the specification of **sequence**. (Below, we only show the part of the expansion set that is used in the derivation.) The specification set consists of the rewrite rules (I1) and (I2), above, snce the nonrecursive implementation is derived from the recursive implementation. The reduction set consists of the rewrite rules in the specification of **sequence**. We employ the inductive strategy by picking the rewrite rule (I1) from the specifictaion set.

Specification Set

(I1) ENQUEUE(e, [ ]) → e + [ ]
(I2) ENQUEUE(e, e1 + v1) → e1 + ENQUEUE(e, v1)

Expansion Set

(7) rotate([ ]) ≡ [ ]
(8) rotate(e + [ ]) ≡ e + [ ]
(9) rotate(e1 + (e2 + v)) ≡ e2 + rotate(e1 + v)

Reduction Set = Expansion Set ∪ Specification Set

(I1) ENQUEUE(e, [ ]) → e + [ ] *Pick (I1)*
ENQUEUE(e, [ ]) → rotate(e + [ ]) *Expansion using (8)*

(G) ENQUEUE(e, s) → rotate(e + s) *Generalize [ ] to get a candidate rewrite rule*

(NR) ENQUEUE(e, s) → rotate(e + s) *Hypothesis-Testing using (G)*
(NR) is the desired nonrecursive implementation.

Example 2: Non-recursive Implementation

The goal of the synthesis in this example is to convert a recursive implementation of a function **Insert–rest** into a non-recursive one. **Reml** and **Insertall** are two functions (for which efficient implementations are assumed to exist) on **List** with generators **nil** and **cons.** **Reml** removes the last element of a list, and **Insertall** inserts the atom 1 before every element in a list. Specifications for **Insertall** (rules 4-5) and **Reml** (rules 1-3) are given below. We employ the inductive strategy to derive the new implementation.

(1) **Reml(nil) → Error**
(2) **Reml(cons(x, nil)) → nil**
(3) **Reml(cons(x, cons(y, L))) → cons(x, Reml(cons(y, L)))**

(4) **Insertall(nil) → nil**
(5) **Insertall(cons(x, L)) → cons(x, cons(1, Insertall(L)))**

Consider the function **Insert_rest** that inserts 1 before every element of a list except the first one. A recursive implementation for **Insert_rest** is:

(6) **Insert_rest(nil) → Error**
(7) **Insert_rest(cons(x, nil)) → cons(x, nil)**
(8) **Insert_rest(cons(x, cons(y, L))) → cons(x, cons(1, Insert_rest(cons(y, L))))**

We want to transform the above implementation of **Insert_rest** into one that uses **Reml** and **Insertall.** The method used is to "guess" an implementation for **Insert_rest** by trying out **Insert_rest** on a few concrete list objects. The concrete objects are determined by computing derived pairs. The guess is generated by generalizing, i.e., replacing subexpressions by variables, in the rewrite rules derived with concrete instances. The **Hypothesis–Testing** inference rule is then used to confirm that our guess is a correct implementation. In this example the expansion set consists of rewrite rules (1) through (5) (which define the functions **Reml** and **Insertall**) and nothing else because the intent is to synthesize an implementation for **Insert–rest** in terms **Reml** and **Insertall.**

Insert_rest(cons(x, nil)) → cons(x, nil)
→ cons(x, Reml(cons(z, nil))) *Derived Pair of (7) on (4)*
→ Reml(Insertall(cons(x, nil))) *Expansion using (2), (4), (5)*

(9) **Insert_rest(cons(x, nil)) → Reml(Insertall(cons(x, nil)))**

(10) **Insert_rest(L) → Reml(Insertall(L))** *Hypothesis-Testing on the rule obtained by replacing* **cons(x, nil)***by* **L** *in (9)*

Note that although this implementation involves pipelining of **Reml** and **Insertall**, it could be more efficient than the recursive implementation since **Insertall** and **Reml** are assumed to be primitive operations.

6. References

[BuD77] R. M. Burstall and J. Darlington, "A Transformation System for Developing Recursive Programs", *Journal of the Association for Computing Machinery*, **24**, 1 (January 1977), 44-67.

[Dar82] J. Darlington, "Program Transformation", in *Functional Programming and its Applications, An advanced course*, J. D. al, (ed.), Cambridge University Press, 1982, 193-209.

[Der82] N. Dershowitz, "Orderings for Term Rewriting Systems", *J.TCS*, **17**, 3 (1982), 279-301.

[DHJ83] N. Dershowitz, J. Hsiang, N. Josephson and D. Plaisted, "Associative-Commutative Rewriting", in *Proc. 8th IJCAI*, Karlsruhe, Germany, 1983.

[Fea82] M. S. Feather, "A System for Assisting Program Transformation", *Transactions on Programming Languages and Systems*, **4**, 1 (January 1982), .

[GoT79] J. A. Goguen and J. Tardo, "An Introduction to OBJ-T", in *Specification of Reliable Software*, IEEE, 1979.

[GTW8.] J. A. Goguen, J. W. Thatcher and E. G. Wagner, "Initial Algebra Approach to the Specification, Correctness, and Implementation of Abstract Data Types", in *Current Trends in Programming Methodology*, vol. IV Data Structuring, R. T. Yeh, (ed.), Prentice Hall (Automatic Computation Series), Englewood Cliffs, NJ, 1978..

[GuH78] J. V. Guttag and J. J. Horning, "The Algebraic Specification of Abstract Data Types", *Acta Informatica*, **10**, 1 (1978), 27-52.

[GKM82] J. V. Guttag, D. Kapur and D. R. Musser, "On Proving Uniform Termination and Restricted Termination of Rewriting Systems", in *Proc. 9th ICALP*, Aarhus, Denmark, 1982.

[Hsi82] J. Hsiang, "Topics in Automated Theorem Proving and Program Generation", UIUCDCS-R-82-1113, U. of Illinois at Urbana Champaign, Urbana Illinios, Dec. 1982.

[HuH80] G. Huet and J. M. Hullot, "Proofs by Induction in Equational Theories with Constructors", in *21st IEEE Symposium on Foundations of Computer Science*, 1980, 96-107.

[Kap80] D. K. Kapur, "Towards a Theory for Abstract Data Types,", Tech. Rep.-237, Lab. for Computer Science, MIT, Cambridge, MA 02139, May 1980.

[KaM82] D. K. Kapur and D. R. Musser, "Rewrite Rule Theory and Abstract Data Type Analysis", in *Computer Algebra, EUROSAM 1982, Lecture Notes in Computer Science 144*, Calmet, (ed.), Springer Verlag, April 1982, 77-90.

[KaS84] D. Kapur and M. K. Srivas, "A Rewrite Rule Based Approach for Synthesizing Abstract Data Types", Tech. Rep. 84/080, Dept. of Computer Science, SUNY at Stony Brook, Stony Brook, NY 11794, July 1984.

[KnB70] D. E. Knuth and P. B. Bendix, "Simple Word Problems in Universal Algebras", in *Computational Algebra*, J. Leach, (ed.), Pergamon Press, 1970, 263-297.

[Kot82] L. Kott, "Unfold/Fold Program Transformations", Research Report No. 155, INRIA, Le Chesnay, France, August 1982.

[LaB77] D. S. Lankford and A. M. Ballantyne, "Decision Procedure for Simple Equational Theories with Commutative-Associative Axioms", Report ATP-39, Univ. of TExas at Austin, 1977.

[LaB79] D. S. Lankford and A. M. Ballantyne, "The Refutation Completeness of Blocked Permutative Narrowing and Resolution", in *4th Conf. on Automated Deduction*, Austin, TX, 1979.

[Lan81] D. S. Lankford, "A Simple Explanation of Inductionless Induction", MTP-14, Louisiana Tech Univ., 1981.

[Les83] P. Lescanne, "Computer Experiments with the REVE Term Rewriting System Generator", in *10th Annual Symposium on Principles of Prgoramming Languages*, Austin, Texas, January 1983.

[MaW80] Z. Manna and R. Waldinger, "A Deductive Approach to Program Synthesis", *ACM Trans. Prog. Lang. and Systems*, **2**, 1 (January 1980), 90-121.

[Mus80a] D. R. Musser, "Abstract Data Types in the AFFIRM System", *Trans. on Software Eng.*, **1**(6), (Jan. 1980), , IEEE.

[Mus80b] D. R. Musser, "On Proving Inductive Properties of Abstract Data Types", in *Conference record of the Seventh Annual ACM Symposium on Principles of Programming Languages*, Las Vegas, Nevada, January 1980, 154-162.

[PeS81] G. E. Peterson and M. E. Stickel, "Complete Sets of Reductions for Some Equational Theories", *J. ACM*, **28**, (1981), 233-264.

[Sri82] M. K. Srivas, "Automatic Synthesis of Implementations for Abstract Data Types from Algebraic Specifications", MIT/LCS/Tech. Rep.-276, Laboratory for Computer Science, MIT, June 1982.

[Sti81] M. E. Stickel, "A Unification Algorithm for Associative-Commutative Functions", *J. ACM*, **28**, (1981), 233-264.

"Delayability" in Proofs of Strong Normalizability in the Typed Lambda Calculus

Michael Karr*
Software Options, Inc.
22 Hilliard Street
Cambridge, Mass 02138

1. Introduction

We consider here the interaction of certain reductions with the β-reduction $(\lambda x.M)N \to [N/x]M$ (substitute N for occurrences of x in M). Given a term in the typed λ-calculus, any sequence of β-reductions must eventually terminate; this is referred to as the *strong normalizability* of the term, or in this paper, simply *normalizability*. It is often useful to consider other reductions in addition to the β-reduction. The most classical example is

$\lambda x.Mx \to M$ when x is not free in M (η-reduction)

Other reductions arise when considering terms extended with the constants left, right and pair, and the reductions:

left (pair x y) $\to x$ (L-reduction)

right (pair x y) $\to y$ (R-reduction)

pair (left z) (right z) $\to z$ (P-reduction)

Most works on the typed lambda calculus consider normalizability for only the β- and η-reductions, for example, [1] and [6]. Using a standard technique for proving strong normalizability [8], it is not difficult to show that applying η-, L-, and R- reductions in addition to β-reduction still leaves all terms strongly normalizable. It is more difficult to show this when P-reductions are added to the set [2], [4]. It is also possible to handle these reductions by generalization of the notion of type [7]. A different approach to proving normalizability has been applied to the L- and R-reductions [3]. The goal here is to have a general theory that allows us to adjoin new reductions to the β-reduction, and by simple structural examination of the new reductions, ascertain that terms remain normalizable in the new system. To do this, we have developed a new strategy for proving strong normalizability, which as a special case yields an independent proof of normalizability of terms under β-reduction. While we use the pairing reductions to exemplify the general results, the emphasis is on clarifying the properties of reductions that allow the proof of normalizability.

In the next section, we develop a general technique for proving that a union of well-founded relations is well-founded, given original relations that are well-founded and whose interaction obeys certain conditions. The third section gives a precise meaning to the term "reduction", and provides results that are helpful in proving the conditions required for the well-foundedness results. In the following section, we show that a broad class of reductions, including the union of the β-, L-, and R-reductions, has the normalizability property. The final section considers reductions having other properties; this class includes the η- and P-reductions.

---

*Part of this work was done at Harvard University, where the author was supported by an IBM graduate fellowship.

2. Well-foundedness of a union of relations

2.1. Delayability

Let $\rightarrow_i$ denote a relation, $\twoheadrightarrow_i$ its reflexive transitive closure, and $\underset{\neq}{\twoheadrightarrow}_i$ its transitive closure. We shall investigate the consequences of the following condition on relations.

Definition A relation $\rightarrow_0$ *is delayable past* $\rightarrow_1$ via $\rightarrow_2 \overset{\text{def}}{\Leftrightarrow}$

$$M \rightarrow_0 M' \rightarrow_1 M'' \Rightarrow \exists N', N'' : M \underset{\neq}{\twoheadrightarrow}_1 N' \twoheadrightarrow_0 N'' \text{ and } M'' \twoheadrightarrow_2 N''$$

□

The following definition is standard.

Definition A pair of relations $\rightarrow_0, \rightarrow_1$ is *commutative* $\overset{\text{def}}{\Leftrightarrow}$ $M \rightarrow_i M_i,\ i = 0 \text{ and } 1 \Rightarrow \exists M_2 : M_i \rightarrow_{1-i} M_2$

□

Lemma 1 Suppose that

$\rightarrow_0$ is delayable past $\rightarrow_1$ via $\rightarrow_2$,

$\twoheadrightarrow_i, \twoheadrightarrow_2$ is commutative for both $i = 0$ and 1,

$\underset{\neq}{\twoheadrightarrow}_k, \twoheadrightarrow_2$ is commutative for either $k = 0$ or 1,

$\rightarrow_i$ is well-founded, $i = 0$ and 1.

Then $\rightarrow_{01} \triangleq \rightarrow_0 \cup \rightarrow_1$ is well-founded.

Proof Suppose we have an infinite path in $\rightarrow_{01}$. Because the $\rightarrow_i$ are individually well-founded, there are an infinity of alternations between $\rightarrow_0$ steps and $\rightarrow_1$ steps. We may write the path in the form:

$$M_0 \twoheadrightarrow_1 M_1 \underset{\neq}{\twoheadrightarrow}_0 M_2 \underset{\neq}{\twoheadrightarrow}_1 M_3 \underset{\neq}{\twoheadrightarrow}_0 \cdots$$

The crux of the proof is that we can shorten $M_1 \underset{\neq}{\twoheadrightarrow}_0 M_2$ by one step, and obtain another infinite sequence. When there is only one $\rightarrow_0$ step between M_1 and M_2, and it is shortened, we have extended $M_0 \twoheadrightarrow_1 M_1$ by one more $\rightarrow_1$ step. The ability to do this indefinitely denies the well-foundedness of $\rightarrow_1$.

Look at the term N_1 just preceding M_2 in the sequence, and the term N_2 just following M_2. Using delayability:

$$N_1 \rightarrow_0 M_2 \rightarrow_1 N_2 \Rightarrow \exists P_1, P_2 : N_1 \underset{\neq}{\twoheadrightarrow}_1 P_1 \twoheadrightarrow_0 P_2 \text{ and } N_2 \twoheadrightarrow_2 P_2$$

Then by using the commutativity conditions starting with $N_2 \rightarrow_1 M_3$ and $N_2 \rightarrow_1 P_2$, we can get a new infinite sequence by extending the "ladder" one step at a time:

$$\begin{array}{ccccccccccccccccc}
M_0 & \twoheadrightarrow_1 & M_1 & \twoheadrightarrow_0 & N_1 & \rightarrow_0 & M_2 & \rightarrow_1 & N_2 & \twoheadrightarrow_1 & M_3 & \underset{\neq}{\twoheadrightarrow}_0 & M_4 & \underset{\neq}{\twoheadrightarrow}_1 & \cdots \\
\| & & \| & & \| & & & & \downarrow & & \downarrow & & \downarrow & & \\
M_0 & \twoheadrightarrow_1 & M_1 & \twoheadrightarrow_0 & N_1 & \underset{\neq}{\twoheadrightarrow}_1 & P_1 & \twoheadrightarrow_0 & P_2 & \twoheadrightarrow_1 & P_3 & \twoheadrightarrow_0 & P_4 & \twoheadrightarrow_1 & \cdots
\end{array}$$

where the relations $\downarrow$ are $\twoheadrightarrow_2$. By positive commutativity, $P_{2i+k-1} \neq P_{2i+k}$ and so the new sequence is indeed not eventually constant. As we claimed, either the first set of $\rightarrow_0$ steps has been shortened, or the first set of $\rightarrow_1$ steps has been lengthened (perhaps from zero to one).

□

Corollary 1 Suppose that $\rightarrow_0$ is delayable past $\rightarrow_1$ via $=$, and that $\rightarrow_i$ is well-founded for $i = 0$ and 1. Then $\rightarrow_0 \cup \rightarrow_1$ is well-founded.

Proof Apply the Theorem, because $\rightarrow, =$ is commutative for any relation $\rightarrow$.

□

The following result is useful in establishing the delayability properties of a union of relations (simple proofs are left to the reader).

Proposition 2.1 If $\rightarrow_0$ and $\rightarrow_1$ are delayable past $\rightarrow_2$ via $\rightarrow_3$ for $i = 0$ and 1, then $\rightarrow_0 \cup \rightarrow_1$ is delayable past $\rightarrow_2$ via $\rightarrow_3$.

□

2.2. Weak Commutativity

When $\rightarrow_1$ is not equality, the commutativity of $\twoheadrightarrow_{\neq 0}$, $\twoheadrightarrow_1$ may be proved by using a property of the relations $\rightarrow_0$, $\rightarrow_1$.

Definition An (ordered) pair of relations $\rightarrow_0$, $\rightarrow_1$ is *positive weak commutative* $\overset{\text{def}}{\Leftrightarrow}$

$$M \rightarrow_i M_i,\ i = 0 \text{ and } 1 \Rightarrow \exists M_2 : M_1 \twoheadrightarrow_{\neq 0} M_2 \text{ and } M_0 \twoheadrightarrow_1 M_2$$

(*Weak commutativity* is the same except that $\twoheadrightarrow_{\neq 0}$ is replaced by $\twoheadrightarrow_0$.)

□

In addition to this property that involves the pair, it is convenient to work with a relation that has the following property.

Definition A relation $\rightarrow$ is *bounded* $\overset{\text{def}}{\Leftrightarrow}$ $\exists \nu\ \forall M$: any $\rightarrow$ path from M has at most $\nu(M)$ steps.

□

Lemma 2 Suppose that

$\rightarrow_0$, $\rightarrow_1$ is positive weak commutative.

$\rightarrow_0$ is bounded.

Then $\twoheadrightarrow_{\neq 0}$, $\twoheadrightarrow_1$ is commutative.

Proof Proofs omitted in this paper that appear in the full paper [5] are indicated by ○.

○

Corollary 2 If $\rightarrow,\rightarrow$ is weak commutative and $\rightarrow$ is bounded, then $\twoheadrightarrow,\twoheadrightarrow$ is commutative.

○

These results show that the issue of commutativity of a transitive closure reduces to the issue of (positive) weak commutativity and boundedness. Proofs of (positive) weak commutativity are often aided by:

Proposition 2.2 If $\rightarrow_0,\rightarrow_1$ and $\rightarrow_0,\rightarrow_2$ are (positive) weak commutative, then $\rightarrow_0,\rightarrow_1 \cup \rightarrow_2$ is (positive) weak commutative. If $\rightarrow_0,\rightarrow_2$ and $\rightarrow_1,\rightarrow_2$ are (positive) weak commutative, then $\rightarrow_0 \cup \rightarrow_1,\rightarrow_2$ is (positive) weak commutative.

□

Boundedness is often proved using the following property of a relation.

Definition A relation $\rightarrow$ *has finite degree* $\overset{\text{def}}{\Leftrightarrow}$ $\forall M$, there are a finite number of N such that $M \rightarrow N$.

□

Proposition 2.3 If a relation is well-founded and has finite degree, then it is bounded.

Proof This is equivalent to the well-known result that an infinite tree with finite branching has an infinite path.

□

Proposition 2.4 If $\rightarrow_0$ and $\rightarrow_1$ have finite degree, then so does $\rightarrow_0 \cup \rightarrow_1$.

□

2.3. Summary

The results of the rest of this paper will be concerned with proving the hypotheses of the result below, for various families of relations.

Definition A family $\{\to_i\}_{i\in I}$ of relations *has finite application* $\overset{\text{def}}{\Leftrightarrow}$

$$\forall M\ \exists \text{ a finite set } I_M : \text{in any path beginning at } M, \text{ each step } \to_i \text{ has } i \in I_M.$$

□

Proposition 2.5 If $\{\to_i\}_{i\in I}$ has finite application and each $\to_i$ has finite degree, then $\to_\infty$ has finite degree.

□

Lemma 3 Let $\{\to_i\}_{i\in I}$ be a family of relations over a well-ordered index set I, and let $\to_\infty \triangleq \cup_i \to_i$. Suppose, with cases a) and b):

$\{\to_i\}_{i\in I}$ has finite application.

$\forall i : \to_i$ is well-founded.

$\forall i : \to_i$ has finite degree.

Case a) $\forall i<j : \to_i$ is delayable past $\to_j$ via =.

Case b) $\forall i<j : \to_i$ is delayable past $\to_j$ via $\to_j$.
$\forall i<j : \to_i, \to_j$ is positive weak commutative.
$\forall i : \to_i, \to_i$ is weak commutative.

Then $\to_\infty$ is well-founded.

○

Definition A relation $\to$ is *Church-Rosser* $\overset{\text{def}}{\Leftrightarrow}$ $\to,\to$ is commutative.

□

Corollary 3 Let $\{\to_i\}_{i\in I}$ be as in Lemma 3, and suppose that in addition to previous hypotheses, we also assume that $\to_i, \to_j$ is weak commutative, $\forall\ i,j$. Then $\twoheadrightarrow_\infty$ is Church-Rosser.

○

Because of this result, we shall also be able to obtain general results for Church-Rosser properties of β-reduction adjoined by other reductions. Again, the emphasis will be on finding simple structural characteristics of the rules which guarantee the property.

3. Reductions

In this section, we will give a technical meaning to the term "reduction", as a type of relation on expressions in the typed λ-calculus. In the first subsection we discuss congruence closure; in particular, we are interested in proving delayability and weak commutativity properties of a relation from related properties on simpler relations. The next subsection discusses substitution and the third uses substitution and congruence closure to define reduction. The final subsection considers a situation in which a delayability condition is easily proved.

3.1. Congruence Closure

We shall work with relations $\to_0$ that have been defined beginning with a "top-level reduction" $\to^0$. Examples of top-level reduction are the β-, η-, and pairing reductions mentioned in the introduction. These relations may extended over expressions as follows:

Definition Given a relation $\to^0$ on expressions, the *congruence closure* $\to_0$ of $\to^0$ is defined inductively on terms, by $M \to_0 N \overset{\text{def}}{\Leftrightarrow}$

- $M \to^0 N$, or

- $M = M_1M_2$, $N = N_1N_2$, $M_i \to_0 N_i$ and $M_j = N_j$ for $i \neq j$, or

- $M = \lambda x.M_0$, $N = \lambda x.N_0$ and $M_0 \to_0 N_0$.

□

Proposition 3.1 The congruence closure of the (reflexive) transitive closure is the same as the (reflexive) transitive closure of the congruence closure.

□

Proposition 3.2 The congruence closure of $\to^0 \cup \to^1$ is $\to_0 \cup \to_1$.

□

By $\twoheadrightarrow^0$ we mean the reflexive transitive closure of $\to^0$; by $\twoheadrightarrow_0$, we mean the relation referred to by the above proposition. The notation $\to_{\bar{0}}$ means $\to_0 - \to^0$. As always, we are concerned with unions, delayability, commutativity, and finite degree.

Proposition 3.3 If $\to^0$ is delayable past $\to_1$ via $\to_2$ and $\to_{\bar{0}}$ is delayable past $\to^1$ via $\to_2$, then $\to_0$ is delayable past $\to_1$ via $\to_2$.

□

Proposition 3.4 If $\to^0,\to_1$ and $\to_{\bar{0}},\to^1$ are (positive) weak commutative, then $\to_0,\to_1$ is (positive) weak commutative.

□

Proposition 3.5 If $\to^0$ has finite degree, so does $\to_0$.

□

The proofs of all of these are by induction on terms.

3.2. Substitution

The definitions in this subsection will be used throughout this paper.

Definition A *substitution* s is a function mapping a finite set of variables to expressions, preserving type.

□

Thus, we may speak of Dom(s) and Range(s). It is often convenient to denote a function using specific variables and expressions. We use the notations

$$[M/x] \triangleq \{\langle x,M\rangle\},\ [M_i/x_i] \triangleq \{\langle x_i,M_i\rangle\}_i, \text{ etc.}$$

We also use the set operations on the functions, such as $\cup$ and $-$. We often need to remove a point from a domain, for which we use the notation $s{-}x \triangleq s{-}\{\langle x,sx\rangle\}$.

We denote the free variables of an expression M by FV(M), and extend this notation to FV(s), meaning the union of free variables of elements of Range(s). This is useful in defining how a substitution acts on an expresion:

Definition Inductively on terms:

$sc \triangleq c$	for constants
$sx \triangleq sx$ if $x \in \text{Dom}(s)$ $\quad x$ otherwise	for variables

$s(M_1M_2) \triangleq (sM_1)(sM_2)$ for applications

$s(\lambda x.M_0) \triangleq \lambda y.((s{-}x \cup [y/x])M_0)$ for abstractions

□

For variables $x \in \mathrm{Dom}(s)$, it is not clear whether sx refers to the result of applying s as a function or of doing the substitution. But the result is the same in both cases and the ambiguity is inconsequential. Note that substitution into abstractions is somewhat arbitrary. We view it as being well-defined only up to $=_\alpha$, the equivalence relation obtained by allowing renaming of bound variables. When we use "=" for expressions, we implicitly mean $=_\alpha$. Using these definitions and notation, top-level β-reduction is correctly defined by the rule:

$$(\lambda x.M)N \to^\beta [N/x]M$$

We shall view β-reduction in a somewhat different way in the next subsection.

3.3. Substitutive Closure

In this subsection, we formalize "reduction". We start with the following notion.

Definition A *reduction set* $\mathfrak{R}$ is a set of pairs of expressions $\langle P,Q \rangle$ where for every such pair in $\mathfrak{R}$:

$$\mathrm{type}(P) = \mathrm{type}(Q) \text{ and } \mathrm{FV}(P) \supseteq \mathrm{FV}(Q).$$

□

For reasons that will become apparent, when $\langle P,Q \rangle \in \mathfrak{R}$, P is called a *pattern* and Q is a *replacement*.

We may view $\mathfrak{R}$ as a relation on expressions. The first step is to get a new relation from it.

Definition Given a reduction set $\mathfrak{R}$, the *substitutive closure* of $\mathfrak{R}$, $\to^{\mathfrak{R}}$ is given by

$$M \to_{\mathfrak{R}} N \overset{\mathrm{def}}{\Leftrightarrow} \exists \text{ a substitution } s \text{ and } \langle P,Q \rangle \in \mathfrak{R} : \mathrm{Dom}(s) = \mathrm{FV}(P),\ M = sP, \text{ and } N = sQ$$

□

Any $\to^{\mathfrak{R}}$ defined in this way is by definition a top-level reduction.

Reduction sets are often parameterized by type. For example, let:

$$\mathfrak{R}_\eta \triangleq \{ \langle \lambda x.qx,\ q \rangle : q \text{ is a variable of type } \sigma,\ \sigma \text{ any type} \} \text{ and } \to^\eta \triangleq \to^{\mathfrak{R}_\eta}$$

(In general, a subscript on a reduction set will be used as a superscript on $\to$ to indicate the corresponding top-level reduction.) The beauty of this approach to the definition of reductions is that the usual side condition for η-reduction, "$x \notin \mathrm{FV}(q)$", is not necessary because substitution renames bound variables to avoid the free variables of a term to which it is applied. Thus $\to^\eta$ is in fact the familiar top-level η-reduction. The reader may define sets $\mathfrak{R}_L$, $\mathfrak{R}_R$, and $\mathfrak{R}_P$ that generate the usual L-, R- and P-reductions.

How does β-reduction fit this scheme? The idea is to parameterize the reduction set by syntax as well as type. For example:

$$\mathfrak{R}_\beta \triangleq \{ \langle (\lambda x.M)q, [q/x]M \rangle : M \text{ is an expression, } q, x \text{ are variables} \}$$

The reader may verify that this corresponds to the definition of $\to^\beta$ given in the previous section. Aside from the unified approach, the reason for such a view of β-reduction will become more apparent when we consider more subtle versions of $\mathfrak{R}_\beta$.

The motivation for the terminology of this section is the following definition and result.

Definition A relation is *substitutive* $\overset{\mathrm{def}}{\Leftrightarrow}$ $M \to N \Rightarrow sM \to sN$, for any substitution s.

□

Proposition 3.6 Given $\mathfrak{R}$, $\to^{\mathfrak{R}}$ is substitutive. If $\mathfrak{R}$ is substitutive, $\to^{\mathfrak{R}} = \mathfrak{R}$.

□

Finally, a *reduction* is a relation that is the congruence closure of a top-level reduction. Reductions are thus substitutive:

Proposition 3.7 If $\to^0$ is substitutive, so is $\to_0$.

□

3.4. Trivial Replacements

One of the conditions for Proposition 3.3 holds for all of the reductions in this paper.

Definition An expression is *trivial* $\overset{\text{def}}{\Leftrightarrow}$ it is a variable.

□

The reduction sets $\mathfrak{R}_\eta$, $\mathfrak{R}_L$, $\mathfrak{R}_R$, $\mathfrak{R}_P$ all have trivial replacements. This property enables a simple proof of one of the delayability conditions.

Proposition 3.8 Let $\mathfrak{R}_0$ have trivial replacements, and $\mathfrak{R}_1$ be arbitrary. Then $\to^0$ can be delayed past $\to_1$ via $=$.

□

4. Stratification

4.1. Positive β-reduction

In this subsection we shall consider a subrelation of β-reduction, as the first and most direct application of the ideas of the previous sections. We shall give all results without proofs, because they are special cases of more general results that we will be presenting in later subsections. The results given here may be proved without much difficulty; the only problem is in defining a relation for which *positive* weak commutativity holds. The key is this:

$$\mathfrak{R}_{\beta+} \triangleq \{ \langle (\lambda x.M)q, [q/x]M \rangle \in \mathfrak{R}_\beta : x \in FV(M) \}$$

The remainder of $\to_\beta$ arises by union with $\to_{\beta 0}$, where:

$$\mathfrak{R}_{\beta 0} \triangleq \{ \langle (\lambda x.p)q, p \rangle : p, q, \text{ and } x \text{ are variables} \}$$

In later subsections, we shall consider the union with $\to_{\beta 0}$. It is useful to *stratify* $\mathfrak{R}_{\beta+}$ as follows:

$$\mathfrak{R}_{\beta+,\sigma} \triangleq \{ \langle (\lambda x.M)q, [q/x]M \rangle \in \mathfrak{R}_{\beta+} : \text{type}(\lambda x.M) = \sigma \}$$

Our strategy is to extend a partial ordering on types to a well-ordering. Specifically:

Definition Inductively on type, $\sigma \subseteq \tau \overset{\text{def}}{\Leftrightarrow}$

$$\sigma = \tau \text{ or } \tau = \tau_1 \to \tau_2 \text{ and } \sigma \subseteq \tau_i,\ i = 1 \text{ or } 2.$$

□

The well-ordering on types becomes the index set I of Lemma 3. The condition $i < j$ in the hypothesis there will translate into $\sigma \not\supseteq \tau$ here, so that any linearization of the partial ordering will suffice.

In the following summary of results, we shall use the convention $\to_\sigma \triangleq \to_{\beta+,\sigma}$ and $\to^\sigma \triangleq \to^{\beta+,\sigma}$.

The family $\{ \to_\tau \}_\tau$ has finite application.

For any type τ, $\to_\tau$ is well-founded and has finite degree.

If $\sigma = \sigma_1 \to \sigma_2$ and $\sigma_1, \sigma_2 \neq \tau$, then $\to_\sigma$ is delayable past $\to_\tau$ via $\to_\tau$.

If $\sigma \neq \tau$, then $\to_\sigma, \to_\tau$ is positive weak commutative.

For any type τ, $\to_\tau, \to_\tau$ is weak commutative.

Since $\sigma_1\to\sigma_2 = \sigma \not\leq \tau \Rightarrow \sigma_1, \sigma_2, \sigma \neq \tau$, we can indeed apply Lemma 3, and conclude that $\to_{\beta+}$ is well-founded and Church-Rosser. The results listed above from an outline for the rest of this section.

4.2. Conjugation

Working through the proof of positive weak commutativity of $\to_\sigma, \to_\tau$ for $\sigma\neq\tau$ reveals that positivity results from the fact that no free variables are dropped by a reduction. This is not true for $\mathfrak{R}_{\beta 0}$, $\mathfrak{R}_L$, and $\mathfrak{R}_R$, and positivity fails in these cases, though we can prove weak commutativity. The trick is to augment these reductions by others, in order to get delayability and positive weak commutativity. As an example, we define $\mathfrak{R}_\gamma$ so that when $y\notin FV(M)\cup FV(Q)$ and $x\in FV(M)$:

$$(\lambda y\lambda x.M)NQ \to^{\gamma} (\lambda y.[Q/x]M)N$$

In other words, we allow the positive β-reduction without doing the $\beta 0$-reduction to expose the abstraction $\lambda x.M$. Note that this rule is sound, because it is given by

$$(\lambda y\lambda x.M)NQ \to_{\beta 0} (\lambda x.M)Q \to^{\beta 0} [Q/x]M\ (\to_{\beta 0})^{-1}\ (\lambda y.[Q/x]M)N,$$

where $^{-1}$ denotes the inverse relation. The sequence $\to_{\beta 0}\to^{\beta+}(\to_{\beta 0})^{-1}$ yields the term conjugation.

The reasoning described above, and the theory that we will develop in this section, apply much more generally than to just $\mathfrak{R}_{\beta 0}$. We will work with a reduction set $\mathfrak{R}_0$ that satisfies certain properties that we will introduce as the section progresses; $\mathfrak{R}_{\beta 0}$ will have the desired properties. The first of these is that $\mathfrak{R}_0$ must have trivial replacements, which we previously saw was true of $\mathfrak{R}_{\beta 0}$.

To understand the next property, review the above expansion of $\to^{\gamma}$. In the forward direction of $\to^{\beta 0}$, the type of the replacement was type($\lambda x.M$); in the reverse direction, the type of the replacement was type($[Q/x]M$)$\subset$type($\lambda x.M$). This causes no difficulty, because $\mathfrak{R}_{\beta 0}$ is parameterized freely over all types. The following definitions capture the essence of this observation.

Definition Given $\mathfrak{R}_0$, its *strata* are given by:

$$\mathfrak{R}_{0,\sigma} \triangleq \{ \langle P, Q\rangle \in \mathfrak{R}_0 : \text{type}(P) = \sigma \}$$

□

Definition A reduction set $\mathfrak{R}_0$ with trivial replacements is *uniform* (over types) $\overset{\text{def}}{\Leftrightarrow}$ for types σ, τ, these is a one-one correspondence between $\mathfrak{R}_{0,\sigma}$ and $\mathfrak{R}_{0,\tau}$, $\langle P_\sigma, p_\sigma \rangle \leftrightarrow \langle P_\tau, p_\tau \rangle$ such that P_σ and P_τ have the same syntactic shape and the same types for free variables other than p_σ and p_τ.

□

In order to make this notion completely rigorous, we would have to set up correspondences between the constants that appear in reductions. For example, we clearly mean that the correspondence in $\mathfrak{R}_L$ be left (pair p_σ y) $\leftrightarrow$ left (pair p_τ y). The two appearances of "left" have different type, but there is only one "left" for each pair of types σ, type(y). In $\mathfrak{R}_{\beta 0}$, all that is necessary is to change the type of "p": $(\lambda x.p_\sigma)q \leftrightarrow (\lambda x.p_\tau)q$. Since well-formedness on types requires only that type(x)=type(q), both these are legitimate typed expressions.

Throughout this section, the types will usually not be affixed to a symbol. Rather, the correspondences will be indicated by $\langle P, p\rangle \leftrightarrow \langle \hat{P}, \hat{p}\rangle$, and the types will be clear from context. In fact, if type(p) = $\sigma_1\to\sigma_2$, we will always have type($\hat{p}$) = σ_2.

Proposition 4.1 Suppose $\mathfrak{R}_0$ has trivial replacements and is uniform. Given $P \twoheadrightarrow_0 p$, and a variable $\hat{p}$. Then $\exists\hat{P}$ with the same syntactic shape as P, where $\hat{P} \twoheadrightarrow_0 \hat{p}$.

Proof We work by induction on the length of $P \twoheadrightarrow_0 p$. If this has length 0, then $P = p$, and we let $\hat{P} \triangleq \hat{p}$. Otherwise, $P \to_0 Q \twoheadrightarrow_0 p$, and by induction there is $\hat{Q} \twoheadrightarrow_0 \hat{p}$. The reduction $P \to_0 Q$ takes place at top level at some subexpression of P, this subexpression may be written in the form sP_0, where $\langle P_0, p_0\rangle \in \mathfrak{R}_0$, and Q has a subexpression of the form sp_0. Compare $\hat{Q}$ with Q. This subexpression will have the same shape, though perhaps different type, and may thus be written $\widehat{sp_0}$, where $\hat{p}_0$ has the appropriate type, $\langle \hat{P}_0, \hat{p}_0\rangle \in \mathfrak{R}_0$ (by uniformity), and $\hat{s}$ has the same shape as s. Form $\hat{P}$ from $\hat{Q}$ by replacing the subexpression $\widehat{sp_0}$ with $\hat{s}\hat{P}_0$; this has the same shape as P.

□

The $\hat{P}$ given by this Proposition may be used whenever we have $P \twoheadrightarrow_0 p$; we do not refer to this result explicitly.

Definition Given a uniform reduction set $\mathfrak{R}_0$ with trivial replacements, its *conjugate* is $\mathfrak{R}_\gamma \triangleq \cup_\sigma \mathfrak{R}_{\gamma,\sigma}$ where

$$\mathfrak{R}_{\gamma,\sigma} \triangleq \{ \langle ([\lambda x.M/p]P)q, [[q/x]M/\hat{p}]\hat{P}\rangle : P \twoheadrightarrow_0 p, x \in FV(M), \text{ and type}(p) = \sigma \}$$

□

Note that $\mathfrak{R}_{\beta+} \subseteq \mathfrak{R}_\gamma$ for any $\mathfrak{R}_0$, because $p \twoheadrightarrow_0 p$. The general soundness argument is that:

$$([\lambda x.M/p]P)q \twoheadrightarrow_0 (\lambda x.M)q \rightarrow^{\beta+} [q/x]M \; (\twoheadrightarrow_0)^{-1} \; [[q/x]M/\hat{p}]\hat{P}$$

With suitable conditions on $\mathfrak{R}_0$, the results for $\mathfrak{R}_{\beta+}$ can be generalized to $\mathfrak{R}_\gamma$, as we shall see.

For the remainder of this section, we shall let $\rightarrow^\tau$ be the top-level reduction induced by the stratum $\mathfrak{R}_{\gamma,\tau}$, and $\rightarrow_\tau$ its congruence closure. This gives rise to the family of relations $\{ \rightarrow_\tau \}_\tau$ where $\rightarrow_\infty$ for the family is $\rightarrow_\gamma$. We begin supplying the hypotheses for Lemma 3 with the tollowing two results.

Proposition 4.2 For any τ, $\{\rightarrow_\tau\}_\tau$ has finite application.

Proof Given M, $\rightarrow_\gamma$ does not produce any new types. Thus, any $\rightarrow_\tau$ steps in a derivation of M will have τ among the types of subexpressions of M, and this set is finite.

□

Proposition 4.3 For any τ, $\rightarrow_\tau$ has finite degree.

Proof By Proposition 3.5, we need prove only that $\rightarrow^\tau$ has finite degree. If $M \rightarrow^\tau M'$, then M is an application M_1M_2, and there are only a finite number of ways to write $M_1 = s([\lambda x.M/p]P)$ for varying s, M, and P, not counting choices of variable names for p and elements of Dom(s).

□

4.3. Linear Expressions

This section introduces one of the restrictions that we shall make on $\mathfrak{R}_0$, and provides two basic results.

Definition An expression P is *linear wrt* a variable $x \overset{\text{def}}{\Leftrightarrow} x$ occurs once in P; P is *linear wrt* a set of variables $\overset{\text{def}}{\Leftrightarrow} P$ is a linear wrt x, $\forall x \in S$; P is *linear wrt* a substitution $s \overset{\text{def}}{\Leftrightarrow} P$ is linear wrt Dom(s). If P is linear wrt FV(P), we say simply that P is linear.

□

We shall be interested in reduction sets with linear patterns. The reader may check that $\mathfrak{R}_L$, $\mathfrak{R}_R$, $\mathfrak{R}_\eta$ and $\mathfrak{R}_{\beta 0}$ all have linear patterns, and that $\mathfrak{R}_P$ does not. As we defined $\mathfrak{R}_\beta$ and $\mathfrak{R}_\gamma$, they do not have linear patterns. We can change this by fiat.

Definition Let $\mathfrak{R}^{\text{old}}_{\gamma,\sigma}$ and $\mathfrak{R}^{\text{old}}_\gamma$ be the previous definitions; redefine $\mathfrak{R}_\gamma \triangleq \cup_\sigma \mathfrak{R}_{\gamma,\sigma}$ where

$$\mathfrak{R}_{\gamma,\sigma} \triangleq \{ \langle P, Q\rangle \in \mathfrak{R}^{\text{old}}_{\gamma,\sigma} : P \text{ is linear} \}$$

□

For the $\mathfrak{R}_0$ in which we are interested, this redefinition makes no difference.

Proposition 4.4 If $\Re_0$ has linear patterns, the old and new definitions of $\Re_\gamma$ yield the same $\to_\gamma$.

○

The importance of the linear property is its connection with substitution. The following is the first of several results concerning reductions into or out of an expression of the form sM.

Lemma 4 Let $\Re_0$ have linear patterns and trivial replacements, and let $N \to_0 sM$ where $\mathrm{Dom}(s) \cap \mathrm{FV}(N) = \phi$ and M is linear wrt s. Then either:

$N = s'M$ where $s' \to_0 s$, or

$N = sM'$ where $M' \to_0 M$ and M' is linear wrt s.

○

For the second result, there are some preliminaries.

Definition Given substitutions s and t, $s \circ t \triangleq [s(tx)/x]_{x \in \mathrm{Dom}(t)}$.

□

Proposition 4.5 For an expression M with $\mathrm{Dom}(t) = \mathrm{FV}(M)$: $(s \circ t) = (s(t(M))$.

□

Definition P_0 is a *factor* of $P \overset{\text{def}}{\Leftrightarrow} \exists\, P_1, p : P = [P_0/p]P_1$. It is called *proper* $\overset{\text{def}}{\Leftrightarrow} P_0 \neq P$.

□

Definition An expression M is *open* $\overset{\text{def}}{\Leftrightarrow} \mathrm{FV}(M) \neq \phi$; otherwise, it is *closed*. A substitution s is *open* (resp., *closed*) $\overset{\text{def}}{\Leftrightarrow}$ each element of Range(s) is open (resp., closed).

□

Definition For an expression P:

$$T(P) \triangleq \{\, \mathrm{type}(P_0) : P_0 \text{ is a non-trivial open proper factor of } P \,\}.$$

For a substitution s:

$$T(s) \triangleq \{\, \mathrm{type}(x) : x \in \mathrm{Dom}(s) \,\}$$

□

Lemma 5 Let $sM = tP$ where s is open, M is non-trivial and P is linear, $\mathrm{Dom}(t) = \mathrm{FV}(P)$, and $T(s) \cap T(P) = \phi$. Then $\exists u : M = uP$ and $t = s \circ u$.

○

4.4. Delayability of $\to_0$

We now have enough machinery to relate $\to_0$, $\to_\beta$, and $\to_\gamma$. Suppose we can prove that $\to_\gamma$ is well-founded. In this subsection, we show how to conclude that $\to_0 \cup \to_\gamma$ is well-founded. Since $\to_\beta \subseteq \to_\gamma$, we will also have $\to_0 \cup \to_\beta$ well-founded.

Proposition 4.6 Let $\Re_0$ have linear patterns and trivial replacements. Then $\to_0$ is delayable past $\to_\sigma$ via $=$.

○

Proposition 4.7 Let $\to_0$ have linear and non-trivial patterns and trivial replacements. Assume that $\to_\gamma$ is well-founded. Then $\to_0 \cup \to_\gamma$ is well-founded.

Proof By non-trivial patterns and trivial replacements, $\to_0$ is well-founded. then use the previous result, the hypothesis for $\to_\gamma$, and Corollary 1.

□

In the remainder of the section, then, our sole concern is the well-foundedness of $\to_\gamma$.

4.5. Prime Expressions

Lemma 5 is very useful, but only indirectly. We need a way to characterize $T(P)$ and $T(s)$ so that we can prove null intersection. The following idea is sufficient for many purposes.

Definition An expression P is *prime* $\overset{\text{def}}{\Leftrightarrow}$ for every non-trivial open proper factor N of P: type(N) $\supset$ type(P).

□

(Note that P is prime $\Leftrightarrow$ ($\sigma \in T(P) \Rightarrow \sigma \supset$type($P$)).) The relations $\Re_{\beta 0}$, $\Re_{\eta}$, $\Re_L$, and $\Re_R$ are all prime; $\Re_P$ is not. As with the linearity, $\Re_\beta$ and $\Re_\gamma$ are not prime, but we may redefine them to be so. We do this after giving the basic result:

Proposition 4.8 Let $sM = tP$, where s is open, M is non-trivial and P is linear and prime. Suppose $\sigma \in T(s)$ has $\sigma \not\supset$type(P). Then $\exists u$: $M = uP$ and $t = s \circ u$.

Proof To use Lemma 5, we must verify only that $T(s) \cap T(P) = \phi$:

$$\sigma \in T(P) \text{ and } \sigma \in T(s) \Rightarrow \sigma \supset \text{type}(P) \text{ and } \sigma \not\supset \text{type}(P), \text{ contradiction.}$$

The reasons are that P is prime, and the hypothesis.

□

We once again redefine $\Re_\gamma$ and its strata.

Definition Given $\Re_0$ with trivial replacements, let $\Re^{\text{old}}_{\gamma,\sigma}$ be the previous definition of $\Re_{\gamma,\sigma}$, and redefine $\Re_\gamma \triangleq \cup_\sigma \Re_{\gamma,\sigma}$, where $\Re_{\gamma,\sigma} \triangleq \{ \langle P, Q \rangle \in \Re^{\text{old}}_{\gamma,\sigma} : P \text{ is prime} \}$ □

Proposition 4.9 Let $\Re_0$ have non-trivial, linear, and prime patterns, and trivial replacements. Then the old and new definitions of $\Re_\gamma$ give rise to the same $\to^\gamma$.

○

Throughout the remainder of this section, we will let $\to^\sigma \triangleq \to^{\gamma,\sigma}$ and $\to_\sigma \triangleq \to_\gamma{}^\sigma$. We close this subsection with a key result and one of its consequences.

Lemma 6 Let $sM \to_\tau N$ where M is linear wrt s and $\sigma \in T(s) \Rightarrow \sigma \not\geq \tau$. Then either:

$N = sM'$ where $M \to_\tau M'$, or
$N = s'M$ where $s \to_\tau s'$.

○

Corollary 4 Let $sM \to_\tau N$ where $\sigma \in T(s) \Rightarrow \sigma \not\geq \tau$. Then either:

$N = sM'$ where $M \to_\tau M'$, or
$N \twoheadrightarrow_\tau s'M$ where $s \to_\tau s'$

○

4.6. Well-foundedness

The main result here not only supplies one of the hypotheses for Lemma 3, it also is necessary in some of the commutativity and delayability results for strata of $\Re_\gamma$. In this subsection, the term *normalizable* will mean not having an infinite $\to_\tau$ derivation and will apply to both expressions and substitutions. We begin with an auxiliary result.

Proposition 4.10 Given τ, s, and M, suppose that $\sigma \in T(s) \Rightarrow \sigma \not\geq \tau$, and that s and M are normalizable. Then sM is normalizable.

○

This provides a simple proof of the desired result.

Proposition 4.11 Let $\Re_0$ be uniform with linear and prime patterns. For any type τ, $\rightarrow_\tau$ is well-founded.

Proof By a simple induction on terms argument, we may reduce the proof to showing the normalizability of an application $N = N_1N_2$, where N_1 and N_2 are normalizable, and at least one step in the derivation is at top level, i.e.:

$$N_1 \twoheadrightarrow_\tau [\lambda x.M/p]P \text{ where } x \in FV(M),\ P \twoheadrightarrow_0 p, \text{ and type}(p) = \sigma$$

$$N_2 \twoheadrightarrow_\tau Q$$

$$N_1N_2 \twoheadrightarrow_\tau ([\lambda x.M/p]P)Q \rightarrow^\tau [[Q/x]M/\hat{p}]\hat{P}$$

Since N_1 and N_2 are normalizable, so are Q, M, and P. Since $T([Q/x]) = \{\text{type}(x)\}$ and $\text{type}(x) \subset \tau$, the previous result says that $[Q/x]M$ is normalizable, and thus, so is $[[Q/x]M/\hat{p}]$. But $T([[Q/x]M/\hat{p}]) = \{\text{type}(\hat{p})\}$ and $\text{type}(\hat{p}) \subset \tau$. Further $\hat{P}$ has the same derivations as P, and is thus normalizable. Again using the previous result, $[[Q/x]M/\hat{p}]\hat{P}$ is normalizable. Thus all derivations from N lead to a normalizable expression, and N is normalizable.

□

As a direct consequence of this result and Proposition 4.6, Lemma 3 yields:

Proposition 4.12 For the same $\Re_0$ as above, and any type τ, $\rightarrow_\tau \cup \rightarrow_0$ is well-founded.

○

4.7. Irreducible Expressions

It remains to prove positive weak commutativity and delayability for the pairs $\rightarrow^\sigma, \rightarrow_\tau$ and $\rightarrow_\sigma, \rightarrow^\tau$. For the pair $\rightarrow^\sigma, \rightarrow_\tau$, primality is the key for both conditions. For the pair $\rightarrow_\sigma, \rightarrow^\tau$, we need a slightly different condition. With apologies for the nomenclature:

Definition An expression P is *irreducible* $\overset{def}{\Leftrightarrow}$ P has no bound variables, and every non-trivial subexpression P_0 of P has $\text{type}(P_0) \supseteq \text{type}(P)$.

□

Neither primality nor irreducibility implies the other; primality considers fewer subexpressions, but requires strict ordering on types. The reduction sets $\Re_{\beta 0}$, $\Re_L$ and $\Re_R$ have irreducible patterns. Those of $\Re_\eta$ are not irreducible, because $\text{type}(px) \not\supseteq \text{type}(\lambda x.px)$; we shall see later that η-reduction can be handled without stratification and conjugation.

Our goal is to show that we may once again modify the definition of $\Re_{\gamma,\sigma}$, this time adding the additional constraint that "P" be irreducible.

Proposition 4.13 Let $\Re_0$ have linear and irreducible patterns and trivial replacements. If $P \twoheadrightarrow_0 p$, then $\exists s, P_0, p_0$:

$$P = sP_0,\ P_0 \twoheadrightarrow_0 p_0, \text{ and } sp_0 = p.$$

$$P_0 \text{ is irreducible and linear wrt } s.$$

○

For the last time:

Definition Given $\Re_0$ with trivial replacements, let $\Re^{old}_{\gamma,\sigma}$ be the previous definition of $\Re_{\gamma,\sigma}$, and redefine $\Re_\gamma \triangleq \cup_\sigma \Re_{\gamma,\sigma}$, where:

$$\Re_{\gamma,\sigma} \triangleq \{ \langle ([\lambda x.M/p]P)q,\ [[q/x]M/\hat{p}]\hat{P} \rangle \in \Re^{old}_{\gamma,\sigma} : P \text{ is irreducible} \}$$

□

Proposition 4.14 The new and old definitions of $\mathfrak{R}_\gamma$ yield the same $\to^\gamma$.

○

4.8. Commutativity

We first examine the simplest of problems—how might we prove that $\to^0, \to^0$ is weak commutative? For the reductions we have been considering, this follows immediately from:

Proposition 4.15 Suppose that $\to^0$ is a function on its domain. Then $\to^0, \to^0$ is weak commutative.

□

Next, we consider ways of proving that $\to_0$ is weak commutative, for example, how might we prove this for $\to_{\beta 0}, \to_{\beta 0}$? Looking at the rule $(\lambda x.p)q \to_{\beta 0} p$, it is evident that no matter what is substituted for p or q, any non top-level $\beta 0$-reduction from the expression will occur within a substituted expression—$\lambda x.P \not\to_0$ anything, for any P. This observation motivates:

Definition A set of expressions P is *independent* $\overset{\text{def}}{\Leftrightarrow}$ $\forall P,Q \in \mathfrak{R}$ and substitutions s and t with $\mathrm{Dom}(s) = \mathrm{FV}(P)$, if M is a non-trivial proper subexpression of P, then $sM \neq tQ$.

□

It is easily checked that $\mathfrak{R}_{\beta 0} \cup \mathfrak{R}_L \cup \mathfrak{R}_R$ has independent patterns; if either $\mathfrak{R}_P$ or $\mathfrak{R}_\pi$ are adjoined to this set, independence is lost.

Proposition 4.16 If $\mathfrak{R}_0$ has linear independent patterns and if $\to^0, \to^0$ is weak commutative, then $\to_0, \to_0$ is weak commutative.

○

In our applications, $\to_0$ will be bounded, and Corollary 2 says that $\twoheadrightarrow_0, \twoheadrightarrow_0$ is commutative.

We next study the commutativity of $\twoheadrightarrow_\beta, \twoheadrightarrow_0$. This not only leads to the commutativity of $\twoheadrightarrow_{\beta+} \cup \twoheadrightarrow_0$, it also figures in the proof of delayability in the next subsection. In order to get the result, we introduce another condition.

Definition An expression is *anchored* $\overset{\text{def}}{\Leftrightarrow}$ no free variable appears as an operator (the first subexpression of an application).

□

Proposition 4.17 Given $\mathfrak{R}_0 \geq \mathfrak{R}_{\beta 0}$ with linear, irreducible, independent and anchored patterns and trivial replacements, and $\to_0, \to_0$ weak commutative. For any τ, $\to_\tau \cup \to_0, \to_0$ is weak commutative.

○

Proposition 4.18 Under the same conditions as above, and for any type τ: $(\to_\tau \cup \to_0)^*$, $\twoheadrightarrow_0$ is commutative.

Proof Use Proposition 4.12 and Corollary 2.

□

Finally, we consider the commutativity relations within $\mathfrak{R}_\gamma$.

Proposition 4.19 Given a uniform reduction set $\mathfrak{R}_0$ with linear and prime patterns. Assume that $\twoheadrightarrow_0, \twoheadrightarrow_0$ is commutative. For types σ, τ with $\sigma \not\geq \tau$, $\to^\sigma, \to_\tau$ is positive weak commutative.

○

Proposition 4.20 Given a reduction set $\mathfrak{R}_0$ with irreducible linear patterns and trivial replacements. If $\sigma \not\geq \tau$, then $\to_\sigma, \to^\tau$ is positive weak commutative.

○

4.9. Delayability

In this section we consider the delayability of $\to_\sigma$ past $\to_\tau$.

Proposition 4.21 Let $\Re_0$ have linear and prime patterns and trivial replacements. Given types σ, τ with $\sigma \not\geq \tau$, $\to^\sigma$ is delayable past $\to_\tau$ via $\to_\tau$.

○

Proposition 4.22 Let $\Re_0$ have linear and irreducible patterns and trivial replacements. Given types σ,τ with $\sigma \not\geq \tau$, $\to_\sigma$ is delayable past $\to^\tau$ via =.

○

4.10. Summary

We now have all the pieces for applying Lemma 3.b. We state the following only to summarize the results.

Theorem 1 Given $\Re_0$, assume:

> The patterns of $\Re_0$ are linear, prime, irreducible, anchored, and form an independent set.
>
> $\Re_0$ has trivial replacements.
>
> $\Re_{\beta 0} \subseteq \Re_0$.

Then $\to_\beta \cup \to_0$ is well-founded and its reflexive transitive closure is Church-Rosser.

□

All of the conditions on the patterns of $\Re_0$, except for independence, are easily checked by looking at individual rules; independence is generally not difficult to verify. Since $\Re_{\beta 0}$ meets all the requirements, this result gives a new proof of the well-foundedness and Church-Rosser properties of $\to_\beta$.

5. Delayability past β-reduction

In this section we consider reductions $\to_0$ that are delayable past β-reduction, and more generally, past reductions $\to_1$ that can be proved well-founded by the methods of the previous section. Examples of $\to_0$ are $\to_\eta$ and $\to_P$. In each case, the delayability of $\to^0$ past $\to_1$ via = follows from Proposition 3.8. By Proposition 3.3, all that we need consider in this section is the delayability of $\to_{\bar{0}}$ past $\to^1$ via =.

5.1. η-reduction

The delayability of $\to_\eta$ past $\to_\beta$ via = is essentially classical; it is called "postponement" in the literature, for example, see [1], p. 382. Let $\to_1 = \to_\beta \cup \to_2$ where $\to_2$ arises from $\Re_2$. We seek a simple condition on the patterns of $\Re_2$ that guarantees the delayability of $\to_\eta$ past $\to_1$.

Definition An expresssion P is *η-delayable* $\overset{\text{def}}{\Leftrightarrow}$ every non-trivial subexpression in operand position has a non-arrow type.

□

Proposition 5.1 Let $\Re_2$ have linear and η-delayable patterns, and $\to_1 \triangleq \to_\beta \cup \to_2$. Then $\to_{\bar{\eta}}$ is delayable past $\to_\beta \cup \to^2$ via =.

○

In our example, $\to_2 = \to_L \cup \to_R$; the only non-trivial subexpression in operand position of a pattern of $\Re_L \cup \Re_R$ is of the form "pair x y", and has type $\sigma\times\tau$, not an arrow type. Hence, we conclude that $\to_\beta \cup \to_L \cup \to_R \cup \to_\eta$ is well-founded.

5.2. λ-free Replacements

The delayability of $\to_P$ past $\to_\beta$ follows from the fact that $\mathfrak{R}_P$ has replacements with the following property.

Definition An expression Q is *λ-free* $\overset{\text{def}}{\Leftrightarrow}$ for every substitution s, sQ is not an abstraction.

□

Proposition 5.2 Let $\mathfrak{R}_0$ have λ-free replacements. Then $\to_{\bar{0}}$ is delayable past $\to^\beta$.

○

This result shows that $\to_P$ can be delayed past $\to_\beta$—the "Q" in a P-reduction has type $\sigma\times\tau$, not the type of an abstraction. Since $\to_\eta$ is also delayable past $\to_\beta$, we have $\to_\eta \cup \to_P$ delayable past $\to_\beta$. The reduction $\to_\eta \cup \to_P$ reduces the size of the expression, and so it is well-founded. This means that $\to_\beta \cup \to_\eta \cup \to_P$ is well-founded. The proposition of this subsection is also useful when $\mathfrak{R}$ has replacements that are applications (which are λ-free regardless of type), but to get delayability of $\to_0$ past $\to_\beta$, we can no longer rely on Proposition 3.8.

5.3. Pairing Interaction

We still have not shown that $\to_\beta \cup \to_\eta$, together with all three pairing reductions, is well-founded. To show this all we must prove is:

Proposition 5.3 $\to_{\bar{P}}$ is delayable past $\to^L \cup \to^R$ via $=$.

Proof Consider $N \to_{\bar{P}}$ left (pair M_1 M_2) $\to^L M_1$. The only interesting case is that $\to_{\bar{P}}$ might supply "pair M_1 M_2". But then $\to^L$ leads to the same expression as $\to_{\bar{P}}$:

$$N = \text{left (pair (left (pair } M_1\ M_2\text{)) (right (pair } M_1\ M_2\text{)))} \to^L \text{left (pair } M_1\ M_2\text{)} \to^L M_1$$

The $\to_P$ reduction simply disappears, and $\to_{\bar{P}}$ is delayable past $\to^L$. Symmetrically, we can prove that it is delayable past $\to^R$, and by Proposition 2.1, $\to_P$ is delayable past $\to^L \cup \to^R$ via $=$.

□

Thus $\to_P$ is delayable past $\to_\beta \cup \to_L \cup \to_R$ via $=$. Since $\to_\eta$ is also, so is $\to_P \cup \to_\eta$, and this relation is well-founded. This finally proves the well-foundedness of $\to_\beta \cup \to_L \cup \to_R \cup \to_P \cup \to_\eta$.

Bibliography

1. Barendregt, H.P. *The Lambda Calculus.* North Holland, 1981.

2. Bercovici, I. Tait's method and strong normalizability of pairing rules. Unpublished

3. Gandy, R.O. Proofs of Strong Normalization. In J.P. Seldin and J.R. Hindley, editors, *To H.B. Curry: Essays on Combinatory Logic, Lambda Calculus and Formalism,* pages 457-477. Academic Press, 1980.

4. Karr, M. Tait's method and surjective pairing. Unpublished

5. Karr, M. "Delayability" in Proofs of Strong Normalizability in the Typed Lambda Calculus. Unpublished

6. Klop, J.W. *Combinatory Reduction Systems.* Mathematisch Centrum Amsterdam, 1980.

7. Nederpelt, R.P. *Strong Normalization in a Typed λ-Calculus with Lambda Structured Types.* Ph.D. Th., The University of Technology, Eindhoven, 1973.

8. Tait, W.W. Intensional Interpretations of Functionals of Finite Type I. *J. of Symbolic Logic 32* (1967), 198-212.

BISIMULATIONS AND ABSTRACTION HOMOMORPHISMS

Ilaria Castellani*

Computer Science Department

University of Edinburgh

Abstract

In this paper we show that the notion of bisimulation for a class of labelled transition systems (the class of *nondeterministic processes*) may be restated as one of "reducibility to a same system" via a simple reduction relation. The reduction relation is proven to enjoy some desirable properties, notably a Church-Rosser property. We also show that, when restricted to finite nondeterministic processes, the relation yields unique minimal forms for processes and can be characterised algebraically by a set of reduction rules.

1. Introduction

Labelled transition systems [K,P] are generally recognised as an appropriate model for nondeterministic computations. The motivation for studying such computations stems from the increasing interest in concurrent programming.

When modelling communication between concurrent programs, some basic difficulties have to be faced. A concurrent program is inherently part of a larger environment, with which it interacts *in the course of* its computation. Therefore a simple input-output function is not an adequate model for such a program. The model should retain some information about the internal states of a program, so as to be able to express the program's behaviour in any interacting environment. Also, *nondeterminacy* arises when abstracting from such parameters as the relative speeds of concurrent programs: as a consequence, we need to regard any single concurrent program as being itself nondeterministic.

The question is then to find a model for nondeterministic programs that somehow accounts for intermediate states. On the other hand, only those intermediate states should be considered which are relevant to the "interactive" (or *external*) behaviour of the program. Now one can think of various criteria for selecting such significant states.

In this respect labelled transition systems provide a very flexible model: by varying the definition of the transition relation one obtains a whole range of different descriptions, going from a full account of the structure of a program to some more interesting "abstract" descriptions. However, even these abstract descriptions still need to be factored by equivalence relations (for a review see [B] or [DeN]).

A natural notion of equivalence, *bisimulation equivalence*, has been recently proposed by D. Park [Pa] for transition systems: informally speaking, two systems are said to *bisimulate* each other if a full correspondence can be established between their sets of

*Supported by a scholarship from the Consiglio Nazionale delle Ricerche (Italy)

states in such a way that from any two corresponding states the two (sub)systems will still bisimulate each other.

In this paper we show that the notion of bisimulation for a class of labelled transition systems (the class of *nondeterministic processes*) may be restated as one of "reducibility to a same system" via a simple reduction relation. The reduction relation is proven to enjoy some desirable properties, notably a Church-Rosser property. We also show that, when restricted to finite nondeterministic processes, the relation yields unique minimal forms for processes and can be characterised algebraically by a set of reduction rules.

The paper is organised as follows. In section 2 we present our computational model, the class of *nondeterministic processes*. In section 3 we argue that this basic model is not abstract enough, particularly when systems are allowed *unobservable* transitions as well as observable ones. We therefore introduce *abstraction homomorphisms* [CFM] as a means of simplifying the structure of a process by merging together some of its states: the result is a process with a simpler description, but "abstractly equivalent" to the original one. We can then infer a *reduction relation* between processes from the existence of abstraction homomorphisms between them. We prove some significant properties of this relation, such as invariance in contexts and the announced Church-Rosser property. Based on the reduction relation, we define an *abstraction equivalence* relation on processes: two processes are equivalent iff they are both reducible to a same (simpler) process.

In sections 4 and 5 we study the relationship between our notions of reduction and abstraction and the notion of *bisimulation* between transition systems. The criterion we use for identifying states of a process via abstraction homomorphisms is similar to the one underlying the definition of bisimulation: we show in fact that any abstraction homomorphism is a *single-valued bisimulation*. We finally prove that the abstraction equivalence is substitutive in contexts and that it coincides with the *largest* (*substitutive*) *bisimulation*. Our equivalence can then be regarded as a simple alternative formulation for bisimulation equivalence.

In section 6 we consider a small *language* for defining *finite* nondeterministic processes: essentially a subset of R. Milner's CCS (Calculus of Communicating Systems) [M1]. We find that our results combine neatly with some established facts about the language. On this language our equivalence is just Milner's *observational congruence*, for which a complete finite axiomatisation has been given in [HM]. So, on the one hand, we get a ready-made algebraic characterisation for the abstraction equivalence; on the other hand, our characterisation proves helpful in working out a complete system of *reduction rules* for that language. We conclude by proposing a denotational *tree-model* for the language, which is isomorphic to the term-model in [HM].

Most of the results will be stated without proof. For the proofs we refer to the complete version of the paper [C].

2. Nondeterministic Systems

In this section we introduce our basic computational model, the class of *nondeterministic systems*. Nondeterministic systems are essentially labelled transition systems with an initial state.

Let A be a set of elementary *actions* or *transitions*, containing a distinguished symbol τ which denotes a hidden or *unobservable* transition. We will use μ, ν... to range over A, and a, b... to range over $A - \{\tau\}$.

Definition 2.1: A *nondeterministic system* (*NDS*) over A is a triple $S = (Q \cup \{r\}, A, \longrightarrow)$, where $Q \cup \{r\}$ is the set of *states* of S, $r \notin Q$ is the *initial* state (or root) of S, and $\longrightarrow \subseteq [(Q \cup \{r\}) \times A \times (Q \cup \{r\})]$ is the *transition relation* on S.

We will use q, q' to range over $Q \cup \{r\}$, and write $q \xrightarrow{\mu} q'$ for $(q, \mu, q') \in \longrightarrow$. We interpret $q \xrightarrow{\mu} q'$ as: S may evolve from state q to state q' via a transition μ.

We will also make use of the transitive and reflexive closure $\longrightarrow^*$ of $\longrightarrow$, which we call the *derivation relation* on S. For an NDS $S = (Q \cup \{r\}, A, \longrightarrow)$, we will use Q_S, r_S, $\longrightarrow_S$ instead of Q, r, $\longrightarrow$ whenever an explicit reference to S is required.

According to our definition, an NDS S is a machine starting in some definite state and evolving through successive states by means of elementary transitions. On the other hand, each state of S may be thought of as the initial state of some NDS: then we might regard S as giving rise to new systems, rather than going through successive states.

In fact, if we consider the class *S* of all NDS's, we may notice that *S* itself can be described as a transition system (although not an NDS, since *S* is obviously not rooted). Let $\longrightarrow^*_{S}$ be the associated derivation relation: we say that S' is a *derivative* of S whenever $S \longrightarrow^*_{S} S'$. Now it is easy to see that, for any $S \in$ *S*, a one-to-one correspondence can be established between the states and the derivatives of S. We shall denote by S_q the derivative corresponding to the state q and by $q_{S'}$ the state corresponding to the derivative S'.

In the following we will often avail of this correspondence between states and (sub)systems.

We assume the class *S* to be closed w.r.t. some simple *operators*: a nullary operator NIL, a set of unary operators $\mu.$ (one for each $\mu \in A$), and a binary operator +. The intended meaning of these operators is the following: NIL represents *termination*, + is a *free-choice* operator, and the μ's provide a simple form of sequentialisation, called *prefixing* by the action μ.

The transition relation of a compound NDS may be inferred from those of the components by means of the rules:

i) $\mu S \xrightarrow{\mu} S$

ii) $S \xrightarrow{\mu} S'$ implies $S + S'' \xrightarrow{\mu} S'$, $S'' + S \xrightarrow{\mu} S'$

The operators will be given a precise definition for a subclass of *S*, the class of *nondeterministic processes* that we will introduce in the next section.

2.1 Nondeterministic processes

As they are, NDS's have an isomorphic representation as (rooted) *labelled directed graphs*, whose nodes and arcs represent respectively the states and the transitions of a system. On the other hand, any NDS may be unfolded into an *acyclic* graph. We shall here concentrate on a class of acyclic NDS's that we call *nondeterministic processes* (NDP's).

Basically, NDP's are NDS's whose derivation relation $\longrightarrow^*$ is a *partial ordering*. Each state of a process is assigned a *label*, that represents the sequence of observable actions leading from the root to that state. To make such a labelling consistent, we only allow two paths to join in the graph if they correspond to the same observable derivation sequence. The labelling is subject to the following further restriction: for any label σ, there are at most finitely many states labelled by σ. As it will be made clear subsequently, this amounts to impose a general *image-finiteness* condition on the systems.

In the formal definition, we will use the following notation: A^* is the set of finite sequences over A, with the usual prefix-ordering, and with empty sequence ε . For simplicity the string $<\mu>$ will be denoted by μ. The *covering relation* $-\!\!\subset$ associated to a partial ordering $\leq$ is given by: $x -\!\!\subset y$ iff $x<y$ and $\not\exists$ z such that $x<z<y$. Also, we make the following convention: τ acts as the identity over A^* and will thus be replaced by ε when occurring in strings.

Definition 2.1.1: A *nondeterministic process* (NDP) over A is a triple $P = (Q \cup \{r\}, \leq, l)$ where:

$(Q \cup \{r\}, \leq)$ is a *rooted* poset: $\forall q, \quad r \leq q$

$l: Q \cup \{r\} \longrightarrow A^*$ is a *monotonic* labelling function, satisfying:

$$l(r) = \varepsilon$$

$$q -\!\!\subset q' \quad \text{implies} \quad l(q') = l(q).\mu, \ \mu \in A^*$$

$$\forall \sigma \in A^*, \quad \{q \mid l(q) = \sigma\} \quad \text{is } \textit{finite}$$

Note that an NDP is very nearly a *labelled tree*: it only differs from a labelled tree in that it might have some confluent paths. The reason we do not directly adopt labelled trees as a model is purely technical (the proof that the model is closed w.r.t. reductions would be rather tricky). However we intend that trees are our real object of interest: in particular, our examples will always be chosen from trees.

As pointed out already, we label nodes with sequences of actions, rather than labelling arcs with single actions: this minor variation w.r.t. the standard notation (see e.g. Milner's *synchronisation trees*) will make it easier to compare different states of a process.

It is easy to see that any NDP P is also an NDS, with $\longrightarrow_P$ given by $-\!\!\subset$. More precisely, for any $\mu \in A$, the relation $\xrightarrow{\mu}_P$ will be given by $\{ (q,q') \mid q -\!\!\subset q' \text{ and } l(q') = l(q).\mu \}$.

Note that, because of our convention that $\tau = \varepsilon$, a τ-transition will be represented in an NDP by the repetition of the same label on the two $\xrightarrow{\tau}$ related nodes. More generally, the label of a node will now represent the sequence of *observable* actions leading to it. For example:

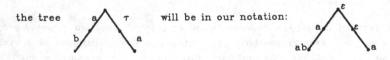

the tree 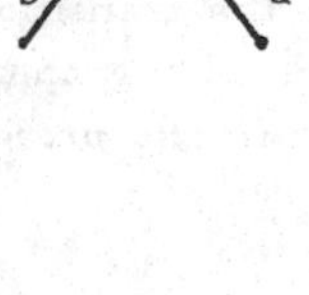

will be in our notation: 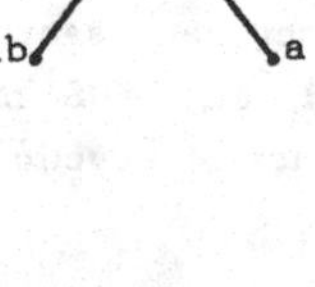

In what follows, nondeterministic processes will always be considered up to isomorphism. Formally, an *isomorphism* between two NDP's: $P_1 = (Q_1 \cup \{r_1\}, \leq_1, l_1)$, $P_2 = (Q_2 \cup \{r_2\}, \leq_2, l_2)$ is a one-to-one correspondence: $\Phi : Q_1 \cup \{r_1\} \longrightarrow Q_2 \cup \{r_2\}$ s.t.

i) $l_2(\Phi(q)) = l_1(q)$

ii) $\Phi(q) \leq_2 \Phi(q')$ iff $q \leq_1 q'$

The operators NIL, μ. and + can be formally defined on NDP's. Let T_i denote the NDP $(Q_i \cup \{r_i\}, \leq_i, l_i)$. Then we have the following :

Definition 2.1.2: *(Operators on NDP's)*

$NIL = (\{r_{NIL}\}, \{(r_{NIL}, r_{NIL})\}, \{(r_{NIL}, \varepsilon)\})$ is the NDP with just a root r_{NIL} and an empty set of subsequent states

μP_1 is the NDP $P = (Q \cup \{r\}, \leq, l)$, where r does not occur in $Q_1 \cup \{r_1\}$, and:

$$Q = Q_1 \cup \{r_1\}$$

$$\leq \; = \; \leq_1 \cup \{(r,q) \mid q \in Q\}$$

$$l(q) = \begin{cases} \varepsilon, \text{ if } q = r \\ \mu.l_1(q) \text{ otherwise} \end{cases}$$

$P_1 + P_2$ is the NDP $P = (Q \cup \{r\}, \leq, l)$, where r does not occur in $Q_1 \cup Q_2$, and:

$$Q = Q_1 \cup Q_2 \quad \text{(disjoint union)}$$

$$\leq \; = \; \leq_1 \upharpoonright Q_1 \cup \leq_2 \upharpoonright Q_2 \cup \{(r,q) \mid q \in Q\}$$

$$l = l_1 \upharpoonright Q_1 \cup l_2 \upharpoonright Q_2 \cup \{(r, \varepsilon)\}$$

Let $P \subseteq S$ denote the class of all NDP's: in what follows our treatment of nondeterministic systems will be confined to P.

3. Abstraction Homomorphisms

The NDP-model, though providing a helpful conceptual simplification, does not appear yet abstract enough. It still allows, e.g., for structural redundancies such as:

Moreover we want to be able, in most cases, to ignore *unobservable* transitions. Such transitions, being internal to a system, should only be detectable indirectly, on account of their capacity of affecting the *observable behaviour* of the system.

We will therefore introduce a *simplification* operation on processes, which we call *abstraction homomorphism.* Essentially an abstraction homomorphism will transform a process in a structurally simpler (but semantically equivalent) process by merging together some of its states.

The criterion for identifying states is that they be *equivalent* in some recursive sense: informally speaking, two states will be equivalent iff they have *equivalent histories* (derivation sequences) and *equivalent futures* or potentials (sets of subsequent states). Formally:

Definition 3.1: Given two NDP's $P_1 = (Q_1 \cup \{r_1\}, \leq_1, l_1)$, $P_2 = (Q_2 \cup \{r_2\}, \leq_2, l_2)$

a function $h: \begin{matrix} r_1 \longrightarrow r_2 \\ Q_1 \longrightarrow Q_2 \end{matrix}$ is an *abstraction homomorphism* (a.h.) from P_1 to P_2 iff:

i) $l_2(h(q)) = l_1(q)$

ii) $succ_2(h(q)) = h(succ_1(q))$

where $succ(q) = \{ q' \mid q \leq q' \}$ is the set of successors of q, *inclusive of q*.

Examples

1)

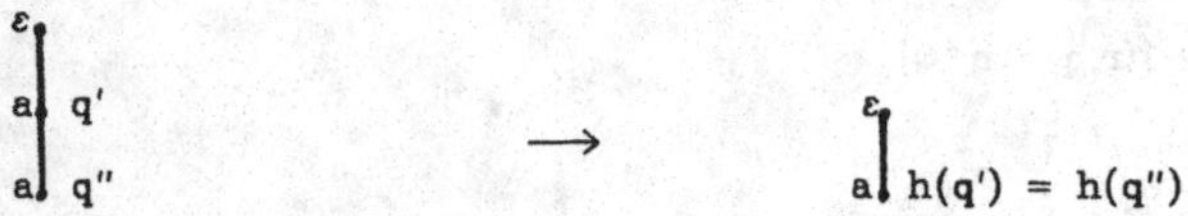

From this example we can see why succ(q) must include q itself: q'' is a *proper* successor of q', whereas h(q'') would not be a proper successor of h(q').

2)

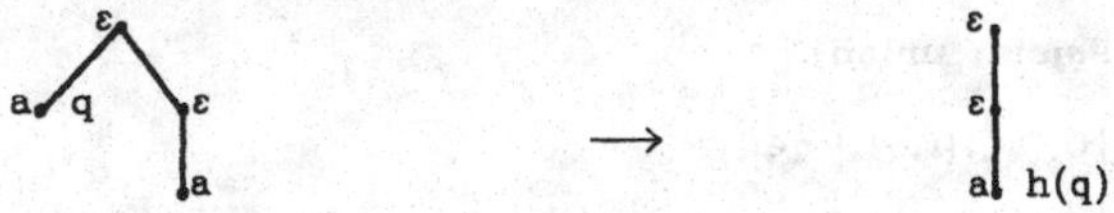

Note that the set of predecessors of q is *not* preserved by the homomorphism.

3)

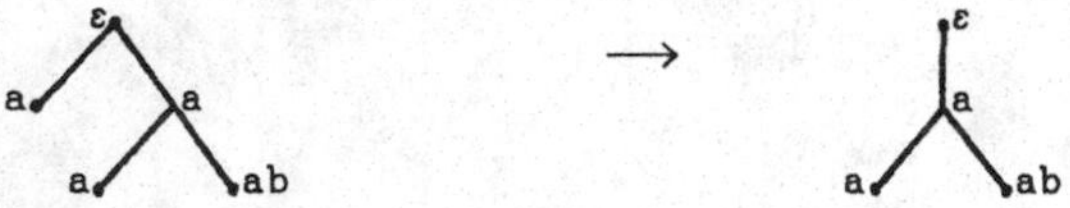

4)

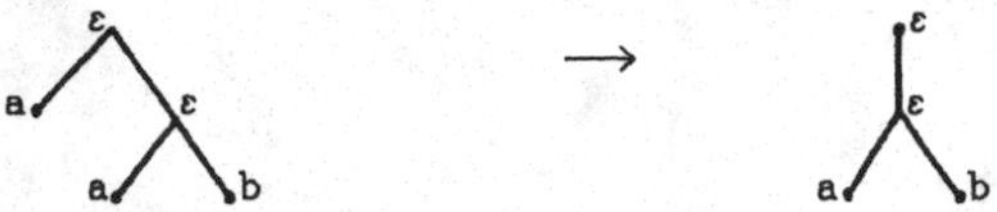

Counterexamples

5)

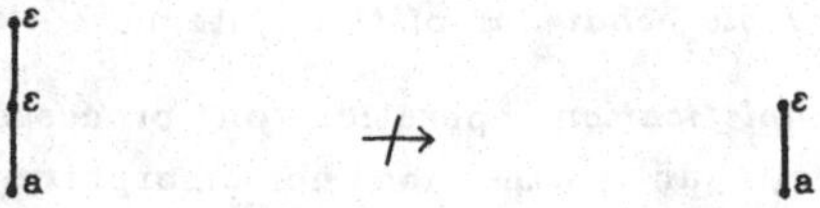

This example shows that a process of the form τP can only be transformed into a process of the same form. This point will subsequently be made more precise.

6) 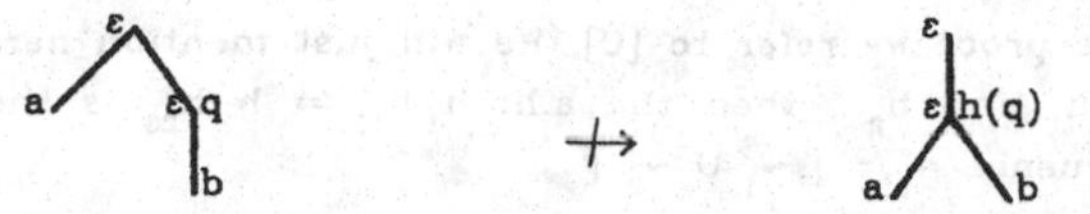

This is not an a.h. because it would increase the set of successors of q.

Abstraction homomorphisms induce the following *reduction relation* $\xrightarrow{abs}$ on processes:

Definition 3.2 : P $\xrightarrow{abs}$ P' iff ∃ a.h. h: P ⟶ P'.

Since the identity function is an a.h. and the composition of two a.h.'s is again an a.h., the relation $\xrightarrow{abs}$ satisfies the following:

Property 1: $\xrightarrow{abs}$ is reflexive and transitive.

Also, it can easily be shown that:

Property 2: $\xrightarrow{abs}$ is preserved by the operators μ. and +.

We turn now to what is perhaps the most interesting feature of our reduction relation, namely its *confluent* behaviour. Confluence of a.h.'s can be proved by standard algebraic techniques, once the notion of *congruence* associated to an a.h. is formalised.

Definition 3.3: Given an NDP P = (Q∪{r}, ≤, l), we say that an equivalence relation ~ on Q is a *congruence* on P iff, whenever q ~ q':

i) l(q) = l(q') (*labels* are preserved)

ii) q ≤ p implies ∃ p' ~ p s.t. q' ≤ p'

(*successors* are preserved)

iii) q ≤ p ≤ q' implies q ~ p ~ q'

(*antisymmetry* of ≤ is preserved)

It can be shown [C] that there is a one-to-one correspondence between congruences and abstraction homomorphisms on a NDP P: any congruence on P is the *kernel* $\sim_h$ of some a.h. h on P, and any a.h. on P is the *natural mapping* $h_\sim$ associated to some congruence ~ on P. Then the following fact is (almost) standard:

Theorem 3.1 : (*Confluence of abstraction homomorphisms*)

If P, P_1, P_2 are NDP's, and $h_1 : P \longrightarrow P_1$, $h_2 : P \longrightarrow P_2$ are a.h.'s, then ∃ NDP P_3, ∃ a.h.'s $h_{13} : P_1 \longrightarrow P_3$, $h_{23} : P_2 \longrightarrow P_3$ s.t. the following diagram commutes:

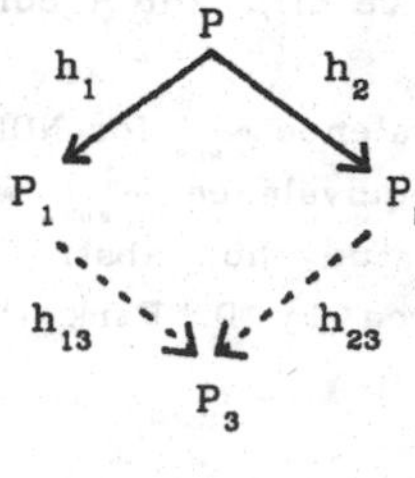

Hint for proof: for the complete proof we refer to [C]. We will just mention here that, if $\sim_1$ and $\sim_2$ are the kernels of h_1 and h_2 , then the a.h. $h_1 h_{13} = h_2 h_{23}$ is the natural mapping associated to the congruence $\sim_3 = [\ \sim_1 \cup \sim_2\ ]^*$ □

Corollary 3.1: ($\xrightarrow{abs}$ is Church-Rosser)

If P, P_1, P_2 are NDP's, and $P_1 \xleftarrow{abs} P \xrightarrow{abs} P_2$
then $\exists$ NDP P_3 s.t. $P_1 \xrightarrow{abs} P_3 \xleftarrow{abs} P_2$

□

3.1 Abstraction equivalence

The relation $\xrightarrow{abs}$ gives us a criterion to regard two processes as "abstractly the same". However, being essentially a simplification, $\xrightarrow{abs}$ is not symmetric and therefore does not, for example, relate the two processes:

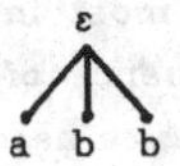

or the processes:

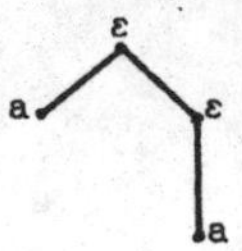

Based on $\xrightarrow{abs}$, we will then define on NDP's a more general relation $\sim_{abs}$, of *reducibility to a same* process:

Definition 3.1.1 : $\sim_{abs} =_{def} \xrightarrow{abs} . \xleftarrow{abs}$

We can immediately prove a few properties for $\sim_{abs}$.

Property 1: $\sim_{abs}$ is an *equivalence*.

Proof: Transitivity follows from the fact that $\xrightarrow{abs}$ is Church-Rosser, which can be restated as: $[\xrightarrow{abs} \cup \xleftarrow{abs}]^* = \xrightarrow{abs} . \xleftarrow{abs}$ □

Property 2: $\sim_{abs}$ is preserved by the operators $\mu.$ and $+$.

Proof: Consequence of $\xrightarrow{abs}$ and $\xrightarrow{abs}^{-1}$ invariance in $\mu.$ and + contexts. □

To sum up, we have now a *substitutive* equivalence $\sim_{abs}$ for NDP's that can be split, when required, in two reduction halves. The equivalence $\sim_{abs}$ will be called *abstraction equivalence*. In the coming section we will study how abstraction equivalence relates to bisimulation equivalence, a notion introduced by D. Park [Pa] for general transition systems.

4. Bisimulation relations

A natural method for comparing different systems is to check to which extent they can *behave like* each other, according to some definition of behaviour.

Now, what is to be taken as the *behaviour* of a system need not be known a priori. One can always, in fact, having fixed a criterion for deriving subsystems, let the behaviour of a system be recursively defined in terms of the behaviours of its subsystems.

Based on such an *implicit* notion of behaviour, one gets an (equally implicit) notion of equivalence of behaviour, or *bisimulation*, between systems: two systems are said to bisimulate each other iff any subsystem of either of the two, selected with some criterion, recursively bisimulates a subsystem of the other, selected with the same criterion.

For an NDS S, the transition relation provides an obvious criterion for deriving a subsystem S': S' is a μ-subsystem of S iff $S \xrightarrow{\mu} S'$ for some μ. However, if we are to abstract from internal transitions, a weaker criterion will be needed. To this purpose the following *weak* transition *relations* $\overset{\mu}{\Longrightarrow}$ are introduced:

$$\overset{a}{\Longrightarrow} \; = \; \xrightarrow{\tau}{}^{n} \xrightarrow{a} \xrightarrow{\tau}{}^{m} \qquad\qquad n, m \geq 0$$

$$\overset{\tau}{\Longrightarrow} \; = \; \xrightarrow{\tau}{}^{n} \qquad\qquad n \geq 0$$

S' is called a *μ-derivative of S* iff $S \overset{\mu}{\Longrightarrow} S'$. We can then formally define bisimulations on NDS's as follows:

Definition 4.1: A (*weak*) *bisimulation* relation is a relation $R \subseteq (\mathcal{S} \times \mathcal{S})$ such that $R \subseteq F(R)$, where $(S_1, S_2) \in F(R)$ iff $\forall\, \mu \in A$:

i) $S_1 \overset{\mu}{\Longrightarrow} S'_1$ implies $\exists\, S'_2$ s.t. $S_2 \overset{\mu}{\Longrightarrow} S'_2$, $S'_1 \; R \; S'_2$

ii) $S_2 \overset{\mu}{\Longrightarrow} S'_2$ implies $\exists\, S'_1$ s.t. $S_1 \overset{\mu}{\Longrightarrow} S'_1$, $S'_1 \; R \; S'_2$

Now we know that F has a *maximal fixed-point* (which is also its maximal postfixed-point) given by $\cup_{R \subseteq F(R)} \{R\}$. We will denote this largest bisimulation by $<\approx>$, and, since $<\approx>$ turns out to be an equivalence, refer to it as *the bisimulation equivalence*.

Unfortunately, $<\approx>$ is not preserved by all the operators. Precisely, $<\approx>$ is not preserved by the operator +, as shown by the example:

 $<\approx>$ NIL , but 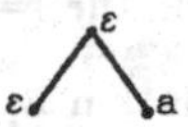 $<\not\approx>$ 

On the other hand the relation $<\approx>^+$, obtained by closing $<\approx>$ w.r.t. the operator +:

$$S_1 <\approx>^+ S_2 \quad \text{iff } \forall\, S\text{: } S + S_1 <\approx> S + S_2$$

can be shown to be a *substitutive* equivalence, and in fact to be the largest such equivalence contained in $<\approx>$. (For more details on $<\approx>$ and $<\approx>^+$ we refer to [M2]).

To conclude, $<\approx>^+$ seems a convenient restriction on $<\approx>$ to adopt when modelling NDS's. We will see in the next section that $<\approx>^+$ coincides, on NDP's, with our abstraction equivalence $\sim_{abs}$.

5. Relating Bisimulations to Abstraction Homomorphisms

Looking back at out relations $\xrightarrow{abs}$ and $\sim_{abs}$, we notice that they rely on a notion of *equivalence of states* which, like bisimulations, is *recursive*. Moreover, the recursion builds up on the basis of a similarity requirement (equality of *labels*) that reminds of the criterion (equality of *observable derivation sequences*) used in bisimulations to derive "bisimilar" subsystems. All this indicates there might be a close analogy between abstraction equivalence and bisimulation equivalence.

In fact, since we know that $\sim_{abs}$ is substitutive, we shall try to relate it with the substitutive bisimulation equivalence $<\approx>^+$. To this purpose, we will need a direct (recursive) definition for $<\approx>^+$.

Note that $<\approx>^+$ only differs from $<\approx>$ in that it takes into account the *preemptive capacities* a system can develop when placed in a sum-context. Such preemptive capacities depend on the system having some silently reachable state where, informally speaking, some of the "alternatives" offered by the sum-context are no more available. This suggests that we should adopt, when looking for a direct definition of $<\approx>^+$, the more *restrictive* transition *relations* $\stackrel{\mu}{\Longmapsto}$:

$$\stackrel{\mu}{\Longmapsto} \; = \; \xrightarrow{\tau}^{n} \xrightarrow{\mu} \xrightarrow{\tau}^{m} \qquad\qquad n,m\geq 0$$

In particular, we will have $\stackrel{\tau}{\Longmapsto} \; = \; \xrightarrow{\tau}^{n}$, $n>0$. Note on the other hand that, for $a\in A$, it will be: $\stackrel{a}{\Longmapsto} \; = \; \stackrel{a}{\Longrightarrow}$.

However, $<\approx>^+$ is *restrictive* with respect to $<\approx>$ only as far as the first $\stackrel{\tau}{\Longrightarrow}$ derivation steps are concerned: at further steps $<\approx>^+$ behaves like $<\approx>$, as it can be seen from the example:

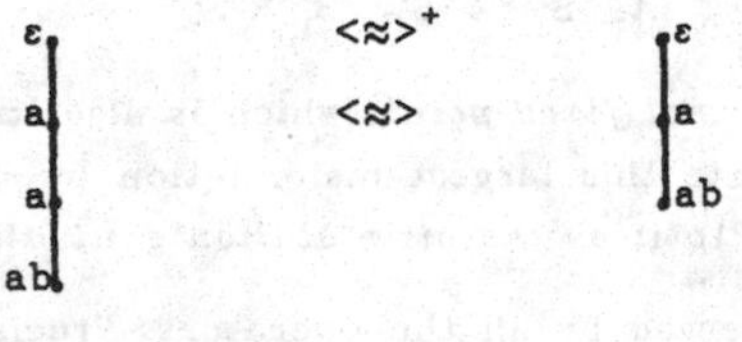

So, if we are to recursively define $<\approx>^+$ in terms of the transitions $\stackrel{\mu}{\Longmapsto}$, we will have to somehow counteract the strengthening effect of the $\stackrel{\mu}{\Longmapsto}$'s at steps other than the first.

To this end, for any relation $R\subseteq (S \times S)$, a relation R_a ("almost" R) is introduced: $(S_1, S_2) \in R_a$ iff $(S_1,S_2) \in R$, or $(\tau S_1,S_2) \in R$, or $(S_1,\tau S_2) \in R$

Then we can define a-bisimulation ("almost" bisimulation) relations on NDS's as follows:

Definition 5.1: A *(weak) a-bisimulation* relation is a relation $R\subseteq (S\times S)$ such that $R\subseteq F_a(R)$, where $(S_1,S_2) \in F_a(R)$ iff $\forall\, \mu \in A$:

i) $S_1 \stackrel{\mu}{\Longmapsto} S_1'$ implies $\exists\, S_2'$ s.t. $S_2 \stackrel{\mu}{\Longmapsto} S_2'$, $S_1' \, R_a \, S_2'$

ii) $S_2 \stackrel{\mu}{\Longmapsto} S_2'$ implies $\exists\, S_1'$ s.t. $S_1 \stackrel{\mu}{\Longmapsto} S_1'$, $S_1' \, R_a \, S_2'$

Again, F_a has a *maximal (post)fixed-point* which is an equivalence, and which we will denote by $<\approx>^a$. The equivalence $<\approx>^a$ has been proven to *coincide* with $<\approx>^+$. Both the definition of $<\approx>^a$ and the proof that $<\approx>^a = <\approx>^+$ are due to M. Hennessy.

It can be easily shown that, if R is an a-bisimulation, then R_a is an ordinary bisimulation. In particular, for the maximal a-bisimulation $<\approx>^a$, it is the case that $<\approx>^a_a = <\approx>$.

Now, it can be proved that:

Theorem 5.1: $\xrightarrow{abs}$ is an a-bisimulation.

The proof relies on the two following lemma's:

Lemma 5.1: If $P_1 \xrightarrow{abs} P_2$ then:

$P_1 \stackrel{\mu}{\Longmapsto} P'_1$ implies $\exists\, P'_2$ s.t. $P_2 \stackrel{\mu}{\Longmapsto} P'_2$ where

either $P'_1 \xrightarrow{abs} P'_2$ *or* $P'_1 \xrightarrow{abs} \tau P'_2$.

Lemma 5.2: If $P_1 \xrightarrow{abs} P_2$ then:

$P_2 \stackrel{\mu}{\Longmapsto} P'_2$ implies $\exists\, P'_1$ s. t. $P_1 \stackrel{\mu}{\Longmapsto} P'_1$ where

either $P'_1 \xrightarrow{abs} P'_2$ *or* $P'_1 \xrightarrow{abs} \tau P'_2$.

Note that in lemma's 5.1 and 5.2 we do not need consider the case $\tau P'_1 \xrightarrow{abs} P'_2$. The reason this case does not arise is that a.h.'s are *single-valued* relations.

Corollary 5.1: $\xrightarrow{abs} \;\subseteq\; <\approx>^a$

Proof: $<\approx>^a$ is the maximal a-bisimulation □

Moreover, we have the following characterisation for a.h.'s:

Terminology: For any NDP P, let $S_P = S_P = \{P' \mid P \longrightarrow^* P'\}$. We say that a bisimulation (a-bisimulation) relation R is *between* P_1 and P_2 iff $(P_1, P_2) \in R$ and $R \subseteq (S_{P_1} \times S_{P_2})$.

Theorem 5.2: An abstraction homomorphism from P_1 to P_2 is a *single-valued* relation which is both a *bisimulation* and an *a-bisimulation* between P_1 and P_2.

We now come to our main result, concerning the relationship between the abstraction equivalence $\sim_{abs}$ and the substitutive bisimulation equivalence $<\approx>^a$. It turns out that these two equivalences coincide:

Theorem 5.3: $\sim_{abs} = <\approx>^a$

Proof of $\subseteq$: From corollary 5.1 we can infer that $\sim_{abs} = [\xrightarrow{abs} \xrightarrow{abs}{}^{-1}] \subseteq <\approx>^a$, since $<\approx>^a$ is symmetrically and transitively closed.

Proof of $\supseteq$: Suppose $P_1 <\approx>^a P_2$. We want to show that $\exists\, P_3$ s.t. $P_1 \xrightarrow{abs} P_3 \xleftarrow{abs} P_2$.

Let R be an a-bisimulation between P_1 and P_2. Then R can be written as:

$$R = (P_1, P_2) \cup R\restriction [(S_{P_1} - P_1) \times (S_{P_2} - P_2)]$$

Now consider:

$$R' = (P_1, P_2) \cup R_a\restriction [(S_{P_1} - P_1) \times (S_{P_2} - P_2)]$$

It is easy to see that R' is both a bisimulation and an a-bisimulation between P_1 and P_2. However R' will not, in general, be single-valued. Let then $\sim$ be the equivalence induced by R' on the states of P_2:

$$q_{P'_2} \sim q_{P''_2}$$

iff $\exists\, P'_1 \in S_{P_1}$ s.t. both (P'_1, P'_2) and $(P'_1, P''_2) \in R'$.

It can be shown that $\sim$ is a congruence on P_2 and therefore $\exists\, P_3$ s.t. $h_\sim : P_2 \longrightarrow P_3$ is an a.h.. So $P_2 \xrightarrow{abs} P_3$.

Also, by theorem 5.2, $h_\sim$ can be regarded as a bisimulation R'' between P_2 and P_3 .Consider now the composition R'·R'': this is by construction a *single-valued* relation contained in $(S_{P_1} \times S_{P_3})$ and containing (P_1, P_3). Moreover R'·R'' is a bisimulation and an a-bisimulation, because both R' and R'' are. So, by theorem 5.2 again, $P_1 \xrightarrow{abs} P_3$.

Summing up, we have $P_1 \xrightarrow{abs} P_3 \xleftarrow{abs} P_2$. □

In view of the last theorem, $\sim_{abs}$ can be regarded as an alternative definition for $<\approx>^a$ = $<\approx>^+$. In the next section, we will see how this new characterisation can be used to derive a set of reduction rules for $<\approx>^+$ on finite processes.

6. A language for finite processes

In this section, we study the subclass of *finite* NDP's, and show how it can be used to model terms of a simple language L.

The language is essentially a subset of R. Milner's CCS (Calculus of Communicating Systems[M1]). In [HM] a set of axioms is presented for L that exactly characterises the equivalence $<\approx>^a$ (and therefore $\sim_{abs}$) on the corresponding transition systems. We show here that the reduction $\xrightarrow{abs}$ itself can be characterised algebraically, by a set of *reduction rules*. These rules yield *normal forms* which coincide with the ones suggested in [HM].

Finally, we establish a notion of *minimality* for NDP's and use it to define a denotational model for L, a class of NDP's that we call *Representation Trees*. The model is shown to be isomorphic with Hennessy and Milner's term-model.

We shall now introduce the language L. Following the approach of [HM], we define L as the term algebra T_Σ over the signature:

$$\Sigma = A \cup \{ NIL, + \}$$

If we assume the operators in Σ to denote the corresponding operators on NDP's (A will denote the set of unary operators μ.), we can use *finite NDP's* to model terms in T_Σ. For a term t, we will use P_t for the corresponding NDP.

We shall point out, however, that the denotations for terms of T_Σ in *P* will always be *trees*, i.e. NDP's $P = (Q \cup \{r\}, \leq, l)$ obeying the further constraint:

confluence-freeness: $\exists\, q''$ s.t. $q \leq q''$ and $q' \leq q''$

implies $q \leq q'$ or $q' \leq q$

Consider now the set of axioms:

E_c

	E1. $x + x' = x' + x$
- sum-laws	E2. $x + (x' + x'') = (x + x') + x''$
	E3. $x + NIL = x$
	E4. $\mu\tau x = \mu x$
- τ-laws	E5. $\tau x + x = \tau x$
	E6. $\mu\,(x + \tau y) + \mu y = \mu\,(x + \tau y)$
- absorption law	E7. $x + x = x$

Let $=^c$ be the equality generated by E_c. It has been proved [HM] that E_c is a sound and complete axiomatisation for Milner's *observational congruence* $\approx^c$ [M1], namely that:

$$t =^c t' \quad \text{iff} \quad P_t \approx^c P_{t'}$$

The relation $\approx^c$ is defined as the closure w.r.t. sum-contexts of the relation (Milner's *observational equivalence*)

$$\approx \; = \; \bigcap_n F^n(P \times P)$$

where $(P \times P)$ is the universal relation on NDP's and F is the function on relations introduced in section 4.

For *image-finite* systems*, the relations $\approx$ and $\approx^c$ have been shown [HM] to coincide with the relations $<\approx>$ and $<\approx>^a$ introduced in the previous sections. In particular, we can assume $\approx^c$ to be defined as $<\approx>^a$ on finite NDP's. Combining these facts together, we have that:

$$t =^c t' \quad \text{iff} \quad P_t \sim_{abs} P_{t'}$$

So $=^c$ is an algebraic analogue for $\sim_{abs}$. Note on the other hand that, although each axiom of E_c could be viewed as a reduction rule (when applied from left to right), the corresponding reduction relation would not characterise $\xrightarrow{abs}$. Consider for example the terms t = aNIL+τ(aNIL+bNIL), t' = τ(aNIL+bNIL). Then the transformation: $t \longrightarrow t'$ would *not* be allowed, whereas we have $P_t \xRightarrow{abs} P_{t'}$.

However, using the axiomatisation E_c as reference, we are able to derive a new system of reduction rules, which characterises $\xRightarrow{abs}$.

We first need to define the relations $\xrightarrow{\mu}$ on terms of T_Σ: $\forall\, \mu \in A^*$, $\xrightarrow{\mu}$ is the *least* relation satisfying the rules:

i) $\mu t \xrightarrow{\mu} t$

ii) $t \xrightarrow{\mu} t'$ implies $t + t'' \xrightarrow{\mu} t'$, $t'' + t \xrightarrow{\mu} t'$

---

*Our restriction on the labelling for NDP's corresponds to the *general image-finiteness* condition: $\forall q, \forall \mu$, $\{q' \mid q \xRightarrow{\mu} q'\}$ is finite

The weak relations $\overset{\mu}{\Longrightarrow}$ are derived from the $\overset{\mu}{\longrightarrow}$'s just as in section 4.

Let now $\longrightarrow^{c}$ be the reduction relation generated by the following set of reduction rules R_c (where $\longleftrightarrow$ stands for $(\longrightarrow \cap \longrightarrow^{-1})$):

R_c

- sumlaws	R1.	$x + x' \longleftrightarrow x' + x$
	R2.	$(x + x') + x'' \longleftrightarrow x + (x' + x'')$
	R3.	$x + NIL \longrightarrow x$
- 1st τ-law	R4.	$\mu\tau x \longrightarrow \mu x$
- generalised absorption law	R5.	$x + \mu x' \longrightarrow x$, whenever $x \overset{\mu}{\Longrightarrow} x'$

Then it can be proved [C] that:

Theorem 6.1: $t \longrightarrow^{c} t'$ iff $P_t \overset{abs}{\Longrightarrow} P_{t'}$

Corollary 6.1: R_c is a rewriting system for the equational theory E_c.

We can make use of our new axiomatisation for $=^{c}$ to characterise normal forms for terms in T_Σ. We say that a term is in *normal form* if no proper reduction (R3, R4 or R5) can be applied to it. It can be shown that:

Theorem 6.2: A term $t = \Sigma_i\, \mu_i t_i$ is a *normal form* iff (Hennessy-Milner characterisation):

i) no t_i is of the form $\tau t'$

ii) each t_i is a normal form

iii) for $i \neq j$, $t_i \longleftrightarrow t'_j \;\; \forall\, t'_j$ s.t. $\mu_j t_j \overset{\mu_i}{\Longrightarrow} t'_j$

Corresponding to normal forms, we have a notion of *minimality* for processes. We say that a process P is *irreducible* or *minimal* iff $P \overset{abs}{\Longrightarrow} P'$ implies $P = P'$. Then the following is trivial:

Theorem 6.3: For any finite NDP P, $\exists\,!$ *minimal* NDP P' s.t. $P \sim_{abs} P'$.

Proof: for uniqueness, use $\overset{abs}{\Longrightarrow}$'s Church-Rosser property □

We shall denote by $\hat{P}$ the unique minimal process corresponding to the NDP P.

Corollary 6.2: $P \sim_{abs} P'$ iff $\hat{P} = \hat{P}'$. □

As we mentioned earlier, the denotation P_t of a term t is always a *tree*. However its "abstract" denotation $\hat{P}_t$ might not be a tree. We shall now propose a *tree-model* for terms of T_Σ, which is isomorphic to the term-model $T_\Sigma/=^{c}$.

Note first that any NDP which is not a tree has a *unique unwinding* into a tree. The tree-unwinding of an NDP P (which is not defined formally here) will be denoted by U(P).

Let now *RT* (*representation trees*) be the class: RT = { U(P) | P is a minimal NDP }. The *denotation* T_t of a term $t \in T_\Sigma$ in RT is defined by: $T_t = U(\hat{P}_t)$.

It can be shown that:

Theorem 6.4: $t =^c t'$ iff $T_t = T_{t'}$ □

We shall finally argue that our model *RT* is *isomorphic* to the term-model $T_\Sigma/=^c$:

RT is a Σ-algebra satisfying the axioms E_c (by theorem 6.4), with the operators defined by:

$$\mu U(P) = U(\widehat{\mu P})$$

$$U(P_1) + U(P_2) = U(\widehat{P_1 + P_2})$$

Therefore, since $T_\Sigma/=^c$ is the initial Σ-algebra satisfying the axioms E_c, we know that:

$$\exists !\ \Sigma\text{-homomorphism } \Psi : T_\Sigma/=^c \longrightarrow RT$$

It is easily seen that Ψ is given by: $\Psi([t]) = U(\hat{P}_t) = T_t$. Also, by theorem 6.4 again, Ψ is a *bijection* between T_Σ and *RT*.

Conclusion

We have proposed an alternative definition for the (substitutive) bisimulation equivalence $<\approx>^+$ for a class of transition systems. Note that the ordinary bisimulation equivalence could be characterised just as easily, by slightly changing the definition of homomorphism: in fact it would be enough to drop the requirement that *proper* states should be preserved. Also, using our definition, we have been able to derive a denotational model for the language L, which is isomorphic to Hennessy and Milner's term model for the same language.

Our approach is intended to extend to richer languages, for programs which are *both* nondeterministic *and* concurrent (meaning that the actual concurrency is not interpreted nondeterministically). Some simple results have already been reached in that direction.

Acknowledgements

The definition of abstraction homomorphism and the idea of using it to characterise Milner's notions of observational equivalence and congruence stems from a joint work with U. Montanari at Pisa University. I would like to thank him for inspiration and for subsequent discussions. I would also like to thank my supervisor M. Hennessy for the substantial help he gave me all along, and R. Milner for helpful suggestions. Many thanks to my colleagues Francis Wai and Tatsuya Hagino for helping me with the wordprocessing of the paper.

References

LNCS stands for Lecture Notes in Computer Science, Springer-Verlag

[BR] S. Brookes, C. Rounds (1983), "Behavioural Equivalence Relations induced by Program Logics", in Proc. ICALP '83, LNCS 154.

[C] Full version of this paper. Contact the author.

[CFM] I. Castellani, P. Franceschi, U. Montanari (1982), "Labelled Event Structures: A Model for Observable Concurrency", in: D. Bjorner (ed.):Proc. IFIP TC2 Working Conference on Formal Description of Programming Concepts II, Garmisch, June 1982: North-Holland Publ. Company 1983

[DeN] R. De Nicola (1984), "Behavioural Equivalences for Transition Systems", Internal Report I.E.I., Pisa, Italy.

[HM] M. Hennessy, R. Milner (1983), "Algebraic laws for Nondeterminism and Concurrency" , Technical Report: CSR-133-83 , University of Edinburgh.

[K] R. Keller (1976), "Formal verification of Parallel Programs", Communications of the ACM n. 19, Vol. 7.

[M1] R. Milner (1980), A Calculus of Communicating Systems, LNCS 92.

[M2] R. Milner (1982), "Calculi for Synchrony and Asynchrony", J. Theoretical Computer Science, Vol. 25.

[Pa] D. Park (1981), "Concurrency and Automata on Infinite Sequences", in LNCS 104.

[P] G. Plotkin (1981), "A Structured Approach to Operational Semantics", DAIMI FN-19, Computer Science Dept, Aarhus University.

A METRIC CHARACTERIZATION OF FAIR COMPUTATIONS IN CCS *

Gerardo COSTA

Istituto di Matematica - Università di Genova

Via L.B. Alberti 4 - I 16132 Genova - Italy

ABSTRACT. We address the problem of characterizing fair (infinite) behaviours of concurrent systems as limits of finite approximations. The framework chosen is Milner's Calculus of Communicating Systems. The results can be summarized as follows. On the set FD of all finite derivations in the calculus we define three distances: da, dw, ds. Then the metric completion of (FD,da) yields the space of all derivations, while the completion of (FD,dw), resp. (FD,ds), yields the space of all finite derivations together with all - and only - the weakly, resp. strongly, fair computations (i.e. non-extendable derivations). The results concerning da and dw are a reformulation of previously known ones, while that concerning ds is - we believe - new.

INTRODUCTION

One of the most challanging problems in modelling fairness properties of concurrent systems is to reconcile infinite fair behaviours with our ideas and wishes about approximations and limits. We would like a framework in which we have some notions of finiteness, approximation and limit such that infinite behaviours are fully determined, via "countable limits", by the sets of their finite approximants. Now, fair infinite behaviours do not seem to fit into this pattern.

In [Pa1] Park provides a fixpoint relational semantics for data-flow networks with a fair merge operator. In doing so he has to consider limits (lub's and inf's) of transfinite chains. The denotational semantics for a while language with a fair parallel operator in [Pl] also appeals to transfinite chains. More recently, De Bakker and Zucker have proposed metric spaces of processes where they can interpret fair parallel operators using the standard notion of limit of (countable) sequences [BZ]. Here, however, the finiteness of the approximants is in question as fairness is achieved using the equivalent of random assignment. We see the same problem in the use of oracles in [Pa3].

Apparently, if there is a way out it is well concealed. A customary solution in such cases is to settle for less. Going in this direction, one realizes that the difficulties are not entirely due to the interaction of fairness on one side and our requirements, countable limits and finite approximations, on the other: our notion of behaviour plays its part. Some of the features that we see as most desirable become stumbling blocks. The most obvious one is abstraction: we are particularly keen to remove information which would allow us to capture fairness more easily. But there is another feature which, though desirable in general, becomes an impediment here: we want to consider global behaviours and handle them as a whole. For instance, in |Pa1| the behaviour of a network is its input-output relation; that of a program in |Pl| is the

* This research was carried out while the author was at the Dept. of Computer Science of the University of Edinburgh.

set of its (possible) results. It is such a global entity that one tries to obtain through the limiting process.

In a recent paper, Degano and Montanari consider a notion of behaviour closely connected to the operational notion of step-wise evolution of a system |DM|. Then they can characterize some liveness properties, including fairness, of the individual entities which constitute the global behaviour of systems in terms of convergence properties in appropriate metric spaces. We omit the details of their approach as here we proceed along the same lines, though in a slightly different setting: that provided by Milner's CCS. Indeed, the starting point for the present work has been to find an analogue of the results in |DM| which concern fairness within the framework proposed in |CS2|.

We consider the kind of concurrent systems that are represented by (closed) expressions in CCS without value-passing. These are systems in which components evolve asynchronously and can interact by exchanging signals on which they synchronize. We shall call them processes in the sequel. We take as behaviour of a process the set of its computations, i.e. non-extendable derivation sequences as defined by the rules of the calculus. As the bare derivation sequences (without mention of how each step is inferred) do not contain enough information, we add labels to the calculus. They allow us to see if a subprocess remains unduly idle in the course of a computation. The setting is rich enough to allow weak and strong fairness |Pa2,Pl| to be distinguished. On the set of all derivations we define three distances: da, dw,ds. Then, infinite computations - taken individually - are fully determined as limits w.r.t. da of (countable) sequences of finite derivations (which play the role of finite approximants). The same result is obtained for infinite computations which are weakly, resp. strongly, fair using dw, resp. ds.

The first two results are not new and are presented for completeness sake and as a stepping stone towards the third. The one concerning infinite computations, without regard to their fairness, is a straightforward adaptation of well known results (see e. g. |AN|). The result concerning weak fairness is a reformulation of one in |DM|. It is the result about strong fairness which is new, or, at least, an improvement on previous ones. Indeed, the characterization of strong fairness given by Degano and Montanari is weaker than ours.

It should be clear that we do not regard our results as a real breakthrough towards a solution of the initial problem. Forfeiting both abstraction and "globality" in the notion of behaviour is, perhaps, too much. On the other hand, the present work adds to our confidence in the framework for handling fairness developed jointly with Colin Stirling |CS1,CS2|. The approach in those papers is operational; the problem addressed is that of giving a finite set of rules for generating fair computations in CCS, without resorting to random assignment. In |CS2| two calculi are presented, one for weak and the other for strong fairness. A starting point is the notion of "fairness at i" (see Sect. 2) which yields, at least in the weak case, a more local characterization of fairness. Here, this same notion provides the basis for defining the distances dw and ds mentioned above.

In Section 1 we give a concise account of Milner's CCS and of the labelled calculus we use. In Section 2 we formalize the notion of fairness in the context of CCS and give a more local characterization of it. Most of the material in this two sections is borrowed from |CS2|. Section 3 provides the basic tools used in defining our distances. The metric characterization of infinite (fair) computations is in Section 4. The final section is devoted to concluding remarks and the Appendix to an example. Proofs are omitted; they appear in the extended version of this paper |C|.

We assume the reader familiar with the elementary notions of the theory of metric spaces.

1. C C S

We give here a concise account of CCS without value-passing (hence communication is reduced to pure synchronization) and renaming. These features have been omitted for simplicity sake; they can be accomodated within the framework we develop here. For a detailed account of CCS see |M1|. Then we introduce the labelled CCS we shall actually use. Finally, we define the notions of derivation and computation. The example at the end of the paper provides an illustration of the labelled calculus and, by suppressing labels, of standard CCS.

Let AA be a set of <u>atomic actions</u> and $\overline{AA}$ be a set of <u>co-actions</u> disjoint from AA and in bijection with it. The bijection is $\bar{\ }$: $\bar{a}$ stands for the co-action of a and $\bar{\bar{a}} = a$. The calculus allows for synchronization of co-actions; this is represented by τ, a silent or internal action not in $AA \cup \overline{AA}$. Let $Act = AA \cup \overline{AA}$ and $Move = Act \cup \{\tau\}$ and a,b,c,... range over Act, m over Move and X over a suitable set of variables. The syntax of <u>CCS expressions</u> is:

$$E ::= X \mid NIL \mid mE \mid E+E \mid E|E \mid fixX.E \mid E\backslash a\ .$$

NIL is the process which does nothing; + represents nondeterministic choice; | concurrency; fix recursion; \a restriction (prevention) of a and $\bar{a}$ actions. We assume that in fixX.E X is <u>guarded</u> in E: every free occurrence of X in E is within a subexpression of the form mG. The <u>rules</u> of the calculus are (where $E \xrightarrow{m} G$ means E becomes G by performing the move m and E[G/X] denotes substitution of G for X in E):

$$mE \xrightarrow{m} E \qquad \frac{E \xrightarrow{m} E'}{E+G \xrightarrow{m} E'} \qquad \frac{G \xrightarrow{m} G'}{E+G \xrightarrow{m} G'} \qquad \frac{E[fixX.E/X] \xrightarrow{m} E'}{fixX.E \xrightarrow{m} E'}$$

$$\frac{E \xrightarrow{m} E'}{E|G \xrightarrow{m} E'|G} \qquad \frac{G \xrightarrow{m} G'}{E|G \xrightarrow{m} E|G'} \qquad \frac{E \xrightarrow{a} E' \quad G \xrightarrow{\bar{a}} G'}{E|G \xrightarrow{\tau} E'|G'} \qquad \frac{E \xrightarrow{m} E' \quad m \notin \{a,\bar{a}\}}{E\backslash a \xrightarrow{m} E'\backslash a}$$

Three points are worth mentioning. Sometimes the + rules do not allow choice: (aE+bG)\b can only become E by performing 'a'. The rules for | do not compel synchronization: $aE|\bar{a}G$ can perform a or $\bar{a}$ as well as τ - on the other hand, $(aE|\bar{a}G)\backslash a$ can only perform τ because of the restriction. Finally, the number of concurrent subprocesses may increase as moves are performed; for instance, if E = fixX.aX|bX, then $E \xrightarrow{a} E|bE$.

In discussing fairness we shall need to know "who is doing what". One way of achieving this is to introduce labels into the calculus. The full definition of the <u>labelled calculus</u> appears in |CS2| (and a similar calculus is in |CS1|). The precise details are needed only in the proofs of the results in the following section (for them see |CS2|). We think it suffices here to point out the essential features of this calculus.

Labels are strings in {1,2}* , with typical elements u,v,w,... and ε denoting the null string. They are assigned systematically following the structure of expressions. Due to recursion the labelling is in part dynamic: the rule for fix generates new labels. The syntax of expressions is unchanged. However, variables, NIL, symbols in Move and the operators +, |, \a, fixX are labelled. Each label occurs at most once in an expression; we call this property unicity of labels. An example is $(a_{11}NIL_{111} +_1 b_{12}NIL_{121}) |_\varepsilon (fixX)_2.a_{21}X_{211}$. In $(fixX)_u.E$ we assume that $(fixX)_u$ binds any (free) labelled X within E. Similarly, in $E(\backslash a)_u$ all labelled a and $\bar{a}$ are restricted. Moreover, a_u and $\bar{a}_v$ are assumed to be complementary, irrespective of u and v. The rules of the calculus are just the ones in standard CCS where we allow expressions to carry labels; moves remain unlabelled. Examples are:

$$m_uE\text{-}m\rightarrow E \ ; \qquad \frac{E\text{-}m\rightarrow E'}{E|_uG\text{-}m\rightarrow E'|_uG} \quad \text{and} \quad \frac{E\text{-}m\rightarrow E' \qquad m \notin \{a,\bar{a}\}}{E(\backslash a)_u\text{-}m\rightarrow E'(\backslash a)_u} .$$

The only real change is in the rule for fix: standard substitution, [-/-] , is replaced by a substitution operation which also changes the labels in the substituted expression by prefixing them with the label(s) of the variable occurrence(s) it replaces. The net result is that, for instance, under the labelling fixX.aX is "equivalent" to the infinite expression $a_{u_1}a_{u_2}...a_{u_i}...$ where each u_i occurs only once.

If E-m→G, then the labels in G are determined by those in E and unicity of labels is preserved. If a label occurs both in E and G it will be attached to the same symbol (and this will indicate that the move has not affected this symbol) or to a variable, say X, in E and to a fixX in G. Once a label is lost via a move, then it is never regained. Finally, if r(-) corresponds to the operation of removing labels: if E-m→G, then r(E)-m→r(G) in standard CCS; if E'-m→G' in standard CCS, then for any E s.t. r(E)=E' there exists G s.t. r(G)=G' and E-m→G.

From now on we assume that we are working within the labelled calculus. However, whenever possible, we leave labels implicit.

A derivation is any finite or infinite sequence of the form $E_0\text{-}m_1\rightarrow E_1\text{-}m_2\rightarrow E_2 \ ...$ A computation is just a non-extendable derivation: if it is finite its last term is unable to make a move; on the other hand any infinite derivation is a computation. It is useful to have a more precise definition and an alternative representation for derivations. If we know that the initial term is E_0, then the sequence above can be represented as a sequence of pairs: $(m_1,E_1)(m_2,E_2)...$ This can be formalized as follows, where

$[1\ k]$ denotes the set $\{i,\ 1\le i\le k\}$, $k\ge 0$, and N^+ denotes the set of positive natural numbers. A finite derivation of lenght k from E_0, $k\ge 0$, is a sequence $Y=(m_1,E_1)\ldots(m_k,E_k)$ s.t. (*) below holds, where $Dom(Y)=[1\ k]$. An infinite derivation from E_0 is an infinite sequence $Y=(m_1,E_1)\ldots(m_k,E_k)\ldots$ s.t. (*) below holds, where $Dom(Y)=N^+$.

$$(*)\quad \forall\ i \in Dom(Y)\ .\ E_{i-1} \ -m_i\!\rightarrow E_i\ .$$

The following notation will also be convenient: $Dom0(Y) = Dom(Y) \cup \{0\}$.

In the sequel, we shall use both representations for derivations, according to which is more appropriate.

2. FAIRNESS IN CCS

Fairness imposes the constraint that each concurrent subprocess always eventually proceeds, unless it is deadlocked or has terminated. In CCS, however, subcomponents may not be able to proceed autonomously, hence weak and strong fairness are distinguishable [Pa2,Pl]. The weak fairness constraint states that if a subcomponent can almost always proceed then eventually it must do so, while the strong fairness contraint states that if a subcomponent can proceed infinitely often then it must proceed infinitely often. Consider $Y = E\text{-}a\!\rightarrow G\text{-}c\!\rightarrow E\text{-}a\!\rightarrow G\text{-}c\!\rightarrow \ldots E\text{-}a\!\rightarrow G\text{-}c\!\rightarrow E \ldots$, where $E=(F|\bar{b}NIL)\backslash b$, $F= fixX.\ a(cX+bNIL)$ and $G= ((cF+bNIL)|\bar{b}NIL)\backslash b$. The subcomponent $\bar{b}NIL$ is blocked in E (as \b prevents it from moving autonomously and it cannot synchronize with F, whose only action is a) while it can synchronize with bNIL in G. Hence Y is not strongly fair: $\bar{b}NIL$ can move infinitely often and never does. On the other hand, all subcomponents which are almost always able to move do so; hence Y is weakly fair.

We stress the fact that we regard fairness as an issue concerning concurrent subcomponents only. For instance, if $H = fixX.\ aX+bX$, then the sequence $H\text{-}a\!\rightarrow H\text{-}a\!\rightarrow H\text{-}a\!\rightarrow \ldots$ is fair. (Indeed, at each step in the above sequence there is a choice between a and b; if a is chosen - performed - then b is discarded. At each step ther is a "new" choice, with "new" a and b, independent from the previous - and future - ones.) So, from now on subprocess, subcomponent, component will mean concurrent subprocess, subcomponent, component.

Now we formalize the two fairness constraints and then give an alternative, more local, characterization for them. The reader can find more details in |CS2|.

We need to define the notion of (top-level) live subprocess of a process E: a (concurrent) subcomponent that can contribute to the performance of a move of E.

First, we let $P(E)$ be the set of (top-level) subprocesses of E irrespective of liveness. This set is defined inductively, letting labels represent processes.

$P(X) = \emptyset$; $P(NIL_u) = P(m_u E) = \{u\}$; $P(fixX.E) = P(E\backslash a) = P(E)$;

$P(E|G) = P(E) \cup P(G)$; $P(E +_u G) = $ if $P(E) \cup P(G) = \emptyset$ then $\emptyset$ else $\{u\}$.

Note that the notion of subcomponent is dynamic: if $E = a_u a_v NIL | b_w NIL$ and E performs 'a' then the resulting subcomponents do not include u. A simpler, static, notion of subprocess is inadequate here because the number of subcomponents may grow under derivation.

We now define Act(E), the set of unlabelled actions in E (excluding τ) which can happen autonomously:

$Act(X) = Act(NIL) = Act(\tau E) = \emptyset$; $Act(aE) = \{a\}$; $Act(fixX.E) = Act(E)$;

$Act(E\backslash a) = Act(E) - \{a,\bar{a}\}$; $Act(E+G) = Act(E|G) = Act(E) \cup Act(G)$.

A simple consequence of this definition is: $a \in Act(E)$ iff $\exists G . E\text{-}a\text{-}\!\!>G$.

Next, we define LP(E,A) to stand for the set of live subprocesses of E when the environment prevents the actions in A, a subset of Act. Clearly, LP(E,A) is a subset of P(E). If $\overline{Act(E)}$ denotes $\{\bar{a} : a \in Act(E)\}$; then:

$LP(X,A) = LP(NIL,A) = \emptyset$; $LP(fixX.E , A) = LP(E,A)$;

$LP(m_u E,A) = $ if $m \notin A$ then $\{u\}$ else $\emptyset$;

$LP(E +_u G , A) = $ if $LP(E,A) \cup LP(G,A) = \emptyset$ then $\emptyset$ else $\{u\}$;

$LP(E\backslash a , A) = LP(E , A \cup \{a,\bar{a}\})$;

$LP(E|G , A) = LP(E , A - \overline{Act(G)}) \cup LP(G , A - \overline{Act(E)})$.

The set of live subprocesses in E is defined as $LP(E,\emptyset)$ which we abbreviate to LP(E).

Looking at our running example (in Appendix) one sees that the labels in $P(E_i)$ and $LP(E_i)$ are the expected ones: the following lemma provides a formal justification for the definitions just given. (Recall that if a subcomponent contributes to a move, then afterwards it no longer exists.)

Lemma 2.1

1. if $u \in P(E) - LP(E)$ and $E\text{-}m\text{-}\!\!>G$ then $u \in P(G)$;
2. if $u \in LP(E)$ then $\exists G, m . E\text{-}m\text{-}\!\!>G$ and $u \notin P(G)$;
3. if $E\text{-}m\text{-}\!\!>G$ then $\exists u \in LP(E) . u \notin P(G)$.

The set P(E) has a "persistency" property - not shared with LP(E) - which will be very useful in dealing with strong fairness.

Fact 2.2 If $E_0\text{-}m_1\text{-}\!\!>E_1\text{-}m_2\text{-}\!\!>\ldots\text{-}m_k\text{-}\!\!>E_k$ and $u \in P(E_0) \cap P(E_k)$ then $u \in P(E_i)$, $1 \le i < k$.

Given these definitions, we can now define admissibility under the two fairness constraints. (Intuitively, E_0 below should be closed - and the reader is free to assume it is so - formally we do not need this condition.)

Definition 2.3. Let $Y = E_0 -m_1\rightarrow E_1 -m_2\rightarrow \ldots$ be a computation.

1. Y is weakly fair iff $\forall u\ \forall i\ \exists n\geq i\ .\ u \notin LP(E_n)$;
2. Y is strongly fair iff $\forall u\ \exists i\ \forall n\geq i\ .\ u \notin LP(E_n)$.

Quantification over i and n is implicitly restricted to DomO(Y).

A computation is weakly fair iff no subcomponent becomes live and then remains live throughout. It is strongly fair iff no subcomponent is live infinitely often. Notice that a component cannot be live infinitely often and proceed infinitely often because of the labelling: as soon as a component contributes to a move then it "disappears". Clearly, if a computation is strongly fair it is also weakly fair; also, any finite computation is strongly fair (its final term will have no live subcomponents).

Our next step is to try and express fairness as a local property and not just as a property of complete computations. This attempt proves successful for weak fairness, through the notion of "w-fairness at i". Not surprisingly, it fails for strong fairness: if Y is "s-fair at i" this can only be known, in general, by inspecting the complete tail of Y after i. Let $Y = E_0 -m_1\rightarrow E_1 -m_2\rightarrow \ldots$ be a derivation (not necessarily a computation).

Definition 2.4

1. Y is w-fair at i iff $\exists k\geq i\ .\ \bigcap\{LP(E_j),\ i\leq j\leq k\} = \emptyset$;
2. Y is s-fair at i iff $\exists k\geq i\ \forall n\geq k\ .\ P(E_i) \cap LP(E_n) = \emptyset$.

Quantification over k and n is implicitly restricted to DomO(Y). Any k satisfying the r.h.s. in 1 (resp. 2) will be called a w-witness (resp. an s-witness) for i.

The following fact is straigthforward when x=w. In the case x=s, it relies on fact 2.2; it would be false if $LP(E_i)$ replaced $P(E_i)$ in Def. 2.4.2.

Fact 2.5 Let $x \in \{w,s\}$. If Y is x-fair at i, then Y is x-fair at j for all j<i, any x-witness for i being also an x-witness for j.

Borrowing the terminology from [M2] we can say that any u in $P(E_i)$ represents an expectation of E_i. This expectation is immediate when $u \in LP(E_i)$ - i.e. there is a move from E_i which involves u, fulfilling the expectation, see Lemma 2.1.2 - otherwise it is quiescent. Then, using Fact 2.5, we can read the definition of w-fairness at i as follows. Upon reaching E_k we know that each of the immediate expectations of E_j, $0\leq j\leq i$, has disappeared completely, or it has become quiescent at least in one of the terms $E_{j+1},\ldots,E_k$. A similar explanation - but taking into account the differences already pointed out - can be given of s-fairness at i. We thus have the intuitive justification for the next result.

Theorem 2.6 If Y is a computation then it is weakly (resp. strongly) fair iff Y is w-fair (resp. s-fair) at i for all i in DomO(Y).

3. THE BASIC TOOLS

Here we introduce some notation and lw, ls, LS: the basic tools we use in defining our distances for weak and strong fairness. We have found it convenient to concentrate on a fixed initial term. So we assume E_0 to be fixed throughout this and the next section. Later (at the end of Sect. 4), we shall outline how to remove this restriction.

In what follows, derivation, computation,... abreviate derivation from E_0, computation from E_0,... ; similarly, FD, D^ω, D,... below abbreviate $FD(E_0)$, $D^\omega(E_0)$, $D(E_0)$,...

Then: FD is the set of finite derivations; D^ω is the set of infinite derivations; $D^{\omega w}$ is the set of infinite weakly fair derivations; $D = D^\omega \cup FD$; $D^w = FD \cup D^{\omega w}$. $D^{\omega s}$ and D^s are defined in a similar way.

If Y is in D and i in Dom(Y), then: Y(i) is the i-th element of Y - i.e. (m_i, E_i) - and Y[i] is the initial segment of Y of lenght i - i.e. Y(1)...Y(i). By convention, $Y[0] = \varepsilon$ i.e. the null string.

Let $Y = E_0 - m_1 \rightarrow E_1 - m_2 \rightarrow \ldots - m_n \rightarrow E_n \ldots$ be in D.

Definition 3.1. For k in DomO(Y), $lw(Y,k) = lub\,\{i : 0 \leq i \leq k \text{ and } LP(E_i) \cap \ldots \cap LP(E_k) = \emptyset\}$.

The intuitive idea behind this definition is to take the one for w-fairness at i and read it backwards. The lub is taken in N -the set of natural numbers. We use lub instead of max just to handle the empty set: lub $\emptyset = 0$. For an example see Appendix. From this definition we obtain immediately fact 3.2. Lemma 3.3 provides yet another characterization of infinite weakly fair computations. Its proof is straightforward given theor. 2.6 and fact 3.2.1.

Fact 3.2. 1. $lw(Y,n) \geq i$, for some $n \geq i$, iff Y is w-fair at i;

2. if $m \leq n$ then $lw(Y,m) \leq lw(Y,n)$.

Lemma 3.3. If Y is in D^ω, then Y is weakly fair iff $\lim\,\{lw(Y,n),\ n \geq 0\} = +\infty$.

We now define an appropriate correlate of s-fairness at i. Clearly, the difference between s- and w-fairness at i will come forward. A definition which mimics that of lw will only provide us with a convenient notation: ls. The true correlate of s-fairness at i is LS, whose definition involves looking ahead in the derivation.

Definition 3.4. If k is in DomO(Y), then:

$ls(Y,k) = lub\,\{ i : 0 \leq i \leq k \text{ and } P(E_i) \cap LP(E_k) = \emptyset \}$;

$LS(Y,k) = lub\,\{ i : 0 \leq i \leq k \text{ and } \forall\, n \geq k,\ n \text{ in DomO}(Y).\ P(E_i) \cap LP(E_n) = \emptyset \}$.

Clearly: $0 \leq LS(Y,k) \leq ls(Y,k) \leq lw(Y,k) \leq k$.

We now state the analogue of fact 3.2 for LS; its proof is immediate. Notice that neither part 1 nor part 2 hold of ls. This can be seen using the example given in the

Appendix (for part 1, consider the derivation Y[6] : ls(Y[6] , 5)=4 but Y[6] is not s-fair at 4). The link between ls and LS is established by lemma 3.6.

Fact 3.5. 1. LS(Y,n) $\geq$ i , for some n$\geq$i, iff Y is s-fair at i;

2. if m $\leq$ n then LS(Y,m) $\leq$ LS(Y,n).

Lemma 3.6. LS(Y,k) = min { ls(Y,n) , n$\geq$k and n is in DomO(Y) }.

As expected, LS provides an alternative characterization of strong fairness. But this is also true of ls: taking the limit we overcome the "weaknesses" of ls. Once more the proof is straightforward.

Lemma 3.7. If Y is in D^{ω}, then Y is strongly fair iff

lim { ls(Y,n) , n$\geq$0 } = + ∞ iff lim { LS(Y,n) , n$\geq$0 } = + ∞ .

4. A METRIC CHARACTERIZATION OF D , D^{w} AND D^{s}

Here we introduce our distances: da,dw,ds. The first allows us to characterize all infinite computations as limits of sequences of finite derivations. The analogue for infinite weakly (resp. strongly) fair computations is achieved using dw (resp. ds). These characterization results are stated in theorem 4.6.

The distance da is essentially a well known distance on strings; also well known is the relative result (see e. g. [AN]). The other two distances are derived from lw and LS, hence from the notions of w- and s-fairness at i. However, an analogue of the result concerning dw can be found in [DM]: the characterization of "globally fair" computations as limits of "histories" w.r.t. d_2. What is new is the result concerning strong fairness; indeed the one by Degano and Montanari is ... weaker. (We shall come back to this point in the concluding section.)

Recall that here we assume that derivations have a fixed initial term, E_0. At the end of this section we outline how to remove this restriction.

Let Y and Z be two (finite or infinite) derivations s.t. Y$\neq$Z and let P be their longest common prefix - as sequences - which is finite even if Y and Z are not. The lenght of P provides a natural basis for a distance between Y and Z without regard to fairness; hence the definition of δa and da below. The intuition behind lw and lemma 3.3 suggest that lw(P,lenght(P)) can play the same role w.r.t. a distance between Y and Z which takes weak fairness into account. This motivates the definition of δw - notice that if k$\leq$lenght(P), then lw(P,k) = lw(Y,k) = lw(Z,k) . Then dw is obtained from δw in the same way as da is obtained from δa. Lemma 4.4 states that da and dw are indeed distances (actually, ultra-distances) and that dw is more "stringent" than da.

<u>Definition 4.1</u>. If Y and Z are in D and $Y \neq Z$ then:

$$\delta a(Y,Z) = \max \{i : Y[i] = Z[i]\} ; \quad \delta w(Y,Z) = lw(Y , \delta a(Y,Z)) .$$

<u>Definition 4.2</u>. If Y and Z are in D , $Y \neq Z$ and x=a,w , then:

$$dx(Y,Y) = 0 ; \quad dx(Y,Z) = [\delta x(Y,Z) + 1]^{-1}.$$

Our distance for strong fairness is defined directly from LS (but, probably, by refining the basic idea one could define some δs, hence derive ds in the same way as dw is). A more fundamental difference with dw is however the one inherited from the difference between LS and lw (and ls). While lw(Y,n) depends entirely upon the initial segment of Y, LS(Y,n) depends also on its tail. So if $\delta a(Y,Z) = n$ the distance dw uses lw to explore the common prefix of Y and Z while ds needs to know both LS(Y,n) and LS(Z,n). For examples concerning ds see Appendix.

<u>Definition 4.3</u>. If Y and Z are in D and $Y \neq Z$ and $n = \delta a(Y,Z)$ then:

$$ds(Y,Y) = 0 ; \quad ds(Y,Z) = [LS(Y,n) + 1]^{-1} + [LS(Z,n) + 1]^{-1}.$$

<u>Lemma 4.4</u>. da , dw and ds are distances on D. Moreover, for any Y and Z:

$$da(Y,Z) \leq dw(Y,Z) \leq \tfrac{1}{2} ds(Y,Z) .$$

The only non-trivial part in the proof of this lemma is to show that the triangular inequality holds for dw and ds.

We can now state the promised results. Theorem 4.5 says that the infinite computations are fully determined by the finite derivations and the distance da. Any computation in D can be obtained as limit of a sequence of finite derivations; conversely, any sequence of derivations which is a Cauchy sequence w.r.t. da has a limit in D. The analog is true of infinite weakly/strongly fair computations and dw/ds. In other words, when completing FD w.r.t. dw/ds, we add all and only the infinite computations which are weakly/strongly fair. Notice that if Y is infinite, then, rather trivially, Y is the limit w.r.t. da of $\{ Y[n] , n \geq 0 \}$ - and of infinitely many other sequences. The "value" of dw/ds is in the following fact:

$\{Y[n] , n \geq 0\}$ converges w.r.t. dw/ds <u>if and only if</u> Y is weakly/strongly fair (and then the limit is Y).

<u>Notation</u>. If D' is a subset of D and d is a distance on D, the restriction of d to D' (which is a distance on D') will also be denoted by d. If x = a,w,s , $\lim_x$ denotes a limit w.r.t. dx; limits are taken in D.

<u>Theorem 4.5</u>. The spaces (D,da) , $(D^w$,dw) , $(D^s$,ds) are (isomorphic to) the metric completion of (FD,da) , (FD,dw) , (FD,ds) , respectively.

The proof of the result concerning da is well known - and, anyway, straightforward. The proof of the other two results relies on the previous one and on lemmas 4.6 and 4.7 below; given these it is straightforward. Part 1 of lemma 4.6 is immediate from lemma 4.4, while lemma 4.7 essentially follows from lemmas 3.3 and 3.7. The crucial proof is that of part 2 in lemma 4.6; not surprisingly, the case x=s is harder than the other.

Lemma 4.6. If x=w,s and S is a Cauchy sequence in (D,dx), then:

1. S is also a Cauchy sequence in (D,da);
2. $\lim_x S$ exists and coincides with $\lim_a S$.

Lemma 4.7. If x=w,s and S is a sequence in FD and $\lim_x S = Y$, then Y is in D^x.

(The non-trivial case is when Y is infinite; then it is fair.)

Let us now outline how to extend the results above to all derivations. Recall that FD, D, ... have been used so far as explicit abbreviations for $FD(E_0)$, $D(E_0)$, ... Implicitly, also dx has been an abbreviation for $dx(E_0)$, x=a,w,s. Now we redefine FD, D,... and dx. If ⊕ denotes disjoint union and EXP denotes the set of (labelled) CCS expressions, then we let:

FD = ⊕ { FD(E) : E in EXP } ;   D = ⊕ {D(E) : E in EXP} ;   and so on.

Therefore, an element of D can be represented as a pair: <E,Y> , with Y in D(E). Now, on D we can define, for x=a,w,s :

dx(<E,Y> , <E,Z>) = dx(E)(<Y,Z>) ;   dx(<E,Y> , <G,Z>) = 2 , when E≠G.

It is immediate to check that dx is a distance on D (notice that the diameter of D(E) w.r.t. dx(E) is at most 1). Moreover, $\{<E_n, Y_n>, n\geq 0\}$ is a Cauchy sequence in (D,dx) iff $\exists k\geq 0\ \exists E$ s.t. $\forall n\geq k$. $E_n = E$ and $\{<E, Y_n>, n\geq k\}$ is a Cauchy sequence in (D(E),dx(E)).

It is clear then that theorem 4.5 and the auxiliary lemmas hold also when FD , D, ..., da, dw and ds are those just defined.

CONCLUDING REMARKS

The final "bringing everything together" should not mislead the reader. What we have shown here is that fair computations, taken individually, can be characterized as limits of finite derivations. We do not know whether our setting allows a similar characterization ot the fairness of global behaviours keeping approximants finite.

The other question is whether we really need all the information we carry around: labels, intermediate steps in derivations (the E_i's) and so on. In |DM|, Degano and

Montanari characterize weakly fair computations as limits of finite "histories", which are more abstract than our derivations. On the other hand, the information contained in these histories seems insufficient to obtain the analogous result for strong fairness ("local fairness" in |DM|). The distance proposed in |DM|, d_3, seems the most stringent that can be defined in that framework, nevertheless it does not give the wanted results. If we translate d_3 in our framework we obtain a distance on finite derivations only; let us call it d_3'. If $(\overline{FD},\bar{d}_3')$ denotes the completion of (FD,d_3'), then $\overline{FD}$ is in bijection with D^w (but $\bar{d}_3'$ is not equivalent to dw). In other words, the completion contains also the infinite computations which are weakly fair but fail to be strongly fair. The infinite strongly fair computations can only be characterized through "canonical" Cauchy sequences (of finite derivations). We do not see an immediate analogue of our distance ds within the framework in |DM|: histories lack the information we use in defining ds. Our feeling is that (part of) this additional information is actually needed. If one could make this conjecture precise and prove it, this result would be - in our opinion - much more interesting than those presented here.

On the positive side, we can say that, as pointed out by M. Hennessy |H|, the present work is not confined to CCS. Our characterization is based on the sets LP(-) and P(-) and more precisely on their properties, rather than their explicit definition. Therefore, everything could be "lifted up" to the more abstract setting of (general labelled) transition systems, axiomatizing the notion of subprocess and "live" subprocess.

ACKNOWLEDGEMENTS.

I would like to thank Pierpaolo Degano and Ugo Montanari for discussions concerning their paper and Edmund Robinson for suggesting a better presentation of some of the results. Robin Milner and Gordon Plotkin should be thanked for their encouragement. Finally, I have relied heavily on the work on fairness done together with Colin Stirling he has also made useful comments on an earlier version of this paper.

REFERENCES (*)

AN A. ARNOLD, M. NIVAT. Metric interpretations of infinite trees and semantics of non-deterministic recursive procedures. T.C.S. 11 (1980) 181-205.

BZ J.W. DE BAKKER, J.I. ZUCKER. Processes and a fair semantics for the ADA rendez-vous. ICALP'83, LNCS 154 (1983) 52-66.

C G. COSTA. A metric characterization of fair computations in CCS. Technical Rep. CSR-169-84, Dept. Comput. Sci. Univ. Edinburgh (1984).

CS1 G. COSTA, C. STIRLING. A fair calculus of communicating systems. To appear in Acta Informatica; shortened version: FCT'83, LNCS 158 (1983) 94-105.

CS2 G. COSTA, C. STIRLING. Weak and strong fairness in CCS. Technical Rep. CSR-167-84, Dept. Comput. Sci. Univ. Edinburgh (1984); shortened version: MFCS'84, LNCS 176 (1984) 245-254.

DM P. DEGANO, U. MONTANARI. Liveness properties as convergence in metric spaces. Proc. 16th ACM STOC (1984).

H M. HENNESSY. Private communication.

M1 R. MILNER. A calculus of communicati_ng systems. LNCS 92 (1980).

M2 R. MILNER. A finite delay operator in synchronous CCS. Technical Rep. CSR-116-82 Dept. Comput. Sci. Univ. Edinburgh (1982).

Pa1 D. PARK. On the semantics of fair parallelism. LNCS 86 (1980) 504-526.

Pa2 D. PARK. A predicate transformer for weak fair iteration. Proc. 6th IBM Symp. on Mathematical Foundat. of Comput. Sci. , Hakone, Japan (1981).

Pa3 D. PARK. The "fairness" problem and nondeterministic computing networks. Foundat. of Comput. Sci. IV, De Bakker - Van Leuven edit. Amsterdam (1982).

Pl G. PLOTKIN. A powerdomain for countable nondeterminism. ICALP'82, LNCS 140 (1982) 418-482.

========================

(*) LNCS n stands for Lecture Notes in Computer Science Vol.n, Springer.

APPENDIX : AN EXAMPLE

We use the following abbreviations: $\underline{n}$ denotes the string composed of n 1's; when u is a label, E^{u} denotes the term obtained by prefixing all labels in E with u.

Now let:

$$F = (\mathrm{fix}X)_{\varepsilon} \cdot a_{\underline{2}} b_{\underline{3}} X_{\underline{4}} +_{1} \bar{c}_{12} X_{121} \quad ; \quad H = (\mathrm{fix}X)_{\varepsilon} \cdot c_{1} X_{11} \quad ; \quad E_0 = (F^{\underline{2}} \,|_{1}\, H^{12})(\backslash c)_{\varepsilon}$$

According to the definitions above we have, say, $\underline{3}21 = 11121$ and

$$F^{\underline{2}} = (\mathrm{fix}X)_{\underline{2}} \cdot a_{\underline{4}} b_{\underline{5}} X_{\underline{6}} +_{\underline{3}} \bar{c}_{\underline{3}2} X_{\underline{3}21} \quad \text{and} \quad H^{12} = (\mathrm{fix}X)_{12} \cdot c_{121} X_{1211} \ .$$

Then, a derivation from E_0 in the labelled calculus is

$$Y = E_0 \xrightarrow{a} E_1 \xrightarrow{b} E_2 \xrightarrow{a} E_3 \xrightarrow{b} E_4 \xrightarrow{a} E_5 \xrightarrow{b} E_6 \xrightarrow{\tau} E_7 \xrightarrow{a} E_8 \ .$$

The expressions for E_i (suppressing some of the labels) are given in the table below, where $u = \underline{15}21$ and $v = 121$. The table also shows the sets $LP(E_i)$ and $P(E_i)$ and the values of $lw(Y,i)$, $ls(Y,i)$ and $LS(Y,i)$, for $0 \le i \le 8$.

i	E_i	$P(E_i)$	$LP(E_i)$	$lw(Y,i)$	$ls(Y,i)$	$LS(Y,i)$
0	$(\ F^{\underline{2}} \,\vert\, H^{12})\backslash c$	$\{\underline{3}, v\}$	$\{\underline{3}, v\}$	0	0	0
1	$(b_{\underline{5}} F^{\underline{6}} \,\vert\, H^{12})\backslash c$	$\{\underline{5}, v\}$	$\{\underline{5}\}$	0	0	0
2	$(\ F^{\underline{6}} \,\vert\, H^{12})\backslash c$	$\{\underline{7}, v\}$	$\{\underline{7}, v\}$	1	0	0
3	$(b_{\underline{9}} F^{\underline{10}} \,\vert\, H^{12})\backslash c$	$\{\underline{9}, v\}$	$\{\underline{9}\}$	2	2	0
4	$(\ F^{\underline{10}} \,\vert\, H^{12})\backslash c$	$\{\underline{11}, v\}$	$\{\underline{11}, v\}$	3	0	0
5	$(b_{\underline{13}} F^{\underline{14}} \,\vert\, H^{12})\backslash c$	$\{\underline{13}, v\}$	$\{\underline{13}\}$	4	4	0
6	$(\ F^{\underline{14}} \,\vert\, H^{12})\backslash c$	$\{\underline{15}, v\}$	$\{\underline{15}, v\}$	5	0	0
7	$(\ F^{u} \,\vert\, H^{v1})\backslash c$	$\{u1, v\underline{2}\}$	$\{u1, v\underline{2}\}$	6	6	6
8	$(b_{u\underline{3}} F^{u\underline{4}} \,\vert\, H^{v1})\backslash c$	$\{u\underline{3}, v\underline{2}\}$	$\{u\underline{3}\}$	7	7	7

Moreover, recalling that $Y[i] = E_0 \xrightarrow{a} \ldots \xrightarrow{\ .\ } E_i$, we have, for instance:

$LS(Y[5],i) = 0$, $0 \le i \le 4$; $\quad LS(Y[5],5) = 4$; $\quad LS(Y[6],i) = 0$, $0 \le i \le 6$.

$$ds(Y_3,Y_4) = [LS(Y_3,3) + 1]^{-1} + [LS(Y_4,3) + 1]^{-1}$$
$$= [ls(Y_3,3) + 1]^{-1} + [ls(Y_4,4) + 1]^{-1} = 1/3 + 1 \ ;$$

$$ds(Y_3,Y_i) = 1/3 + 1 \ , \quad i=5,6,7,8 \ ;$$

$$ds(Y_7,Y_8) = [ls(Y_7,7) + 1]^{-1} + [ls(Y_8,7) + 1]^{-1} = 1/(6+1) + 1/(6+1).$$

A COMPLETE MODAL PROOF SYSTEM FOR A SUBSET OF SCCS *

Colin Stirling
Dept. of Computer Science
Edinburgh University
Edinburgh, U.K.

Abstract

Logical proof systems for concurrent programs are notoriously complex, often involving arbitrary restrictions. One of the main reasons for this is that unlike other major programming concepts parallelism does not appear to have a logical correlate. Here using a simple semantic strategy we tentatively propose one and offer an example modal proof system for a subset of Milner's SCCS. The proof rules are reminiscent of Gentzen introduction rules except that there are also introduction rules for the operators of the program language.

Introduction

Logical proof systems for concurrent programs are notoriously complex, often involving arbitrary restrictions: a representative sample is [AFR,L,LG,OG,ZBR]. There does not seem to be a clean way of reasoning from the parts of a system to the whole. It is very hard to deduce even a weak property of a concurrent system without knowing a good deal about its subcomponents: unexpected interactions can arise and are all too common in practice.

Suppose p is a program and $p \models A$ means that p satisfies (the property expressed by) the formula A. For a standard binary commutative and associative parallel operator $\|$ there is little hope of finding an interesting binary function * on arbitrary formulas which validates an unrestricted implication of the form:

$$\text{if } p \models A \text{ and } q \models B \text{ then } p \| q \models A * B$$

No general method is available for deriving information about a concurrent program from little bits of information about its parts. Unlike other major programming notions parallelism does not appear to have a logical correlate. Here we tentatively propose one and offer an example proof system. The suggestion is a rationalization of ideas in [BK,BKP], and arises from a simple semantic strategy.

Let $\models_B$ be a semantic relation relativized to a formula B which (partially) describes an 'environment': $p \models_B A$ is stipulated to mean

$$\text{for any program } q, \text{ if } q \models B \text{ then } q \| p \models A$$

From $p \models_B A$ and $q \models B$ we can, therefore, derive information about $p \| q$. An immediate consequence of associativity of $\|$ is:

---

* This work was supported by the Science and Engineering Research Council of the U.K.

$$\text{if } p \models_A B \text{ and } q \models_B C \text{ then } p \| q \models_A C$$

For suppose r is any program satisfying A then $(r\|p)\|q \models C$ and so also $r\|(p\|q) \models C$. In principle, we have a method of reasoning about concurrent programs with arbitrary numbers of components. Again, in principle, we have a compositional semantics of $\|$: in terms of the pair of semantic relations the meaning of a concurrent system can be built from the meanings of its components.

This semantic ploy suggests we introduce two proof-theoretic consequence relations $\vdash$ and $\vdash_B$ to coincide with the semantic pair. Resulting proof rules (introduction rules) for $\|$ are then straightforward:

$$\frac{p \vdash B \qquad q \vdash_B C}{p \| q \vdash C} \qquad\qquad \frac{p \vdash_A B \qquad q \vdash_B C}{p \| q \vdash_A C}$$

Such rules, unlike most logics for parallel programs, will not presuppose (as a proof proceeds) either a fixed or a bounded number of potential concurrent subcomponents. They allow one to treat $\|$ as a first class program operator on a par with sequential composition ';'. From a program point of view the formula B in $\models_B$ suggests an environment description. Logically, it suggests an assumption : $p \models_B A$ could be written as $B, p \models A$. The proposed proof rules for $\|$ are, therefore, analogous to Gentzen's cut rule [G] in the form:

$$\frac{A, \Gamma \vdash B \qquad\qquad B, \Delta \vdash C}{A, \Gamma, \quad \vdash C}$$

(They are also analogous to the sequential composition rule of Hoare logic.)

Using this strategy we offer a sound and complete modal proof system for a subset of Milner's Synchronous Calculus of Communicating Systems, SCCS [Mi3]. The binary parallel operator is a synchronous parallel, a tight coupling. The modal language used is Hennessy-Milner logic [HM1,HM2] which has the virtue, unlike more standard program logics, that its expressiveness is tied to a logic independent criterion, namely bisimulation equivalence. This together with the theoretical simplicity of SCCS considerably aids the development of the modal proof system: the proof rules are introduction rules not only for the logical operators but also for the combinators of the program language. The parallel introduction rules include the pair offered above. This work extends results presented in [St1].

The paper is in five sections followed by a conclusion. Proofs of the results will be contained in a fuller version. The first three sections are introductory: sections 1 and 2 describe Hennessy-Milner logic and its logic independent criterion of expressibility; section 3 outlines the subset of SCCS we build a proof theory for. In section 4 we examine the kind of proof rules we would like and in section 5 the proof system together with example proofs is given.

1. Transition Systems and Bisimulation Equivalence

A nondeterministic or concurrent program may communicate repeatedly with its environment. A simple input/output function is, therefore, too austere as a model of such a program: two programs determining the same input/output function may behave very differently in the same environment. Transition systems have long been recognized as a richer model [K,P1,Si1]. More recently they have been used extensively as models of concurrent programs within the framework of tense (temporal) logic [EH, HS,MP1,MP2,QS]. Here our interest is transition systems whose transitions are labelled.

<u>Definition 1.1</u> A transition system is a triple $\langle P, Act, \longrightarrow\rangle$ where

i. P is a set (of processes)

ii. Act is a set (of actions)

iii. $\longrightarrow$ is a mapping which associates with each $a \in Act$ a relation $\xrightarrow{a} \subseteq P \times P$.

A transition system T is <u>finite branching</u> provided that each relation $\xrightarrow{a}$, $a \in Act$, is image finite: $\xrightarrow{a}$ is image finite if for each $p \in P$ the set $\{q \mid p \xrightarrow{a} q\}$ is finite.

In [HM1,HM2] the authors offer, in effect, a very intuitive understanding of a transition system. The set Act is viewed as a set of atomic experiments. An atomic experiment on a process (program) $p \in P$ is understood as an attempt to communicate with p. Communication may change a process depending on its internal structure. The relations $\xrightarrow{a}$ $a \in Act$, are intended to capture the effect of experimentation : $p \xrightarrow{a} q$ means that p can evolve to q in response to an a experiment, or q is the result of a successful a experiment on p. A computation can be viewed as a successful sequence of experiments (communications). Similar ideas are also contained in [DeH,HBR,Ho,Mo].

Hennessy and Milner propose that two processes (programs) should be equivalent (have the same meaning) when no amount of finite experimentation distinguishes them. A formal criterion is offered which is the same as bisimulation equivalence when T is finite branching.

<u>Definition 1.2</u> A relation $R \subseteq P \times P$ on T is a bisimulation just in case

pRq iff i. $\forall a \forall p'.$ if $p \xrightarrow{a} p'$ then $\exists q'.\ q \xrightarrow{a} q'$ and $p'Rq'$

ii. $\forall a \forall q'.$ if $q \xrightarrow{a} q'$ then $\exists p'.\ p \longrightarrow p'$ and $p'Rq'$

This definition characterizes a property a relation may or may not have on T. (The identity relation, for instance, is a bisimulation.) Such relations give rise to a natural equivalence, bisimulation equivalence, on processes in T:

$$p \sim_T q \text{ iff there exists a bisimulation R such that pRq}$$

It is straightforward to check that $\sim_T$ is an equivalence and, moreover, that it is also the maximal bisimulation under inclusion. Bisimulation equivalence is a very fine equivalence. For instance, consider the transition system given by example 1.3.

Example 1.3

Note that $p_2 \not\sim_T q_2$ and $p_2 \not\sim_T q_3$ because p_2 can respond successfully to both b and c experiments whereas q_2 and q_3 each fail one of the pair. Consequently, $p_1 \not\sim_T q_1$. Thus, if two processes have the same computations (may respond to the same sequences of experiments) this does not guarantee their equivalence. Strong connections between their respective intermediate states are also required. It is precisely these sorts of connections which are, in general, needed for comparing the behaviours of concurrent programs.

Bisimulations have been investigated in [Mi3,Pa,Si2]. Alternative equivalences based on experimental indistinguishability can be found in [Ab,DeH,He2,HBR,Mo,RB]. There is a need to allow infinite experiments when modelling fairness. Bisimulation equivalence is then no longer sufficient. The result is even finer equivalences [He1,Mi2]. However, fairness does not arise in the process language we examine later because its parallel operator is that of tight coupling.

2. Hennessy-Milner Logic

Hennessy and Milner present a modal logic which characterizes bisimulation equivalence on finite branching transition systems [HM1,HM2]. We offer here a negation free version of their logic: the avoidance of negation aids the development of the modal proof theory provided in the sequel. Let $T = (P, Act \longrightarrow)$ be a transition system and L_T the modal language:

$$A ::= Tr \mid False \mid A \wedge A \mid A \vee A \mid \langle a \rangle A \mid [a]A \quad \text{where } a \in Act$$

L_T is reminiscent of propositional dynamic logic [Ha]. Here, however, only atomic actions appear within the modal operators. Furthermore, unlike dynamic logic, the satisfaction relation $\models$ is defined between processes and formulas. $\models \subseteq P \times L_T$ is the least relation such that:

$$
\begin{array}{lll}
p \models Tr & \text{for all } p \in P \\
p \not\models False & \text{for all } p \in P \\
p \models A \wedge B & \text{iff } p \models A \text{ and } p \models B \\
p \models A \vee B & \text{iff } p \models A \text{ or } p \models B \\
p \models \langle a \rangle A & \text{iff } \exists q, p \xrightarrow{a} q \text{ and } q \models A \\
p \models [a]A & \text{iff } \forall q. \text{ if } p \xrightarrow{a} q \text{ then } q \models A
\end{array}
$$

The only atomic formulas are Tr and False: Tr stands for true which every process satisfies whereas False does not hold of any process. $p \models \langle a \rangle A$ means that p can evolve under some successful a experiment to a process satisfying A. Likewise, $p \models [a]A$ means that every process which is the result of a successful a experiment on p satisfies A. In particular, $p \models [a]False$ means that p is a-deadlocked: no a experiment on p can be successful. This modal language is, therefore, expressively rich; it can say of processes not only what they can do but also what they can't do. This power of distinguishing is required for the characterization of bisimulation equivalence.

Let $L_T(p) = \{A | p \models A\}$ then Hennessy and Milner (in effect) prove the following theorem:

Theorem 2.1 If T is finite branching then $L_T(p) = L_T(q)$ iff $p \sim_T q$

The properties expressible in L_T are tied to the distinguising powers of bisimulation equivalence: the logic L_T cannot differentiate processes which are bisimulation equivalent and vice versa. For instance, the processes p_1, q_1 of example 1.3 are distinguishable by L_T formulas:

$$p_1 \models [a](\langle b\rangle Tr \wedge \langle c\rangle Tr) \qquad q_1 \not\models [a](\langle b\rangle Tr \wedge \langle c\rangle Tr)$$
$$p_1 \not\models \langle a\rangle [b]False \qquad q_1 \models \langle a\rangle [b]False$$

Thus, a virtue of L_T, unlike programming logics in general, is that its criterion for expressiveness lies completely outside logic. However, we may wonder how this expressiveness can be translated into expression of particular process properties. If Act in T is finite than $\bigwedge_{a\in Act} [a]False$ expresses deadlock or termination, and its dual $\bigvee_{a\in Act} \langle a\rangle Tr$ may be said to express a form of liveness. More generally, the formula or set of formulas expressing a particular property like absence of deadlock will depend upon the particular process under consideration. The assumption that T be finite branching in theorem 2.1 can be discarded if infinite disjunctions and conjunctions are allowed in L_T [HS,Mi1]. Further modal characterizations of equivalences can be found in [BR,GS,HS,Mi1].

The description of Hennessy-Milner logic, L_T, here is semantic. Our aim is to develop a proof system on L_T for a particular process language containing a binary commutative and associative parallel operator. The process languages we choose is a subset of Milner's SCCS.

3. A Subset of SCCS

Milner developed the Synchronous Calculus of Communicating Systems, SCCS, as a tractable model of systems which interact synchronously [Mi3]. It is a transition system $T = (P, Act, \longrightarrow)$ whose set of processes P is built up from as few combinators or operators as possible, each of which is intended to capture a distinctive intuitive concept. This makes SCCS ideal for the sort of proof system we wish to develop. Here we only consider a subset of SCCS, namely SCCS with only finite summation, and without restriction and renaming.

Processes in SCCS evolve relative to some universal discrete time. If $p \xrightarrow{a} p'$ and $q \xrightarrow{b} q'$ then the synchronous parallel of p and q responds to the product of the experiments a and b, $a \times b$, and evolves to the parallel of p' and q'. Product of actions is captured by a structure on Act. Milner assumes that (Act,$\times$,1) is an abelian monoid: $\times$ is both commutative and associative with 1 as identity. For simplicity we further assume the left cancellation law:

$$\text{if } a \times b = c \times b \text{ then } a = c$$

The right cancellation law also holds because $\times$ is commutative. We abbreviate $a \times b$ to ab. If ab = c then we let $c\backslash a = b$ (and $c\backslash b = a$): by the cancellation laws if $d\backslash e$ exists it is unique.

The set of processes P of SCCS we consider is given by the <u>closed</u> <u>expressions</u> of the following process language where Z ranges over process variables

$$p ::= Z \mid \mathbb{0} \mid a.p \mid \text{fix } Z.p \mid p + p \mid p \times p \quad \text{with } a \in \text{Act}$$

$\mathbb{0}$ stands for 'disaster': it is a process which cannot respond to any experiment. The process a.p responds to a, and evolves to p. Potentially infinite computations are allowed by the recursion combinator fix Z which binds free occurrences of Z in p in the process fix Z.p. We impose a syntactic restriction on fix Z.p, that Z is guarded in p: that is, every free occurrence of Z in p is within a subexpression a.q of p. Without this restriction the resulting transition system would not be finite branching. The operator + represents external nondeterministic choice: the experimenter may resolve the choice. Finally, $\times$ represents synchronous parallelism.

The remaining undefined feature of the transition system T is the transition relation $\longrightarrow$. It is defined as the least set such that:

$a.p \xrightarrow{a} p$

fix $Z.p \xrightarrow{a} p'$ whenever $p[\text{fix } Z.p/Z] \xrightarrow{a} p'$ where $[\cdot/\cdot]$ denotes substitution

$p + q \xrightarrow{a} r$ whenever $p \xrightarrow{a} r$ or $q \xrightarrow{a} r$

$p \times q \xrightarrow{bc} p' \times q'$ whenever $p \xrightarrow{b} p'$ and $q \xrightarrow{c} q'$

The process a.p can only respond to an a experiment and in so doing evolves to p. In a.p + b.q, where $a \neq b$, the experimenter may resolve the choice: the offer of an a experiment results in p whereas the offer of a b experiment results in q. This is not true of a.p + a.q : the experimenter has no control on whether p or q is the result of an a experiment. The number of concurrent subprocesses may increase in response to an experiment. This only happens when the concurrent combinator $\times$ occurs within the scope of a fix Z; for instance, if $p = \text{fix } Z(aZ \times bZ)$ then $p \xrightarrow{ab} p \times p$. This possibility of growth must be reflected in any logical proof system for this language of processes. For a full discussion of SCCS with examples see [Mi3].

Bisimulation equivalence is not only an intuitively natural equivalence on the transition system T it is also a congruence: process contexts preserve equivalence [Mi3]. The following implies that T is finite branching.

<u>Fact 3.1</u> $\forall p.\ \{q \mid \exists a.\ p \overset{a}{\Rightarrow} q\}$ is finite

By theorem 2.1 we know that the modal logic L_T characterizes bisimulation equivalence, $\sim_T$, on T. The following fact states that the parallel operator $\times$ is both commutative and associative up to $\sim_T$ [Mi3].

<u>Fact 3.2</u> i. $p \times q \sim_T q \times p$

ii. $p \times (q \times r) \sim_T (p \times q) \times r$

This means that formulas of L_T cannot distinguish between these equivalent processes.

The process $\mathbb{0}$ cannot respond to any experiment. It is, therefore, deadlocked: $\mathbb{0} \models [a]\mathrm{False}$ for every $a \in \mathrm{Act}$. Moreover so is $\mathbb{0} \times p$ for any p. As remarked in section 1 processes which may respond to the same sequences of experiments need not be bisimulation equivalent. The following example illustrates this where $\bar{a}$ is the inverse of a; that is $a\bar{a} = 1$.

<u>Example 3.3</u> $p_i \times q \not\sim_T p_j \times q$ for $1 \leq i < j \leq 3$ where

$q = \mathrm{fix}\ Z.\ \bar{a}.\ Z$

$p_1 = \mathrm{fix}\ Z.\ a.\ Z$

$p_2 = a(a.\mathbb{0} + \mathrm{fix}\ Z.a.Z)$

$p_3 = \mathrm{fix}\ Z.(a.\mathbb{0} + a.Z)$

The three processes $p_i \times q$, $1 \leq i \leq 3$ satisfy every formula in the set $\{\langle 1\rangle Tr, \langle 1\rangle\langle 1\rangle Tr, \ldots\}$. (Note the only experiment they can ever respond to is 1.) However, they are not bisimulation equivalent because of differences in possibilities of deadlock. The process $p_1 \times q$ is deadlock free whereas $p_2 \times q$ can only deadlock in one circumstance unlike $p_3 \times q$ which can always deadlock. The process $p_3 \times q$ satisfies every formula in the set $\{\langle 1\rangle[1]\mathrm{False}, \langle 1\rangle\langle 1\rangle[1]\mathrm{False}, \ldots\}$ whereas $p_2 \times q$ only satisfies $\langle 1\rangle\langle 1\rangle[1]\mathrm{False}$ and fails the rest; $p_1 \times q$, on the other hand, fails them all.

4. <u>Towards a Modal Proof Theory: A Relativized Satisfaction Relation</u>

Our aim is to offer a sound and complete modal proof system on L_T for the subset of SCCS outlined. We want, therefore, to define a proof-theoretic consequence relation $\vdash$ which coincides with $\models$. Ideally, the proof rules will be Gentzen style introduction rules [G]. But, we also need to take account of the structure of processes. The theoretical simplicity of the process language suggests that we also offer introduction rules for the combinators. The question then arises as to what, if anything, is the logical correlate of the combinators. In [St1] we provided proof rules for an even more restricted process language, a language devoid of both concurrency and recursion.

Introduction rules for . in a.p. are straightforward. The following schemas suffice:

$$\frac{p \vdash A}{a.p \vdash \langle a\rangle A} \qquad \frac{p \vdash A}{a.p \vdash [a]A}$$

Their justification is that a.p. evolves to p under any a experiment. The logical correlate of . is, therefore, modal iteration. Consequently, these are also <a> and [a] introduction rules.

A global + introduction rule of the form

$$\frac{p \vdash A \qquad q \vdash B}{p + q \;\vdash f(A,B)}$$

where f is truth-functional and not the constant true function is always unsound: this is shown in [St1]. Restricted versions of such a rule, however, which depend on the form of A and B can be found:

$$\frac{p \vdash \langle a\rangle A}{p + q \vdash \langle a\rangle A} \qquad \frac{q \vdash \langle a\rangle A}{p + q \vdash \langle a\rangle A} \qquad \frac{p \vdash [a]A \qquad q \vdash [a]A}{p + q \vdash [a]A}$$

Their justification is that p + q only evolves to a process which either p evolves to or q evolves to. A restricted metalogical 'or' or 'and' is the correlate of + : if <a>A is true of p or of q it is true of p + q; if [a]A is true of p and of q it is true of p + q.

A global × introduction rule suffers the same fate as a + global rule. Unlike the + case, however, restricted versions which depend on the forms of A and B are inadequate. Such rules would also need to take into account most, if not all, the modal subformulas of A and B. Even if such rules could be found they would be in opposition to the style of proof rules we are suggesting. An alternative approach, a rationalization of [BK,BKP], which fits in with the style of rules suggested already, is now offered. This approach was outlined in the introduction.

We complicate the semantics of L_T, when T is the subset of SCCS outlined, by introducing a relativized satisfaction relation $\models_A$ where A is a formula of L_T. We stipulate that

$$p \models_A B \;\text{ iff }\; \forall q.\ \text{if } q \models A \;\text{ then }\; q \times p \models B$$

The pair of semantic relations, $\models$, $\models_A$ gives a compositional semantics for concurrency: the samentics of a concurrent system are built up from the semantics of the components. Recall Fact 3.2 that × is both commutative and associative (up to $\sim_T$). By commutativity if $q \models A$ and $p \models_A B$ then also $p \times q \models B$. By associativity the following holds

<u>Fact 4.1</u> If $p \models_A B$ and $q \models_B C$ then $p \times q \models_A C$

Consequently, if q is the concurrent process with components p_j $0 \leq j \leq n$, in any order and $p_0 \models A_0$ and $p_i \models_{A_{i-1}} A_i$ $1 \leq i \leq n$ then $q \models A_n$.

This semantic ploy suggests that we introduce a second proof-theoretic consequence relation $\vdash_A$. Natural introduction rules for $\times$ then arise:

$$\frac{q \vdash A \quad p \vdash_A B}{q \times p \vdash B} \qquad \frac{q \vdash A \quad p \vdash_A B}{p \times q \vdash B}$$

and

$$\frac{q \vdash_A B \quad p \vdash_B C}{q \times p \vdash_A C} \qquad \frac{q \vdash_A B \quad p \vdash_B C}{p \times q \vdash_A C}$$

Computationally, A in $p \models_A B$ can be viewed as an environment description, a (possibly partial) summary of any process q such that $q \times p$ satisfies B. Logically, A can be viewed as an assumption : $p \models_A B$ could be rewritten $A, p \models B$. This suggests a logical correlate of $\times$ introduction - in fact, a logical analogy - namely Gentzen's cut rule [G] in the form

$$\frac{A, \Gamma \vdash B \qquad B, \Delta \vdash C}{A, \Gamma, \Delta \vdash C}$$

There is also a similarity to Hoare's (introduction) rule for sequential composition (when $p \vdash_A B$ is written A{p}B). From now on $A,p \models B$ ($A,p \vdash B$) is written instead of $p \models_A B$($p \vdash_A B$).

The additional semantic relation means that introduction rules for two consequence relations $\vdash$ and $\vdash_B$ need to be offered. These relations will be connected by the first pair of $\times$ introduction rules above. Introduction rules for the . and + combinators in the context of $\vdash_B$ are straightforward and not dissimilar from above:

$$\frac{A, p \vdash B}{\langle b \backslash a\rangle A,\ a.p \vdash \langle b\rangle B} \text{ if } b\backslash a \text{ exists} \qquad \frac{A, p \vdash B}{[b \backslash a]A,\ a.p \vdash [b]B} \text{ if } b\backslash a \text{ exists}$$

$$\frac{A, p \vdash \langle a\rangle B}{A, p+q \vdash \langle a\rangle B} \qquad \frac{A, q \vdash \langle a\rangle B}{A, p+q \vdash \langle a\rangle B} \qquad \frac{A, p \vdash [a]B \quad A, q \vdash [a]B}{A, p+q \vdash [a]B}$$

Left unmentioned are introduction rules for fix. The behaviour of fix Z.p is however, fully determined by repeated 'unfolding' : an unfolding of fix Z.p is p[fix Z.p/Z]. (Recall that fix Z.p $\xrightarrow{a}$ q whenever p[fix Z.p] $\xrightarrow{a}$ q.) Contextual introduction rules for fix Z are offered which appeal to this unfolding. The rules depend upon the modal degree of a formula A, m(A), which is inductively defined as the maximum depth of modal operators in A:

$$\begin{aligned} m(Tr) &= m(False) &&= 0 \\ m(A \vee B) &= m(A \wedge B) &&= \max(m(A), m(B)) \\ m(\langle a\rangle A) &= m([a]A) &&= 1 + m(A) \end{aligned}$$

If $p \models A$ and m(A) = n then A is a property of p's evolution through at most n processes; a property of computations from p of length at most n. If p = fix Z.q then A is at most a property of the nth 'unfolding' of p given that Z is guarded in q. Consequently, we can appeal to standard approximation techniques: when p = fix Z.q then p^n, $n \geq 0$, is defined inductively:

$$p^0 = \mathbb{0}$$
$$p^{n+1} = q[p^n/Z]$$

For instance, if p = fix Z. a.Z then $p^0 = \mathbb{0}$ and $p^n = \overbrace{a.a.\ \ldots\ a.}^{n \text{ times}}\mathbb{0}$, $n > 0$. Hence, p^n, $n \geq 0$, can respond in the same way as fix Z.q to any sequence of experimenting whose length is less than or equal to n: such experimenting is summed up in L_T by formulas whose modal degree is less than or equal to n. The outcome is the following pair of rules for p = fix Z.q

$$\frac{p^n \vdash A}{p \vdash A}\ m(A) \leq n \qquad \frac{A,\ p^n \vdash B}{A,\ p \vdash B}\ m(B) \leq n$$

This method of dealing with fix was suggested by Gerardo Costa.

5. A Complete Modal Proof System for T

The full proof system on L_T for T, the subset of SCCS, is now given.

Axioms $p \vdash Tr$ $\quad \mathbb{0} \vdash [a]A$ $\quad A,p \vdash Tr$ $\quad False, p \vdash A$ $\quad A,\mathbb{0} \vdash [a]B$

$a.p \vdash [b]A$ if $a \setminus b$ $\quad A, a.p \vdash [b]B$ if $b \setminus a$ doesn't exist

$$\vee I \quad \frac{p \vdash A}{p \vdash A \vee B} \qquad \frac{p \vdash B}{p \vdash A \vee B} \qquad \frac{A,p \vdash B}{A,p \vdash B \vee C} \qquad \frac{A,p \vdash C}{A,p \vdash B \vee C} \qquad \frac{A,p \vdash C \quad B,p \vdash C}{A \vee B, p \vdash C}$$

$$\wedge I \quad \frac{p \vdash A \quad p \vdash B}{p \vdash A \wedge B} \qquad \frac{A,p \vdash C}{A \wedge B, p \vdash C} \qquad \frac{B,p \vdash C}{A \wedge B, p \vdash C} \qquad \frac{A,p \vdash B \quad A,p \vdash C}{A,p \vdash B \wedge C}$$

$$\langle a\rangle I \quad \frac{p \vdash A}{a.p \vdash \langle a\rangle A} \qquad \frac{A,p \vdash B}{\langle a\setminus b\rangle A, b.p \vdash \langle a\rangle B}\ a\setminus b \text{ exists}$$

$$[a]I \quad \frac{p \vdash A}{a.p \vdash [a]A} \qquad \frac{A,p \vdash B}{[a\setminus b]A, b.p \vdash [a]B}\ a\setminus b \text{ exists}$$

$$+I\langle\rangle \quad \frac{p \vdash \langle a\rangle A}{p+q \vdash \langle a\rangle A} \qquad \frac{q \vdash \langle a\rangle A}{p+q \vdash \langle a\rangle A} \qquad \frac{A,p \vdash \langle a\rangle B}{A,p+q \vdash \langle a\rangle B} \qquad \frac{A,q \vdash \langle a\rangle B}{A,p+q \vdash \langle a\rangle B}$$

$$+I[\,] \quad \frac{p \vdash [a]A \quad q \vdash [a]A}{p+q \vdash [a]A} \qquad \frac{A,p \vdash [a]B \quad A,q \vdash [a]B}{A,p+q \vdash [a]B}$$

$$\times I \quad \frac{p \vdash A \quad A,q \vdash B}{p \times q \vdash B} \qquad \frac{p \vdash A \quad A,q \vdash B}{q \times p \vdash B} \qquad \frac{A,p \vdash B \quad B,q \vdash C}{A,p \times q \vdash C} \qquad \frac{A,p \vdash B \quad B,q \vdash C}{A,q \times p \vdash C}$$

$$\text{fix } I \quad \frac{p^n \vdash A}{p \vdash A}\ \begin{array}{l} p = \text{fix } Z.q \\ n \geq m(A) \end{array} \qquad \frac{A,p^n \vdash B}{A,p \vdash B}\ \begin{array}{l} p = \text{fix } Z.q \\ n \geq m(B) \end{array}$$

The axioms and rules are reasonably straightforward. The axioms $p \vdash Tr$ and $A,p \vdash Tr$ hold because every process satisfies Tr. In contrast, the axiom $False,p \vdash A$

holds because no process satisfies False. The axioms for $\mathbf{0}$ depend upon its inability to respond to any experiment. (Recall that $p \models [a]\mathrm{False}$ says that p is a-deadlocked.) The final three axioms are justified by noting that a.p can only respond to a. The $\vee$I and $\wedge$I rules are as expected. Note, however, that these include introduction rules for the 'environment' or assumption formula. The presence of the assumption complicates the $\langle a\rangle$I and [a]I rules as expected. Most of the rest of the rules were mentioned in the previous section. We illustrate the use of some of the rules with an example proof where, as before, $\bar{d}$ for any $d \in \mathrm{Act}$, is the inverse of d; $d\bar{d} = 1$.

Example 5.1 $\quad a(b.\mathbf{0} + c.\mathbf{0}) \times (\mathrm{fix}\, Z(\bar{a}.1.z) \times 1.\bar{b}.\mathbf{0}) \vdash [1]\langle 1\rangle Tr$

$$\times I\;\dfrac{[a]I\;\dfrac{+\langle\rangle I\;\dfrac{\langle b\rangle I\;\dfrac{\mathbf{0}\vdash Tr}{b.\mathbf{0}\vdash \langle b\rangle Tr}}{b\mathbf{0}+c\mathbf{0}\vdash \langle b\rangle Tr}}{a(b\mathbf{0}+c\mathbf{0})\vdash [a]\langle b\rangle Tr} \qquad x\, I\;\dfrac{\mathrm{fix}\, I\;\dfrac{[1]I\;\dfrac{\langle b\rangle I\;\dfrac{Tr, \quad \bar{a}.1.\mathbf{0}\vdash Tr}{\langle b\rangle Trm \quad 1.\bar{a}.1.\mathbf{0}\vdash \langle b\rangle Tr}}{[a]\langle b\rangle Tr,\; \bar{a}.1.\bar{a}.1.\mathbf{0}\vdash [1]\langle b\rangle Tr}}{[a]\langle b\rangle Tr,\; \mathrm{fix}\, Z.\bar{a}.1.Z\vdash [1]\langle b\rangle Tr} \quad [1]I\;\dfrac{\langle 1\rangle I\;\dfrac{Tr, \quad \mathbf{0}\vdash Tr}{\langle b\rangle Tr, \quad \bar{b}.\mathbf{0}\vdash \langle 1\rangle Tr}}{[1]\langle b\rangle Tr, 1.\bar{b}.\mathbf{0}\vdash [1]\langle 1\rangle Tr}}{[a]\langle b\rangle Tr, \quad \mathrm{fix}\, Z(\bar{a}.1.Z) \times 1.\bar{b}.\mathbf{0}\vdash [1]\langle 1\rangle Tr}}{a(b.\mathbf{0}+c.\mathbf{0}) \times (\mathrm{fix}\, Z(\bar{a}.1.Z) \times 1.\bar{b}.\mathbf{0}) \vdash [1]\langle 1\rangle Tr.}$$

In this example proof the final formula $[1]\langle 1\rangle Tr$ is proved of the process $1.\bar{b}.\mathbf{0}$ relative to the assumption $[1]\langle b\rangle Tr$. The proof, however, could have proceeded instead by proving the final formula of either of the other two subprocesses (relative to assumptions). For instance, it is straightforward to show the following pair.

$$[\bar{a}]\langle\bar{b}\rangle Tr, \quad a(b.\mathbf{0} + c.\mathbf{0}) \vdash [1]\langle 1\rangle Tr$$
$$\mathrm{fix}\, Z(\bar{a}.1.Z) \times 1.\bar{b}\mathbf{0} \vdash [\bar{a}]\langle\bar{b}\rangle Tr$$

The conclusion follows by $\times$ I. All concurrent subprocesses then are to this extent on an equal footing.

In the next example we offer a proof of a process whose number of subcomponents grows as it responds to experiments.

Example 5.2 $\quad \mathrm{fix}\, Z(a.Z \times \bar{a}.Z) \vdash \langle 1\rangle\langle 1\rangle Tr$

$$\mathrm{fix}\, I\;\dfrac{\times I\;\dfrac{\langle a\rangle I\;\dfrac{\times I\;\dfrac{\langle\bar{a}\rangle I\;\dfrac{\mathbf{0}\vdash Tr}{\bar{a}.\mathbf{0}\vdash \langle\bar{a}\rangle Tr} \qquad \langle 1\rangle I\;\dfrac{Tr, \quad \mathbf{0}\vdash Tr}{\langle\bar{a}\rangle Tr, \quad a.\mathbf{0}\vdash \langle 1\rangle Tr}}{(a.\mathbf{0}\times\bar{a}\mathbf{0})\vdash \langle 1\rangle Tr}}{a.(a.\mathbf{0}\times\bar{a}.\mathbf{0})\vdash \langle a\rangle\langle 1\rangle Tr} \qquad \langle 1\rangle I\;\dfrac{\times I\;\dfrac{\langle a\rangle I\;\dfrac{Tr, \quad \mathbf{0}\vdash Tr}{\langle 1\rangle Tr, \quad a.\mathbf{0}\vdash \langle a\rangle Tr} \qquad \langle 1\rangle I\;\dfrac{Tr, \quad \mathbf{0}\vdash Tr}{\langle a\rangle Tr, \; \bar{a}.\mathbf{0}\vdash \langle 1\rangle Tr}}{\langle 1\rangle Tr, \quad a.\mathbf{0}\times\bar{a}.\mathbf{0}\vdash \langle 1\rangle Tr}}{\langle a\rangle\langle 1\rangle Tr, \quad \bar{a}(a.\mathbf{0}\times\bar{a}.\mathbf{0})\vdash \langle 1\rangle\langle 1\rangle Tr}}{a.(a.\mathbf{0}\times\bar{a}.\mathbf{0}) \times \bar{a}(.a.\mathbf{0}\times\bar{a}.\mathbf{0})\vdash \langle 1\rangle\langle 1\rangle Tr}}{\mathrm{fix}\, Z(a.Z\times\bar{a}.Z)\vdash \langle 1\rangle\langle 1\rangle Tr}$$

In the final example we prove, in effect, the possibility of deadlock. The process is $p_2 \times q$ of example 3.3. This process can only ever respond to 1 experiments. Thus, $\langle 1\rangle\langle 1\rangle[1]\mathrm{False}$ says that it can become deadlocked.

<u>Example 5.3</u> $a(a.\mathbb{0} + \text{fix } Z.a.z) \times \text{fix } Z.\bar{a}.Z \vdash \langle 1\rangle\langle 1\rangle[1]\text{False}$

$$
\times I\;\dfrac{
\text{fix}I\,\dfrac{\langle\bar a\rangle I\,\dfrac{\langle\bar a\rangle I\,\dfrac{\mathbb{0}\vdash \text{Tr}}{\bar a.\mathbb{0}\vdash\langle\bar a\rangle \text{Tr}}}{\bar a.\bar a.\mathbb{0}\vdash\langle\bar a\rangle\langle\bar a\rangle \text{Tr}}}{\text{fix}Z.\bar a.Z\vdash\langle\bar a\rangle\langle\bar a\rangle \text{Tr}}
\qquad
\langle 1\rangle I\,\dfrac{+\langle\rangle I\,\dfrac{\langle 1\rangle I\,\dfrac{\text{Tr},\quad \mathbb{0}\vdash[1]\text{False}}{\langle\bar a\rangle \text{Tr},\ a.\mathbb{0}\vdash\langle 1\rangle[1]\text{False}}}{\langle\bar a\rangle \text{Tr},\ a.\mathbb{0}+\text{fix}Z.aZ\vdash\langle 1\rangle[1]\text{False}}}{\langle\bar a\rangle\langle\bar a\rangle \text{Tr},\ a(a.\mathbb{0}+\text{fix}Z.a.Z)\vdash\langle 1\rangle\langle 1\rangle[1]\text{False}}
}{a(a.\mathbb{0}+\text{fix }Z.a.Z)\times\text{fix }Z.\bar a.Z\vdash\langle 1\rangle\langle 1\rangle[1]\text{False}}
$$

<u>Soundness and Completeness</u>

The modal proof system above is sound. This is the content of the following theorem:

<u>Theorem 5.4</u> i. if $p \vdash B$ then $p \models B$

ii. if $A, p \vdash B$ then $A,p \models B$

Its proof is standard: the axioms are all valid and the introduction rules preserve validity.

Completeness can be understood in two ways:

weak completeness : if $p \models B$ then $p \vdash B$

strong completeness : weak completeness and if $A,p \models B$ then $A,p \vdash B$

The modal system is weak complete. Constructing a strongly complete system is problematic. One difficulty is that if A is SCCS-valid, true of every SCCS process, then $B,p \models A$ for any B. In fact, there is not a simple dichotomy between weak and strong completeness: a weak completeness result depends on a corresponding strong result to justify $p\times q \models A$ implies $p\times q \vdash A$. The strong completeness result we offer is modulo formula equivalence on SCCS in the case of the relativized relation. Let $|A| = \{p : p \models A\}$.

<u>Theorem 5.5</u> i. if $p \models A$ then $p \vdash A$

ii. if $A,p \models B$ then $\exists C.\ |C|=|A|$ and $C,p \models B$

The most difficult part of the proof is the following expressibility (or parallel decomposition) lemma

<u>Lemma 5.6</u> i. if $p\times q \models A$ then $\exists C.\ p \models C$ and $C,q \models A$

ii. if $A,p\times q \models B$ then $\exists C,D.\ |A|=|C|$ and $C,p \models D$ and $D,q \models B$.

<u>Conclusion</u>

A complete compositional modal proof system for a subset of Milner's SCCS has been offered. The parallel introduction rules were suggested by a simple semantic strategy. There is a number of directions for further work. The semantic strategy involves an appeal to a notion of environment. More general work on environments for SCCS type languages is contained in [La]. The current work is also closely related to [Wi]. More generally, the strategy needs to be tested against other concurrent programming languages and other logical languages. A small step in this direction is [St2] where we show that the strategy works for the more intractable asynchronous parallel operator of Milner's CCS.

There is also the issue of extending the current modal proof system to cover full SCCS. Allowing infinitary summation means the modal logic must become infinitary [HS,Mi1]. Renaming can be dealt with straightforwardly. Restriction is the operator which causes most problems. Preliminary results suggest that only a subset of restriction contexts can be dealt with by a simple extension of the logic.

Acknowledgements

I would like to thank Bob Constable, Matthew Hennessy, Robin Milner, Gordon Plotkin and in particular Gerardo Costa for discussions and helpful comments on the topic of this paper. I would also like to thank Dorothy McKie for typing.

References

LNCS abbreviates Lecture Notes in Computer Science, Springer-Verlag.

[Ab] S. Abramsky. 'Experiments, powerdomains and fully abstract models for applicative multiprogramming', LNCS Vol.158, pp.1-13 (1983).

[AFR] K. Apt, N. Francez and W.de Roever. 'A proof system for communicating sequential processes', TOPLAS pp. 359-385 (1980).

[BK] H. Barringer and R. Kuiper. 'Towards the hierarchical, Temporal logic, specification of concurrent systems',presented at STL/SERC Workshop on the Analysis of Concurrent Systems, Cambridge. (1983).

[BKP] H. Barringer, R. Kuiper and A. Pnueli, 'Now you may compose temporal logic specifications', Proceedings STOC (1984).

[BR] S. Brookes and W. Rounds. 'Behavioural equivalence relations induced by programming logics', LNCS Vol.154 pp. 97-108 (1983).

[DeH] R. de Nicola and M. Hennessy. 'Testing equivalences for processes', in LNCS Vol. 154 pp. 548-560 (1983).

[EH] E. Emerson and J. Halpern. 'Sometimes and not never revisited: on branching versus linear time', pp. 127-140 POPL Proceedings (1983).

[G] G. Gentzen. 'Investigations into logic deduction', in '<u>The Collected Works of Gerhard Gentzen</u>' ed. Szabo, North-Holland (1969).

[GS] S. Graf and J. Sifakis. 'A modal characterization of observational congruence on finite terms of CCS', IMAG Technical Report No. 402 (and to appear in ICALP '84) (1983).

[Ha] D. Harel. '<u>First-Order Dynamic Logic</u>' LNCS Vol.68 (1979).

[HBR] C. Hoare, S. Brookes and A. Roscoe. 'A theory of communicating sequential processes', Technical Monograph Prg-16, Computing Lab, University of Oxford (1981).

[He1] M. Hennessy. 'Axiomatizing finite delay operators', Acta Informatica 21, pp. 61-88 (1984).

[He2] M. Hennessy. 'Modelling finite delay operators'. Technical Report CSR-153-83 Dept. of Computer Science, Edinburgh (1983).

[HM1] M. Hennessy and R. Milner. 'On observing nondeterminism and concurrency', LNCS Vol.85, pp. 299-309 (1980).

[HM2] M. Hennessy and R. Milner. 'Algebraic laws for nondeterminism and concurrency' Technical Report CSR-133-83 (and to appear in JACM) (1983).

[Ho] C. Hoare. 'A model for communicating sequential processes'. Technical Monograph, Prg-22, Computing Lab University of Oxford (1982).

[HS] M. Hennessy and C. Stirling. 'The power of the future perfect in program logics', LNCS Vol.176 pp.301-311 (1984).

[K] R. Keller. 'A fundamental theorem of asynchronous parallel computation', in Parallel Processing ed. T. Feng, Springer-Verlag (1975).

[L] L. Lamport. 'The 'Hoare logic' of concurrent programs', Acta Informatica pp. 21-37 (1980).

[La] K. Larsen. 'A context dependent equivalence between processes'. To appear.

[LG] G. Levin and D. Gries. 'A proof technique for communicating sequential processes', Acta Informatica pp. 281-302 (1981).

[Mi1] R. Milner. 'A modal characterisation of observable machine-behaviour', LNCS Vol. 112 pp. 25-34 (1981).

[Mi2] R. Milner. 'A finite delay operator in synchronous CCS', Technical Report CSR-116-82, Dept. of Computer Science, Edinburgh (1982).

[Mi3] R. Milner. 'Calculi for synchrony and asynchrony', Theoretical Computer Science, pp. 267-310 (1983).

[Mo] E. Moore. 'Gedanken-experiments on sequential machines', in 'Automata Studies' ed. C. Shannon and J. McCarthy, Princeton University Press, pp. 129-153 (1956).

[MP1] Z. Manna and A. Pnueli. 'Temporal verification of concurrent programs : the temporal framework for concurrent programs', in 'The Correctness Problem in Computer Science', ed. R. Boyer and J. Moore, Academic Press, pp. 215-273 (198).

[MP2] Z. Manna and A. Pnueli. 'How to cook a temporal proof system for your pet language', POPL Proceedings pp. 141-154 (1983).

[OG] S. Owicki and D. Gries. 'An axiomatic proof technique for parallel programs I' Acta Informatica pp. 319-340 (1976).

[Pa] D. Park. 'Concurrency and automata on infinite sequences', LNCS Vol.104 (1981).

[P] G. Plotkin. 'A structural approach to operational semantics'. Lecture Notes, Aarhus University (1981).

[QS] J. Queille and J. Sifakis. 'Fairness and related properties in transition systems - a temporal logic to deal with fairness', Acta Informatica 19, pp. 195-220 (1983).

[RB] W. Rounds and S. Brookes. 'Possible futures, acceptances, refusals and communicating processes', in Proc. FOCS pp. 140-149 (1981).

[Si1] J. Sifakis. 'Unified approach for studying the properties of transition systems', Theoretical Computer Science, pp. 227-258 (1982).

[Si2] J. Sifakis. 'Property preserving homomorphisms of transition systems', Technical Report, IMAG (1982).

[St1] C. Stirling. 'A proof theoretic characterization of observational equivalence' in Procs. FCT-TCS Bangalore (1983) (and to appear in TCS).

[St2] C. Stirling. 'A compositional modal proof system for a subset of CCS'. To appear.

[ZBR] J. Zwiers, A. de Bruin and W. de Roever. 'A proof system for partial correctness of dynamic networks of processes', Technical Report RUU-CS-83-15, Dept. of Computer Science, University of Utrecht (1983).

[Wi] G. Winskel. 'Complete proof systems for SCCS with modal assertions'. To appear.

Amalgamation of Graph Transformations with Applications to Synchronization

Paul Boehm

Harald-Reto Fonio

Annegret Habel

Technische Universität Berlin

D-1000 Berlin 10, Franklinstr. 28/29

Abstract: In the present paper we generalize the well-known PARALLELISM THEOREM for graph derivations to the AMALGAMATION THEOREM. In this theorem the assumption of "parallel independence" is dropped. For each pair of productions together with a relational production (allowing productions to be associated with each other) we construct a single "amalgamated" production. The AMALGAMATION THEOREM states that graph derivations which respect the given associations can be amalgamated to a single derivation via the "amalgamated" production.
The amalgamation mechanism can be used to handle synchronization phenomena. The amalgamation concept is applied to synchronization of graph manipulations in a simplified railway control system as well as in GDS, a graph grammar formalism for distributed systems.

1. Introduction

Graphs and transformations of graphs are important in many areas of computer science (see /CER 79/ and /ENR 83/). There are many different ways how to generalize string productions and derivations to graphs. A survey over several approaches including an extensive bibliography is given in /Na 79/. This paper is based on the gluing approach defined in /EPS 73/ and /Ro 75/ and extensively described in /Eh 79/.

Some of the most fundamental concepts of graph transformation theory are graph productions and graph derivations. One important class of problems and phenomena can be characterized by the following situation:
Given an initial graph, there can be several productions applicable. What to do now? There are several interesting possibilities, e.g. to use some productions in parallel, or to use a well-defined sequence of productions, or to use some specific "composed" productions, or.... These and similar phenomena are investigated under the headings parallelism, sequentialization, concurrency (see e.g. /ER 79a/, /EK 80/, /JKRE 82/, /Eh 83/, /Pe 80/), e.t.c.. One further concept belonging to this class is amalgamation which is the topic of this paper.

In the present paper we generalize the well-known PARALLELISM THEOREM for graph derivations to the AMALGAMATION THEOREM. In both theorems we consider a graph G and productions p1 and p2 which are applicable to G. If the corresponding graph derivations $G \Longrightarrow H1$ via p1 and $G \Longrightarrow H2$ via p2 are "parallel independent", i.e. the occurrences of p1 and p2 in G are allowed to overlap in items which remain preserved in each derivation, the PARALLELISM THEOREM states that the productions can be applied one after the other or "in parallel". In general p1 and p2 may have common items which remain not preserved. Then the productions p1 and p2 only can be applied "synchronously" , provided that the derivations via p1 and p2 are "amalgamable", i.e. that the occurrences of p1 and p2 in G are allowed to overlap in items which shall be preserved - or deleted - by both productions. For each pair of productions (which are allowed to possess a common part) we construct a single "amalgamated" production. The AMALGAMATION THEOREM states that "amalgamable" graph derivations can be amalgamated to a single derivation. Applying the amalgamated production of p1 and p2 has essentially the same effect as applying first the common part of p1 and p2 and then the remainders of p1 and p2.

Let us illustrate our AMALGAMATION CONCEPT by a simplified example of a railway control system: States in the railway system are represented by graphs, tracks and trains are some types of nodes. "Blowprints" for changes of states are described by graph productions and the actual changes of states by graph derivations. Here the main problem is whether planned changes for a subnet are consistent with those of other subnets, i.e., with respect to the topic of this paper, whether plans for subnets can be amalgamated to a single plan for the whole railway net.
Consider the elementary productions MOVE and HALT given in Fig. 1.1 which are part of the small railway system studied in /MW 82/. More precisely MOVE and HALT are production rules where the node labels have to be recolored by actual parameters.

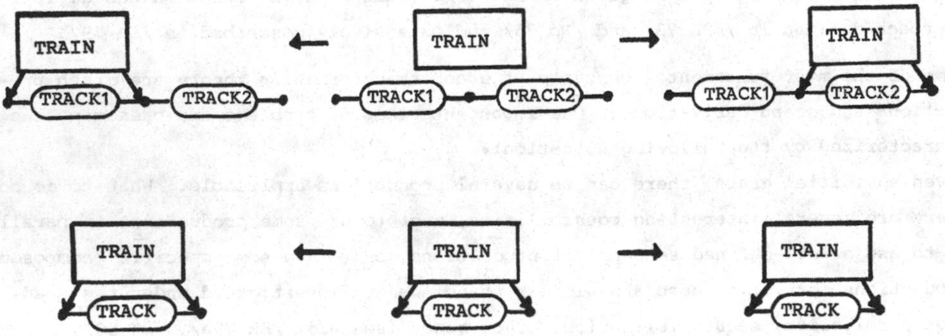

Figure 1.1: Productions for moving and halting of trains

Moving a train means to apply the production rule MOVE with suitable actual parameters to the current state of the railway system which is represented as a graph.

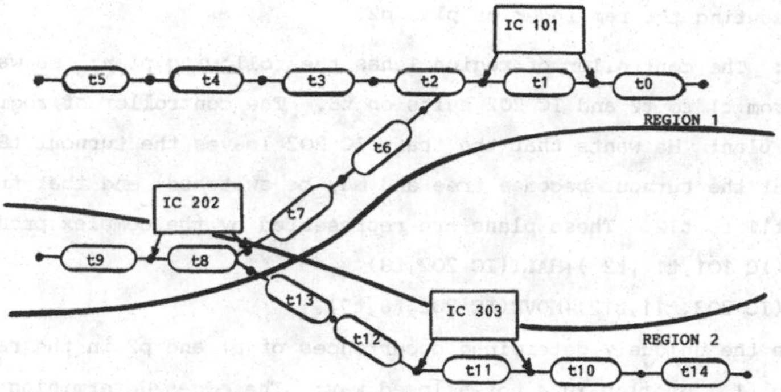

Figure 1.2: Graph representing the current state of the railway system

The railway net is covered by regions which are controlled by controllers. Each controller designs a plan for his region (controller is meant to be a person, not to be a system). A plan is represented as a complex production p consisting of an elementary production for each train in the region. A plan may be executed if it harmonizes with the plans for the adjacent regions. Now there may be one of the following situations:

Situation 1: The controller of region 1 has the plan that the train IC 101 moves from the track t1 to the turnout t2 (we assume that the turnout t2 is directed to t3) and IC 202 halts on t8. The controller of region 2 wants that IC 202 halts on t8 and that IC 303 moves from t11 to t12. In this case the plans for region 1 and region 2 harmonize and can be executed one after the other or "in parallel".

Situation 2: The controller of region 1 has the plan that IC 101 moves from t1 to t2 and IC 202 moves from t8 to t9. The controller of region 2 has a plan which harmonizes with the first plan: He wants that IC 202 moves from t8 to t9 and that IC 303 moves from t11 to t12. These plans cannot be executed one after the other, but only "synchronously".

The plans for the single regions are represented by

p1=MOVE (IC 101,t1 ,t2)+MOVE (IC 202,t8,t9),

p2=MOVE (IC 303,t11,t12)+MOVE (IC 202,t8,t9).

The plan for the region of common control is represented by

r=MOVE (IC 202,t8,t9).

The plan for the whole region is an amalgamation of the single plans with respect to r. It is represented by the r-amalgamated production

$p\oplus_r p'$=MOVE(IC 101,t1,t2)+MOVE(IC 202,t8,t9)+MOVE(IC 303,t11,t12).

Applying this production to the railway graph G we obtain a derivation $G \Longrightarrow X$ via $p\oplus_r p'$ which can be seen as an amalgamation of the derivations $G \Longrightarrow H1$ via p1 and $G \Longrightarrow H2$ via p2. The AMALGAMATION THEOREM applied to this example states that executing the plan $p1\oplus_r p2$ has essentially the same effect as first executing the plan p1 and than executing the remainder of plan p2.

Situation 3: The controller of region 1 has the following plan: He wants that IC 101 moves from t1 to t2 and IC 202 halts on t8. The controller of region 2 has a conflicting plan: He wants that the train IC 202 leaves the turnout t8 and moves to t7 (such that the turnout becomes free and may be switched) and that train IC 303 moves from t11 to t12. These plans are represented by the complex productions

p1=MOVE(IC 101,t1 ,t2)+HALT(IC 202,t8)

p2=MOVE(IC 303,t11,t12)+MOVE(IC 202,t8,t7).

In this case the uniquely determined occurrences of p1 and p2 in the railway graph G given in Fig. 1.2 overlap in a not-allowed way: The edges determining the location of the train IC 202 at turnout t8 shall be preserved by production p1 and deleted by production p2, i.e. the occurrences of p1 and p2 in G are in "conflict". Such conflict cases will not be treated in the present paper.

The AMALGAMATION THEOREM, the main result of this paper, is formulated and proved in the framework of the algebraic theory of graph grammars using pushout and pullback constructions in the category of labeled graphs.

PARALLELISM

AMALGAMATION CONCURRENCY

The AMALGAMATION THEOREM generalizes the well-known PARALLELISM THEOREM considerably. The CONCURRENCY THEOREM presented e.g. in /ER 79a/ and /EHR 83/ can be seen as another type of generalization of the PARALLELISM THEOREM. Although the amalgamation and the concurrency concepts are different it is shown how we can profit from the CONCURRENCY THEOREM in the proof of the AMALGAMATION THEOREM.

Some basic notions from the theory of graph grammars are reviewed in Section 2. The main result is stated in Section 3 and proved in Section 4. Finally an application of the amalgamation concept to special synchronization problems in distributed systems is discussed in Section 5. Due to the limitation of space the proofs of the technical lemmata are omitted. A complete formal treatment can be found in /BFH 84/.

Acknowledgements

We would like to thank Hartmut Ehrig and Hans-Jörg Kreowski for fruitful discussions. For excellent typing we are most grateful to H. Barnewitz.

2. Preliminaries

This section provides some basic notions concerning graphs, graph productions and graph derivations used in the following. For more details we refer to the tutorial survey /Eh 79/.

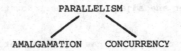

The object of our considerations are directed, labeled graphs over a fixed pair of labeling alphabets $C=(C_N,C_A)$. A graph $G=(G_N,G_A,s,t,m_N,m_A)$ over (C_N,C_A) consists of a set of nodes G_N, a set of arcs G_A, maps $s,t:G_A \to G_N$ assigning source and target to each arc of G, and labeling maps $m_N:G_N \to C_N$, $m_A:G_A \to C_A$ assigning a node labeling to each node and an arc labeling to each arc of G.

Given two graphs G and G' a graph morphism $f:G \to G'$ is a pair of maps $(f_N:G_N \to G'_N, f_A:G_A \to G'_A)$ satisfying $f_N s=s'f_A$, $f_N t=t'f_A$, $m'_N f_N=m_N$, and $m'_A f_A=m_A$ (composition of maps). f is called injective if the maps f_N and f_A both are injective. The composition of graph morphisms $f:G \to G'$ and $f':G' \to G''$ is defined by the composition of their components.

Following /Eh 79/ a graph production $p=(B_1 \xleftarrow{b_1} K \xrightarrow{b_2} B_2)$ consists of a pair of graphs (B_1,B_2) and an auxiliary graph K called gluing graph, which is related to B_1 and B_2 by injective graph morphisms $b_1:K \to B_1$, $b_2:K \to B_2$.

Given a graph G, a graph production $p=(B_1 \longleftarrow K \longrightarrow B_2)$ and a graph morphism $g:B_1 \to G$, p is applicable to G with respect to $g:B_1 \to G$ if the gluing condition

$$\mathrm{BOUNDARY}(B_1 \longrightarrow G) \subseteq b_1K$$

is satisfied. [$\mathrm{BOUNDARY}(B_1 \to G)$ consists of dangling items, i.e. items of $\mathrm{DANGLING}(B_1 \to G)=\{n \in B_{1N} \mid \exists a \in (G-gB_1)_A : gn=sa \text{ or } gn=ta\}$ and identification items, i.e. items of $\mathrm{IDENTIFICATION}(B_1 \to G)=\{x \in B_1 \mid \exists y \in B_1 : x \neq y \text{ and } gx=gy\}$].

In this case the "context" graph D can be constructed in such a way that G becomes the gluing of D and B_1 along the gluing graph K. Now the result of the application H of p to G with respect to $g:B_1 \to G$ is the gluing of D and B_2 along the gluing graph K.

$$\begin{array}{ccccc}
B_1 & \xleftarrow{b_1} & K & \xrightarrow{b_2} & B_2 \\
{\scriptstyle g}\downarrow & (1) & \downarrow{\scriptstyle d} & (2) & \downarrow{\scriptstyle h} \\
G & \xleftarrow[c_1]{} & D & \xrightarrow[c_2]{} & H
\end{array}$$

We write $G \overset{}{\underset{p}{\Longrightarrow}} H$ and say that $g:B_1 \to G$ is the occurrence of p in G. $G \Longrightarrow H$ is also called direct derivation via p based on g. (Note that G and H are pushout objects in the diagrams (1) and (2) constructed in the category of labeled graphs. We will often use the short notations PO and PB for pushouts and pullbacks (/AM 75/). Moreover, we will use a linear notation for squares if the notion of morphisms is not essential, e.g. PO (1) above will be written as KB_1DG or KDB_1G.

Two direct derivations $G \Rightarrow H$ via p based on g and $G \Rightarrow H'$ via p' based on g' are parallel independent if the intersection of B_1 and B'_1 in G (which are the occurrences of p and p' in G) consists of common gluing items only, that means

$$gB_1 \cap g'B'_1 \subseteq gb_1K \cap g'b'_1K'$$

On the other hand we can construct the parallel production of p and p'

$$p+p'=(B_1+B'_1 \leftarrow K+K' \to B_2+B'_2)$$

built up by componentwise disjoint union from the single productions

$p=(B_1 \leftarrow K \rightarrow B_2)$ and $p'=(B_1' \leftarrow K' \rightarrow B_2')$. Derivations via parallel productions are called parallel derivations.

The connection between parallel independent derivations and parallel derivations is established in the PARALLELISM THEOREM (see /ER 79a/):

PARALLELISM THEOREM

Let p and p' be productions and p+p' the corresponding parallel production. Then we have

1. SYNTHESIS Given parallel independent derivations $G \Rightarrow H$ via p and $G \Rightarrow H'$ via p' then there are a graph X, direct derivations $H \Rightarrow X$ via p' and $H' \Rightarrow X$ via p and a parallel derivation $G \Rightarrow X$ via p+p'.
2. ANALYSIS Given a parallel derivation $G \Rightarrow X$ via p+p' then there are derivation sequences $G \Rightarrow H \Rightarrow X$ via (p,p') and $G \Rightarrow H' \Rightarrow X$ via (p',p) such that the direct derivations $G \Rightarrow H$ via p and $G \Rightarrow H'$ via p' are parallel independent.

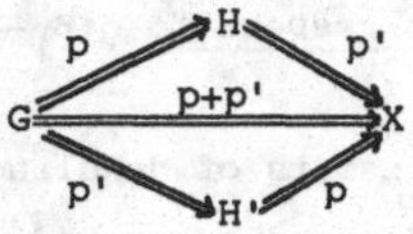

3. The operations SYNTHESIS and ANALYSIS are inverse to each other in the following sense: there is a bijective correspondence between parallel independent derivations and parallel derivations.

3. Amalgamation of Transformations

In this section we will study the problem of "amalgamating" direct transformations to a single transformation using only one "amalgamated" production. Applying the amalgamated production the original productions shall be executed at the same time and hence synchronously.

We introduce the notion of a relational production r for productions, r-amalgamable derivations, and the construction of r-amalgamated productions and derivations. The connection between r-amalgamable derivations and r-amalgamated derivations is established in the Amalgamation Theorem which will be stated together with a number of corollaries.

3.1 DEFINITION

1. Given a production $p=(B_1 \leftarrow K \rightarrow B_2)$ a production $r=(R_1 \leftarrow R \rightarrow R_2)$ together with graph morphisms $R_1 \rightarrow B_1$, $R \rightarrow K$, $R_2 \rightarrow B_2$ is called a subproduction of p if the diagrams RR_1KB_1 and RR_2KB_2 commute and the conditions

(1) $\mathrm{BOUNDARY}(R_1 \rightarrow B_1) \subseteq r_1R$ and $\mathrm{IDENTIFICATION}(R_2 \rightarrow B_2) \subseteq r_2R$

(2) $f_1^{-1}\, b_1K \subseteq r_1R$ and $f_2^{-1}\, b_2K \subseteq r_2R$

are satisfied.

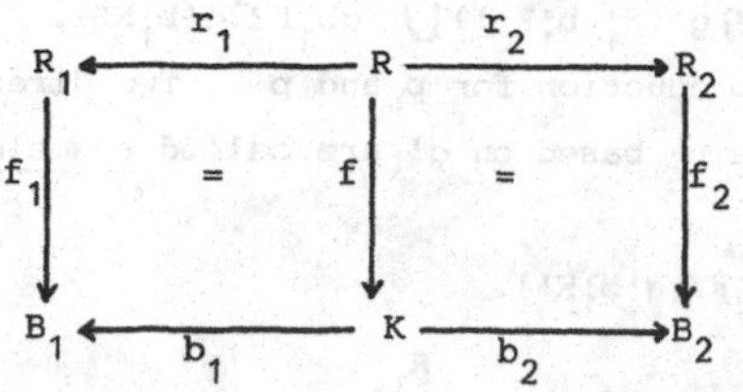

2. Given two productions $p=(B_1 \leftarrow K \rightarrow B_2)$, $p'=(B_1' \leftarrow K' \rightarrow B_2')$ a production $r=(R_1 \leftarrow R \rightarrow R_2)$ together with graph morphisms $B_1 \leftarrow R_1 \rightarrow B_1'$, $K \leftarrow R \rightarrow K'$, $B_2 \leftarrow R_2 \rightarrow B_2'$ is called a <u>relational production</u> for p and p' if r is a subproduction of p as well as p'.

3. Let p and p' be productions and r be a relational production for p,p'. Then the r-amalgamated production $\bar{p}$ is defined by the following construction:

(1) Let $\bar{B}_1$ be the gluing of B_1 and B_1' along R_1,

(2) let $\bar{K}$ be the gluing of K and K' along R,

(3) and $\bar{B}_2$ be the gluing of B_2 and B_2' along R_2.

Moreover let $\bar{K} \rightarrow \bar{B}_1$ and $\bar{K} \rightarrow \bar{B}_2$ be the uniquely existing morphisms.

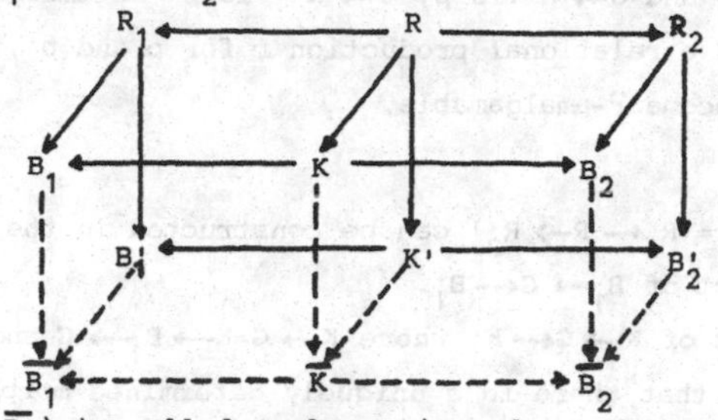

Then $\bar{p}=(\bar{B}_1 \leftarrow \bar{K} \rightarrow \bar{B}_2)$ is called <u>amalgamation of p and p' with respect to r</u>, written $p \oplus_r p'$. $\bar{p}$ also is called <u>r-amalgamated production</u>. A direct derivation $G \Rightarrow X$ via $\bar{p}$ is called <u>r-amalgamated derivation</u>.

REMARKS

1. Let p be a production and r be a subproduction of p. Then there is a uniquely determined production p_o and a relation S for (r,p_o) such that $p=r*_S p_o$. The production p_o is called <u>remainder of p with respect to r</u>.

2. The relational production r relates the productions p and p'. In the case $r=\emptyset$, i.e. the empty production $(\emptyset \leftarrow \emptyset \rightarrow \emptyset)$, the amalgamation of p and p' w.r.t. r is equal to the parallel production $p+p'=(B_1+B_1' \leftarrow K+K' \rightarrow B_2+B_2')$.

3. The construction of r-amalgamated productions and derivations can be iterated.

3.2 DEFINITION

1. Given two productions $p=(B_1 \leftarrow K \rightarrow B_2)$ and $p'=(B_1' \leftarrow K' \rightarrow B_2')$ two direct derivations $G \Rightarrow H$ via p based on g and $G \Rightarrow H'$ via p' based on g' are called <u>amalgamable</u> if the intersection of B_1 and B_1' in G (which are the occurrences of p and p' in G) consists of common gluing items or items which shall be deleted by p as well as by p'. That means precisely

$gB_1 \cap g'B_1' \subseteq (g(B_1-b_1K) \cap g'(B_1'-b_1'K')) \cup (gb_1K \cap g'b_1'K')$.

2. Let r be a relational production for p and p'. Two direct derivations $G \Rightarrow H$ via p based on g and $G \Rightarrow H'$ via p' based on g' are called <u>r-amalgamable</u> if the diagram $R_1B_1B_1'G$ commutes and

$gB_1 \cap g'B_1' \subseteq gf_1R_1 \cup (gb_1K \cap g'b_1'K')$.

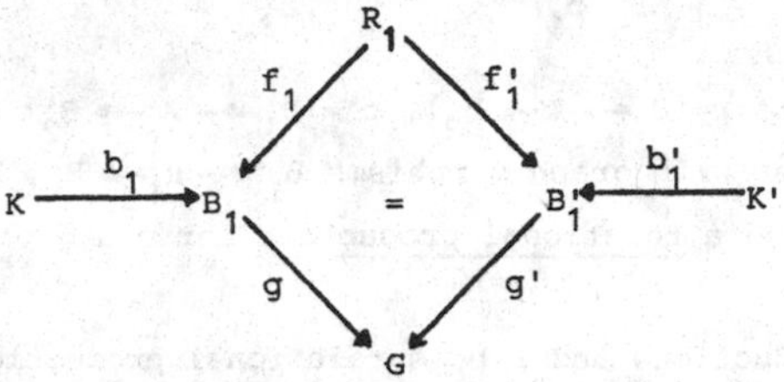

<u>REMARKS</u>: 1. Amalgamability generalizes parallel independency. 2. r-amalgamable derivations are amalgamable.

3.3 LEMMA

Let p and p' be productions and $G \Rightarrow H$ via p, $G \Rightarrow H'$ via p' be amalgamable direct derivations. Then there is a relational production r for p and p', such that the given direct derivations become r-amalgamable.

<u>CONSTRUCTION</u>

The relational production $r=(R_1 \leftarrow R \rightarrow R_2)$ can be constructed in the following way:

(1) Let R_1 be the PB-object of $B_1 \rightarrow G \leftarrow B_1'$.

(2) Let R be the PB-object of $K \rightarrow G \leftarrow K'$ where $K \rightarrow G = K \rightarrow B_1 \rightarrow G$ and $K' \rightarrow G = K' \rightarrow B_1' \rightarrow G$. (Note that there is a uniquely determined morphism $R \rightarrow R_1$ such that the diagrams RR_1KB_1 and RR_2KB_2 commute.)

(3) Let $R_2=R$ and $R \rightarrow B_2 = R \rightarrow K \rightarrow B_2$, $R \rightarrow B_2' = R \rightarrow K' \rightarrow B_2'$.

Now we will state the main theorem of this paper:

3.4 AMALGAMATION THEOREM

Let r be a relational production for productions p,p' and $\bar{p}=p\oplus_r p'$ the corresponding r-amalgamated production. Moreover, let p_o, p_o' be the remainders of p resp. p' w.r.t. r. Then we have

1. <u>SYNTHESIS</u> Given r-amalgamable direct derivations $G \Rightarrow H$ via p and $G \Rightarrow H'$ via p' then there are a graph X, direct derivations $H \Rightarrow X$ via p_o' $H' \Rightarrow X$ via p_o, and an r-amalgamated derivation $G \Rightarrow X$ via $p\oplus_r p'$.

2. <u>ANALYSIS</u> Given an r-amalgamated derivation $G \Rightarrow X$ via $p\oplus_r p'$ then there are derivation sequences $G \Rightarrow H \Rightarrow X$ via (p,p_o') and $G \Rightarrow H' \Rightarrow X$ via (p',p_o) such that the direct derivations $G \Rightarrow H$ via p and $G \Rightarrow H'$ via p' are r-amalgamable.

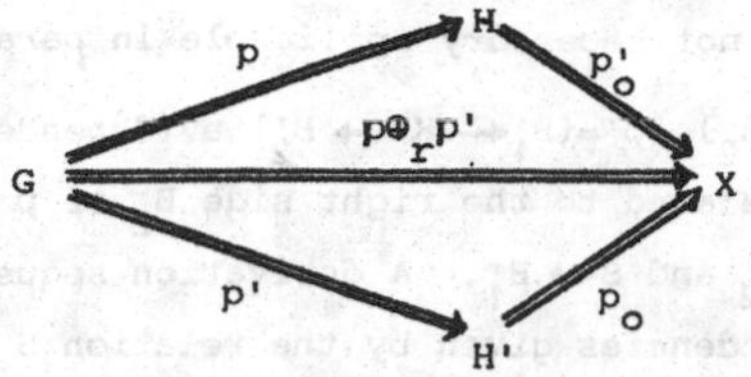

3. The operations SYNTHESIS and ANALYSIS are inverse to each other in the following sense: There is a bijective correspondence between r-amalgamable and r-amalgamated derivations.

The proof of the AMALGAMATION THEOREM and some useful lemmata are given in Sect.4. We conclude the present section with some corollaries.

3.5 COROLLARY 1

Taking the empty production as relational production in the AMALGAMATION THEOREM we obtain the PARALLELISM THEOREM.

3.6 COROLLARY 2

Given an r-amalgamated derivation G$\Longrightarrow$ X via $p\oplus_r p'$ there is a "complete" analysis into derivation sequences G$\Longrightarrow$ Y$\Longrightarrow$ H$\Longrightarrow$ X via (r,p_o,p_o') and G$\Longrightarrow$ Y$\Longrightarrow$ H'$\Longrightarrow$ X via (r,p_o',p_o).

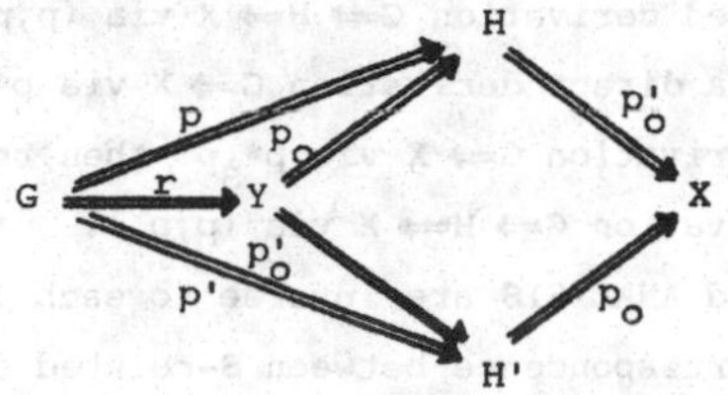

3.7 COROLLARY 3

Let G$\Longrightarrow$ H via p, G$\Longrightarrow$ H' via p' be r-amalgamable derivations and G$\Longrightarrow$ Y the corresponding derivation via r. Moreover, let G$\Longrightarrow$ X be the r-amalgamated derivation via $p\oplus_r p'$. Then X can be constructed from H and H' by gluing of H and H' along Y, provided that RR_1KB_1 and RR_2KB_2 are gluing diagrams (i.e. PO's).

4. The Proof of the Amalgamation Theorem

In this Section we give the proof of the AMALGAMATION THEOREM 3.4 together with some lemmata used in the proof. Since we can profit from the CONCURRENCY THEOREM presented in /ER 79a/ and /EHR 83/ in the proof of the AMALGAMATION THEOREM the CONCURRENCY THEOREM is stated first.

Let us review the problem of simulating a transformation sequence by a single transformation using only one "concurrent" production instead of a sequence of productions (see /ER 79a/ and /Ha 80/). Applying the concurrent production we can execute the relevant parts of the productions at the same time and hence concurrently, although the productions themselves are not necessary applicable in parallel.

Given productions $p=(B_1 \leftarrow K \rightarrow B_2)$, $p'=(B_1' \leftarrow K' \rightarrow B_2')$ a (dependency) <u>relation</u> for (p,p') is a graph S which is related to the right side B_2 of p and the left side B_1' of p' by graph morphisms $S \rightarrow B_2$ and $S \rightarrow B_1'$. A derivation sequence $G \Rightarrow H \Rightarrow X$ via (p,p') which reflects the dependencies given by the relation S is called <u>S-related.</u>

On the other hand we can construct the <u>S-concurrent production</u> of p and p'

$$p*_S p' = (B_1^* \leftarrow K^* \rightarrow B_2^*)$$

where - roughly speaking - the left side B_1^* consists of B_1 and the non-S-related parts of B_1' and similar the right side B_2^* consists of B_2' and the non-S-related parts of B_2. A direct derivation $G \Rightarrow X$ via a S-concurrent production $p*_S p'$ is called <u>S-concurrent</u>. (For formal versions of the notions above we refer to /ER 79a/ and /EHR 83/.)

The connection between S-related sequences and S-concurrent derivations is established in the Concurrency Theorem.

CONCURRENCY THEOREM

Let S be a relation for a pair of productions p,p' and $p*_S p'$ the corresponding S-concurrent production.

1. <u>SYNTHESIS</u> Given a S-related derivation $G \Rightarrow H \Rightarrow X$ via (p,p') then there is a canonical synthesis leading to a direct derivation $G \Rightarrow X$ via $p*_S p'$.
2. <u>ANALYSIS</u> Given a direct derivation $G \Rightarrow X$ via $p*_S p'$ then there is a canonical analysis into a S-related derivation $G \Rightarrow H \Rightarrow X$ via (p,p').
3. The operations SYNTHESIS and ANALYSIS are inverse to each other in the following sense: There is a bijective correspondence between S-related derivations $G \Rightarrow H \Rightarrow X$ via (p,p') and S-concurrent derivations $G \Rightarrow X$ via $p*_S p'$.

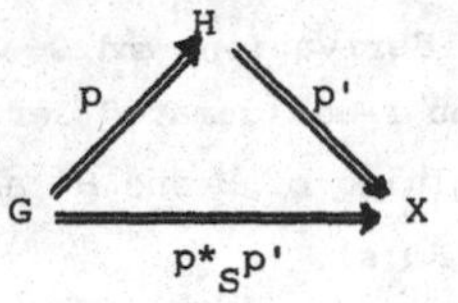

Using the CONCURRENCY THEOREM and some technical lemmata the proof of the AMALGAMATION THEOREM becomes very simple:

<u>**PROOF OF THE AMALGAMATION THEOREM**</u>

1. <u>SYNTHESIS</u>: Let $G \Rightarrow H$ via p, $G \Rightarrow H'$ via p' be r-amalgamable direct derivations.

1.1 Then there are derivation sequences

$G \Rightarrow Y \Rightarrow H$	via (r,p_o)	(S-related)
$G \Rightarrow Y \Rightarrow H'$	via (r,p_o')	(S'-related)

because p,p' can be decomposed into $p=r*_S p_o$, $p'=r*_{S'} p_o'$ (Lemma 1, Concurrency Theorem).

1.2 The direct derivations $Y \Rightarrow H$ via p_o, $Y \Rightarrow H'$ via p_o' are parallel independent, because $G \Rightarrow H$ via p, $G \Rightarrow H'$ via p' are r-amalgamable (Lemma 3). Hence there are a graph X and direct derivations $H \Rightarrow X$ via p_o', $H' \Rightarrow X$ via p_o such that the derivation sequences $Y \Rightarrow H \Rightarrow X$ via (p_o, p'_o), $Y \Rightarrow H' \Rightarrow X$ via (p'_o, p_o) become sequentially independent (see CHURCH-ROSSER-PROPERTY I e.g. in /Eh 79/).

1.3 The derivation sequence $G \Rightarrow H \Rightarrow X$ via (p, p'_o) is S'-related, because $G \Rightarrow Y \Rightarrow H'$ via (r, p'_o) is S'-related (Lemma 4). Moreover, the S'-concurrent production $p *_{S'} p'_o$ is equal to the r-amalgamated production $p \oplus_r p'$ (Lemma 2). Hence there is a canonical synthesis leading to a direct derivation $G \Rightarrow X$ via $p \oplus_r p'$.

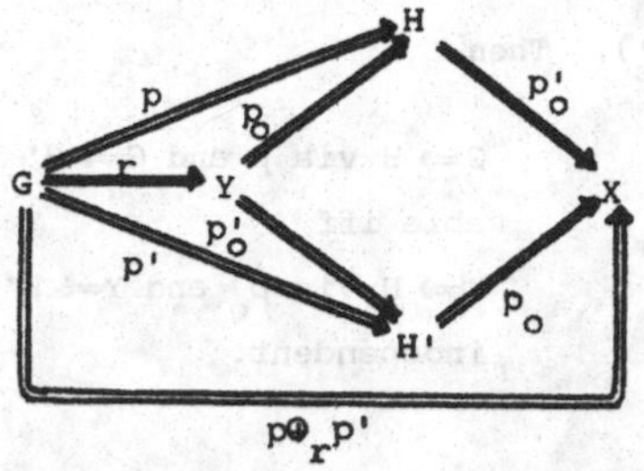

2. **ANALYSIS**: Let $G \Rightarrow X$ via $p \oplus_r p'$ be an r-amalgamated derivation.

2.1 There are derivation sequences

$G \Rightarrow H \Rightarrow X$ via (p, p'_o) (S'-related)

$G \Rightarrow H' \Rightarrow X$ via (p', p_o) (S-related)

because $p \oplus_r p'$ possesses decompositions $p *_{S'} p'_o$ and $p' *_S p_o$ (Lemma 2,Concurrency Theorem).

2.2 The direct derivations $G \Rightarrow H$ via p, $G \Rightarrow H'$ via p' are r-amalgamable, which can be seen as follows: There are derivation sequences

$G \Rightarrow Y \Rightarrow H$ via (r, p_o) (S-related)

$G \Rightarrow Y \Rightarrow H'$ via (r, p'_o) (S'-related)

because p and p' possess decompositions $p = r *_S p_o$ and $p' = r *_{S'} p'_o$ (Lemma 1, Concurrency Theorem). Now the S'-relatedness of $G \Rightarrow H \Rightarrow X$ implies the sequential independency of $Y \Rightarrow H \Rightarrow X$ (Lemma 5) and the parallel independency of $Y \Rightarrow H$, $Y \Rightarrow H'$ (CHURCH-ROSSER-PROPERTY II, /Eh 79/). By Lemma 3 $G \Rightarrow H$, $G \Rightarrow H'$ become r-amalgamable.

3. The bijective correspondence between r-amalgamable derivations and r-amalgamated derivations is an immediate consequence of the bijective correspondence between related derivation sequences and concurrent derivations.

Finally we state the lemmata used in the proof. For the proofs of these lemmata we refer to /BFH 84/ and /Fo 84/.

LEMMA 1 (SEPARATION OF PRODUCTIONS)

Let r be a relational production for p and p'. Then there are uniquely determined productions p_o and p'_o and relations S for (r, p_o) and S' for (r, p'_o) such that

$p = r *_S p_o$ and $p' = r *_{S'} p'_o$.

LEMMA 2 (DECOMPOSITION OF AMALGAMATED PRODUCTIONS)

Let r be a relational production for p and p' and $\bar{p}$ the corresponding r-amalgamated production. Then there are uniquely determined decompositions

$\bar{p}=p*_{S'}p'_o$ and $\bar{p}=p'*_{S}p_o$,

where p'_o and p_o are the remainders of p' and p and S' and S are uniquely determined relations for (p,p'_o) and (p',p_o) with $S'\rightarrow B_2=S'\rightarrow R_2\rightarrow B_2$ and $S\rightarrow B'_2=S\rightarrow R_2\rightarrow B'_2$.

LEMMA 3

Let G⟹H, G⟹H' be direct derivations via p resp. p' and r be a relational production for p and p'. Let G⟹Y⟹H, G⟹Y⟹H' be the corresponding derivation sequences via (r,p_o) resp. (r,p'_o). Then

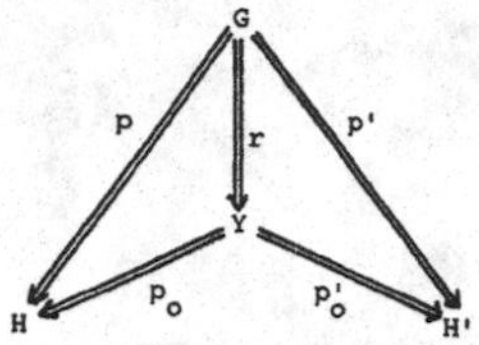

G⟹H via p and G⟹H' via p' are r-amalgamable iff

Y⟹H via p_o and Y⟹H' via p'_o are parallel independent.

LEMMA 4

Let G⟹H be a direct derivation via p and G⟹Y⟹H be the corresponding derivation sequence via (r,p_o). Let Y⟹H⟹X be a sequentially independent derivation sequence via (p_o,p'_o) and Y⟹H'⟹X the corresponding derivation sequence (p'_o,p_o). Then

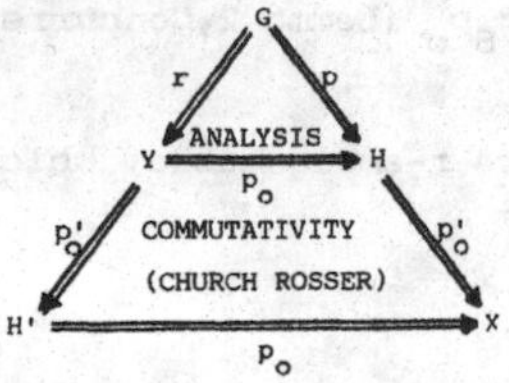

G⟹H⟹X via (p,p'_o) is S'-related if G⟹Y⟹H' via (r,p'_o) is S'-related.

LEMMA 5

Let G⟹H⟹X be a derivation sequence via (p,p'_o) and G⟹Y⟹H via (r,p_o) the analysis of G⟹H via p. Then S'-relatedness of G⟹H⟹X via (p,p'_o) implies sequential independency of Y⟹H⟹X via (p_o,p'_o).

5. Application to Synchronization in Distributed Systems

In this section we exemplify the application of the amalgamation mechanism to synchronization problems within Montanari's graph grammar formalism GDS (Grammars for Distributed Systems) /CM 83/, /DM 83/, /CDM 84/, a formalism for modelling the behaviour of nondeterministic dynamic process nets with distributed synchronization. Following Kung /Ku 80/, who classified the class of concurrent systems with respect to their module granularity, communication geometry and concurrency control (synchronization), GDS supports essential features of synchronization and distributedness. For details, concerning the link to other specification techniques, we refer to /DM 83/, where the relationship of GDS with Petri-Nets /Pe 80/, Milner's SCCS /Mi 82/, and Hoare's CSP /HBR 81/ is discussed. The graph grammar formalism GDS, introduced in /CM 83/, is based on labeled, partially ordered hypergraphs, called distributed systems. A distributed system models both the spatial and temporal aspects of a real system through the relations of adjacency and causality. The productions of a grammar represent the possible stand-alone evaluations of system components. For modelling synchronized evaluation of adjacent system components the productions have to be synchronized. The synchronization of productions as well as the application of synchronized productions is described by two procedures A and B. The (terminal) distributed systems derived within a given grammar represent the alternative deterministic, concurrent computations of a single nondeterministic system which is completely modeled by the grammar.

In the following we will show that the algorithmic procedures A and B, given in /CM 83/, easily can be expressed in terms of amalgamation and application of graph productions. (Note that our considerations are based on graphs instead of hypergraphs.)

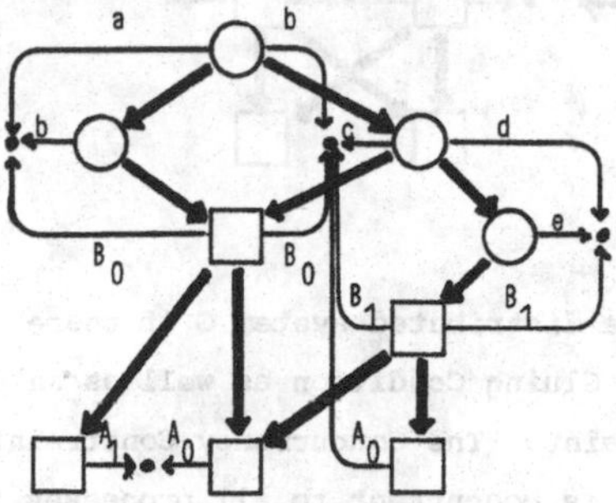

Representation Conventions:

Processes, events and ports are special colored nodes whereby boxes (□) denote processes, circles (○) events and bullets (•) ports.

Lowercase letters (a,b,c,d,e) correspond to terminal labels (actions) and uppercase letters (A_o,A_1,B_o,B_1) to nonterminal labels (process-types).

Fig. 5.1: distributed system G

A distributed system consists of three different kinds of nodes, processes, events and ports and three different kinds of arcs, terminal labeled arcs from events to ports, nonterminal labeled arcs from processes to ports and causal arcs (bold arcs) between subsystems, i.e. processes and events. A terminal label marks the action that happened, a nonterminal label marks the activated process type. The causal arcs induce a partial ordering on the set of subsystems. If two subsystems are related with respect to the partial ordering, they are called causally or sequentially dependent, otherwise they are concurrent. If subsystems are linked by a common port, they are called adjacent. Adjacency and concurrency may be interpreted as spatial and temporal overlapping. Events protocolize evaluation steps in the past while processes represent the possible nondeterministic future of a distributed system. Ports correspond to common storage, channels etc. Various requirements are formulated for distributed systems which ensure technical and logical consistency. The most important ones are, that no process may preceed an event with respect to the partial ordering and that an event cannot be concurrent with an adjacent subsystem.

A (GDS-)production $p=(B_1 \xleftarrow{b_1} K \xrightarrow{b_2} B_2)$ consists of distributed systems B_1,K,B_2 and injective graph morphisms b_1,b_2, such that certain technical conditions are satisfied. In general p describes the evaluation of processes with corresponding process types and ports together with the local modification of the partial ordering.

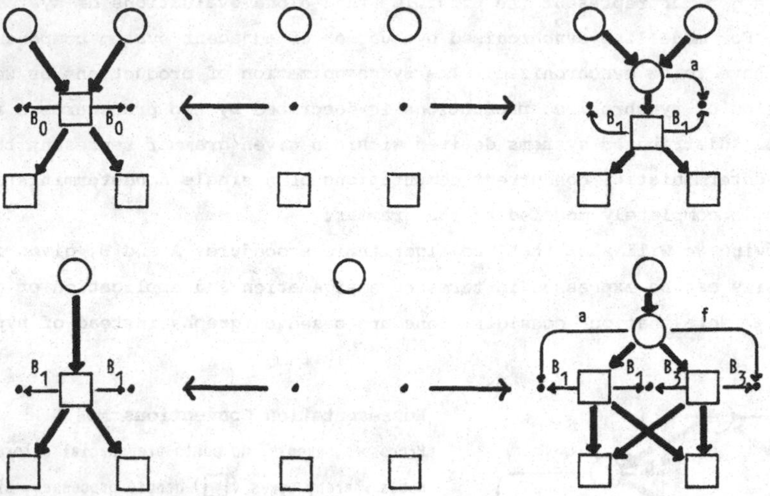

Fig. 5.2: productions p,p'

A production $p=(B_1 \longleftarrow K \longrightarrow B_2)$ can be applied to a distributed system G if there is an embedding morphism $g:B_1 \rightarrow G$, satisfying the Gluing Condition as well as an Injectivity, a Locallity and a Concurrency Constraint. The Concurrency Constraint requires that no process or event in $G-gB_1$, which is concurrent to all processes in B_1-b_1K is connected to an image g(s) of a synchronization port s of B_1 (synchronization ports s of B_1 are those ports, whose images $b_2b_1^{-1}(s)$ are connected to an event in B_2). The result H of applying a production p to a distributed system G w.r.t. an embedding morphism g is defined by the direct derivation $G \xRightarrow{p,g} H$.

In our example neither p nor p' can be applied to G: each embedding morphism does not satisfy the Concurrency Constraint. This calls for synchronization of p,p'; the underlying two processes have to be evaluated synchronously.

Synchronization of two GDS-productions $p=(B_1 \leftarrow K_1 \rightarrow B_2)$, $p'=(B_1' \leftarrow K' \rightarrow B_2')$ w.r.t. embedding morphisms $g:B_1 \rightarrow G$, $g':B_1' \rightarrow G$ and a distributed system G is done as follows: First we try to construct a relational production r for p,p' w.r.t. g,g'. If there is a relational production, then p,p' are synchronizable w.r.t. g,g', otherwise other productions or embedding morphisms have to be chosen. In the case of synchronizability, we can construct the amalgamation $p \oplus_r p'$ of p,p' w.r.t. r, which we call the synchronized production or synchronization of p,p' w.r.t. g,g'.

REMARK: An interpretation of synchronizability is, that processes which overlap in time and space must generate same actions at common ports. The procedure mentioned above, can be iterated. Because of that, synchronization of more than two productions is possible. This is necessary, if the synchronized production is not applicable, i.e. there are no embedding morphisms which satisfy the applicability constraints.

In our example p,p' are not applicable but synchronizable w.r.t. the distributed system G. The constructed relational production r for p,p' is given in Fig. 5.3. The amalgamated production $p\oplus_r p'$ of p,p' with respect to r is presented in Fig. 5.4.

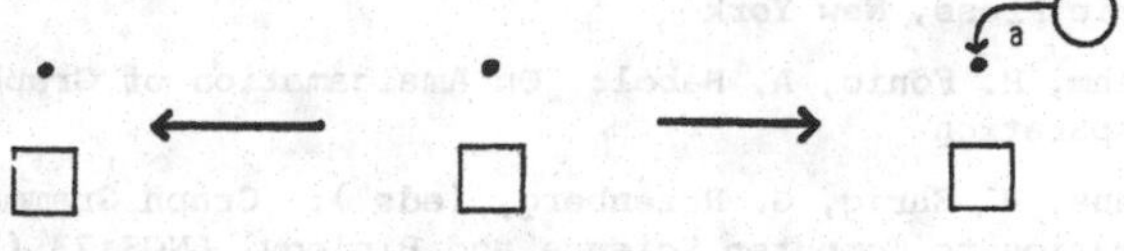

Fig. 5.3: relational production r for p,p'

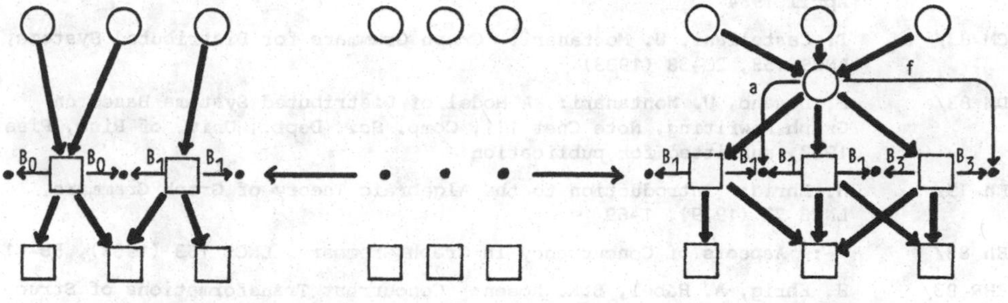

Fig. 5.4: amalgamated production $p\oplus_r p'$ of p,p' w.r.t. r

The amalgamated production $p\oplus_r p'$ of p,p' with respect to the relational production r is applicable, i.e. the uniquely determined embedding morphism into G satisfies all applicability constraints. Application of the amalgamated production $p\oplus_r p'$ to G leads to the distributed system H given in Fig. 5.5.

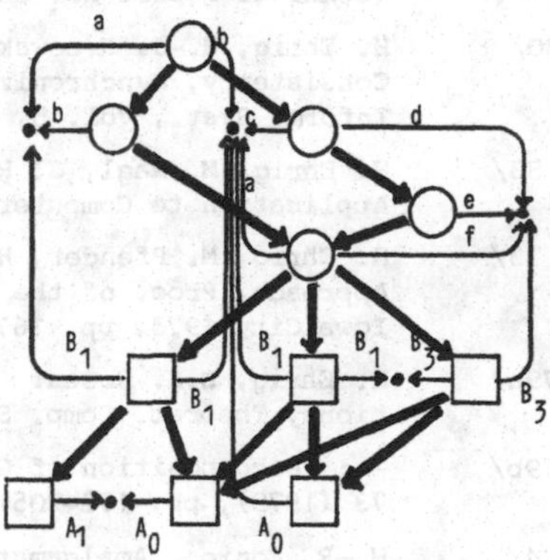

Fig. 5.5: derived distributed system H

The amalgamation concept, introduced in Section 3, can be used to describe the synchronization of productions in the sense of /CM 83/. This approach avoids the algorithmic procedures for synchronization and application of productions. As a consequence we get easy proofs with respect to iterated synchronization and consistency:

(1) The synchronization of two GDS-productions w.r.t. a relational GDS-production leads to a GDS-production.

(2) The application of a GDS-production to a distributed system leads to a distributed system.

With regard to a formal treatment we refer to /BFH 84/. We can state that the parallelism and concurrency concept together with the amalgamation concept give us excellent possibilities for modelling the behaviour of distributed systems.

References

/AM 75/ M.A. Arbib, E.G. Manes: Arrows, Structures, and Functors, Academic Press, New York

/BFH 84/ P. Boehm, H. Fonio, A. Habel: On Amalgamation of Graph Manipulations; in preparation

/CER 79/ V. Claus, H. Ehrig, G. Rozenberg, (eds.): Graph Grammars and Their Application to Computer Science and Biology, LNCS 73 (1979)

/CDM 84/ A. Corradini, P. Degano, U. Montanari: Specifying Highly Concurrent Data Structure Manipulation; Comp. Sci. Dept., Univ. of Pisa, Pisa, April 1984

/CM 83/ I. Castellani, U. Montanari: Graph Grammars for Distributed Systems, LNCS 153, 20-38 (1983)

/DM 83/ P. Degano, U. Montanari: A Model of Distributed Systems Based on Graph Rewriting, Note Cnet 111, Comp. Sci. Dept., Univ. of Pisa, Pisa 1983, submitted for publication

/Eh 79/ H. Ehrig: Introduction to the Algebraic Theory of Graph Grammars, LNCS 73 (1979), 1-69

/Eh 83/ --: Aspects of Concurrency in Graph Grammars, LNCS 153 (1983), 58-81

/EHR 83/ H. Ehrig, A. Habel, B.K. Rosen: Concurrent Transformations of Structures: From Graphs to Relational Data Structures; TU Berlin, FB 20, Technical Report No. 83-01, January 1983

/EK 80/ H. Ehrig, H.-J. Kreowski: Applications of Graph Grammar Theory to Consistency, Synchronization and Scheduling in Data Base Systems, Inform. Syst., Vol. 5, pp. 225-238, Pergamon Press Ltd., 1980

/ENR 83/ H. Ehrig, M. Nagl, G. Rozenberg (eds.): Graph Grammars and Their Application to Computer Science, LNCS 153 (1983)

/EPS 73/ H. Ehrig, M. Pfender, H.J. Schneider: Graph-Grammars: An Algebraic Approach, Proc. of the IEEE Conf. on Automata and Switching Theory, Iowa City 1973, pp. 167-180

/ER 79a/ H. Ehrig, B.K. Rosen: Parallelism and Concurrency of Graph Manipulations, Theoret. Comp. Sci. 11 (1980), pp. 247-275

/ER 79b/ --: Decomposition of Graph Grammar Productions and Derivations, LNCS 73 (1979), pp. 192-205

/Fo 84/ H.-R. Fonio: Amalgamation of Graph Transformations with Application to Synchronization in Distributed Systems, to appear as Techn. Report, TU Berlin, FB 20

/Ha 80/ A. Habel: Concurrency in Graph-Grammatiken, TU Berlin, FB 20, Technical Report No. 80-11, March 1980

/HBR 81/ C.A.R. Hoare, S.D. Brookes, A.W. Roscoe: A Theory of Communicating Sequential Processes, Techn. Monograph PRG-16, Progr. Research Group, Oxford Univ., 1981

/JKRE 82/ D. Janssens, H.-J. Kreowski, G. Rozenberg, H. Ehrig: Concurrency of Node-label Controlled Graph Transformations, Techn. Report No. 82-38, Univ. of Antwerp, U.I.A. (1982)

/Ku 80/ H.T. Kung: The Structures of Parallel Algorithms, Advances in Computers, Vol. 19, 65-108, Academic Press Inc. 1980

REFERENCES (cont'd)

/MW 82/ B. Mahr, A. Wilharm: Graph Grammars as a Tool for Description in Computer Processed Control: A Case Stude, Proc. 8th Conf. on Graph-theoretic Concepts in Comp. Sci. (WG'82), 165-176 (1982)

/Mi 83/ R. Milner: Calculi for Synchrony and Asynchrony, Theoret. Comp. Sci. 25 (1983), pp. 267-310

/Na 79/ M. Nagl: A Tutorial and Bibliographical Survey on Graph Grammars, LNCS 73 (1979), 70-126

/Pe 80/ C.A. Petri: Concurrency, Proc. Net Theory and Applications, LNCS 84 (1980), 251-260

/Ro 75/ B.K. Rosen: Deriving Graphs from Graphs by Applying a Production, Acta Informatica, 4 (1975), pp. 337-357

DECOMPILATION OF CONTROL STRUCTURES
BY MEANS OF GRAPH TRANSFORMATIONS *

Ulrike Lichtblau
Lehrstuhl Informatik II
Universität Dortmund
Postfach 500500
D-4600 Dortmund 50

ABSTRACT

Decompilation denotes the translation from lower level into higher level programming languages. Here we deal with the aspect of detecting higher level control structures, including loops with any number of exits, in line-oriented programs.
The detection is carried out on the control flow graph of the source program by means of so called wellstructuring transformations. We show that the iteration of these transformations always terminates in a time linearly depending on the number of vertices of the underlying control flow graph.

1. INTRODUCTION

Decompilation denotes the reverse of the compilation process, i.e. the translation from lower level into higher level programming languages. Simply embedding the source language into the target language is not considered to be sufficient. Rather, the higher target language should be exhausted. Therefore, the main problem of decompilation is the detection of high level structures in - at first sight - unstructured programs.
Decompilers are investigated in the literature since the middle of the sixties. Hopwood gave a survey of the subject [4].

Here we deal merely with one aspect of decompilation, namely the analysis of the control flow of source programs and its translation into semantically equivalent higher control structures of the target language.
The target languages which we consider are higher procedural ones like Ada or Modula-2. Their control structures include loops with any number of exits. The source languages

---

* This work was supported by Deutsche Forschungsgemeinschaft (grant Cl 53/3-2).

are line-oriented such as assembler languages or BASIC.
Programs written in one of these source languages can easily be represented by control flow graphs. We use this representation here.
The detection of higher control structures is described in terms of graph transformation rules. By application of a transformation rule a pattern of vertices and edges is replaced by a single vertex. Each pattern corresponds directly to a control structure of the target languages. Fig. 1 illustrates the idea of these transformation rules.

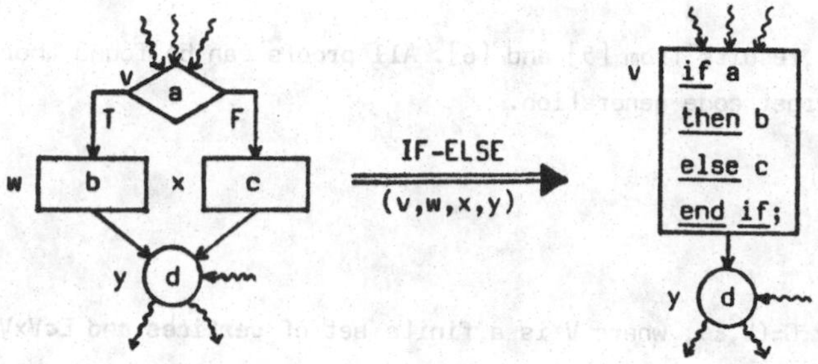

Fig. 1

Iteration of transformations based on such rules causes a control flow graph to decrease in size. The more higher structures it contains, the smaller it becomes.
We show that the process of reducing a control flow graph as far as possible always terminates in a time linearly depending on the number of vertices of the graph. Note that this holds although the size and the number of the patterns to be detected grow with the size of the input graph. The result can be obtained because it is possible to recognize the nesting of structures before starting the reduction process and because the system of wellstructuring transformation rules has the finite Church-Rosser property.

We do not explain the generation of target code here. However, this task can easily be solved by assigning labels to the vertices and edges of control flow graphs. In Fig. 1 such a labeling was assumed to exist. The edge labels should indicate 'true'- and 'false'-branches. The vertex labels should consist of simple conditions or simple statements at the beginning, and later on of target code which corresponds to the structure already detected. It is important to note that the transformation rules introduced here support the generation of code.
There are, of course, control flow graphs that by means of these transformation rules cannot be reduced to a single vertex. For code generation purposes they can be treated in the following way. Whenever there is no rule applicable, an edge is removed from the graph and a goto-statement is generated at the appropriate position. Afterwards, the process of reduction is restarted.

Rosendahl and Mankwald [8] use a similar idea of decompiling control structures by reducing the control flow graph and constructing the target program in the labels of its vertices. However, they only support the detection of loops with one exit. Further, they do not construct an efficient transformation algorithm.
The set of graph transformation rules introduced by Farrow, Kennedy and Zucconi [1] considers multiple exits from loops in an even more general way than is done in this paper. The derived reduction algorithm is of linear time complexity. These rules, however, do not support the translation of the recognized structure as they do not directly correspond to control structures of a higher programming language.

This paper states results from [5] and [6]. All proofs can be found there, as well as the details of target code generation.

2. PRELIMINARIES

A **graph** is a pair $G=(V,E)$, where V is a finite set of **vertices** and $E\subseteq V\times V$ is a set of directed **edges**.
Let $G=(V,E)$, $H=(V',E')$ be graphs.
For each vertex $v\in V$ we call **pred(v)**$:=\{w\in V \mid (w,v)\in E\}$ the set of all predecessors and **suc(v)**$:=\{w\in V \mid (v,w)\in E\}$ the set of all successors of v.
$\mathbf{d^+(v)}:=\#\mathrm{pred}(v)$ is the **in-degree** and $\mathbf{d^-(v)}:=\#\mathrm{suc}(v)$ is the **out-degree** of v.
A finite sequence of vertices $(v_0,\ldots,v_n)$, $n\geq 0$, is a **path** in G if $v_i\in V$ for $0\leq i\leq n$ and $(v_i,v_{i+1})\in E$ for $0\leq i<n$.
A vertex $r\in V$ is called a **root** of G if there exists a path $(r,\ldots,v)$ in G for every $v\in V$.
$v\in V$ **dominates** $w\in V$ **with respect to** a root r of G if v occurs in every path $(r,\ldots,w)$ in G.
H is a **subgraph** of G if $V'\subseteq V$ and $E'=E\cap(V'\times V')$.
For each $V''\subseteq V$ the graph $(V'',E\cap(V''\times V''))$ is called the **subgraph** of G **induced by** V''.
An **isomorphism** i from G to H, denoted by $i\colon G\rightarrow H$, is a bijective mapping $i\colon V\rightarrow V'$ with the property $(i(v),i(w))\in E'$ iff $(v,w)\in E$ for all $v,w\in V$.

Let X be a set and $R\subseteq X\times X$ a binary relation on X.
We denote by $\mathbf{R^m}$, $m\in\mathbb{N}_0$, the m-fold iteration of R and by $\mathbf{R^*}$ the reflexive transitive closure of R.

3. CONTROL FLOW GRAPHS

In this section we introduce control flow graphs. We use them to carry out the detection of higher control structures since they are normal forms of line-oriented programs.

Definition :

A **control flow graph** is a 4-tupel $k=(V,E,r,s)$, where

(i) (V,E) is a graph

(ii) $V=\{r\}$ or $V=V_A+V_C+\{r,s\}$ and $V_A\neq\emptyset$

(iii) r is a root of (V,E)

(iv) $d^+(r)=0$ and $d^-(s)=0$

(v) $\forall v\in V_A\cup\{r\}: d^-(v)\leq 1$

(vi) $\forall v\in V_C \quad : d^-(v)=2$.

K denotes the set of all control flow graphs and **SK** the subset of all graphs in K consisting of exactly one vertex.

There are four kinds of vertices in a control flow graph: a vertex r representing the start-statement of a program, a vertex s representing the stop-statement, vertices V_A corresponding to other statements and vertices V_C corresponding to conditions.

Example :

Fig. 2 shows a control flow graph $k\in K$.

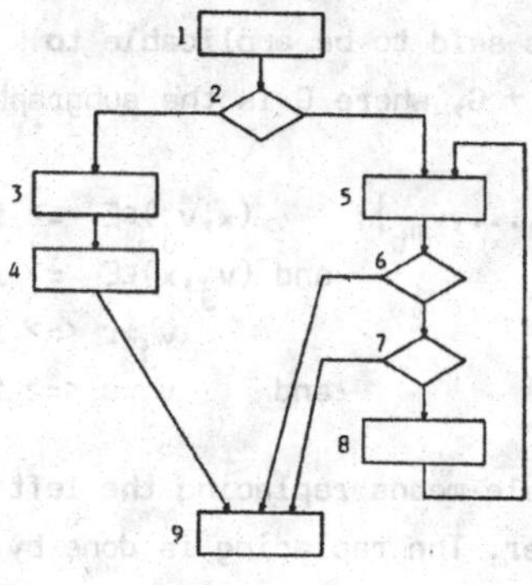

Fig. 2

4. A GRAPH TRANSFORMATION SYSTEM FOR STRUCTURE RECOGNITION PURPOSES

Here we introduce a system of transformation rules on the set of all control flow graphs. Each of these rules is applicable if and only if the given graph contains a pattern of vertices and edges that corresponds to a boolean operation on conditions or to one of the control structures 'sequence', 'if-then', 'if-then-else', 'loop' or 'while-loop', where a loop may have any number of exits. On application of a transformation rule the detected pattern is reduced to a single vertex.

Definition :

The **system of wellstructuring transformation rules** T is the collection of rules depicted in Fig. 3.

Notation :

Each transformation rule $t \in T$ consists of two graphs, the **left hand side** of t $L(t)=(V_{L(t)},E_{L(t)})$ and the **right hand side** $R(t)=(V_{R(t)},E_{L(t)})$ and two sets of vertices $I(t) \subset V_{L(t)}$ and $O(t) \subset V_{L(t)}$, the **in-vertices** and the **out-vertices** of t.
In Fig. 3 the arrow-heads mark the in-vertices and the circles symbolize the out-vertices.

The in-vertices and the out-vertices describe the allowed embedding of the left hand side of a rule into a control flow graph. In-vertices are the only vertices which may be connected to the outside by incoming edges, out-vertices are those which may be connected by outgoing edges.
Note that the out-vertices of a transformation rule do not belong to the pattern of the detected control structure. Rather, they are used to define the context.

Definition :

Let be $k=(V,E,r,s) \in K$.
A transformation rule $t \in T$ is said to be **applicable to** k **in** $v_1,\dots,v_{m_t} \in V$ if there exists an isomorphism i: $L(t) \to G$, where G is the subgraph of (V,E) induced by $\{v_1,\dots,v_{m_t}\}$ and

$$\text{(i)}\quad \forall v_j,\ 1 \le j \le m_t:\ \forall x \in V \setminus \{v_1,\dots,v_{m_t}\}:\quad (x,v_j) \in E \Rightarrow i^{-1}(v_j) \in I(t) \ \text{and}\ (v_j,x) \in E \Rightarrow i^{-1}(v_j) \in O(t)$$

$$\text{(ii)}\quad \forall v_j,\ 1 \le j \le m_t:\quad v_j=r \Leftrightarrow i^{-1}(v_j)=r \ \text{and}\ v_j=s \Leftrightarrow i^{-1}(v_j)=s \quad .$$

Applying a transformation rule means replacing the left hand side with the right hand side and embedding the latter. The replacing is done by removing certain vertices and adding certain edges. This is possible because $V_{R(t)} \subset V_{L(t)}$ for each $t \in T$. The embedding is straightforward since for every $t \in T$ $I(t) \cup O(t) \subset V_{R(t)}$.

Definition :

Let be $k=(V,E,r,s) \in K$ and $t \in T$ such that t is applicable to k in $v_1,\dots,v_{m_t} \in V$ via the isomorphism i: $L(t) \to G$.
The **result of applying** t **to** k **in** $v_1,\dots,v_{m_t}$ is defined by $\mathbf{t(v_1,\dots,v_{m_t})(k)}:=$ (V',E',r',s'), where

- $V'=(V \setminus \{v_1,\dots,v_{m_t}\}) \cup \{i(x) \mid x \in V_{R(t)}\}$
- $E'=(E \cap (V' \times V')) \cup \{(i(x),i(y)) \mid (x,y) \in E_{R(t)}\}$
- $r'=r$
- $s'=\begin{cases} r, & \text{if } t=BLK \\ s & \text{else} \end{cases}$.

The binary relation on K induced by T is denoted by $\mathbf{T_R}$.

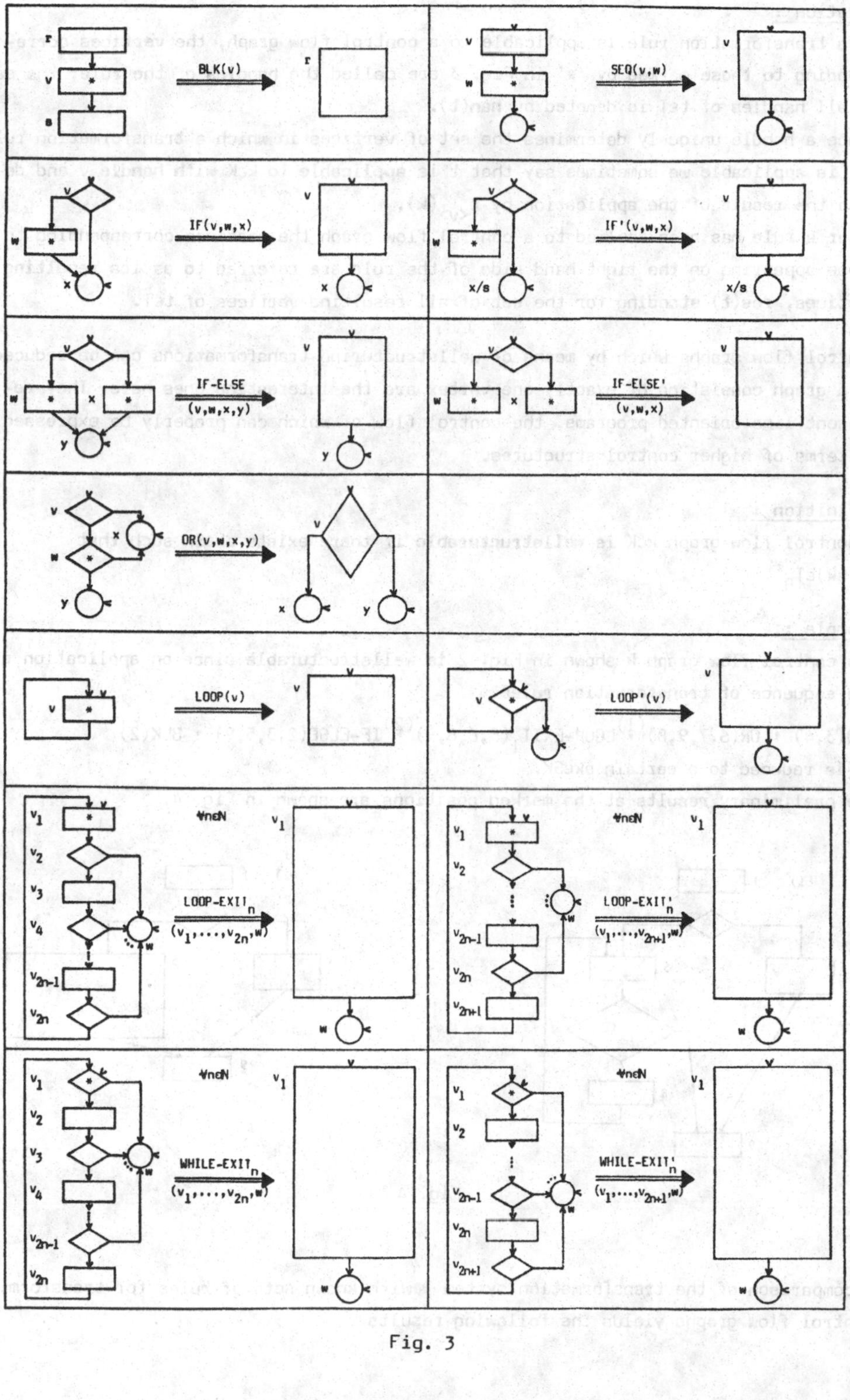
BLK(v)
SEQ(v,w)
IF(v,w,x)
IF'(v,w,x)
IF-ELSE
(v,w,x,y)
IF-ELSE'
(v,w,x,y)
OR(v,w,x,y)
LOOP(v)
LOOP'(v)
∀n∈N
LOOP-EXIT$_n$
(v_1,...,v_{2n},w)
LOOP-EXIT'$_n$
(v_1,...,v_{2n+1},w)
WHILE-EXIT$_n$
(v_1,...,v_{2n},w)
WHILE-EXIT'$_n$
(v_1,...,v_{2n+1},w)

Fig. 3

Notation :

If a transformation rule is applicable to a control flow graph, the vertices corresponding to those marked by '*' in Fig. 3 are called the **handles** of the rule; the set of all handles of $t \epsilon T$ is denoted by **han(t)**.

Since a handle uniquely determines the set of vertices in which a transformation rule $t \epsilon T$ is applicable we sometimes say that t is **applicable to** $k \epsilon K$ **with handle** v and denote the result of the application by $\mathbf{t}_{<v>}(\mathbf{k})$.

After a rule has been applied to a control flow graph the vertices corresponding to those appearing on the right hand side of the rule are referred to as its **resulting vertices, res(t)** standing for the set of all resulting vertices of $t \epsilon T$.

Control flow graphs which by means of wellstructuring transformations can be reduced to a graph consisting of exactly one vertex are the interesting ones here. They represent line-oriented programs, the control flow of which can properly be expressed in terms of higher control structures.

Definition :

A control flow graph $k \epsilon K$ is **wellstructurable** if there exists $sk \epsilon SK$ such that $(k,sk) \epsilon T_R{}^*$.

Example :

The control flow graph k shown in Fig. 2 is wellstructurable since on application of the sequence of transformation rules

$SEQ(3,4) \cdot OR(6,7,9,8) \overset{(1)}{\cdot} LOOP\text{-}EXIT_1(5,6,8,9) \overset{(2)}{\cdot} IF\text{-}ELSE(2,3,5,9) \cdot BLK(2)$

it is reduced to a certain $sk \epsilon SK$.

The preliminary results at the marked positions are shown in Fig. 4.

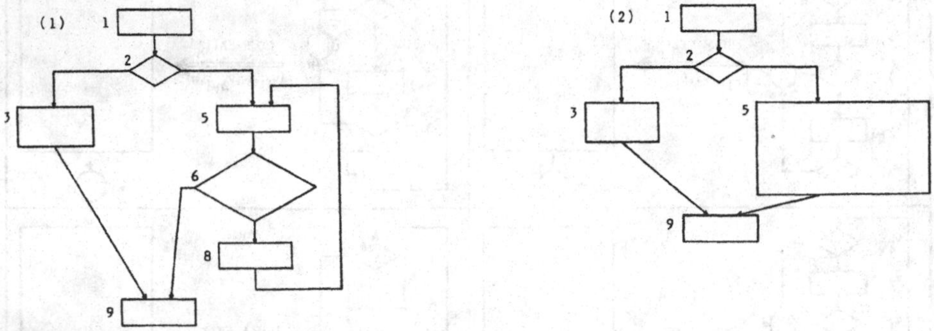

Fig. 4

A comparison of the transformation system T with known sets of rules for transforming control flow graphs yields the following results.

Each wellstructurable control flow graph is collapsible with respect to the definition of Hecht and Ullman [3]; the reverse does not hold.
The set of acyclic wellstructurable control flow graphs is properly covered by the set of acyclic SSFG-reducible ones defined by Farrow, Kennedy and Zucconi [1]. Considering arbitrary control flow graphs the inclusion does no longer hold.
The fully well structured control flow graphs of Rosendahl and Mankwald [8] are a proper subset of the wellstructurable ones.

5. PROPERTIES OF THE TRANSFORMATION SYSTEM

Here we collect some properties of the system of wellstructuring transformation rules T which prove to be useful in designing a linear time reduction algorithm based on T.

We start with the finite Church-Rosser property. It says that any sequence of transformation rules applicable to a control flow graph is of bounded length, and that the limit graph which can be derived from a given control flow graph is uniquely determined - and thus not dependent on the decision which of several applicable rules is chosen in a given situation.
The following formulation of the finite Church-Rosser property is due to Rosen [7].

Definition :
Let X be a set and $R \subset X \times X$ a binary relation on X.
(X,R) is **finite Church-Rosser (FCR)** if
(i) $\forall x \in X\ \exists m_x \in \mathbb{N}_0\ \forall y \in X, m \in \mathbb{N}_0 : (x,y) \in R^m \Rightarrow m \le m_x$
(ii) $\forall x,y,y' \in X: ((x,y) \in R \text{ and } (x,y') \in R) \Rightarrow \exists z \in X: (y,z) \in R^* \text{ and } (y',z) \in R^*$.

Lemma 1 :
(K, T_R) is FCR.

Proof :
By considering all possible interferences of any two wellstructuring transformation rules (cf. [5]). ∎

It is an easy consequence of lemma 1 that one is free to assign priorities to the wellstructuring transformation rules. We do this in the following way.

Definition :
Let be **prio_1** := {BLK, SEQ, IF, IF', IF-ELSE, IF-ELSE', OR} ⊂ T,
prio_2 := {LOOP, LOOP'}
∪ $\{\text{LOOP-EXIT}_n \mid n \in \mathbb{N}\}$ ∪ $\{\text{LOOP-EXIT}_n' \mid n \in \mathbb{N}\}$
∪ $\{\text{WHILE-EXIT}_n \mid n \in \mathbb{N}\}$ ∪ $\{\text{WHILE-EXIT}_n' \mid n \in \mathbb{N}\}$ ⊂ T .

Prio_i, i=1,2, is called the **class of wellstructuring transformation rules of priority i**.

When testing whether a transformation rule is applicable it is reasonable to try to find a handle first. Therefore, we next state easy to check conditions which are necessary for a vertex being a handle of a transformation rule of a certain priority.

Lemma 2 :

Let be $k=(V,E,r,s)\in K$ and $t\in T$ such that t is applicable to k with handle $v\in V$.

If $t\in prio\_1$, then $d^+(v)=1$.

If $t\in prio\_2$, then $d^+(v)>1$.

Proof :

By definition. ∎

We now turn to the question how the set of vertices which are potential handles of transformation rules changes during the reduction process.

Concerning transformation rules of priority 1 we get the following result: at most the resulting vertices of the rule applied last are new potential handles.

Lemma 3 :

Let be $k=(V,E,r,s)$, $k'=(V',E',r',s')\in K$, $v\in V\cap V'$, $t\in T$ and $t'\in prio\_1$ such that t is applicable to k, $t(k)=k'$ and t' is not applicable to k with handle v.

If t' is applicable to k' with handle v, then $han(t')\cap res(t)\neq\emptyset$.

Proof :

By verifying the following statement for each $t'\in prio\_1$ and each $t\in T$:

There are certain properties of a vertex v which are relevant to the decision whether t' is applicable with handle v. An application of t changes relevant properties of some vertices, but only of the resulting vertices of t. (cf. [6]) ∎

For transformation rules of priority 2 the answer is even easier: each vertex which may be used as a handle at any time during a reduction can be detected in the original control flow graph. This follows from lemma 2.

Furthermore, it is possible to sort the potential handles of transformation rules of priority 2 in such a way that for each nested loop the handle of the corresponding transformation rule appears before the handle of the rule detecting the outer loop. To show this we use a wellknown ordering on the vertices of a control flow graph. The definition is taken from Hecht [2].

Definition :

Let be $k=(V,E,r,s)\in K$.

A mapping rPOSTORDER: $V \rightarrow \{1,\ldots,\#V\}$ is called a **vertex numbering** of k if the ordering induced on V equals the reverse of the order in which each vertex was last visited during a depth-first search of k.

A depth-first search algorithm that computes vertex numberings is given in [2]. Note that there exist different vertex numberings of a given control flow graph.

The reverse of any vertex numbering is an ordering of potential handles of transformation rules of priority 2 with the property described above.

Lemma 4 :

Let be $k=(V,E,r,s)$, $k_i=(V_i,E_i,r_i,s_i)\epsilon K$, $i=1,2$, and $t_1,t_2 \epsilon prio\_2$ such that t_1 is applicable to k_1 with handle $v_1 \epsilon V_1$, t_2 is applicable to k_2 in $v_{21},\ldots,v_{2m},w \epsilon V_2$, $m\geq 2$, with handle $v_2=v_{21}$ and (k,k_1), $(k_1,k_2)\epsilon T_R{}^*$.
If there exists a subgraph $G'=(V',E')$ of (V_1,E_1) such that $v_1 \epsilon V'$ and G' is transformed into v_{2j}, $1\leq j\leq m$, during the transition from k_1 to k_2, then $\{v_1,v_2\}\subset V$ and $rPOSTORDER(v_1)\geq rPOSTORDER(v_2)$ for any vertex numbering $rPOSTORDER$: $V \rightarrow \{1,\ldots,\#V\}$ of k.

Proof :

By two claims:

v_2 dominates v_1 with respect to r. (By definition of the transformation rules.)

If x dominates y with respect to r, then $rPOSTORDER(x)\leq rPOSTORDER(y)$ for any vertex numbering rPOSTORDER. (By definition of a vertex numbering.) (cf. [6]) ∎

6. THE TIME COMPLEXITY OF THE REDUCTION PROCESS

In this last section we are concerned with the process of reducing a control flow graph by means of wellstructuring transformations as far as possible and, in particular, contracting each wellstructurable graph to a single vertex.

As our main result we show that this process always terminates in time linear in the number of vertices of the input control flow graph.

The algorithm which we present is based on the fact that, since (K,T_R) is FCR (lemma 1), transformation rules may be applied in any order. It uses two work-lists, each corresponding to a priority class and containing all vertices of the current control flow graph which can possibly be used as handles of transformation rules belonging to that class. These potential handles are determined by means of lemma 2. The potential handles of rules of priority 2 are ordered in such a way that handles of inner loops appear in the work-list before handles of outer loops (see lemma 4). During each iteration of the algorithm the first entry - say vertex v - is removed from the current work-list. The lists are treated in order of the priorities they correspond to. If there exists a transformation rule in the corresponding priority class that is applicable to the current control flow graph with handle v, then this rule is applied. After a successful application the work-lists are updated according to lemmas 3, 2 and 4.

Theorem :

Let be $k=(V,E,r,s)\varepsilon K$.

The problem of reducing k as far as possible by means of wellstructuring transformations can be solved in time $O(\#V)$.

Proof :

Consider the following algorithm REDUCE .

```
Input   :  k=(V,E,r,s)εK
Output  :  k'=(V',E',r',s')εK such that
           (i)    (k,k')εT_R*
           (ii)   ∀tεT: t is not applicable to k'
           (iii)  ∀k"=(V",E",r",s")εK: if k" fulfills (i) and (ii),  then #V"≥#V'
Method  :  Compute a vertex numbering rPOSTORDER: V → {1,...,#V};
           prio_1_handle_1 := list of all xεV with d+(x)=1,
                                        in any order;
           prio_2_handle_1 := list of all xεV with d+(x)>1,
                                        in order of decreasing values of rPOSTORDER;
           kk := k;
           while  there exists 1≤i≤2 such that prio_i_handle_1 is nonempty
           loop   if    prio_1_handle_1 is nonempty
                  then  perform actions on kk corresponding to priority 1;
                  else  perform actions on kk corresponding to priority 2;
                  end if;
           end loop;
           k' := kk .
Refine-
ments   :  (1)  perform actions on kk corresponding to priority i, 1≤i≤2 :
                v := first entry of prio_i_handle_1;
                remove v from prio_i_handle_1;
                if    there exists tεprio_i such that t is applicable to kk with
                      handle v
                then  kk := t_<v>(kk);
                      insert all xεres(t) with d+(x)=1 into prio_1_handle_1
                      at any position;
                      if    i=2 and d+(v)>1
                      then  insert v into prio_2_handle_1
                                  at the new first position;
                      end if;
                end if;
```

```
(2)  there exists tεprio_i such that t is applicable to kk with handle v,
     1≤i≤2 :
     case  i  is
     when  1  =>  verified := false;
                  for   all tεprio_1
                  loop  if    t is applicable to kk with handle v
                        then  verified := true;
                              exit;
                        end if;
                  end loop;
                  there exists := verified;
     when  2  =>  verified := false;
                  if    there is a vertex w that is a proper successor of
                        the potential loop at v
                  then  falsified := false;
                        x := v;
                        while  not verified and not falsified
                        loop   if     suc(x)\{w}={v}
                               then   verified := true;
                               elsif  yεsuc(x)\{w} is acceptable as a succes-
                                      sor of x in a loop
                               then   x := y;
                               else   falsified := true;
                               end if;
                        end loop;
                  end if;
                  there exists := verified;
     end case;
(3)  there is a vertex w that is a proper successor of the potential loop
     at v :
     falsified := false;
     if     d⁻(v)=2
     then   x := v;
     elsif  there exists yεsuc(v)
     then   x := y;
     else   falsified := true;
     end if;
     if     not falsified
     then   if     there exists yεsuc(x) with d⁺(y)>1
            then   w := y;
            elsif  there exists yεsuc(x) with d⁻(y)≠1
            then   w := y;
```

```
            elsif there exists y∈suc(x) with (y,v)∉E
            then  w := y;
            else  falsified := true;
            end if;
      end if;
      there is := not falsified;
```

Claim 1 :

Algorithm REDUCE is correct, i.e. it terminates and has the specified input/output-behaviour.

The main ideas of the proof of claim 1 are contained in the remarks at the beginning of this section.

Claim 2 :

Algorithm REDUCE can be implemented in such a way that its time complexity is O(#V).

Proof of claim 2 :

The computation of rPOSTORDER can be done in time O(#E) (cf. [2]). Since the out-degree of all vertices of a control flow graph is bounded by 2 we have O(#E)=O(#V) here.
Obviously, the initialization of the work-lists needs O(#V) steps.
Further, it is fairly clear that the number of iterations of the main while-loop appearing in REDUCE is in O(#V).
One yet has to look at the nested loops. There are two of them, both appearing in refinement (2).
The first is iterated a constant number of times.
For the second it is possible to show that the total number of its iterations (over all iterations of the main loop) is in O(#V). To prove this we observe that each vertex of the input control flow graph is only visited a constant number of times during all iterations of the inspected loop. The result depends on lemma 4.
So far we have seen that the total number of executions of any test and any simple statement appearing in REDUCE is in O(#V). The proof is completed by showing that each of these atomic components can be executed in constant time. This requires the development of an appropriate internal data structure, which for each vertex allows direct access to its predecessors and to its entries in the lists of predecessors of its successors as well as to its entry in a work-list. (cf. [6]) ∎

ACKNOWLEDGEMENT

The author would like to thank Prof. Dr. V. Claus for initiating this work and for helpful discussions.

REFERENCES

1. Farrow,R., Kennedy,K. and Zucconi,L. Graph grammars and global data flow analysis. Proceedings of the Seventeenth Annual IEEE Symposium on Foundations of Computer Science, Houston, Texas, Oct. 1976, 42-56.

2. Hecht,M.S. Flow Analysis of Computer Programs. North-Holland, New York, 1977, Chapt. 3.

3. Hecht,M.S. and Ullman,J.D. Flow graph reducibility. SIAM J. Comput. 1, 2 (June 1972), 188-202.

4. Hopwood,G.L. Decompilation. Ph.D. dissertation, University of California, Irvine, Feb. 1978.

5. Lichtblau,U. Graphtransformationen zur Erkennung Ada-ähnlicher Kontrollstrukturen in maschinennahen Programmen. Techn. Rep. 142, Abteilung Informatik, Universität Dortmund, 1982.

6. Lichtblau,U. Ein Algorithmus zur Erkennung höherer Kontrollstrukturen durch Graphtransformationen. Techn. Rep. 183, Abt. Informatik, Universität Dortmund, 1984

7. Rosen,B. Tree manipulating systems and Church-Rosser theorems. Journ. ACM 20, 1 (Jan. 1973), 160-187

8. Rosendahl,M. and Mankwald,K.P. Analysis of programs by reduction of their structure. In Graph-Grammars and Their Application to Computer Science and Biology, Claus,V., Ehrig,H. and Rozenberg,G. (Eds.). Springer-Verlag, Berlin, 1979, 409-417.

Synchronized Bottom-Up Tree Automata and L-Systems

E. Fachini and M. Napoli
Dipartimento di Informatica e Applicazioni
Universita' di Salerno
84100 Salerno, Italy

Introduction.

In this paper a new type of bottom-up tree automaton, called synchronized bottom-up tree automaton, is considered. This automaton processes a tree in a bottom-up way and one level at a time. Moreover, more than one transition function is allowed, but only one of them at a time can be applied to nodes at the same level of a tree.

The tree language recognized by these automata are the images, under projection, of the set of derivation trees of EPTOL languages.

The model introduced in this paper is a generalization of the bottom-up tree automaton. Its behaviour, relative to ETOL systems, is the same as the bottom-up tree automaton behaviour relating to context free grammars (7).

Furthermore, many properties of the bottom-up tree automata continue to hold for the class of automata here introduced. In fact, in the case that one transition function is allowed, the class of recognized tree languages is a boolean algebra and has a decidable equivalence problem. In the general case, the membership, the emptiness and the finiteness problems turn out to be decidable.

As it has been observed in the case of context-free languages, the introduction of tree automata recognizing sets of derivation trees of L-languages allows to state properties or to give simpler proofs of already known properties about the corresponding classes of L-languages.

We consider a subclass of trees in which a special symbol e labels a node representing an aborting computation, and we introduce a particular synchronized bottom-up tree automaton, called e-synchronized bottom-up tree automaton, which recognizes languages of this kind of trees.

The language recognized by these automata are the images, under a proper projection, of the sets of derivation trees of ETOL systems.

We are also able to construct, for every tree language L recognized by an e-synchronized bottom-up tree automaton, a synchronized bottom-up tree automaton which recognizes the set of trees obtained from L by pruning the dead branches. This result supplies a different method to construct an EPTOL system equivalent to a given ETOL system.

In section 1 preliminary concepts and definitions are introduced.

Section 2 contains the definitions of the considered classes of synchronized bottom-up tree automata and the proofs of their properties.

In section 3 the relationships between the synchronized bottom-up tree automata and L-systems are pointed out.

In section 4 some decision problems are dealt with.

1. Terminology.

We suppose the principal notions of tree languages and L-systems theories. We just give a definition of the set of derivation trees of an ETOL system, which is a slight modification of the one given by Engelfriet in (2).

Given a ranked alphabet Σ, we will consider the tree language over Σ containing only trees whose paths from the root to the leaves have the same lenght.

Definition 1.1. Let Σ be a ranked alphabet, consider the tree language $H_\Sigma = \bigcup_{k>=0} H_\Sigma^k$ where H_Σ^k is defined recursively as follows:
$H_\Sigma^0 = \Sigma_0$
$H_\Sigma^{k+1} = \{a(t_1 \ldots t_r) \mid a \in \Sigma_r \text{ and } t_i \in H_\Sigma^k \text{ for } 1<=i<=r\}$.

Definition 1.2. Let $G=(\Sigma,\mathcal{P},S,\Delta)$ be an ETOL system. Given a symbol $e \notin \Sigma$, define Ω to be the ranked alphabet $\Sigma \cup \{e\}$ such that $\Omega_0 = \Delta \cup \{e\}$, $\Omega_1 = \{a \in \Sigma$ such that there exist $b \in \Sigma \cup \{\varepsilon\}$ and $P \in \mathcal{P}$ such that $a \rightarrow b$ is in $P\} \cup \{e\}$ and for $r>=2$, $\Omega_r = \{a \in \Sigma$ such that there exist $P \in \mathcal{P}$ and $w \in \Sigma^+$ such that $a \rightarrow w$ is in P and $|w|=r\}$ ($|x|$ is the lenght of the word x). For $a \in \Sigma$ and $\pi \in \mathcal{P}^*$, the set of derivation trees with top a and control word π, denoted $D_\pi^a(G)$, is defined recursively as follows:
1. for $a \in \Delta$, $a \in D_\lambda^a(G)$ (where λ is the empty word of $\mathcal{P}^*$);
2. for $a \in \Sigma$ and $P \in \mathcal{P}$, if $a \rightarrow \varepsilon$ is in P then $a(e^n)e$ is in $D_{\pi P}^a(G)$, for every $n>=0$ and for every $\pi \in \mathcal{P}^*$ such that $|\pi|=n$;
3. for $n>=1$, $a, a_1, \ldots, a_n \in \Sigma$, $P \in \mathcal{P}$ and $t_1, \ldots, t_n \in T_\Sigma$, if $a \rightarrow a_1 \ldots a_n$ is in P and $t_i \in D_\pi^{a_i}(G)$ for $1<=i<=n$, then $a(t_1 \ldots t_n) \in D_{\pi P}^a(G)$.

The set of derivation trees of G, denoted $D(G) \subseteq H_\Sigma$, is defined by $D(G) = \bigcup_{a \in S, \pi \in \mathcal{P}^*} D_\pi^a(G)$.

Note that if G is propagating then we do not need the symbol e and the clause 2. of the above definition.

For a given ranked alphabet Σ, besides the usual frontier function $fr: H_\Sigma \rightarrow \Sigma_0^*$, we will introduce the e-free frontier function for a symbol e, defined as follows.

Definition 1.3. Let Σ be a ranked alphabet. Chosen a symbol $e \in \Sigma_0$ we define a mapping $fr_e : H_\Sigma \rightarrow \Sigma_0^*$ recursively as follows:
i. for $a \in \Sigma_0$, $fr_e(a) = \varepsilon$ if $a=e$, otherwise $fr_e(a)=a$
ii. for $k>=1$, $a \in \Sigma_k$ and $t_i \in H_\Sigma$, for $1<=i<=k$,

$$fr_e(a(t_1 \ldots t_k)) = fr_e(t_1) fr_e(t_2) \ldots fr_e(t_k).$$

It is easy to see that, for every ETOL system G, $fr_e(D(G)) = L(G)$ and that for every EPTOL system G $fr(D(G)) = L(G)$.

Definition 1.4. Given two ranked alphabets Σ and Δ such that $\Sigma_j \neq \emptyset \Rightarrow \Delta_j \neq \emptyset$ and given $R_j \subseteq \Sigma_j \times \Delta_j$, for every $j>=0$ such that $\Sigma_j \neq \emptyset$, a relabeling R with domain T_Σ and range T_Δ is the relation $R \subseteq T_\Sigma \times T_\Delta$ defined as follows:
$(a,b) \in R$ if $(a,b) \in R_0$;
$(a(t_1 \ldots t_k), b(t_1' \ldots t_k')) \in R$ if $(t_i, t_i') \in R$ for $1<=i<=k$ and $(a,b) \in R_k$.

Given a relabeling R with domain T_Σ and range T_Δ and $t \in T_\Sigma$ let $R(t) = \{t' \in T_\Delta \mid (t,t') \in R\}$. If $L \subseteq T_\Sigma$, let $R(L) = \bigcup_{t \in L} R(t)$.

Definition 1.5. A projection P with domain T_Σ and range T_Δ is a relabeling such that for every $a \in \Sigma_j$ there exists a unique $b \in \Delta_j$ such that $(a,b) \in P$, for every j such that $\Sigma_j \neq \emptyset$.

In such a case we will write $t'=P(t)$ instead of $(t,t') \in P$.

A projection P with domain T_Σ and range T_Δ is said to be frontier preserving if for every $t \in T_\Sigma$ $fr(t)=fr(t')$ for every $t'=P(t)$.

2. Synchronized bottom-up tree automata.

In this section we introduce the k-synchronized bottom-up tree automata, abbreviated k-SBUTA, where k is an integer greater than zero, and we prove that the deterministic and non deterministic versions of such automata are equivalent.

We state that the class of tree languages recognized by any k-SBUTA is closed with respect to set theoretical union and intersection, whereas the closure with respect to complementation holds for the class of tree languages recognized by a 1-SBUTA.

We introduce a particular kind of k-synchronized bottom up tree automaton, called e-k-synchronized bottom-up tree automaton, in order to recognize trees in which some paths, labelled by words in $\{e\}^*$, represent aborted computations. Moreover, we prove that, given an e-k-synchronized bottom up tree automaton recognizing a tree language L, it is possible to construct a k-synchronized bottom-up tree automaton which recognizes the language obtained from L by cutting the dead subtrees from each tree.

Definition 2.1. Let $k \in N^+$. A k-NSBUTA is a 5-tuple $A=(\Sigma,Q,\rho_0,\rho,F)$ where

Σ is a ranked alphabet,

Q is a finite set of states,

$F \subseteq Q$ is a set of final states,

$\rho_0 \subseteq \Sigma_0 \times Q$ and

$\rho = \{\rho_1,\dots,\rho_k\}$ where $\rho_i = \{\rho_{i,j} \subseteq (Q^j \times \Sigma_j) \times Q \mid j>=1, \Sigma_j \neq \emptyset\}$.

The binary relation ρ_0 assignes initial states to the leaves; besides, if $(q_1,\dots,q_j,a,q) \in \rho_{i,j}$ then the state q can be assigned to a node labelled a if $q_1,\dots,q_j$ have been assigned to its sons. If k=1 then we will write ρ instead of ρ_1.

A k-NSBUTA $A=(\Sigma,Q,\rho_0,\rho,F)$ is deterministic if $\rho_0: \Sigma_0 \rightarrow Q$ and $\rho_{i,j}: Q^j \times \Sigma_j \rightarrow Q$ are partial functions. A deterministic k-NSBUTA will be called k-SBUTA.

Let us define a binary relation $\vdash^{A,t,h}_{(N)SB} \subseteq (\bigcup_{n \in N} (\Sigma \times Q)^n)^2$ to describe the computation step of a k-(N)SBUTA A on $t \in T_\Sigma$, by using the h-th function (relation):

$(a_{1,1},q_{1,1},\dots,a_{1,r_1},q_{1,r_1},\dots,a_{s,1},q_{s,1},\dots,a_{s,r_s},q_{s,r_s}) \vdash^{A,t,h}_{(N)SB}$

$(a_1,q_1,\dots,a_s,q_s)$ if $1<=h<=k$ and for every $1<=i<=s$ it holds that $a_i \in \Sigma_{r_i}$, $r_i>0$, $a_{i,1},\dots,a_{i,r_i}$ are the labels of the sons of the nodes labelled a_i and $\beta_{h,r_i}(q_{i,1},\dots,q_{i,r_i},a)=q_i$ $((q_{i,1},\dots,q_{i,r_i},a,q_i)\ \beta_{h,r_i})$.

We define $x_1 \vdash^{A,t,\pi +} x_n$ if $\pi = j_1 \dots j_{n-1}$ and there exist $x_2,\dots,x_{n-1} \in \bigcup_{n\in N} (\Sigma \times Q)^n$ such that $x_i \vdash^{A,t,j_i}$ (N)SB x_{i+1} for $1<=i<=n-1$, and $x1 \vdash^{A,t,\pi *}$ (N)SB x_n if $x_1=x_n$ and $\pi=\lambda$ or $x_1 \vdash^{A,t,\pi +}$ (N)SB x_n for some $\pi \in \{1,\dots,k\}^+$.

We say that $t \in T_\Sigma$ is q-π-accepted by A if $q \in Q$ and $(a_1,q_1,\dots,a_n,q_n) \vdash^{A,t,\pi *}$ (N)SB (a,q) where $a_1,\dots,a_n$ are the labels of the leaves and a is the label of the root of t, $(a_i,q_i) \in \beta_0$ $(\beta_0(a_i)=q_i)$ for $1<=i<=n$ and $\pi \in \{1,\dots k\}^*$.

We say that $t \in T_\Sigma$ is q accepted by A if it is q-π-accepted by A for some $\pi \in \{1,\dots,k\}^*$. Moreover, $t \in T_\Sigma$ is accepted by A if it is q-accepted by A for some $q \in F$.

Let L(A) the set of trees accepted by a k-(N)SBUTA A. Note that for every k-NSBUTA $A=(\Sigma,Q,\beta_0,\beta,F)$, $L(A) \subseteq H_\Sigma$.

A language L is said to be k-(N)SB-recognizable if there exists a k-(N)SBUTA A such that L=L(A).

Let k-NSB-RECOG (k-SB-RECOG) be the set of the k-NSB-recognizable (k-SB-recognizable) languages and NSB-RECOG (SB-RECOG) the set of 1-NSB-recognizable (1-SB-recognizable) languages.

Theorem 2.1. For every $k>=1$, k-NSB-RECOG = k-SB-RECOG.

Proof. The proof exploits the usual subset construction.

In the following we will prove that $\bigcup_{k>=1}$ k-SB-RECOG and SB-RECOG are closed with respect to union and intersection, and that the closure with respect to the complementation holds for SB-RECOG.

Theorem 2.2. $\bigcup_{k>=1}$ k-SB-RECOG is closed with respect to intersection and union.

Proof. Let $A=(\Sigma,Q,\beta_0,\beta,F)$ be a k-SBUTA and $A'=(\Sigma',Q',\beta_0',\beta',F')$ be a k'-SBUTA. We can suppose, without loss of generality, that $Q \cap Q' = \emptyset$ and that $k>=k'$.

We now construct a k-NSBUTA A" such that $L(A")=L(A)\cup L(A')$, according to theorem 2.1 there exists a k-SBUTA equivalent to A".

Let $A"=(\Sigma",Q",\beta_0",\beta",F")$ be the k-NSBUTA defined as follows:

$\Sigma"=\Sigma \cup \Sigma'$, $Q"=Q \cup Q'$, $F"=F \cup F'$,

$\beta_0"=\{(a,q) \mid a \in \Sigma_0 \cup \Sigma'_0,\ q \in Q \cup Q'$ and $\beta_0(a)=q$ or $\beta_0'(a)=q\}$,

$\beta"_{j,r}=\{(q_1,\dots,q_r,a,q) \mid a \in \Sigma_r \cup \Sigma'_r,\ q_1,\dots,q_r,q \in Q \cup Q'$ and $\beta_{j,r}(q_1,\dots,q_r,a)=q$ or $\beta'_{j,r}(q_1,\dots,q_r,a)=q\}$, for every $1<=j<=k$.

Consider now the k"-SBUTA A" such that $L(A")=L(A)\cap L(A')$, defined as follows:

$k"=kk'$, $\Sigma"=\Sigma \cap \Sigma'$, $Q"=Q \cap Q'$, $F"=F \cap F'$,

$\beta_0"(a)=(q,p)$ if $\beta_0(a)=q$ and $\beta_0'(a)=p$ for $a \in \Sigma_0 \cap \Sigma'_0$ and

$\beta"_{(j-1)k'+l,r}((q_1,p_1),\dots,(q_r,p_r),a)=(q,p)$ if $\beta_{j,r}(q_1,\dots,q_r,a)=q$ and $\beta'_{l,r}(p_1,\dots,p_r,a)=p$, for every $1<=j<=k$, $1<=l<=k'$ and $a \in \Sigma_r \cap \Sigma'_r$.

In both cases the proof is carried out by induction on the input tree.

Let ^L=H -L be the complement of L in H .

Theorem 2.3. The class of SB-recognizable subset of H is a boolean algebra.

Proof. From the constructions given in the proof of theorem 2.2, it immediately follows that SB-RECOG is closed with respect to union and intersection. Let $A=(\Sigma,Q,\rho_0,\rho,F)$ be a SBUTA recognizing a language L, we construct a SBUTA $A'=(\Sigma,Q',\rho_0',\rho',F')$ recognizing ^L. Let $Q'=Q\cup\{p\}$, where $p \notin Q$, $F'=(Q-F)\cup\{p\}$, $\rho_0'(a)=\rho_0(a)$ if $\rho_0(a)$ is defined, otherwise $\rho_0'(a)=p$, $\rho'_r(q_1,\ldots,q_r,a)=\rho_r(q_1,\ldots,q_r,a)$ if $q_i\in Q$ and $\rho_r(q_1,\ldots,q_r,a)$ is defined, otherwise $\rho'_r(q_1,\ldots,q_r,a)=p$.

In the following we will prove that k-SB-RECOG is closed with respect to relabeling. As a consequence we obtain the closure with respect to projection and inverse projection.

Theorem 2.4. For every k>=1, k-SB-RECOG is closed with respect to relabeling.

Proof. Given a k-SBUTA $A=(\Sigma,Q,\rho_0,\rho,F)$ and a relabeling R with domain T_Σ and range T_Δ we will construct a k-NSBUTA $A'=(\Delta,Q,\rho_0',\rho',F)$, such that $L(A')=R(L(A))$, defined as follows: $\rho_0'=\{(b,q) \mid (a,b)\in R_0$ and $\rho_0(a)=q$ for $a\in\Sigma_0\}$, $\rho'_{j,r}=\{(q_1,\ldots,q_r,b,q)$ if $(a,b)\in R_r$ and $\rho_{j,r}(q_1,\ldots,q_r,a)=q$ for $a\in\Sigma_r\}$.

By induction on t it is easy to see that

i. for every $t\in H_\Sigma$, if t is q-π-accepted by A and there exists $t'\in H_\Delta$ such that $(t,t')\in R$ then t' is q-π-accepted by A';

ii. for every $t\in H_\Delta$, if t is q-π-accepted by A' then there exists $t'\in H_\Sigma$ such that $(t',t)\in R$ and t' is q-π-accepted by A.

From i. and ii. the thesis follows.

Corollary 2.1. For every k>=1, k-SB-RECOG is closed with respect to (frontier preserving) projection and inverse projection.

Proof. Obvious.

Since a tree can represent a computation process, it is quite natural to consider that some computation paths may abort. In order to represent trees with aborting paths in H_Σ, we will introduce a special symbol e to label the nodes of a chain which completes an aborting path. For example the tree t = (a(b(d(d)b(bb))c) $\in T_\Sigma$ can be represented by the tree t' = (a(b(d(d)b(bb))c(e(e))) $\in H_{\Sigma\cup\{e\}}$.

Definition 2.2. Given a ranked alphabet Σ, we will call e-ranked alphabet the alphabet $\Sigma'=\Sigma\cup\{e\}$ such that $e\notin\Sigma$, $\Sigma'_0=\Sigma_0\cup\{e\}$, $\Sigma'_1=\Sigma_1\cup\{e\}$ and $\Sigma'_j=\Sigma_j$ for every j>=1.

Definition 2.3. Let $k\in N^+$. A non deterministic synchronized bottom-up tree automaton over an e-ranked alphabet Σ, abbreviated e-k-NSBUTA, is a k-NSBUTA $A=(\Sigma,Q,\rho_0,\rho,F)$ such that there is a special state $q_e\in Q$ satisfying the following conditions:

i. if $(a,q)\in\rho_0$ then $a=e \Leftrightarrow q=q_e$,

ii. if $(q,a,q')\in\rho_{i,1}$ then $a=e \Leftrightarrow q=q'=q_e$ and $q'=q_e \Rightarrow a=e \wedge q=q_e$,

iii. for j>=2, if $(q_1,\ldots,q_j,a,q_{j+1})\in\rho_{i,j}$ then $q_h\neq q_e$ for every $1\le h\le j+1$

iv. for every $1\le i\le k$ $(q_e,e,q_e)\in\rho_{i,1}$.

We will call e-k-SBUTA a deterministic bottom-up tree automaton over an e-ranked alphabet. The set of languages recognized by any e-k-(N)SBUTA will be called e-k-(N)SB-RECOG.

Lemma 2.1. For every e-k-NSBUTA A there exists an e-k-SBUTA A' such that L(A)=L(A').

Proof. A slight modification of the usual subset construction suffices to take in account the special symbol e.

Definition 2.4. Given two e-ranked alphabets Σ and Δ, a relabeling R with domain T_Σ and range T_Δ is said to be an e-relabeling if for every $(a,b) \in R_0 \cup R_1$ we have that a=e iff b=e. A projection which is an e-relabeling is called e-projection.

Lemma 2.2. For every k>=1, e-k-SB-RECOG is closed with respect to e-relabeling.

Proof. The construction given in the theorem 2.4 applies to the case of an e-k-SBUTA as well.

Given an e-ranked alphabet Σ, let Σ_{pe} be the ranked alphabet defined by: $(\Sigma_{pe})_0=\Sigma_0$ and $(\Sigma_{pe})_k=\bigcup_{r>=k}\Sigma_r$. Let pe be such a function that pe : $H \rightarrow H_{pe}$ and pe(a)=a, for every $a \in \Sigma_0$,
$pe(a(t_1 \ldots t_k))=e$ if $pe(t_i)=e$ for every $1<=i<=k$,
$pe(a(t_1 \ldots t_k))=a(pe(t_{l_1}) \ldots pe(t_{l_m}))$ if $pe(t_{l_s})=e$ for $1<=s<=m$,
$0<l_1<\ldots<l_m<=k$ and $pe(t_{i_r})=e$ for every $i_r=l_s$, $1<=i_r<=k$.

In the following the concept of state diagram will be useful.

Definition 2.5. Given a k-SBUTA $A=(\Sigma,Q,\rho_0,\rho,F)$, let the state diagram G=(V,E) be the directed labelled graph such that $V=2^Q$ and $(U,U') \in E$ with label j iff $1<=j<=k$ and $U'=\{\rho_{j,r}(q_1,\ldots,q_r,a) \mid a \in \Sigma_r$ and $q_i \in U$ for $1<=i<=r\}$.

Let $S_j(U)$ be the node U' such that $(U,U') \in E$ with label j.
Let $P(G) \subseteq 2^Q \times \{1,\ldots,k\}^* \times 2^Q$ such that $(U,\pi,U') \in P(G)$ iff $\pi \in \{1,\ldots,k\}^+$ and there exists a path from U to U' in G labelled π or $\pi=\lambda$ and U=U'. Let $C^U(G) = \{U' \in 2^Q \mid$ there exists $\pi \in \{1,\ldots,k\}^*$ and $(U,\pi,U') \in P(G)\}$.

Theorem 2.5. For every e-k-SBUTA A, there exists a k-SBUTA A' such that $L(A')=pe(L(A))-\{e\}$.

Proof. Given an e-k-SBUTA $A=(\Sigma,Q,\rho_0,\rho,F)$, let $G=(2^Q,E)$ be its state diagram. In the k-NSBUTA which we will construct, the states will be couples (q,U) where $q \in Q$ and U is a subset of Q containing the states p such that there exists a p-π-accepted tree, for a proper π, whose frontier belongs to $\{e\}^*$. Consider the k-NSBUTA $A'=(\Sigma',Q',\rho_0',\rho',F')$ where $\Sigma'=\Sigma_{pe}$, $Q'=Q \times 2^Q$, $F'=F \times 2^Q$, $\rho_0'=\{(a,(\rho_0(a),\{q_e\})) \mid a \in \Sigma_0-\{e\}\}$ and ρ' is defined as follows:
if $\rho_{j,r}(q_1,\ldots,q_r,a)=q$ then $((q_{i_1},U),\ldots,(q_{i_l},U),a,(q,S_j(U))) \in \rho'_{j,l}$ for $1<=l<=r$, $1<=i_1<\ldots<i_l<=r$ and $U \in C^{\{q_e\}}(G)$ iff there exist $h_1,\ldots,h_{r-l} \in$

$\{1,\ldots,r\} - \{i_1,\ldots,i_l\}$ such that $q_{h_n} \in U$, for every $1 \leq n \leq r-1$.

By induction on $|\pi|$ it is easy to prove that, for every $q \in Q$, $U \in C^{\{q_e\}}(G)$ and $\pi \in \{1,\ldots,k\}^*$ such that $(\{q_e\},\pi,U) \in P(G)$ it holds that $q \in U$ iff there exists $t \in H_\Sigma$ such that $pe(t)=e$ and t is q-π-accepted by A. By induction on t and by exploiting this result it is easy to prove that:

1. if $t \in H_\Sigma$ is q-π-accepted by A and $pe(t)=e$, then $pe(t)$ is (q,U)-π-accepted by A' for $(\{q_e\},\pi,U) \in P(G)$.
2. if $t \in H_{\Sigma pe}$ is (q,U)-π-accepted by A' then $(\{q_e\},\pi,U) \in P(G)$ and there exists $t' \in H_\Sigma$, such that $pe(t)=t'$, which is q-π-accepted by A.

So that $pe(L(A))-\{e\} = L(A')$. According to theorem 2.1 there exists a k-SBUTA A" such that L(A")=L(A').

3. L-systems and synchronized bottom-up tree automata.

In this section we relate the considered classes of synchronized bottom-up tree automata with classes of L-systems. We will prove that a language L is an ETOL language iff it coincides, up to the empty word, with the set of frontiers of trees belonging to a language recognized by a synchronized bottom-up tree automaton.

As an example of a result which can be obtained by exploiting the above correspondence between $\bigcup_{k\geq 1}$ k-SB-RECOG and ETOL languages, we show that for every ETOL system there exists a structurally equivalent ETOL system such that in every table the right hand side of a rule uniquelly determines its left hand side.

Theorem 3.1. For every ETOL system G, there exists an e-k-SBUTA A such that L(A)=D(G).

Proof. Given an ETOL system $G=(\Sigma,\wp,S,\Delta)$, consider the e-k-SBUTA $A=(\Omega,Q,\beta_o,\beta,F)$ such that $k=|\wp|$, Ω is the e-ranked alphabet $\Sigma\cup\{e\}$, $Q=\{q_a \mid a \in \Sigma\} \cup \{q_e\}$, $F=\{q_a \mid a \in \Delta\}$,
$\beta_o(a)=q_a$ if $a \in \Delta\cup\{e\}$,
$\beta_{P,1}(q_e,e)=q_e$ for every $P \in \wp$,
$\beta_{P,1}(q_e,a)=q_a$ for every $a \in \Omega_1$ such that $a \rightarrow \varepsilon \in P$ and finally
$\beta_{P,r}(q_{a_1},\ldots,q_{a_r},a)=q_a$ for every $a \in \Sigma_r$ such that $a \rightarrow a_1\ldots a_r \in P$, for $r>0$.

It is easy to see that L(A)=D(G).

Theorem 3.2. For every EPTOL system G, there exists a k-SBUTA A such that L(A)=D(G).

Proof. By eliminating the construction rules regarding the special symbol e, the construction given in the previous proof supplies the wanted k-SBUTA.

Theorem 3.3. Every e-k-SB-recognizable subset of H_Σ is the image, under a frontier preserving e-projection, of the set of derivation trees of an ETOL system $G=(\Omega,\wp,S,\Delta)$ for some alphabet Ω.

Proof. Let $A=(\Sigma,Q,\beta_0,\beta,F)$ be an e-k-SBUTA. Let us consider the alphabet $\Omega=\bigsqcup_{r>=1}(\Sigma_r-\{e\})\times(Q\cup\{R\})\cup(\Sigma_0-\{e\})$ where $R\notin\Sigma\cup Q$. Consider the ETOL $G=(\Omega,\mathcal{P},S,\Delta)$ defined as follows:
$\Delta=\{a\in\Sigma_0-\{e\}\mid\beta_0(a)\text{ is defined}\}$,
$S=\{(a,q)\mid a\in\Sigma_r,\ r>=0,\ q\in F\}\cup\{a\in\Sigma_0\mid\beta_0(a)\in F\}$,
$\mathcal{P}$ contains the tables $\mathfrak{J}=\{(a,q)\rightarrow(a,q_1),\ldots,(a,q_r)\mid\beta_{\mathfrak{J},r}(q_1,\ldots,q_r,a)=q$ for every $a\in\Sigma_r$, $r>0$ and $q\neq q_e$ and $q_j\neq q_e$ for $1<=j<=r\}\cup\{(a,q)\rightarrow\varepsilon\mid\beta_{\mathfrak{J},1}(q_e,a)=q\}\cup\{(a,q)\rightarrow(a,R)\ldots(a,R)\mid q\in Q,\ a\in\Sigma_r$ and $\beta_{\mathfrak{J},r}(q_1,\ldots,q_r,a)$ is not defined for every $q_i\in Q$, $1<=i<=r$ or $a\in\Sigma_r$ and $q=R\}\cup\{a\rightarrow R$ for every $a\in\Sigma_0\cup\{R\}\}$. Let Ω' be the ranked alphabet $\Omega\cup\{e\}$, hence $D(G)\subseteq H_{\Omega'}$. Consider the frontier preserving e-projection P with domain $T_{\Omega'}$ and range T_Σ such that $P(a)=a$ for every $a\in\Omega'_0$ and $P((a,q))=a$ for every $a\in\Sigma_r$ and $q\in Q\cup\{R\}$.

By induction on $t\in H_\Sigma$ it is easy to prove that $t\in L(A)$ iff there exists $t'\in D(G)$ such that $P(t')=t$, hence the thesis holds.

Theorem 3.4. Every k-SB-recognizable subset of H_Σ is the image, under a frontier preserving projection, of the set of derivation trees of an EPTOL system $G=(\Omega,\mathcal{P},S,\Delta)$ for some alphabet Ω.

Proof. The construction of theorem 3.3 applies to a k-SBUTA by giving an EPTOL system.

Note that the above results supply a new proof of the existence of an EPTOL system equivalent to a given ETOL system. In fact given an ETOL system $G=(\Sigma,\mathcal{P},S,\Delta)$ generating a language U, from theorem 3.1 there exists an e-k-SBUTA A such that $L(A)=D(G)$. From theorem 2.5 there exists a k-SBUTA A' such that $L(A')=pe(L(A))-\{e\}$ and $U-\{\varepsilon\}=fr_e(L(A))-\{\varepsilon\}=fr(L(A'))$. According to theorem 3.2 there exists an EPTOL system G' and a frontier preserving projection P such that $L(A')=P(D(G'))$, so that G' generates $U-\{\varepsilon\}$.

Corollary 3.1. A subset $U\subseteq\Delta^*$ is an E(T)OL language iff $U-\{\varepsilon\}=fr(V)$ for some (k)-SB-recognizable $V\subseteq H_\Delta$.

Proof. Given an ETOL language U, consider the EPTOL system $G=(\Sigma,\mathcal{P},S,\Delta)$, generating $U-\{\varepsilon\}$ and the k-SBUTA A such that $L(A)=D(G)$. Chosen a symbol a_0 in Δ, consider the following ranking of Δ: $a_0\in\Delta_n$ for every n such that $\Sigma_n\neq\emptyset$, $a\in\Delta_0$ for every $a\in\Delta$, and the frontier preserving projection P with domain T_Σ and range T_Δ such that $(a,a)\in P_0$ for every $a\in\Sigma_0$ and $(a,a_0)\in P_n$, for every $a\in\Sigma_n$, $n>0$.

In accordance to the corollary 2.1, there exists a k-SBUTA A' such that $L(A')=P(L(A))$. Furthermore $U-\{\varepsilon\}=fr(P(D(G)))=fr(L(A'))$.

Viceversa, let us consider a k-SB-recognizable subset $V\subseteq H_\Delta$. From theorem 3.4, there exists an EPTOL system $G=(\Omega,\mathcal{P},S,\Theta)$ and a frontier preserving projection P with domain T_Ω and range T_Δ such that $P(D(G))=V$. Hence, $fr(P(D(G)))=fr(D(G))=fr(V)=U$ and so U is an ETOL language.

From the constructions given in theorems 3.1 and 3.4, it immediately follows that $U\subseteq\Delta^*$ is an EOL language iff $U-\{\varepsilon\}=fr(V)$ for some SB-recognizable $V\subseteq H_\Delta$.

Definition 3.1. Given an e-ranked alphabet Ω consider the e-ranked alphabet Ω^* defined by $(\Omega^*)_0=\Omega_0$ and $(\Omega^*)_k=\{*\}$ for every k such that $\Omega_k\neq\emptyset$. Moreover consider the frontier preserving e-projection R^* with domain T_Ω and range T_{Ω^*} such that $(a,a)\in(R^*)_0$ for every $a\in\Omega_0$ and

$(a,*) \in (R^*)_r$ for every $a \in \Omega_r$, $r>0$. Two ETOL systems are called structurally equivalent if $R^*(D(G_1))=R^*(D(G_2))$.

Definition 3.2. An ETOL system $G=(\Sigma,\mathcal{P},S,\Delta)$ is invertible if for every $P \in \mathcal{P}$ the right hand side of a rule in P uniquelly determines its left hand side.

Theorem 3.5. For every ETOL system G there exists a structurally equivalent invertible ETOL system G'.

Proof. Let $G=(\Sigma,\mathcal{P},S,\Delta)$ be an ETOL system with $|\mathcal{P}|=k$. Let us consider the e-k-SBUTA $A=(\Omega,Q,\beta_0,\beta,F)$ such that $L(A)=D(G)$ whose construction is given in theorem 3.1. Note that β_0 is an injective function. Consider the frontier preserving e-projection R^* above defined, according to lemma 2.2 there exists an e-k-SBUTA $A'=(\Omega^*,Q',\beta_0',\beta',F')$ such that $R^*(D(G))=L(A')$. Note that $\beta_0'=\beta_0$. Let us consider the ETOL system $G'=(\{*\}X(Q\cup\{R\})\cup\Delta,\mathcal{P}'$, $\{(*,q) \mid q \in F'\},\Delta)$ constructed in the proof of theorem 3.3. Since A' is deterministic then G' is invertible, furthermore it is obvious that G' and G are structurally equivalent.

4. Decision problems.

In this section some decision problems about k-SB-RECOG are dealt with. In particular the membership, the emptiness and the finiteness problems for $\bigcup_{k>=1}$ k-SB-RECOG and the equivalence problem for SB-RECOG are shown to be decidable. As a consequence of these results we prove that the property of structural ambiguity is decidable for ETOL systems and the structural equivalence is decidable for EOL systems.

Theorem 4.1. The membership, the emptiness and the finiteness problems are decidable in the class $\bigcup_{k>=1}$ k-SB-RECOG.

Proof. Since every k-SB-recognizable tree language $L \subseteq H_\Sigma$, viewed as language over the alphabet $\Sigma \cup \{(,)\}$, is an EPTOL language, the thesis follows from the decidability of the considered problems for EPTOL languages (see (6)).

Definition 4.1. A set of trees U is structurally ambiguous if U contains two different trees t_1 and t_2 such that $R^*(t_1)=R^*(t_2)$.

Corollary 4.1. It is decidable whether a k-SB-recognizable tree language is structurally ambiguous.

Proof. The proof is analogous to the one given by Paull and Unger in (5) for the class of tree languages recognized by finite state bottom-up tree automata. It suffices to consider in $H_{\Sigma\times\Sigma}$ the tree language W which contains all the trees such that the frontiers are labelled with pairs (a,a) and at least one internal node is labelled with a pair (a,a') with a=a'.
Clearly W∈SB-RECOG. Consider the relabelings R_1 and R_2 with domain T_Σ and range $T_{\Sigma\times\Sigma}$ defined as follows:
$(a,(a,a)) \in (R_i)_0$ for every $a \in \Sigma_0$, $i=1,2$,

$(a,(a,a')) \in (R_1)_r$ for every $a,a' \in \Sigma_r$ and $r>0$ and
$(a,(a',a)) \in (R_2)_r$ for every $a,a' \in \Sigma_r$ and $r>0$.

If $U \in$ k-SB-RECOG then, according to the theorems 2.2 and 2.4 it holds that $R_1(U) \cap R_2(U) \cap W \in k^2$-SB-RECOG. Finally, U is structurally ambiguous iff $R_1(U) \cap R_2(U) \cap W$ is non empty.

Theorem 4.2. The equivalence problem for SB-RECOG is decidable.

Proof. It immediately follows from theorems 2.3 and 4.1.

Corollary 4.2. It is decidable whether two EOL systems are structurally equivalent or not.

Proof. Given two EOL systems G_1 and G_2, $R^*(D(G_1))$ and $R^*(D(G_2))$ are SB-recognizable and then, according to theorem 4.2, it is decidable if $R^*(D(G_1))=R^*(D(G_2))$.

We would like to thank professor Jost Engelfriet for his valuable and helpful comments on this paper.

References.

(1) J. Engelfriet, Bottom-up and top-down tree transformations. A comparison, Math. Syst. Theory 9, 3(1975), 198-231.
(2) J. Engelfriet, Surface tree languages and parallel derivation trees, Theor. Comp. Sc. 2(1976), 9-27.
(3) J. Engelfriet, G. Rozenberg, G. Slutzki, Tree transducers, L-systems and two machines, J. C. S. S. 20(1980), 150,202.
(4) J. van Leeuwen, Variations of a new machines model, Proc. 17th Annual Symp. Found. Comput. Sci., Houston, Texas (1976), 228-235.
(5) M. C. Paull, S. H. Unger, Structural equivalence of context-free grammars, J. C. S. S. 2(1968), 427-463.
(6) G. Rozenberg, A. Salomaa, The mathematical theory of L-systems, Academic Press, New York (1980).
(7) J. W. Tatcher, Generalized sequential machines maps, J. C. S. S. 4 (1970), 339-367.

ON OBSERVATIONAL EQUIVALENCE AND ALGEBRAIC SPECIFICATION

— Extended abstract[1] —

Donald Sannella and Andrzej Tarlecki[2]
Department of Computer Science
University of Edinburgh

Abstract
The properties of a simple and natural notion of observational equivalence of algebras and the corresponding specification-building operation (observational abstraction) are studied. We begin with a definition of observational equivalence which is adequate to handle reachable algebras only, and show how to extend it to cope with unreachable algebras and also how it may be generalised to make sense under an arbitrary institution. Behavioural equivalence is treated as an important special case of observational equivalence, and its central role in program development is shown by means of an example.

1 Introduction

Probably the most exciting potential application of formal specifications is to the formal development of programs by gradual refinement from a high-level specification to a low-level "program" or "executable specification" as in HOPE [BMS 80]. Each refinement step embodies some design decisions (such as choice of data representation) under the requirement that behaviour must be preserved. If each refinement step can be proved correct, then the program which results is guaranteed to satisfy the original specification.

This paper studies what is meant by "behaviour" in the context of algebraic specifications. Intuitively, the behaviour of a program is determined just by the answers which are obtained from computations the program may perform. We may say (informally) that two Σ-algebras are *behaviourally equivalent* with respect to a set OBS of *observable sorts* if it is not possible to distinguish between them by evaluating Σ-terms which produce a result of observable sort. For example, suppose Σ contains the sorts *nat*, *bool* and *bunch* and the operations *empty*: $\rightarrow$ *bunch*, *add*: *nat*,*bunch* $\rightarrow$ *bunch* and $\in$: *nat*,*bunch* $\rightarrow$ *bool* (as well as the usual operations on *nat* and *bool*), and suppose A and B are Σ-algebras with

$|A|_{bunch}$ = the set of finite sets of natural numbers
$|B|_{bunch}$ = the set of finite lists of natural numbers

with the operations and the remaining carriers defined in the obvious way (but B does *not* contain operations like *cons*, *car* and *cdr*). Then A and B are behaviourally equivalent with respect to {bool} since every term of sort *bool* has the same value in both algebras (the interesting terms are of the form $m \in add(a_1, \ldots, add(a_n, empty)\ldots)$). Note that A and B are not isomorphic.

In the above we assume that the only observations (or experiments) we are allowed to perform are to test whether the results of computations are equal. In this paper we deal

---

[1] The full version of this paper is available as report CSR-172-84, Department of Computer Science, University of Edinburgh.

[2] On leave from Institute of Computer Science, Polish Academy of Sciences, Warsaw.

with the more general situation in which observations may be arbitrary logical formulae. We discuss a notion of *observational equivalence* in which two algebras are observationally equivalent if they both give the same answers to any observation from a prespecified set.

Observational equivalence (or more specifically, behavioural equivalence) seems to be a concept which is fundamental to programming methodology. For example:

Data abstraction

A practical advantage of using abstract data types in the construction of programs is that the implementation of abstractions by program modules need not be fixed. A different module using different algorithms and/or different data structures may be substituted without changing the rest of the program provided that the new module is behaviourally equivalent to the module it replaces (with respect to the non-encapsulated types). ADJ [ADJ 76] have suggested that "abstract" in "abstract data type" means "up to isomorphism"; we suggest that it really means "up to behavioural equivalence".

Program specification

One way of specifying a program is to describe the desired input/output behaviour in some concrete way, e.g. by constructing a very simple program which exhibits the desired behaviour. Any program which is behaviourally equivalent to the sample program with respect to the primitive types of the programming language satisfies the specification. This is called an *abstract model specification* [LB 77]. In general, specifications under the usual algebraic approaches are not abstract enough; it is either difficult, as in Clear [BG 80] or impossible, as in the initial algebra approach of [ADJ 76] and the final algebra approach of [Wand 79] to specify sets of natural numbers in such a way that both A and B above are models. The kernel specification language ASL [SW 83] provides a specification-building operation abstract which when applied to a specification SP relaxes interpretation to all those algebras which are observationally equivalent to a model of SP with respect to the given set of "equational" observations. With a properly chosen set of observations, this gives *behavioural abstraction*.

Stepwise refinement

A formalisation of stepwise refinement requires a precise definition of the notion of refinement, i.e. of the *implementation* of one specification by a lower-level specification. In the context of a specification language which includes an operation like behavioural abstraction, it is possible to adopt a very simple definition of implementation (see section 5 for details). This notion of implementation has two very desirable properties (vertical and horizontal composability, see [GB 80]) which permit the development of programs from specifications in a gradual and modular fashion. An alternative approach which illustrates the same point is to use a definition of implementation which implicitly involves behavioural equivalence, as in [GM 82] and [Sch 83].

This paper establishes a number of basic definitions and results concerning observational equivalence in an attempt to provide a sound foundation for its application to problems such as those indicated above. We begin by treating in section 2 the case in which observations are logical formulae containing no free variables. We define observational equivalence of algebras and a specification-building operation (abstract) which performs observational abstraction and explore their basic properties. We generalise this material in two different dimensions; section 3 discusses observations which contain free variables (to handle "junk" in unreachable algebras without resorting to infinitary logic) and we also mention how the definitions can be generalised to make sense under an arbitrary logical system (or *institution* [GB 83]). Section 4 deals with the problem of proving theorems about structured specifications in the context of observational abstraction. Section 5 discusses behavioural equivalence as an important special case of observational equivalence. A simple notion of implementation is defined, and we demonstrate the role of behavioural equivalence in program development by carrying out one refinement step in the development of a fragment of an optimising compiler from its specification.

We assume that the reader is familiar with the basic algebraic notions presented in e.g. [ADJ 76] (cf. [BG 82]) as well as basic notions of logic as in e.g. [End 72] including some infinitary logic, see [Karp 64].

2 Observational equivalence: the ground case

What is an observation on an algebra? In the axiomatic framework, the most natural choice is to take logical formulae as observations; the result of an observation on an algebra is just the truth or falsity of the formula in the algebra. The kind of formulae we use dictates the kinds of observations we are allowed to make on algebras. On the other hand, the kinds of observations we want to make on algebras dictates the kind of formulae we need, that is the logic we should use.

For example, if we want only to examine results of computations, the natural choice is equations which allow us to compare the values of terms. Another natural choice is first-order predicate calculus which allows us to distinguish between e.g. closed and open intervals of rationals (the observation/formula $\forall x.\exists y.x<y$ yields true in the latter and false in the former). Another choice is an infinitary logic such as $L_{\omega_1\omega}$ which allows us to check e.g. reachability of algebras (that is, whether all elements of the algebra are values of ground terms). Note that the latter two kinds of observations are not computationally-based; they are at a more abstract level, i.e, they describe algebras rather than computations in algebras. Still another kind of formulae are necessary if we want to deal with e.g. problems of concurrency, but we disregard such issues in this paper.

For the moment, we do not want to commit ourselves to any particular logic, and so we leave the notion of "formula" undefined. (In fact, all our definitions work in an even more general setting.) The reader may feel more comfortable to imagine that we are talking about first-order logic.

We use the term "formula" rather than "sentence" to indicate the possible presence of free variables to name elements which are not values of ground terms ("junk"). Free variables introduce some complications which we postpone to the next section. We will assume for the remainder of this section that formulae contain no free variables; we call these *ground observations* or *sentences*. The following definition corresponds directly to the definition of *elementary equivalence* in [Pep 83].

Definition: Let Σ be a signature, Φ a set of Σ-sentences, and let A,B be Σ-algebras. A and B are *observationally equivalent with respect to* Φ, written $A\equiv_\Phi B$, if for any $\varphi\in\Phi$, $A\models\varphi$ iff $B\models\varphi$.

<u>Easy facts</u>

Fact 1: For any signature Σ and set Φ of Σ-sentences, $\equiv_\Phi$ is an equivalence relation on the class of Σ-algebras. □

Fact 2: For any signature Σ, sets Φ,Φ' of Σ-sentences, and Σ-algebras A,B, $\Phi\supseteq\Phi'$ and $A\equiv_\Phi B$ implies $A\equiv_{\Phi'}B$. □

Fact 3: For any signature Σ, family $\{\Phi_i\}_{i\in I}$ of sets of Σ-sentences, and Σ-algebras A,B, $A\equiv_{\Phi_i}B$ for all $i\in I$ implies $A\equiv_\Phi B$ where $\Phi = \bigcup_{i\in I}\Phi_i$. □

Two algebras are observationally equivalent wrt Φ if they satisfy exactly the same sentences of Φ. Note that this remains true if we consider not only the sentences of Φ but also their negations and conjunctions, possibly infinite or empty ($\bigwedge\phi$ is true). We can also add to Φ sentences equivalent to the ones already in Φ, and so everything which is definable in terms of negation and conjunction as well (disjunctions, implications etc.). For any set Φ of Σ-sentences, let $Cl(\Phi)$ denote the closure of Φ under negation, conjunction and equivalence, insofar as the logic in use allows.

Fact 4: $\equiv_{\Phi} = \equiv_{Cl(\Phi)}$ □

Note that this implies that the premise $\Phi \supseteq \Phi'$ in fact 2 may be replaced by the weaker condition $Cl(\Phi) \supseteq \Phi'$.

A *signature morphism* $\sigma:\Sigma \to \Sigma'$ is a renaming of the sorts and operations in Σ to those of Σ' which preserves the argument and result sorts of operations. This induces in a natural way a translation of Σ-terms to Σ'-terms and of Σ-sentences to Σ'-sentences; if φ is a Σ-sentence, then $\sigma(\varphi)$ denotes its translation to a Σ'-sentence. A signature morphism $\sigma:\Sigma \to \Sigma'$ also induces a *σ-reduct* functor translating any Σ'-algebra A' to a Σ-algebra $A'|_{\sigma}$. For the exact definitions of these notions see e.g. [ST 84, section 2]. These translations satisfy the following condition (see [GB 83]):

For any Σ-sentence φ and Σ'-algebra A', $A'|_{\sigma} \models \varphi$ iff $A' \models \sigma(\varphi)$ (Satisfaction condition)

This gives immediately the following fact:

Fact 5: For any signature morphism $\sigma:\Sigma \to \Sigma'$, set Φ of Σ-sentences and Σ'-algebras A',B', $A' \equiv_{\sigma(\Phi)} B'$ iff $A'|_{\sigma} \equiv_{\Phi} B'|_{\sigma}$ where $\sigma(\Phi) = \{\sigma(\varphi) \mid \varphi \in \Phi\}$. □

This says that observational equivalence is coherent with translation along signature morphisms. We can also show that observational equivalence is preserved under combination of "independent" algebras.

Let $\Sigma1$ and $\Sigma2$ be disjoint signatures and let $\Sigma1+\Sigma2$ be their (disjoint) union. For any $\Sigma1$-algebra $A1$ and $\Sigma2$-algebra $A2$, let $\langle A1,A2\rangle$ be the unique $\Sigma1+\Sigma2$-algebra such that $\langle A1,A2\rangle|_{\iota 1}=A1$ and $\langle A1,A2\rangle|_{\iota 2}=A2$, where $\iota 1$ and $\iota 2$ are the inclusions of $\Sigma1$ and $\Sigma2$ (respectively) into $\Sigma1+\Sigma2$. Note that all $\Sigma1+\Sigma2$-algebras are of this form.

Fact 6: For any disjoint signatures $\Sigma1,\Sigma2$, sets of $\Sigma1$-sentences $\Phi1$ and $\Sigma2$-sentences $\Phi2$, $\Sigma1$-algebras $A1,B1$ and $\Sigma2$-algebras $A2,B2$, $\langle A1,A2\rangle \equiv_{\iota 1(\Phi 1)\cup \iota 2(\Phi 2)} \langle B1,B2\rangle$ iff $A1 \equiv_{\Phi 1} B1$ and $A2 \equiv_{\Phi 2} B2$. □

A specification describes a collection of models of the same signature. To formalise this, for any specification SP let Sig[SP] denote its signature and Mod[SP] denote the class of its models, which are Sig[SP]-algebras. The notion of observational equivalence gives rise to a very powerful specification-building operation:

Definition: For any specification SP and set Φ of Sig[SP]-sentences

Sig[**abstract** SP **wrt** Φ] = Sig[SP]
Mod[**abstract** SP **wrt** Φ] = $\{ A \mid A \equiv_{\Phi} B$ for some $B \in$ Mod[SP] $\}$

Informally, **abstract** SP **wrt** Φ is a specification which admits any model which is observationally equivalent to some model of SP. This provides a way of abstracting away from certain details of a specification (see [SW 83], [ST 84]).

<u>Easy facts</u>

Fact 7: For any specification SP and set Φ of Sig[SP]-sentences, Mod[SP] $\subseteq$ Mod[**abstract** SP **wrt** Φ]. □

Fact 8: For any specification SP and set Φ of Sig[SP]-sentences,
Mod[**abstract** (**abstract** SP **wrt** Φ) **wrt** Φ] = Mod[**abstract** SP **wrt** Φ]. □

Fact 9: For any specification SP and sets Φ,Φ' of Sig[SP]-sentences,
$Cl(\Phi)\supseteq\Phi'$ implies Mod[**abstract** SP **wrt** Φ] $\subseteq$ Mod[**abstract** SP **wrt** Φ']. □

Fact 10: For any specifications SP,SP' such that Sig[SP]=Sig[SP'] and set Φ of Sig[SP]-sentences, Mod[SP]$\subseteq$Mod[SP'] implies Mod[**abstract** SP **wrt** Φ] $\subseteq$ Mod[**abstract** SP' **wrt** Φ]. □

Using the above facts we may derive simple identities which allow us to transform specifications involving **abstract**. For example:

Fact 11: For any specification SP and sets Φ,Φ' of Sig[SP]-sentences,

a. Mod[**abstract** SP **wrt** $\Phi\cup\Phi'$] $\subseteq$ Mod[**abstract** (**abstract** SP **wrt** Φ) **wrt** Φ']
$\subseteq$ Mod[**abstract** SP **wrt** $Cl(\Phi)\cap Cl(\Phi')$]

b. $Cl(\Phi)\supseteq\Phi'$ implies
Mod[**abstract** SP **wrt** Φ'] = Mod[**abstract** (**abstract** SP **wrt** Φ) **wrt** Φ']
= Mod[**abstract** (**abstract** SP **wrt** Φ') **wrt** Φ] □

Note that the second equality in (b) need not hold if $Cl(\Phi)\not\supseteq\Phi'$, and that the inclusions in (a) may be proper.

Every algebraic specification language provides an operation for specifying the class of models of a given signature which satisfy a given set of axioms, that is:

Definition: For any signature Σ and set Δ of Σ-sentences, $\langle\Sigma,\Delta\rangle$ is a *basic specification* and

$Sig[\langle\Sigma,\Delta\rangle] = \Sigma$
$Mod[\langle\Sigma,\Delta\rangle] = \{ A \mid A \text{ is a } \Sigma\text{-algebra and } A\models\Delta \}$

For any signature Σ and class K of Σ-algebras, let Th(K) denote the set of all Σ-sentences which hold in K. Note that $K\subseteq Mod[\langle\Sigma,Th(K)\rangle]$ but the converse inclusion is true only for classes K definable by basic specifications.

Fact 12: For any specification SP with Sig[SP]=Σ and set Φ of Σ-sentences,
Mod[**abstract** SP **wrt** Φ] $\subseteq$ $Mod[\langle\Sigma,Th(Mod[SP])\cap Cl(\Phi)\rangle]$. □

In the following, we try to further characterise how **abstract** works for classes of models definable by basic specifications. For any signature Σ and set Δ of Σ-sentences, let $\Delta^{\bullet} = Th(Mod[\langle\Sigma,\Delta\rangle])$ be the *closure of* Δ *under consequence*.

Fact 13: In first-order logic, for any signature Σ and sets Δ,Φ of Σ-sentences,
Mod[**abstract** $\langle\Sigma,\Delta\rangle$ **wrt** Φ] = $Mod[\langle\Sigma,\Delta^{\bullet}\cap Cl(\Phi)\rangle]$.

Proof sketch (for $\supseteq$): For unsatisfiable Δ the containment holds; assume Δ has a model. Let $A\in Mod[\langle\Sigma,\Delta^{\bullet}\cap Cl(\Phi)\rangle]$, and $\Psi = \{\varphi\in Cl(\Phi) \mid A\models\varphi\}$. Assume that $\Delta^{\bullet}\cup\Psi$ is not satisfiable. Then by the compactness theorem of first-order logic and since Ψ is closed under conjunction there is a $\psi\in\Psi$ such that $\Delta^{\bullet}\cup\{\psi\}$ has no model. Hence $\Delta^{\bullet}\models\neg\psi$, which, since $\Delta^{\bullet}$ is closed under logical consequence implies that $\neg\psi\in\Delta^{\bullet}$. Thus $\neg\psi\in\Delta^{\bullet}\cap Cl(\Phi)$ and so $A\models\neg\psi$, which contradicts $\psi\in\Psi$. This proves that $\Delta^{\bullet}\cup\Psi$ has a model, say B. It is easy to check that $A\equiv_{\Phi}B$. □

An examination of the above proof shows that the fact holds for $L_{\omega_1\omega}$ as well as for first-order logic, and in fact for any infinitary logic (any logic which admits negation and conjunction of sets of formulae of any cardinality less than the cardinality in which the logic is compact).

Ground observations are powerful enough if we are only interested in reachable (subalgebras of) algebras and we do not want to distinguish between isomorphic algebras, provided that our logic is at least capable of expressing ground equations (i.e. equations between terms without variables).

Fact 14: For any signature Σ and Σ-algebras A,B, $A \equiv_{GEQ(\Sigma)} B$ iff A and B have isomorphic reachable subalgebras, where GEQ(Σ) is the set of all *ground* Σ-equations. □

3 Observational equivalence: the general case

In the last section we dealt with observational equivalence based on ground observations only (formally, on formulae without free variables). As fact 14 indicates, this is quite satisfactory when we restrict our considerations to reachable algebras. If we want to deal with algebras containing "junk" things become more complicated.

Why do we bother about non-reachable algebras? First, when dealing with parameterised specifications it is usual to consider examples in which some sorts have no generators at all, but where we are interested in algebras having the associated carriers non-empty. This is shown by standard examples such as Stack-of-X, where X is an arbitrary set. Second, when we view algebras from different levels of abstraction we view them with respect to different sets of operations. It is then natural that an algebra which is reachable at a certain level of abstraction becomes non-reachable when viewed from a higher level. A more technical but related point is that there is no natural definition of the specification-building operation **derive** [BG 80], [ST 84] (which can be used to "forget" operations) if models of the result are required to be reachable. Finally, there are examples [SW 83] in which unreachable elements can be useful in constructing specifications; an element which is unreachable at one stage of the construction can become reachable and useful at a later stage.

It should be noted that if the logic we are working in is sufficiently powerful then we can identify algebras up to isomorphism using only ground observations. According to Scott's Theorem [Scott 65] this may be achieved using $L_{\omega_1\omega}$ for countable algebras. Using an even more powerful logic the same may be done for arbitrary algebras.

Fact 15: For any signature Σ and Σ-algebra A, there is a Σ-sentence $\zeta(A)$ of $L_{\infty\infty}$ such that for any Σ-algebra B, $B \models \zeta(A)$ iff $A \cong B$.

Proof sketch: Assume for notational convenience that Σ has only one sort. Consider the formula $\zeta(A) =_{def} \exists|A|.(\bigwedge\mathcal{D}(A) \;\&\; \forall x.\bigvee\{x=a \mid a\in|A|\})$ where $|A|$ is the carrier of A, and $\mathcal{D}(A)$ is the first-order diagram of the expansion of A to a $\Sigma(|A|)$-algebra with the natural interpretation of the new constants. If $B\cong A$ then obviously $B\models\zeta(A)$.
Conversely, assume that B satisfies $\zeta(A)$. Thus, there is a valuation $v{:}|A|\to|B|$ such that $B\models_v \bigwedge\mathcal{D}(A) \;\&\; \forall x.\bigvee\{x=a \mid a\in|A|\}$ ($B\models_v\varphi$ means B satisfies φ under the valuation v; we are going to use this notation throughout the paper.) It is easy to see that v is an isomorphism between A and B. □

Note from the construction that in order to handle algebras of cardinality α it is enough to consider formulae with quantifiers binding α variables. It seems to be possible to sharpen this result (requiring only quantifiers binding less than α variables) using some kind of "back-and-forth" construction as in the proof of Scott's theorem in [Bar 73].

In practice it is desirable to avoid use of infinitary logic (although [MSV 83] argue for an approach to specification in which infinitary logic is central). What we are trying to do in the following is to obtain a balance between the power of the logic in use (the simpler the logic, the better) and the simplicity of the definition of observational equivalence.

It is obvious that using ground equations as observations we are not able to talk about junk at all. If we use equations with universally quantified variables, although we are able to say something about junk we cannot always distinguish between algebras which are intuitively not equivalent. For example, the two algebras

cannot be distinguished by any equation with universally quantified variables (neither of them satisfy $\forall x.f(x)=x$) although if we go to first-order logic then the formula $\exists x.f(x)=x$ distinguishes between them. But even in the framework of first-order logic using only closed sentences, we are not able to deal with junk in a satisfactory way. We cannot even express such a basic property as its existence. For example, it is well-known that the standard model of arithmetic (the natural numbers) and non-standard models (the natural numbers with junk) satisfy exactly the same set of first-order sentences. Thus, there is no set of ground first-order observations which can distinguish between standard and non-standard models of arithmetic. To distinguish between these models using ground observations we need $L_{\omega_1\omega}$.

We are going to extend the definitions of the previous section by allowing free variables in observations. The idea is that these provide a way of referring to otherwise unnameable values. For example, it should be intuitively clear (and will be formalised below) that the observation $f(x)=x$ with free variable x distinguishes between the two algebras A and B above, and the set of observations $\{x=succ^n(0) \mid n\geq 0\}$ with free variable x distinguishes between standard and non-standard models of arithmetic. As in logic, we need a valuation of the free variables into the algebra under consideration to provide these names with interpretations.

Given a signature Σ, a set X of variables (of sorts in Σ), a set $\Phi(X)$ of Σ-formulae with free variables in X, and two Σ-algebras A,B there are a number of possible ways to define $A\equiv_{\Phi(X)}B$. For example, in [SW 83] and [ST 84] $A\equiv_{\Phi(X)}B$ was defined as follows:

> $A\equiv_{\Phi(X)}B$ if there exist surjective valuations $v_A:X\to|A|$ and $v_B:X\to|B|$ such that for all $\varphi\in\Phi(X)$, $A\models_{v_A}\varphi$ iff $B\models_{v_B}\varphi$.

The justification for this definition is that v_A and v_B identify "matching parts" of A and B; each part of A must match some part of B and vice versa. But there are some problems with this definition. Technically, this relation is restricted to comparing algebras of cardinality less than or equal to that of X because of the surjectivity requirement on v_A and v_B. Also, we have to exclude algebras with empty carriers, (at least) on sorts in which X is non-empty; otherwise the valuations v_A and/or v_B cannot exist. Finally, in the "general" case in which models and the logic are arbitrary (see [ST 84]) this definition is rather messy and inelegant because of the difficulty of formulating in abstract terms the requirement of surjectivity.

We are going to concentrate on a different definition of observational equivalence. We define the observational equivalence relation in terms of a preorder.

Definition: For any signature Σ, set X of variables of sorts in Σ, set $\Phi(X)$ of Σ-formulae with free variables in X, and Σ-algebras A,B, A is *observationally reducible to* B *wrt* $\Phi(X)$, written $A\leq_{\Phi(X)}B$, if for any valuation $v_A:X\to|A|$ there exists a valuation $v_B:X\to|B|$ such that for all $\varphi\in\Phi(X)$, $A\models_{v_A}\varphi$ iff $B\models_{v_B}\varphi$.

Fact 16: For any signature Σ, set X of variables of sorts in Σ, and set $\Phi(X)$ of Σ-formulae with free variables in X, $\leq_{\Phi(X)}$ is a preorder on the class of Σ-algebras. □

Definition: For any signature Σ, set X of variables of sorts in Σ, set $\Phi(X)$ of Σ-formulae with free variables in X, and Σ-algebras A,B, A and B are *observationally equivalent wrt* $\Phi(X)$, written $A\equiv_{\Phi(X)}B$, if $A\leq_{\Phi(X)}B$ and $B\leq_{\Phi(X)}A$.

Although we are not going to restate all of them formally here again, facts 2-6 of section

2 hold for the preorder $\leq_{\Phi(X)}$ and facts 1-6 hold for the equivalence $\equiv_{\Phi(X)}$. For example, fact 3 may be reformulated for $\leq_{\Phi(X)}$ here as follows:

Fact 3': For any signature Σ, family of mutually disjoint sets $\{X_i\}_{i\in I}$ of variables of sorts in Σ, family $\{\Phi_i\}_{i\in I}$ of sets of Σ-formulae such that for $i\in I$, Φ_i has free variables in X_i, and Σ-algebras A,B, $A\leq_{\Phi_i(X_i)}B$ for all $i\in I$ implies $A\leq_{\Phi(X)}B$ where $\Phi = \bigcup_{i\in I}\Phi_i$ and $X = \bigcup_{i\in I}X_i$. □

However, because of the problems which empty carriers may cause, we have to be careful with the opposite direction of this implication, that is when discharging variables. Fact 2 should be reformulated as follows:

Fact 2': For any signature Σ, set X of variables of sorts in Σ, sets $\Phi(X)$ and $\Phi'(X)$ of Σ-formulae with free variables in X, and Σ-algebras A,B, $\Phi(X)\supseteq\Phi'(X)$ and $A\leq_{\Phi(X)}B$ implies $A\leq_{\Phi'(X)}B$. □

Note that Φ and Φ' must formally have the same set X of free variables, even if the formulae in the smaller set Φ' do not use all of them. We can discharge such unnecessary variables only if X contains other variables of the same sorts, or if the algebras we are dealing with are guaranteed to have non-empty carriers of these sorts.

As in the previous section, we can define a specification-building operation **abstract** in terms of observational equivalence with exactly the same semantics:

Definition: For any specification SP, set X of variables of sorts in Sig[SP] and set $\Phi(X)$ of Sig[SP]-formulae with free variables in X

Sig[**abstract** SP **wrt** $\Phi(X)$] = Sig[SP]
Mod[**abstract** SP **wrt** $\Phi(X)$] = { A | $A\equiv_{\Phi(X)}B$ for some B∈Mod[SP] }

Facts 7-12 still hold under this more general definition.

Note that we can give a sharper formulation of the facts which involve forming the closure $Cl(\Phi)$ of a set of formulae Φ. In the presence of free variables, besides conjunctions and negations it is tempting to allow the introduction of quantifiers here. We can redefine $Cl(\Phi(X))$ to be the closure of $\Phi(X)$ under negation, conjunction (possibly infinite), equivalence and uniform quantification, that is, $\varphi\in Cl(\Phi(X))$ implies $\forall X.\varphi\in Cl(\Phi(X))$ and $\exists X.\varphi\in Cl(\Phi(X))$. To prove that all facts are still true with this new definition of $Cl(\Phi(X))$, we have to show the following:

Fact 17: $\equiv_{Cl(\Phi(X))} \supseteq \equiv_{\Phi(X)}$ □

Note that only uniform quantification is allowed above. The above fact does not hold if we allow quantification over a proper subset of the set of free variables. For example, suppose Σ = **sorts** rat,bool **opns** <:rat,rat→bool. Let A and B be Σ-algebras corresponding to, respectively, open and closed intervals of rational numbers. Now consider $\Phi(X) = \{x<y \mid x,y\in X\}$. Obviously $A\equiv_{\Phi(X)}B$ but $A\models\forall x.\exists y.x<y)$ while $B\not\models\forall x.\exists y.x<y)$. Note also that even under this definition of closure, fact 13 does not hold for non-ground observations.

4 Proofs in structured specifications

An important issue connected with specifications is theorem proving. We would like to be able to prove theorems about a specification, that is, that certain sentences of the underlying logic hold in every model of a specification. As suggested by Guttag and Horning [GH 80] by proving that selected theorems hold we can understand specifications and gain confidence that they express what we want. Moreover, in order to do any kind of formal program development or verification (or even specification building, if parameterised specifications with requirements can be used) a theorem-proving capability is necessary.

In the context of structured specifications, we have to cope with two separate problems. First is how to prove theorems in theories of the underlying logic. Note that this task may be eased by the fact that our theories have structure, as this allows us to naturally disregard information which is probably irrelevant to what we are trying to prove. The other problem is dealing with the structure itself. What we need are inference rules for every specification-building operation which allow us to derive theorems about a combined specification from theorems about the components from which it was built. Note that the latter problem is not automatically reducible to the former because not all specifications are equivalent to (have the same class of models as) theories of the underlying logic [ST 84], let alone theories with finite presentations as required for use by a theorem prover.

For simple specification-building operations appropriate inference rules are given in [SB 83], for example

thm in SP $\Rightarrow$ thm in SP + SP'

where "thm in SP" means thm∈Th(Mod[SP]), that is that the sentence *thm* holds in all models of the specification *SP*. The **abstract** specification-building operation defined in sections 2 and 3 is more difficult to handle. One problem is that in contrast to other specification-building operations it is not monotonic, in the sense that

thm in SP $\not\Rightarrow$ thm in **abstract** SP **wrt** . . .

However, fact 12 and its analogue for observations with free variables (see section 3) says that the following inference rule is sound.

Inference rule: For any set $\Phi(X)$ of open formulae with variables in X,
thm in SP and thm∈Cl($\Phi(X)$) $\Rightarrow$ thm in **abstract** SP **wrt** $\Phi(X)$

Moreover, for the case of ground observations (i.e. when X is the empty set), fact 13 shows that in some standard logics (first-order logic, infinitary logics) the above rule is in a sense complete when used together with inference rules for the underlying logic and the other specification-building operations. Note also that facts 7-12 provide us with some subsidiary inference rules; for example, fact 9 implies

thm in **abstract** SP **wrt** Φ and $\Phi\subseteq Cl(\Phi')$ $\Rightarrow$ thm in **abstract** SP **wrt** Φ'

5 Behavioural equivalence — an example

In sections 2 and 3 we defined a very general and powerful notion of observational equivalence. In this section we look at a very important special case and we consider an example of its use. Namely, we restrict observations to equations between terms from some specified set; this gives an equivalence corresponding to the one used in the ASL specification language [SW 83]. A proper choice of the set of terms gives *behavioural equivalence* as informally discussed in the introduction.

Suppose that Σ is a signature and IN and OUT are subsets of the sorts of Σ. Now, consider all computations which take input from sorts IN and give output in sorts OUT; this set of computations corresponds to the set of Σ-terms of sorts OUT with variables of sorts IN. Consider the set $EQ_{OUT}(X_{IN})$ of equations between terms of the same sort in OUT having variables X_{IN} of sorts in IN. Two algebras are observationally equivalent with respect to $EQ_{OUT}(X_{IN})$ if they are behaviourally equivalent, that is they have matching input/output relations. Note that this covers the notions of behavioural equivalence with respect to a single set OBS of *observable* sorts which appear in the literature. For example, in [Rei 81] and [GM 82] we have IN=sorts(Σ), OUT=OBS; in [Sch 83], [SW 83] and [GM 83] IN=OUT=OBS; and in [GGM 76] and [Kam 83] IN=ϕ and OUT=OBS. To denote the corresponding special case of abstract we use the following notation:

behaviour SP **with in** IN **out** OUT $=_{def}$ **abstract** SP **wrt** $EQ_{OUT}(X_{IN})$

This corresponds to *behavioural abstraction* as defined in ASL [SW 83].

As an example we are going to consider a simple language of expressions for arithmetical computation over the integers. This may be imagined as a small piece of a real programming language. We believe that the approach used below may be applied to other programming language constructs as well, leading towards the possible formal development of a compiler.

We assume that we are given some standard specifications of identifiers (Ident) with a sort *ident* and of the integers (Int) with the usual arithmetic operations. The (abstract) syntax of expressions is given by the following specification (we use the notation of the Clear specification language [BG 80][3]):

Expr = **enrich** Int + Ident **by data sorts** expr
opns const : int → expr
var : ident → expr
plus, times : expr, expr → expr
cond : expr, expr, expr → expr

The use of **data** above means that any model of Expr is a free extension of a model of Int + Ident. That is, the sort *expr* contains expressions built up using the newly-introduced operations. We could achieve the same effect using a *hierarchy constraint* [Bau 81] (cf. [SW 82] and [EWT 83]) together with the appropriate inequations.

To describe the semantics of expressions we need the additional concept of an environment from which the values of variables may be retrieved. This is described by the following (loose) specification:

Env = **enrich** Int + Ident **by sorts** env
opns lookup : env, ident → int

For the purpose of our example, no more than the existence of an operation *lookup* is required.

Eval = **enrich** Expr + Env **by**
opns eval : expr, env → int
axioms ∀n:int, ρ:env. eval(const(n),ρ) = n
∀x:ident, ρ:env. eval(var(x),ρ) = lookup(ρ,x)
∀e,e':expr, ρ:env. eval(plus(e,e'),ρ) = eval(e,ρ) + eval(e',ρ)
∀e,e':expr, ρ:env. eval(times(e,e'),ρ) = eval(e,ρ) × eval(e',ρ)
∀e,e',e'':expr, ρ:env. eval(cond(e,e',e''),ρ) = eval(e'',ρ) **if** eval(e,ρ) = 0
= eval(e',ρ) **otherwise**

(We use an obvious notation to simplify the syntax of conditional axioms.)

The models of Eval are just the models of Expr with the expected semantics provided by the operation *eval*. The *cond* construct has the semantics of if _ then _ else _, where 0 (as the value of the first argument) is interpreted as false and any other value is interpreted as true. Note that the models of Eval are pretty well determined; in fact they are determined up to isomorphism given models of Ident and Env. Now imagine that we want to build a compiler which performs some source-level optimisation; for example, recognising that *times*(*const*(0),e) is just *const*(0). Such optimisations are not permitted by the

[3] But for the semantics of **derive**, see [SW 83].

specification above.

Two solutions to this dilemma are offered in the literature. First, [Wand 79] and [Kam 83] advocate the use of *final* models; if we adopt this approach (modifying the above specification appropriately) then every (final) model of Eval would satisfy $e = e'$ iff it satisfies $\forall \rho{:}env.\ eval(e,\rho) = eval(e',\rho)$, for all expressions e and e'. But this disallows non-optimal implementations since it requires that all possible optimisations are performed. Much worse, the specified models are actually not attainable since the optimisation required is not computable (this follows from a result in [Chu 36]).

Second, as advocated in e.g. [Ehr 79] and [EKMP 82] the notion of *implementation* of one specification by another should take care of this problem. Algebras with some optimisations are not models of the specification above but models of a specification which implements it. Unfortunately, the formal notions of implementation which have been suggested are rather complicated, and especially so in the context of loose and parameterised specifications. (Note that the specification above may be viewed as parameterised by Ident.)

We adopt neither of these solutions. Instead, we argue that the specification Eval as given above is not really what we intend. When we specify a program what we are really interested in is its behaviour, that is the answers which we obtain when the program is applied to the various possible inputs. The specification Eval says more than that; it dictates the structure of internal data. We can obtain the class of models having the behaviour which Eval specifies (rather concretely) by applying the **behaviour** operation for the appropriate choice of input and output sorts:

Eval-we-really-want = **behaviour** Eval **with in** {int,ident,env} **out** {int}

The inference rule for **abstract** given in section 4 may be applied here to show e.g. that

$$\forall e,e'{:}expr,\ \rho{:}env.\ eval(plus(e,e'),\rho) = eval(plus(e',e),\rho)$$

is a theorem in Eval-we-really-want, since it is a theorem of Eval and is in the closure of the set of observations we are using here.[4]

The ability to specify classes of algebras up to behavioural equivalence (as in Eval-we-really-want) allows us to greatly simplify our formal view of what an implementation is. Proceeding from a specification to a program means making a series of design decisions, each of which amounts to a restriction on the class of models. Such design decisions are choice of data structures, choice of algorithms, and choice between alternatives which the specification leaves open.

Thus, a simple but natural notion of implementation is as follows.

Definition: A specification SP is *implemented by* a specification SP', written SP⇝SP', if Mod[SP'] ⊆ Mod[SP].

It is easy to see that the above implementation relation is transitive (SP⇝SP' and SP'⇝SP'' implies SP⇝SP''), i.e. that it can be composed *vertically* (see [GB 80]). This means that a specification can be refined gradually. Furthermore, this implementation relation can be composed *horizontally* [GB 80] as well[5] [SW 83] (SP1⇝SP1' and SP2⇝SP2' implies SP1+SP2⇝SP1'+SP2' and similarly for the other specification-building operations). This means that specifications can be refined in a modular fashion. This is in contrast to the more complicated notions of implementation mentioned earlier for which these

---

[4] For technical reasons (see [GM 81]) we assume that there are constants of sort *ident*.

[5] provided that all specification-building operations are monotonic (with respect to model classes), which is the case for the speci-cation-building operations defined in e.g. Clear [BG 80], LOOK [ETLZ 82], ASL [SW 83], and for **abstract** and **behaviour** as defined abo

properties do not hold in general.

The following specification is an implementation of Eval-we-really-want:

```
Eval' =
  let Ev0 = enrich Eval by
          opns    optplus, opttimes : expr, expr → expr
                  optcond : expr, expr, expr → expr
          axioms ∀e,e':expr. optplus(e,e')
                           = e'                    if e = const(0)
                           = e                     if e' = const(0)
                           = opttimes(const(2),e)  if e = e'
                           = plus(e,e')            otherwise
                 ∀e,e':expr. opttimes(e,e')
                           = const(0)              if e = const(0) or e' = const(0)
                           = e'                    if e = const(1)
                           = e                     if e' = const(1)
                           = times(e,e')           otherwise
                 ∀e,e',e":expr. optcond(e,e',e")
                           = e'                    if e = const(n) and n ≠ 0
                           = e"                    if e = const(0)
                           = e'                    if e' = e"
                           = cond(e,e',e")         otherwise
    in derive signature Eval
       from Ev0
       by const  is const
          var    is var
          plus   is optplus
          times  is opttimes
          cond   is optcond
          eval   is eval
```

Eval' specifies the syntax and semantics of our expression language, requiring that certain source-level optimisations (constant folding) be carried out.

In order to prove that Eval' implements Eval-we-really-want we have to show:

Claim: Mod[Eval-we-really-want] $\supseteq$ Mod[Eval']

To prove this we have to show that any model of Eval' is behaviourally equivalent to a model of Eval (with respect to input sorts {int,ident,env} and output sort {int}). This boils down to showing that the value of an expression (as given by *eval*) is the same as the value of its optimisation in any environment (see the long version of this paper for details). □

A different way of proving that two algebras are behaviourally equivalent is suggested in [Sch 83]; in this approach, a relation (called a *correspondence*) between the corresponding carriers is set up explicitly and proved to satisfy a kind of homomorphism property.

6 Concluding remarks

In the previous sections we have been rather vague about what we mean by a "formula". We have mentioned formulae of equational logic, first-order logic and infinitary logic. Moreover, although we have been using the standard notion of many-sorted algebra as in [ADJ 76], this was mostly in order to take advantage of the reader's intuition; in fact, we made use of very few formal properties of algebras. This means that in place of the standard notion we could have used for example partial or continuous algebras. We could even change both the notions of signature and of algebra to deal with errors or coercions.

The notion of an *institution* [GB 83] provides a tool for dealing with any of these different notions of a logical system for writing specifications. An institution comprises definitions of signature, model (algebra), sentence and a satisfaction relation satisfying a few minimal consistency conditions. (For a similar but more logic-oriented approach see [Bar 74].) By basing our definitions (of observational equivalence etc.) on an arbitrary institution we can avoid choosing particular definitions of these underlying notions and do everything at an adequately general level. It is possible to define the semantics of a specification language in an arbitrary institution; see [BG 80] and [ST 84].

We encounter no problems at all in generalising the contents of section 2 (on ground observations) to an arbitrary institution. Moreover, facts 1-12 still hold. (Fact 13 holds for institutions with some simple closure properties. Fact 14 may be generalised if we equip institutions with some notion of reachability along the lines of [Tar 84].)

In order to deal with the general case of observations containing free variables we have first of all to provide a notion of an open formula and a valuation of free variables in the framework of an arbitrary institution. Although sentences as they are used in the definition of an institution above are always closed, this may be done (see [ST 84]). Then the contents of section 3 may be generalised as well; see the longer version of this paper for details.

By exploring the properties of a primitive but powerful and general notion such as observational equivalence and then deriving the more directly useful concept of behavioural equivalence as a special case, we are following in the footsteps of earlier work on kernel specification-building operations [Wir 82,83], [SW 83], [ST 84]. Our ultimate interest is not in the primitive notions themselves but rather in the useful higher-level constructs which can be expressed in their terms. By carefully investigating the primitives we hope to gain insights which can be applied to the derived constructs.

The material in this paper could provide the basis for high-level specification languages such as one in which every specification is surrounded by an implicit (and invisible) application of **behaviour** with respect to input and output sorts appropriate to the context. Such a language is presented in [ST 85]. An issue we have not discussed is the connection between behavioural equivalence/abstraction and parameterisation of specifications. A different approach to the problem of specifying software modules which integrates parameterisation and implementation is given in [Ehrig 84]. We have not yet investigated thoroughly the interaction between **behaviour** and other specification-building operations, although a start in this direction is given by facts 5 and 6.

Acknowledgements

Our thanks to Rod Burstall for many instructive discussions and encouragement and to Martin Wirsing for the collaboration which started us on this line of work and for helpful comments on an earlier version. Support was provided by the Science and Engineering Research Council.

7 References

[ADJ 76] Goguen, J.A., Thatcher, J.W. and Wagner, E.G. An initial algebra approach to the specification, correctness, and implementation of abstract data types. IBM research report RC 6487. Also in: Current Trends in Programming Methodology, Vol. 4: Data Structuring (R.T. Yeh, ed.), Prentice-Hall, pp. 80-149 (1978).

[Bar 73] Barwise, J. Back and forth through infinitary logic. In: Studies in Mathematics, Vol. 8: Studies in Model Theory (M.D. Morley, ed.), Mathematical Assoc. of America, pp. 5-34.

[Bar 74] Barwise, J. Axioms for abstract model theory. Annals of Math. Logic 7, pp. 221-265.

[Bau 81] Bauer, F.L. *et al* (the CIP Language Group) Report on a wide spectrum language for program specification and development (tentative version). Report TUM-

I8104, Technische Univ. München.

[BG 80] Burstall, R.M. and Goguen, J.A. The semantics of Clear, a specification language. Proc. of Advanced Course on Abstract Software Specifications, Copenhagen. Springer LNCS 86, pp. 292-332.

[BG 82] Burstall, R.M. and Goguen, J.A. Algebras, theories and freeness: an introduction for computer scientists. Proc. 1981 Marktoberdorf NATO Summer School, Reidel.

[BMS 80] Burstall, R.M., MacQueen, D.B. and Sannella, D.T. HOPE: an experimental applicative language. Proc. 1980 LISP Conference, Stanford, California, pp. 136-143.

[Chu 36] Church, A. An unsolvable problem of elementary number theory. American Journal of Mathematics 58, pp. 345-363.

[Ehr 79] Ehrich, H.-D. On the theory of specification, implementation, and parametrization of abstract data types. Report 82, Abteilung Informatik, Univ. of Dortmund. Also in: JACM 29, 1, pp. 206-227 (1982).

[Ehrig 84] Ehrig, H. An algebraic specification concept for modules (draft version). Report 84-04, Institut für Software und Theoretische Informatik, Technische Univ. Berlin.

[EKMP 82] Ehrig, H., Kreowski, H.-J., Mahr, B. and Padawitz, P. Algebraic implementation of abstract data types. Theoretical Computer Science 20, pp. 209-263.

[ETLZ 82] Ehrig, H., Thatcher, J.W., Lucas, P. and Zilles, S.N. Denotational and initial algebra semantics of the algebraic specification language LOOK. Draft report, IBM research.

[EWT 83] Ehrig, H., Wagner, E.G. and Thatcher, J.W. Algebraic specifications with generating constraints. Proc. 10th ICALP, Barcelona. Springer LNCS 154, pp. 188-202.

[End 72] Enderton, H.B. A Mathematical Introduction to Logic. Academic Press.

[GGM 76] Giarratana, V., Gimona, F. and Montanari, U. Observability concepts in abstract data type specification. Proc. 5th MFCS, Gdansk. Springer LNCS 45.

[GB 80] Goguen, J.A. and Burstall, R.M. CAT, a system for the structured elaboration of correct programs from structured specifications. Technical report CSL-118, Computer Science Laboratory, SRI International.

[GB 83] Goguen, J.A. and Burstall, R.M. Introducing institutions. Proc. Logics of Programming Workshop, Carnegie-Mellon. Springer LNCS 164, pp. 221-256.

[GM 81] Goguen, J.A. and Meseguer, J. Completeness of many-sorted equational logic. SIGPLAN Notices 16(7), pp. 24-32; extended version to appear in Houston Journal of Mathematics.

[GM 82] Goguen, J.A. and Meseguer, J. Universal realization, persistent interconnection and implementation of abstract modules. Proc. 9th ICALP, Aarhus, Denmark. Springer LNCS 140, pp. 265-281.

[GM 83] Goguen, J.A. and Meseguer, J. An initiality primer. Draft report, SRI International.

[GH 80] Guttag, J.V. and Horning, J.J. Formal specification as a design tool. Proc. ACM Symposium on Principles of Programming Languages, Las Vegas, pp. 251-261.

[Kam 83] Kamin, S. Final data types and their specification. TOPLAS 5, 1, pp. 97-121.

[Karp 64] Karp, C.R. Languages with Expressions of Infinite Length. North-Holland.

[LB 77] Liskov, B.H. and Berzins, V. An appraisal of program specifications. Computation Structures Group memo 141-1, Laboratory for Computer Science, MIT.

[MSV 83] Maibaum, T.S.E., Sadler, M.R. and Veloso, P.A.S. Logical implementation. Technical report, Department of Computing, Imperial College.

[Pep 83] Pepper, P. On the correctness of type transformations. Talk at 2nd Workshop on Theory and Applications of Abstract Data Types, Passau.

[Rei 81] Reichel, H. Behavioural equivalence — a unifying concept for initial and final specification methods. Proc. 3rd Hungarian Computer Science Conf., Budapest, pp. 27-39.

[SB 83] Sannella, D.T. and Burstall, R.M. Structured theories in LCF. Proc. 8th Colloq. on Trees in Algebra and Programming, L'Aquila, Italy. Springer LNCS 159, pp. 377-391.

[ST 84] Sannella, D.T. and Tarlecki, A. Building specifications in an arbitrary institution. Proc. Intl. Symposium on Semantics of Data Types, Sophia-Antipolis. Springer LNCS 173, pp. 337-356.

[ST 85] Sannella, D.T. and Tarlecki, A. Program specification and development in Standard ML. Proc. 12th ACM Symp. on Principles of Programming Languages, New

Orleans.

[SW 82] Sannella, D.T. and Wirsing, M. Implementation of parameterised specifications. Report CSR-103-82, Dept. of Computer Science, Univ. of Edinburgh; extended abstract in: Proc. 9th ICALP, Aarhus, Denmark. Springer LNCS 140, pp. 473-488.

[SW 83] Sannella, D.T. and Wirsing, M. A kernel language for algebraic specification and implementation. Report CSR-131-83, Dept. of Computer Science, Univ. of Edinburgh; extended abstract in: Proc. Intl. Conf. on Foundations of Computation Theory, Borgholm, Sweden. Springer LNCS 158, pp. 413-427.

[Sch 83] Schoett, O. A theory of program modules, their specification and implementation (extended abstract). Report CSR-155-83, Dept. of Computer Science, Univ. of Edinburgh.

[Scott 65] Scott, D. Logic with denumerably long formulas and finite strings of quantifiers. In: Theory of Models. North-Holland, pp. 329-341.

[Tar 84] Tarlecki, A. Free constructions in abstract algebraic institutions. Draft report, Dept. of Computer Science, Univ. of Edinburgh.

[Wand 79] Wand, M. Final algebra semantics and data type extensions. JCSS 19, pp. 27-44.

[Wir 82] Wirsing, M. Structured algebraic specifications. Proc. AFCET Symp. on Mathematics for Computer Science, Paris, pp. 93-107.

[Wir 83] Wirsing, M. Structured algebraic specifications: a kernel language. Habilitation thesis, Technische Univ. München.

Parameter Preserving Data Type Specifications

Peter Padawitz

Universität Passau
Fakultät für Informatik
Postfach 2540
D-8390 Passau
F.R.G.

Abstract

Term rewriting methods are used for solving the persistency problem of parameterized data type specifications. Such a specification is called persistent if the parameter part of its algebraic semantics agrees with the semantics of the parameter specification. Since persistency mostly cannot be guaranteed for the whole equational variety of the parameter specification, the persistency criteria developed here mainly concern classes of parameter algebras with "built-in" logic.

1. Persistency, extensions and inductive theories

Starting from a many-sorted signature <S,OP> with sorts S and operation symbols OP an algebraic specification in this sense of ADJ /1/ is given by a triple SPEC = <S,OP,E> where E is a set of equations between OP-terms. Algebras with signature <S,OP> which satisfy E are called SPEC-algebras. For reasons discussed extensively in the literature (e.g. in /1/) the isomorphism class of initial SPEC-algebras plays a dominant role.
A parameterized specification PAR is a pair of two specifications PSPEC and SPEC where the **parameter** PSPEC is part of the **target** SPEC. The role of initial algebras is taken over by a class of target algebras each of which is "freely generated" over some algebra in a given class K of parameter algebras (cf. /2/). Such a class of target algebras is called a parameterized data type. /10/ deals with the proof-theoretical characterization of the equational variety of parameterized data types. This variety turned out to be a certain "inductive" theory of the target specification.
In many cases this characterization works only if PAR is persistent, i.e. if each algebra in the corresponding data type "preserves" the parameter algebra where it is "freely generated" upon. Persistency is also a sufficient criterion for the "passing compatibility" of PAR with actual parameter specifications (cf. ADJ /3/). So this paper is devoted to decidable and powerful criteria for persistency. The first step towards such conditions is the decomposition of PAR into a "base" specification BPAR and the remaining operations & equations of PAR. BPAR is supposed to contain those operations & equations of PAR that are necessary for the "construction" of data. Following this strategy it

is mostly simple to show that BPAR is persistent. Then PAR is persistent, too, if BPAR is a conservative extension of PAR which means that the "base" part of the data type specified by PAR agrees with the data type specified by BPAR.
The extension property is separated into two parts: completeness and consistency. So the tools for solving the persistency problem are the criteria for persistency of BPAR given by Theorem 2.12 and the completeness and consistency conditions of Thms. 3.4/4.7 and 3.5/5.14, respectively. They involve normalization and confluence properties of term reductions and are tailor-made for parameter algebras with "built-in" logic where the proof-theoretical characterization of parameterized data types given in /10/, section 3, is based upon, too.
Besides well-known notions in term rewriting theory like "confluence" and "critical pair" we use some recently introduced ones like "coherence" (cf. /8/), "contextual reductions" (cf. /12/) and "recursive critical pairs" (cf. /11/). They should support the reader's intuition, although their definitions sometimes deviate from their meaning in the cited papers. Moreover, the corresponding results presented here are different from those given there.
The paper is organized as follows: Section 2 contains basic definitions and proof-theoretical characterizations of conservative extensions (2.9) and persistency (2.12). In section 3 general completeness and consistency theorems (3.4/5) are given that refer to term reductions. Sections 4 and 5 focus on parameter algebras with "built-in" logic and adapt the notions of section 3 to this case. Decidable criteria for the crucial confluence criteria of Thm 3.5 are developed in section 5 that culminates in the Critical Pair Thm. 5.12. The main results of sections 4 and 5 are summarized by Completeness Thm. 4.8, Consistency Thm. 5.14 and Persistency Thm. 5.16.
Former versions of these results are part of the author's Ph.D.thesis /9/.
id, inc and nat denote identity, inclusion and natural mappings, respectively.
The first occurrences of notions used throughout the whole paper are printed in boldface.

2. The syntax and semantics of parameterized specifications

Let SIG = <S,OP> be a many-sorted signature with a set S of sorts and an ($S^* \times S$)-sorted set OP of operation symbols. If $\sigma \in OP_{w,s}$, then arity(σ) = w, sort(σ) = s, and we often write σ:w $\rightarrow$ s. If w = ε (empty word), σ is called a constant. T(SIG) denotes the free S-sorted algebra of OP-terms over a fixed infinite S-sorted set X of variables. If t,t' $\in$ T(SIG) and x $\in$ X, then t[t'/x] is t with x replaced by t'.
For every S-sorted set A and all s1,...,sn $\in$ S, $A_{s1...sn} := A_{s1} \times ... \times A_{sn}$. Let w $\in S^*$, s $\in$ S, $\sigma \in OP_{w,s}$ and t $\in T(SIG)_w$. Then root(σt) = σ, arg(σt) = t, sort(σt) = s, and op(σt) resp. var(σt) denote the set of operation symbols resp. variables of σt. Size(t) is the number of operation symbol occurrences in t. A SIG-equation l=r is a pair of SIG-terms l and r with sort(l) = sort(r). Let A be a SIG-algebra. Z(A) denotes the S-sorted set of functions from X to A. The unique homomorphic extension of f $\in$ Z(A) to T(SIG) is also written f. If f $\in$ Z(T(SIG)), t $\in$ T(SIG) and x $\in$ X, then f[t/x] $\in$ Z(T(SIG)) is defined by f[t/x](x) = t and f[t/x](y) = fy for all y $\in$ X-{x}.
A satisfies a SIG-equation l=r if for all f $\in$ Z(A) fl=fr. (This definition extends to classes of algebras and sets of equations as usual.)

2.1. Definitions (specification & semantics)

An (equational) **specification** SPEC = <S,OP,E> consists of a many-sorted signature SIG = <S,OP> and a set E of SIG-equations. **Alg(SPEC)** denotes the class of SIG-algebras that satify E. The **free SPEC-congruence** $=_{SPEC}$ is the smallest SIG-congruence on T(SIG) that contains all pairs <fl,fr> with l=r in E and $f \in Z(T(SIG))$. $=_{SPEC}$ is also called the **free theory of SPEC.**
G(SIG) denotes the free S-sorted algebra of OP-terms over the empty set. **Gen(SPEC)** is the class of "finitely generated" SIG-algebras that satisfy E, i.e. every $a \in A$ is the interpretation of some $t \in G(SIG)$. The **inductive SPEC-congruence** $\equiv_{SPEC}$ is given by all pairs $<t,t'> \in T(SIG)^2$ such that for all $f \in Z(G(SIG))$ ft $=_{SPEC}$ ft'. $\equiv_{SPEC}$ is also called the **inductive theory of SPEC.** Note that the restriction of $\equiv_{SPEC}$ to $G(SIG)^2$ coincides with $=_{SPEC}$.

Two facts are well-known (cf. /4/ resp. /1/):

2.2. Theorem

1. Alg(SPEC) satisfies t=t' iff $t =_{SPEC} t'$.
2. Gen(SPEC) satisfies t=t' iff $t \equiv_{SPEC} t'$. □

2.3. Definitions (parameterized data types)

A **parameterized specification PAR** is a pair of two specifications PSPEC and SPEC. The forgetful functor from Alg(SPEC) to Alg(PSPEC) is denoted by $\mathbf{U_{PAR}}$, while $\mathbf{F_{PAR}}$ stands for its left adjoint. For every class K of PSPEC-algebras the **parameterized data type specified** by **<PAR,K>** is given by

  PDT(PAR,K) = $\{F_{PAR}(A) \mid A \in K\}$.

Let **PSIG** = <PS,POP>, **PSPEC** = <PS,POP,PE>, **SPEC** = <S,OP,E> and **PX** = $\{x \in X \mid sort(x) \in PS\}$. Regarding PX as constants we obtain the signature **SIGX** = <S,OP ∪ PX> and the specification **SPECX** = <S,OP ∪ PX,E>. $\equiv_{SPECX}$ is called the **inductive theory of PAR.**

Analogously to Thm. 2.2, there is the following proof-theoretical characterization of the data type specified by <PAR,Alg(PSPEC)>:

2.4. Theorem (/10/,1.7)

PDT(PAR,Alg(PSPEC)) satisfies t = t' iff $t \equiv_{SPECX} t'$. □

2.5. Definitions (persistency & extension)

Let ID be the identity functor on Alg(PSPEC). We recall from category theory that there is a functor transformation η_{PAR}: ID $\rightarrow U_{PAR}F_{PAR}$ such that for all $B \in$ Alg(SPEC) each homomorphism h: A $\rightarrow U_{PAR}(B)$ uniquely extends to a homomorphism $h^{\#}: F_{PAR}(A) \rightarrow B$ such that $U_{PAR}(h^{\#}) \circ \eta_{PAR}(A) = h$.
Let K be a class of PSPEC-algebras. <PAR,K> is **persistent** if for all $A \in K$ $\eta_{PAR}(A)$ is bijective.
Let **BPAR** = <PSPEC,BSPEC> be a parameterized subspecification of PAR, i.e. **BSPEC** = **<BS,BOP,BE>** is componentwise included in SPEC. Let **BSIG** = <BS,BOP> and **EXT** = <BSPEC,SPEC>. Since $U_{PAR} = U_{BPAR} \circ U_{EXT}$, $\eta_{PAR}(A)$: A $\rightarrow U_{PAR}F_{PAR}(A)$ uniquely

extends to $\eta_{PAR}(A)^*$: $F_{BPAR}(A) \to U_{EXT}F_{PAR}(A)$ such that $U_{BPAR}(\eta_{PAR}(A)^*) \circ \eta_{BPAR}(A) = \eta_{PAR}(A)$:

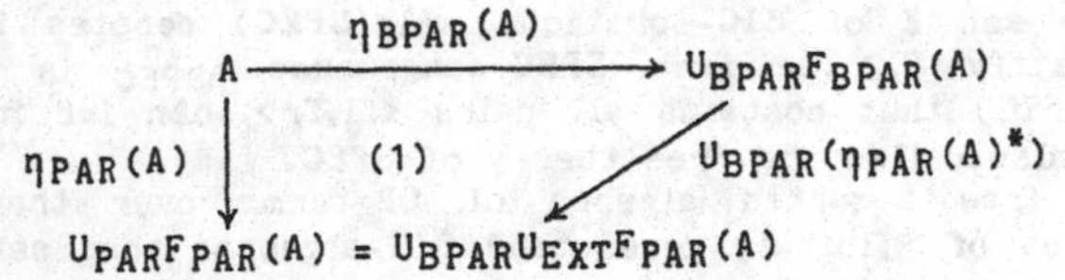

PAR is **complete** (**consistent**) w.r.t. <BPAR,K> if for all $A \in K$ $\eta_{PAR}(A)^*$ is surjective (injective). PAR is a **conservative extension** of <BPAR,K> if PAR is complete and consistent w.r.t. <BPAR,K>.

An immediate consequence of these definitions is the following

2.6. Decomposition Lemma for Persistency

Let <BPAR,K> be persistent. PAR is a conservative extension of <BPAR,K> iff <PAR,K> is persistent. □

2.7. Example

Let BOOL be a specification of Boolean algebras, i.e. BOOL consists of a sort bool, constants true and false, operation symbols $\neg$, $\wedge$, $\vee$, $\Rightarrow$, $\Leftrightarrow$ and the Boolean algebras axioms. Moreover,

```
DATA = BOOL +
   sorts: entry
   opns:  eq: entry entry --> bool
   eqns:  eq(x,x) = true                                   (e1)
          eq(x,y) = eq(y,x)                                (e2)
          (eq(x,y) ∧ eq(y,z)) ⇒ eq(x,z) = true             (e3)

BSET = DATA +
   sorts: set
   opns : Ø: --> set
   ins: set entry --> set
   eqns : ins(ins(s,x),x) = ins(s,x)                       (e4)
          ins(ins(s,x),y) = ins(ins(s,y),x)                (e5)

SET = BSET +
   opns: has: set entry --> bool
         del: set entry --> set
         if-bool: bool bool bool --> bool
         if-set: bool set set --> set
   eqns: has(Ø,x) = false                                  (e6)
         has(ins(s,x),y) = if-bool(eq(x,y),true,has(s,y))  (e7)
         del(Ø,x) = Ø                                      (e8)
         del(ins(s,x),y) = if-set(eq(x,y),del(s,y),ins(del(s,y),x))  (e9)
         if-bool(true,b,b') = b                            (e10)
         if-bool(false,b,b') = b'                          (e11)
         if-set(true,s,s') = s                             (e12)
         if-set(false,s,s') = s'                           (e13)
```

Following the strategy developed in this paper we will show that for a certain class Log of DATA-algebras, which will be given in section 4, <SET,Log(DATA)> is persistent. □

We proceed with the representation of parameterized data types by classes of initial algebras which is essential for the proof-theoretical characterization of conservative extensions (2.9).

2.8. Definition and Theorem (/10/,1.5)

Let $A \in Alg(PSPEC)$. The specification

$$SPEC(A) = \langle S,\ OP \cup A,\ E \cup \triangle(A)\rangle$$

has all operation symbols of SPEC together with all elements of A as constants, while the set of equations of SPEC is extended by the **equational diagram** of A, $\triangle(A)$, that consists of all equations $\sigma(a) = \sigma_A(a)$ with $\sigma \in POP$ and $a \in A_{arity(\sigma)}$.
$F_{PAR}(A)$ gets an $(OP \cup A)$-algebra by interpreting each constant $a \in A$ by $\eta_{PAR}(A)(a)$. Moreover, $F_{PAR}(A)$ is an initial object in Alg(SPEC(A)). □

Using the well-known quotient term algebra representation of initial algebras (cf. ADJ/1/) we can formulate completeness and consistency as free theory properties:

2.9. Theorem

Let $BSIG(A) = \langle BS, BOP \cup A\rangle$ and $SIG(A) = \langle S, OP \cup A\rangle$.

1. PAR is complete w.r.t. <BPAR,K> iff for all $A \in K$, $s \in BS$ and $t \in G(SIG(A))_s$ some $t' \in G(BSIG(A))$ satisfies $t =_{SPEC(A)} t'$.
2. PAR is consistent w.r.t. <BPAR,K> iff for all $A \in K$ and $t,t' \in G(BSIG(A))$ $t =_{SPEC(A)} t'$ implies $t =_{BSPEC(A)} t'$. □

Besides persistency Thms. 2.8 & 2.9 provide a useful criterion for the validity of equations in parameterized data types: PDT(PAR,K) satisfies a set E' of SIG-equations if PAR is complete w.r.t. <BPAR,K> and $\langle PSPEC, \langle S,OP,E \cup E'\rangle\rangle$ is consistent w.r.t. <BPAR,K>.

The proof-theoretical conditions "maximal completeness" and "maximal consistency" defined below deal with variables instead of elements of a particular parameter algebra. Hence they characterize persistency of PAR with respect to all parameter algebras (Thm. 2.12).

2.10. Definitions

PAR is **maximally complete** if for all $s \in PS$ and $t \in G(SIGX)_s$ $t =_{SPECX} t'$ for some $t' \in T(PSIG)$ (cf.2.3.). PAR is **maximally consistent** if for all $t,t' \in T(PSIG)$ $t =_{SPECX} t'$ implies $t =_{PSPEC} t'$.

2.11. Definition

The **simple reduction relation generated** by E, $\xrightarrow[E]{}$, is the smallest relation on T(SIG) resp. Z(T(SIG)) such that

(i) for all $l{=}r$ in E and $f \in Z(T(SIG))$ $fl \xrightarrow[E]{} fr$,

(ii) for all $\sigma \in OP$ $\sigma(t_1,\ldots,t_i,\ldots,t_n) \xrightarrow[E]{} \sigma(t_1,\ldots,t_i',\ldots t_n)$ if $t_i \xrightarrow[E]{} t_i'$,

(iii) for all $f,g \in Z(T(SIG))$ $f \xrightarrow[E]{} g$ if for all $x \in X$ $fx \xrightarrow[E]{} gx$.

$\xrightarrow[E]{=}$, $\xleftrightarrow[E]{}$ and $\xrightarrow[E]{*}$ denote the reflexive, symmetric and reflexive-transitive closures of $\xrightarrow[E]{}$, respectively.

2.12. Persistency Theorem I

<PAR,Alg(PSPEC)> is persistent iff PAR is maximally complete and maximally consistent.

Proof:

"only if": By assumption, $\eta_{PAR(T_{PSIG})/\equiv_{PSPEC}}$ is an isomorphism. Since

$$F_{PAR}(T_{PSIG})/\equiv_{PSPEC} = G(SIGX)/\equiv_{SPECX}$$

(/10/,1.6), we conclude

$$T_{PSIG}/\equiv_{PSPEC} \cong U_{EXT}(G(SIGX)/\equiv_{SPECX}).$$

The surjective resp. injective part of this isomorphism is maximal completeness resp. consistency of PAR.

The "if"-part is more tedious and given in the extended version of this paper. Its main idea is due to H. Ganzinger (cf. /6/, Thm. 5). ◻

2.13. Corollary

<PAR,Alg(PSPEC)> is persistent if for all $\sigma \in OP$, $sort(\sigma) \in PS$ implies $\sigma \in POP$ and if for all $l=r$ in E $sort(l) \in PS$ implies that $l=r$ is in PE. ◻

2.14. Example (cf. 2.7)

Using Corollary 2.13 we immediately observe that <<DATA,BSET>,Alg(DATA)> is persistent. ◻

3. Extension Proofs by term rewriting

Assuming that the "base" <BPAR,K> is persistent we now turn to tools for extension proofs which are based on the proof-theoretical characterization of conservative extensions given in the last section (2.9).

From now on we suppose that S=BS and for all $A \in Alg(PSPEC)$ and $s \in S$ $G(SIG(A))_s$ is nonempty.

A first step towards extension criteria is the decomposition of the free SPEC(A)-congruence into simple reductions and the free BSPEC(A)-congruence:

3.1. Definition

Let $A \in Alg(PSPEC)$. A set R of SIG(A)-equations is **Church-Rosser w.r.t. A** if for all $s \in BS$ and $t,t' \in G(SIG(A))_s$

t $=_{SPEC(A)}$ t' implies

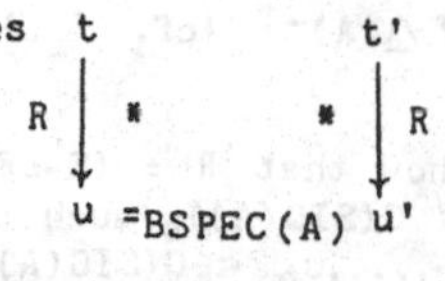

for some u,u' $\in$ G(BSIG(A)).

3.2. Lemma

Suppose that for each l=r in E-BE op(l) contains at least one operation symbol of OP-BOP. For all A $\in$ K let E(A) be a subset of $=_{BSPEC(A)}$. If (E-BE) $\cup$ E(A) is Church-Rosser w.r.t. A, then PAR is consistent w.r.t. <BPAR,K>. □

Localizing the Church-Rosser property by "confluence" and "coherence" conditions goes along with restricting equations to "normalizing" ones:

3.3. Definitions

Let A $\in$ Alg(PSPEC) and R be a set of SIG(A)-equations. t' $\in$ G(BSIG(A)) is an **R-normal form** of t $\in$ G(SIG(A)) if $t \xrightarrow[R]{*} t'$. R is **normalizing** w.r.t. A if for all t $\in$ G(SIG(A)) t has an R-normal form. R is **confluent** w.r.t. A if for all t $\in$ G(SIG(A)) all R-normal forms t_1,t_2 of t satisfy $t_1 =_{BSPEC(A)} t_2$. $\langle t_1,t_2\rangle \in$ G(SIG(A)) is **uniformly** R-**convergent** w.r.t. A if some R-normal forms t_1', t_2' of t_1 resp. t_2 satisfy $t_1' =_{BSPEC(A)} t_2'$, written: $t_1 \downarrow_{R,A} t_2$.
R is **coherent** w.r.t. A if for all $t,t_1,t_2 \in$ G(SIG(A))

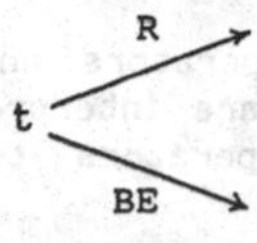

implies $t_1 \downarrow_{R,A} t_2$.

R **commutes with** another set R' of SIG(A)-equations if for all $t,t_1,t_2 \in$ G(SIG(A))

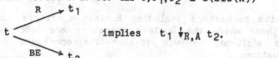

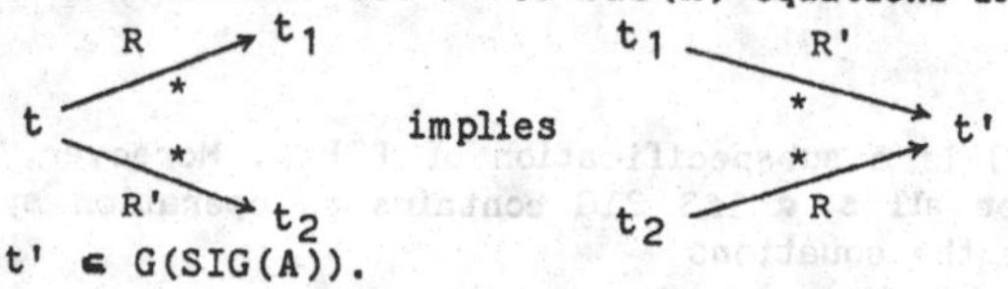

for some t' $\in$ G(SIG(A)).

3.4. Completeness Theorem I

For all A $\in$ K let E(A) be a subset of $=_{SPEC(A)}$. If for all A $\in$ K E $\cup$ E(A) is normalizing w.r.t. <BPAR,K>, then PAR is complete w.r.t. <BPAR,K>.
Proof:
The statement immediately follows from Thm. 2.9.1. □

3.5. Consistency Theorem I

Suppose that for each l = r in BE var(r) $\subseteq$ var(l) and for each l = r in E-BE l contains at least one operation symbol of OP-BOP. For all A $\in$ K let E(A) be a subset of $=_{BSPEC(A)}$. If (E-BE) $\cup$ E(A) is normalizing, confluent and coherent w.r.t. A and commutes with $\triangle(A) \cup \triangle(A)^{-1}$ (cf. 2.8), then PAR is consistent w.r.t. <BPAR,K>.

Proof:

By Lemma 3.2, it is sufficient to show that R = (E-BE) ∪ E(A) is Church-Rosser w.r.t. A. So let s ∈ BS and t,t' ∈ $G(SIG(A))_s$ such that t $=_{SPEC(A)}$ t'. There are a least number n and $t_1,\ldots,t_n$, $u_1,\ldots,u_n$ ∈ G(SIG(A)) with t_1=t, u_n=t' and for all 1 ≤ i<n $u_i \xrightarrow[E-BE]{*} t_i$ and

(i) $u_i \xrightarrow[E-BE]{} t_{i+1}$

or (ii) $u_i \xrightarrow[BE]{} t_{i+1}$

or (iii) $t_{i+1} \xrightarrow[BE]{} u_i$

or (iv) $u_i \xleftrightarrow[\triangle(A)]{} t_{i+1}$.

We prove t $\downarrow_R$ t' by induction on n. n = 1 implies $t' \xrightarrow[R]{*} t$, and t $\downarrow_R$ t' follows from normalization and confluence of R w.r.t. A. Since E(A) ⊆ $=_{BSPEC(A)}$ and for each l=r in E-BE op(l) ∩ (OP-BOP) ≠ ∅,

(*) for all u ∈ G(BSIG(A)) $u \xrightarrow[R]{} u'$ implies u $=_{BSPEC(A)}$ u'.

Let n>1. By induction hypothesis, $t_2 \downarrow_{R,A}$ t'. Hence by confluence of R, it remains to show $t_1 \downarrow_{R,A} t_2$.

The proof proceeds by deriving $t_1 \downarrow_{R,A} t_2$ in each of the cases (i)-(iv) for i=1. □

4. Parameters with "built-in" logic

From now on we deal with parameters including Boolean operators and restrict parameter algebras to those where the Boolean operators are interpreted as in propositional logic. In addition, we use if-then-else operators to simulate conditional axioms by equations.

General Assumption

Suppose that BOOL (cf. 2.7) is a subspecification of PSPEC. Moreover, let ifS be a subset of S such that for all s ∈ ifS SIG contains an operation symbol if-s: bool s s → s and E includes the equations

if-s(true,x,y) = x and if-s(false,x,y) = y.

Vice versa, for each l = r in E

(i) sort(l) = bool implies l ∉ {true, false},

(ii) sort(l) ≠ bool implies t ∈ {true, false} for all bool-sorted subterms t of l.

4.1. Definitions

Let PEXT = <BOOL,PSPEC>. The class **Log(PSPEC)** is given by all PSPEC-algebras A such that $U_{PEXT}(A)$ is the Boolean algebra {true, false}. Hence we drop the equations true=true and false=false from the equational diagram of A (cf. 2.8). For all A ∈ Log(PSPEC) **LE(A)** denotes the set of BSIG(A)-equations l=r with l ∈ $G(BSIG(A))_{bool}$-{true,false}, r ∈ {true, false} and l $=_{BSPEC(A)}$ r.

4.2. Lemma

Let <BPAR,Log(PSPEC)> be persistent. Then for all $A \in Log(PSPEC)$ and $t \in G(BSIG(A))_{bool}$ either t=true or t=false is in LE(A).

Proof:

Let $A \in Log(PSPEC)$. By assumption, $A = U_{BPAR}F_{BPAR}(A)$. By Thm. 2.8, $F_{BPAR}(A) = G(BSIG(A))/{=_{BSPEC(A)}}$. Hence the statement follows from $U_{PEXT}(A) = \{true, false\}$. □

Next we define a reduction relation with conditions (contexts) in order to simulate reductions via LE(A).

4.3. Definition

Let $LT = T(SIG)_{bool}$. The **contextual reduction relation generated by E**, $\{\xrightarrow[E;p]{*}\}_{p \in LT}$, is the family of smallest relations on T(SIG) such that

(i) for all $t \in T(SIG)$ and $p \in LT$ $t \xrightarrow[E;p]{*} t$,

(ii) for all l=r in E, $f \in Z(T(SIG))$ and $p \in LT$ $fl \xrightarrow[E;p]{*} fr$,

(iii) for all $\sigma \in OP$
$\sigma(t_1,\ldots,t_i,\ldots,t_n) \xrightarrow[E;p]{*} \sigma(t_1,\ldots,t_i',\ldots,t_n)$
if $t_i \xrightarrow[E;p]{*} t_i'$,

(iv) for all $s \in ifS$ and $t_1,t_2 \in T(SIG)_s$
$if\text{-}s(p,t_1,t_2) \xrightarrow[E;p]{*} t_1$ and $if\text{-}s(p,t_1,t_2) \xrightarrow[E;\neg p]{*} t_2$,

(v) $t \xrightarrow[E;p \wedge q]{*} t''$ if $t \xrightarrow[E;p]{*} t'$ and $t' \xrightarrow[E;q]{*} t''$,

(vi) $t \xrightarrow[E;p \vee q]{*} t'$ if $t \xrightarrow[E;p]{*} t'$ and $t \xrightarrow[E;q]{*} t'$,

(vii) for all $t,t' \in T(SIG)$ $t \xrightarrow[E;false]{*} t'$.

The following lemma draws the connection between contextual and LE(A)-reductions. Contexts are now restricted to "base" terms so that contextual reductions can be regarded as "hierarchical" ones.

4.4. Lemma

Let <BPAR,Log(PSPEC)> be persistent. Then all $A \in Log(PSPEC)$, $f \in Z(G(SIG(A)))$ and $f' \in Z(G(BSIG(A)))$ with $f \xrightarrow[E \cup LE(A)]{*} f'$ satisfy

(*) $ft \xrightarrow[E \cup LE(A)]{*} ft'$ if $t \xrightarrow[E;p]{*} t'$ and f'p=true is in LE(A). □

Contextual reduction properties that correspond to 3.3 are defined by 4.5 and 4.11 below.

4.5. Definition

Let $t \in G(SIGX)$ (cf. 2.3). t **has contextual E-normal forms** $t_1,\ldots,t_n \in G(BSIGX)$ if there are $n \in \mathbb{N}$ and $p_1,\ldots,p_n \in LT$ such that $p_1 \vee \ldots \vee p_n =_{SPEC} true$ and for all $1 \leq i \leq n$ $t \xrightarrow[E;p_i]{*} t_i$. E is **contextually normalizing** if all $t \in G(SIGX)$ have contextual E-normal forms.

To reduce normalization of $E \cup LE(A)$ to contextual normalization of E we have to guarantee that SPEC(A) does not identify true and false:

4.6. Definition

PAR is **logically consistent** if for all $A \in Log(PSPEC)$ some $B \in Alg(SPEC(A))$ has different interpretations of true and false.

4.7. Lemma

Suppose that <BPAR,Log(PSPEC)> is persistent and PAR is logically consistent. Let $A \in Log(PSPEC)$. $E \cup LE(A)$ is normalizing w.r.t. A if E is contextually normalizing. □

4.8. Completeness Theorem II

Suppose that <BPAR,Log(PSPEC)> is persistent and PAR is logically consistent. If E is contextually normalizing, then PAR is complete w.r.t. <BPAR,Log(PSPEC)>.

Proof:

The statement immediately follows from Lemma 4.7 and Completeness Theorem 3.4. □

4.9. Example (cf. 2.7)

Let E={e6,...,e13}. One easily observes that E is contextually normalizing if

(*) for all $t,t' \in G(BSIGX)_{set}$ has(t,x), del(t,x) and if-set(x,t,t') have contextual E-normal forms.

(*) follows by induction on size(t)+size(t') because we obtain

$$has(\emptyset,x) \underset{e8}{\longrightarrow} false,$$

$$has(ins(t,x),y) \overset{*}{\underset{e7;eq(x,y)}{\longrightarrow}} true,$$

$$has(ins(t,x),y) \overset{*}{\underset{e7;\neg eq(x,y)}{\longrightarrow}} has(t,y),$$

$$del(\emptyset,x) \underset{e8}{\longrightarrow} \emptyset,$$

$$del(ins(t,x),y) \overset{*}{\underset{e9;eq(x,y)}{\longrightarrow}} del(t,y),$$

$$del(ins(t,x),y) \overset{*}{\underset{e9;\neg eq(x,y)}{\longrightarrow}} ins(del(t,y),x),$$

$$if\text{-}set(x,t,t') \underset{E;x}{\longrightarrow} t,$$

$$if\text{-}set(x,t,t') \underset{E;\neg x}{\longrightarrow} t'$$

for all $t,t' \in G(BSIGX)_{set}$. Since <<DATA,BSET>,Log(DATA)> is persistent (cf. Ex. 2.14), we conclude from Thm. 4.8 that <DATA,SET> is complete w.r.t. <<DATA,BSET>,Log(DATA)>. □

Local criteria for confluence and coherence require the "new" equations E-BE to be normalizing (cf. Thm. 3.5). "Base" equations (BE) are often not normalizing. Hence we can use Noetherian induction - to lift local criteria - only with respect to E-BE. But BE must be considered, too. The lack of normalization of BE is circumvented by working with parallel BE-reductions which combine independent simple reductions in one step.

4.10. Definition

The **parallel reduction relation generated** by E, $\underset{E}{\Rightarrow}$, and its reflexive closure $\overset{=}{\underset{E}{\Rightarrow}}$ are the smallest relations on T(SIG) resp. Z(T(SIG)) such that

(i) for all $t \in T(SIG)$ $t \overset{=}{\underset{E}{\Rightarrow}} t$,

(ii) for all $f,g \in Z(T(SIG))$ $f \overset{=}{\underset{E}{\Rightarrow}} g$ if for all $x \in X$ $fx \overset{=}{\underset{E}{\Rightarrow}} gx$,

(iii) for all l=r in E $fl \underset{E}{\Rightarrow} gr$ if $f \overset{=}{\underset{E}{\Rightarrow}} g$,

(iv) $t \overset{=}{\underset{E}{\Rightarrow}} t'$ if $t \underset{E}{\Rightarrow} t'$,

(v) for all $\sigma \in OP$ $\sigma(t_1,..,t_n) \underset{E}{\Rightarrow} \sigma(t_1',..,t_n')$

if $\exists\ 1 \leq i \leq n : t_i \underset{E}{\Rightarrow} t_i'$ and $\forall\ 1 \leq i \leq n : t_i \overset{=}{\underset{E}{\Rightarrow}} t_i'$.

4.11. Definition

$\langle t_1,t_2\rangle \in T(SIG)^2$ is **contextually E-convergent** if there are $n \in \mathbb{N}$, $p_1,\ldots,p_n,q_1,\ldots,q_n \in LT$, $t_1^1,..,t_n^1,t_1^2,..,t_n^2 \in T(SIG)$ such that
(i) $(p_1 \wedge q_1) \vee \ldots \vee (p_n \wedge q_n) =_{SPEC}$ true ,
(ii) for all $1 \leq i \leq n$

$$\begin{array}{ccc} t_1 & \overset{E;p_i}{\underset{*}{\rightarrow}} & t_i^1 \\ & & \Big\Downarrow{=}\ BE \\ t_2 & \overset{*}{\underset{E;q_i}{\rightarrow}} & t_i^2. \end{array}$$

4.12. Lemma

Let <BPAR,Log(PSPEC)> be persistent, let PAR be logically consistent and E be contextually normalizing. Let $A \in Log(PSPEC)$. If $\langle t_1,t_2\rangle \in T(SIG)$ is contextually E-convergent, then for all $f,g \in Z(G(SIG(A)))$ $f \overset{=}{\underset{BE}{\Rightarrow}} g$ implies

$$\begin{array}{ccc} & E \cup LE(A) & \\ ft_1 & \overset{*}{\longrightarrow} & u_1 \\ & & \Big\Vert = \; BE \\ & & \Big\Downarrow \\ gt_2 & \overset{*}{\longrightarrow} & u_2 \\ & E \cup LE(A) & \end{array}$$

for some $u_1,u_2 \in G(SIG(A))$. □

5. Critical pair conditions for consistency

This section is the most technical one. We show that contextual convergence of certain critical pairs is sufficient for confluence, coherence and commutativity of (E-BE) ∪ LE(A) (cf. 3.3/5). The assumptions of section 4 are still valid.

To prepare the critical pair conditions we introduce superposition relations (5.1 and 5.8) as those reductions where the lefthand side of the applied equation l=r overlaps a given prefix t of the term to reduce, pictorially:

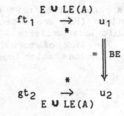

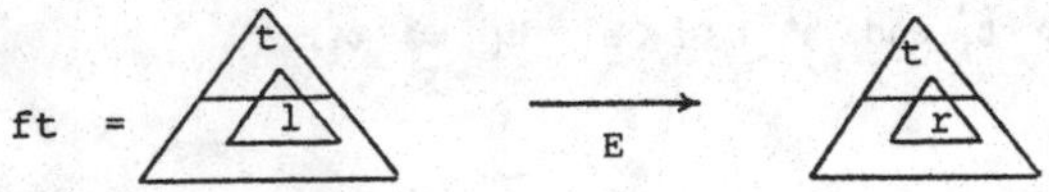

5.1. Definition

The **simple superposition relation generated by E**, $\{\xrightarrow[E;f;t]{}\}_{f \in Z(T(SIG)),\, t \in T(SIG)-X}$, is the family of smallest relations on T(SIG) such that

(i) for all l=r in E $ft \xrightarrow[E;f;t]{} fr$ if ft=fl,

(ii) for all $\sigma \in OP$

$$f\sigma(t_1,..,t_i,..,t_n) \xrightarrow[E;f;\sigma(t_1,..,t_i,..,t_n)]{} f\sigma(t_1,..,t_i',..,t_n)$$

if $ft_i \xrightarrow[E;f;t_i]{} ft_i'$.

An instance of 5.1(i) is **minimal** if f is a most general unifier of t and l.

Let $n(ft \xrightarrow[E;f;t]{} t')$ resp. $n(t \xrightarrow[E]{} t')$ denote the least number of derivation steps 5.1(i)&(ii) resp. 2.11(i)&(ii) that lead to $ft \xrightarrow[E;f;t]{} t'$ resp. $t \xrightarrow[E]{} t'$.

5.2. Proposition

If $ft \xrightarrow[E;f;t]{} t'$, then there are l=r in E, $t_0 \in T(SIG)$, $t_1 \in T(SIG)-X$ and $x \in X$ such that $t=t_0[t_1/x]$, $ft_1=fl$ and $t'=f[fr/x](t_0)$, i.e. l "overlaps" t in ft.

Proof: Straightforward induction on $n(ft \xrightarrow[E;f;t]{} t')$. □

5.3. Proposition

Let t, t' $\in$ T(SIG) and f $\in$ Z(T(SIG)) such that $ft \xrightarrow[E]{} t'$, but not $ft \xrightarrow[E;f;t]{} t'$. Then there are $x \in var(t)$ and $t_x \in T(SIG)$ such that $fx \xrightarrow[E]{} t_x$, $n(fx \xrightarrow[E]{} t_x) \leq n(ft \xrightarrow[E]{} t')$ and

(i) $t' = f[t_x/x](t)$ if t has unique variable occurrences,

(ii) $t' \xrightarrow[E]{*} f[t_x/x](t)$ otherwise.

Proof: Straightforward induction on $n(ft \xrightarrow[E]{} t')$. □

5.4. Definition

E is **linear** if for each l=r in E each variable occurs at most once in l.

The next lemma provides a syntactical criterion for the commutativity property in Consistency Theorem 3.5.:

5.5. Lemma

Suppose that <BPAR, Log(PSPEC)> is persistent, E is linear and for each l=r in E l does not contain operation symbols of POP-{true,false}. Then for all A $\in$ Log(PSPEC) E $\cup$ LE(A) commutes with $\triangle(A) \cup \triangle(A)^{-1}$.

Proof:

Let R = E $\cup$ LE(A), R' = $\triangle(A) \cup \triangle(A)^{-1}$ and

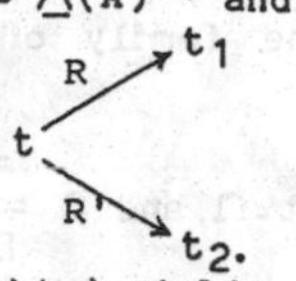

Induction on $n(t \xrightarrow[R]{} t_1) + n(t \xrightarrow[R']{} t_2)$ yields

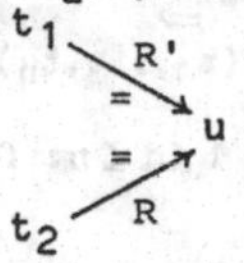

for some u. □

Using the superposition relation we can easily define a **critical pair** of E into l=r as a pair of

(i) the substituted righthand side fr and

(ii) the result t of reducing fl by some equation l'=r' where l' overlaps l,

pictorially:

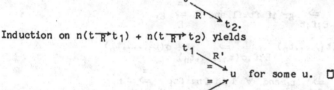

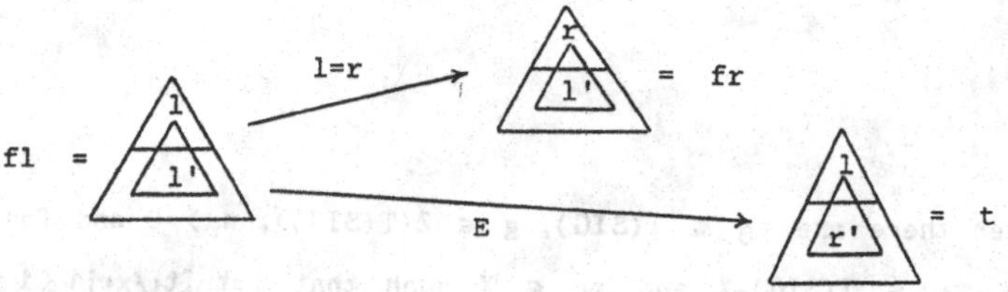

5.6. Definition

Let $fl \xrightarrow[E;f;l]{} t$ and l=r be a SIG-equation. <fr,t> is called a **critical pair of E into l=r**.

5.7. Lemma

Let $A \in Log(PSPEC)$. Suppose that for each l=r in E-BE op(l) contains at least one operation symbol of OP-BOP. Then there are no critical pairs of E-BE into LE(A) or of LE(A) into E-BE. □

In parallel reductions we may have several equations $l_i=r_i$ applied to the same term u. If all outermost l_i overlap a given prefix t of u, we get a "superposing" parallel reduction, pictorially:

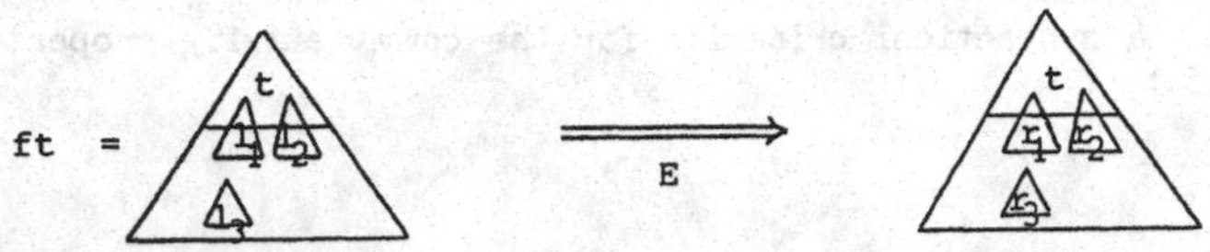

5.8. Definition

The **parallel superposition relation generated by E,**

$\{\underset{E;f;t}{\Rightarrow}\}_{f \in Z(T(SIG)), t \in T(SIG)-X}$, is the family of smallest relations on T(SIG) resp. Z(T(SIG)) such that

(i) for all l=r in E $ft \underset{E;f;t}{\Rightarrow} gr$ if ft=fl and $f \overset{=}{\underset{E}{\Rightarrow}} g$,

(ii) for all $\sigma \in OP$ $f\sigma(t_1,\ldots,t_n) \underset{E;f;\sigma(t_1,\ldots,t_n)}{\Rightarrow} \sigma(t_1',\ldots,t_n')$

if $\exists\ 1 \le i \le n$: $ft_i \underset{E;f;t_i}{\Rightarrow} t_i'$ and $\forall\ 1 \le i \le n$: $ft_i \overset{=}{\underset{E;f;t_i}{\Rightarrow}} t_i'$.

Let $n(ft \underset{E;f;t}{\Rightarrow} t')$ resp. $n(t \underset{E}{\Rightarrow} t')$ denote the least number of derivation steps 5.8.(i)&(ii) resp. 4.9.(i)-(v) that lead to $ft \underset{E;f;t}{\Rightarrow} t'$ resp. $t \underset{E}{\Rightarrow} t'$.

5.9. Proposition

If $ft \underset{E;f;t}{\Rightarrow} t'$, then there are $t_0 \in T(SIG)$, $g \in Z(T(SIG))$, $n > 0$ and for all $1 \le i \le n$ $l_i=r_i$ in E, $t_i \in T(SIG)-X$ and $x_i \in X$ such that $t=t_0[t_i/x_i | 1 \le i \le n]$, $ft_i=fl_i$, $f \overset{=}{\underset{E}{\Rightarrow}} g$ and $t'=f[gr_i/x_i | 1 \le i \le n](t_0)$, i.e. $l_1,\ldots,l_n$ "overlap" t in ft.

Proof: Straightforward induction on $n(ft \underset{E;f;t}{\Rightarrow} t')$. □

5.10. Proposition

Let $t,t' \in T(SIG)$ and $f \in Z(T(SIG))$ such that t has unique variable occurrences. $ft \underset{E}{\Rightarrow} t'$, but not $ft \underset{E;f;t}{\Rightarrow} t'$. Then there are $n > 0$, $x_1,\dots,x_n \in var(t)$ and $t_1,\dots,t_n \in T(SIG)$ such that $fx_i \underset{E}{\Rightarrow} t_i$ and $t' = f[t_i/x_i \mid 1 \le i \le n](t)$.

Proof: Straightforward induction on $n(ft \underset{E}{\Rightarrow} t')$. □

Parallel critical pairs of E into l=r arise in situations like the following one where $l_1=r_1$, $l_2=r_2$ and $l_3=r_3$ are in E:

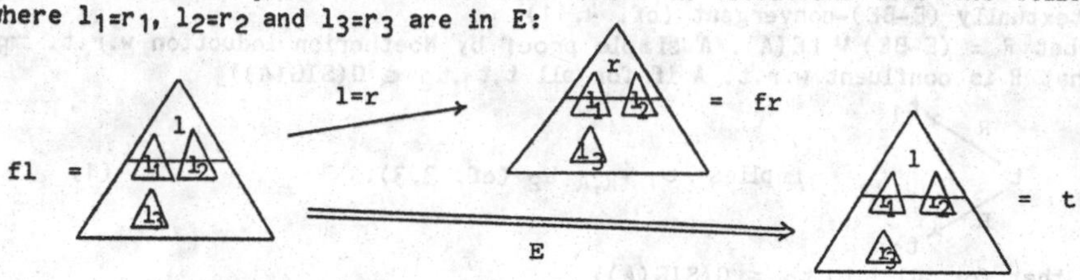

A more complicated case of a parallel overlapping can occur if l_1 shares a subterm of l **and** a prefix of l_3, e.g.

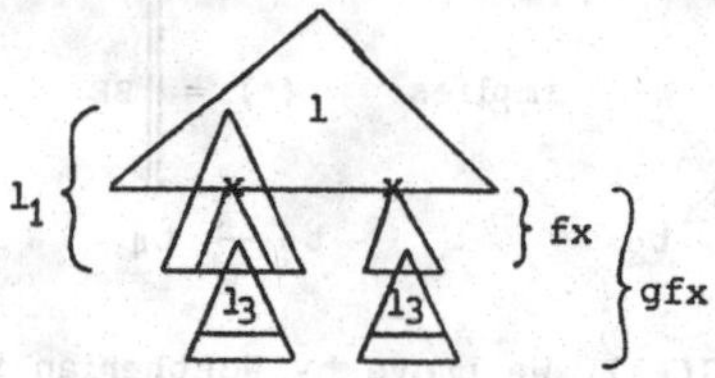

Applying $(l_1=r_1) \in E$ on one hand and $(l=r),(l_3=r_3) \in R$ on the other hand leads to a **recursive critical pair** of E into R.

5.11. Definitions

Let $fl \underset{E;f;l}{\Rightarrow} t$ and l=r be a SIG-equation. <fr,t> is called a **parallel critical pair of E into** l=r.

Let R be a set of SIG-equations and l=r in R. Suppose that $fl \xrightarrow[E;f;l]{} ft$ is derived from a minimal instance of 5.1(i). If for all $x \in X$ $gfx = hx$ or $gfx \underset{R;g;fx}{\Rightarrow} hx$, then <gft,hr> is a **recursive critical pair of E into** R.

E is **terminating** if there are no infinite sequences $t_1 \xrightarrow[E]{} t_2 \xrightarrow[E]{} t_3 \xrightarrow[E]{} \ldots$

t is **E-reducible** if $t \xrightarrow[E]{} t'$ for some t'.

5.12. Critical Pair Theorem

Suppose that <BPAR,Log(PSPEC)> is persistent and PAR is logically consistent. Let E-BE be linear, terminating and contextually normalizing (cf. 4.5), for each l=r in BE var(r) $\subseteq$ var(l) and for each l=r in E-BE l contains at least one operation symbol of OP-BOP.
Let A $\in$ Log(PSPEC). (E-BE) $\cup$ LE(A) is confluent and coherent w.r.t. A (cf. 3.3) if

(i) all critical pairs of E-BE into E-BE,
(ii) all parallel critical pairs of BE into E-BE,
(iii) all recursive critical pairs of E-BE into BE

are contextually (E-BE)-convergent (cf. 4.11).

Proof: Let R = (E-BE) $\cup$ LE(A). A simple proof by Noetherian induction w.r.t. $\xrightarrow[R]{}$ shows that R is confluent w.r.t. A if for all t,t_1,t_2 $\in$ G(SIG(A))

$$t_1 \xleftarrow[R]{} t \xrightarrow[R]{} t_2 \quad \text{implies} \quad t_1 \downarrow_{R,A} t_2 \text{ (cf. 3.3).} \qquad (1)$$

Suppose that for all t,t_1,t_2 $\in$ G(SIG(A))

$$t_1 \xleftarrow[R]{} t \xrightarrow[R]{} t_2 \quad \text{or} \quad t_1 \xleftarrow[R]{} t \overset{BE}{\Longrightarrow} t_2 \quad \text{implies} \quad \begin{array}{ccc} t_1 & \xrightarrow[*]{R} & t_3 \\ (*) = & & \Big\Downarrow BE \\ t_2 & \xrightarrow[R]{*} & t_4 \end{array} \qquad (2)$$

for some t_3,t_4 $\in$ G(SIG(A)). We prove by Noetherian induction w.r.t. $\xrightarrow[R]{}$ that (2) implies (1) and coherence of R w.r.t. A.
By Lemma 4.6, R is normalizing w.r.t. A. Hence if t_3 is not R-reducible, we have t_3 $\in$ G(BSIG(A)) and thus t_4 $\in$ G(BSIG(A)) so that $t_1 \downarrow_{R,A} t_2$. If t_3 is R-reducible, then $t_3 \xrightarrow[R]{} t_5$ for some t_5 $\in$ G(SIG(A)). We obtain $t_5 \downarrow_{R,A} t_4$ by induction hypothesis and thus $t_1 \downarrow_{R,A} t_2$.
Hence it remains to show (2).

a) Let $t \xrightarrow[R]{} t_1$, $t \xrightarrow[R]{} t_2$.

Induction on $n(t \xrightarrow[R]{} t_1) + n(t \xrightarrow[R]{} t_2)$ and a tedious case analysis leads to (*).

b) Let $t \xrightarrow[R]{} t_1$, $t \overset{BE}{\Longrightarrow} t_2$.

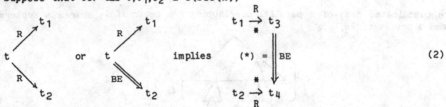

Induction on $n(t \xrightarrow[R]{} t_1) + n(t \underset{BE}{\Rightarrow} t_2)$ and a tedious case analysis leads to (*). □

5.13. Example (cf. 2.7)

Let PAR = <DATA,SET> and BPAR = <DATA,BSET>. One immediately verifies all assumptions of Thm. 5.12 (cf. Exs. 2.14 and 4.9) except for termination of E-BE and the critical pair conditions. For termination we refer to the recursive path ordering method (cf. /5/, /7/), which applied to E-BE = {(e6),...,(e13)} provides a straightforward termination proof.

Assume that there is a critical pair of E-BE into E-BE or a recursive critical pair of E-BE into BE. In both cases we would have l=r in E, $f \in Z(T(SIG))$ and $t \in T(SIG)$ such that $fl \xrightarrow[E\text{-}BE;f;l]{} t$. By Prop. 5.2, there would be l'=r' in E-BE, $t_0 \in T(SIG)$, $t_1 \in T(SIG)-X$ and $x \in X$ such that $l = t_0[t_1/x]$, $ft_1 = fl'$ and $t = f[fr'/x](t_0)$. Since for all l=r in E-BE $op(l) \cap (OP\text{-}BOP) = \{root(l)\}$, we conclude $t_0=x$, $l=l'$ and $r=r'$. Thus we have no recursive critical pair of E-BE into BE, and if <fr,t> is a critical pair of E-BE into E-BE, then fr=fr'=t.

Let $\langle t_1,t_2\rangle$ be a parallel critical pair of BE into E-BE. Then t_1=fr and

$fl \underset{BE;f;l}{\Rightarrow} t_2$ for some l=r in E-BE and $f \in Z(T(SIG))$. By Prop. 5.9, there are $t_0 \in T(SIG)$, $g \in Z(T(SIG))$, n>0 and for all $1 \le i \le n$ $l_i=r_i$ in BE, $u_i \in T(SIG)-X$ and $x_i \in X$ such that $l = t_0[u_i/x_i \mid 1 \le i \le n]$, $fu_i=fl_i$, $f \underset{BE}{\overset{=}{\Rightarrow}} g$ and $t_2 = f[gr_i/x_i \mid 1 \le i \le n](t_0)$.

Case 1: t_0 = has(x_1,y), u_1 = ins(s,x) and l=r is e7.

Case 1.1: fs = ins(fs',fx) and $l_1=r_1$ is e4. Then

t_1 = ifb(eq(fx,fy),true,has(fs,fy)),

t_2 = has(ins(gs',gx),fy)

so that

$t_1 \xrightarrow{E\text{-}BE,eq(fx,fy)}$ true and $t_2 \xrightarrow{E\text{-}BE,eq(gx,fy)}$ true; $t_1 \xrightarrow{E\text{-}BE,\neg eq(fx,fy)}$ has(fs,fy) $\overset{=}{\Downarrow}_{BE}$ t_2.

Since

(eq(fx,fy) $\wedge$ eq(gx,fy)) $\vee$ ¬eq(fx,fy)

$=_{BSPEC}$ eq(fx,fy) $\vee$ ¬eq(fx,fy) $=_{BSPEC}$ true,

$\langle t_1,t_2\rangle$ is contextually (E-BE)-convergent.

Case 1.2: fs = ins(fs',fx') and $l_1=r_1$ is e5. Then

t_1 = ifb(eq(fx,fy),true,has(fs,fy)),

t_2 = has(ins(ins(gs',gx),gx'),fy)

so that

$t_1 \xrightarrow{E\text{-}BE,\ eq(fx,fy) \vee (\neg eq(fx,fy) \wedge eq(fx',fy))}$ true

$t_2 \xrightarrow{E\text{-}BE,\ eq(gx',fy) \vee (\neg eq(gx',fy) \wedge eq(gx,fy))}$ true

and

$t_1 \xrightarrow{E\text{-}BE,\ \neg eq(fx,fy) \wedge \neg eq(fx',fy)}$ has(fs',fy)

$\overset{=}{\Downarrow}_{BE}$

$t_2 \xrightarrow{E\text{-}BE,\ \neg eq(gx',fy) \wedge \neg eq(gx,fy)}$ has(gs',fy).

Since

$$
\begin{aligned}
&((eq(fx,fy) \vee (\neg eq(fx,fy) \wedge eq(fx',fy))) \\
&\quad \wedge (eq(gx',fy) \vee (\neg eq(gx',fy) \wedge eq(gx,fy)))) \\
&\vee (\neg eq(fx,fy) \wedge \neg eq(fx',fy) \wedge \neg eq(gx',fy) \wedge \neg eq(gx,fy)) \\
&=_{BSPEC} ((eq(fx,fy) \vee eq(fx',fy)) \wedge (eq(gx',fy) \vee eq(gx,fy))) \\
&\qquad \vee (\neg eq(fx,fy) \wedge \neg eq(fx',fy)) \\
&=_{BSPEC} eq(fx,fy) \vee eq(fx',fy) \vee (\neg eq(fx,fy) \wedge \neg eq(fx',fy)) \\
&=_{BSPEC} true,
\end{aligned}
$$

$\langle t_1,t_2\rangle$ is contextually (E-BE)-convergent.

Case 2: $t_0 = del(x_1,y)$, $u_1 = ins(s,x)$ and l=r is e9.
Analogously to case 1 we can deduce that $\langle t_1,t_2\rangle$ is contextually (E-BE)-convergent.

Hence all parallel critical pairs of BE into E-BE are contextually (E-BE)-convergent, and we conclude from Thm. 5.12 that for all A ∊ Log(PSPEC) (E-BE) ∪ LE(A) is confluent and coherent w.r.t. A. □

Thms. 3.5 & 5.12 and Lemmata 4.7 & 5.5 imply the

5.14. Consistency Theorem II

Suppose that <BPAR,Log(PSPEC)> is persistent and PAR is logically consistent.
Let E-BE be linear, terminating and contextually normalizing, for each l=r in BE var(r) ⊆ var(l) and for each l=r in E-BE l contains at least one operation symbol of OP-BOP, but no operation symbols of POP-{true,false}.
If all critical pairs of E-BE into E-BE, all parallel critical pairs of BE into E-BE and all recursive critical pairs of E-BE into BE are contextually (E-BE)-convergent, then PAR is consistent w.r.t. <BPAR,Log(PSPEC)>. □

5.15. Example (cf. 2.7)

Let PAR = <DATA,SET> and BPAR = <DATA,BSET>. Using Thm. 5.14 we conclude from Ex. 5.13 that PAR is consistent w.r.t. <BPAR,Log(PSPEC)>. Hence by Ex. 2.14, PAR is a conservative extension of <BPAR,Log(PSPEC)>. Thus the Decomposition Lemma for Persistency (2.6) implies that <PAR,Log(PSPEC)> is persistent. □

Putting together all "syntactical" criteria developed in this paper we obtain the

5.16. Persistency Theorem II

<PAR,Log(PSPEC)> is persistent if PAR is logically consistent and contains a "base" specification BPAR such that

(i) for all σ ∊ BOP sort(σ) ∊ PS implies σ ∊ POP,
(ii) for all l=r in BE var(r) ⊆ var(l), and sort(l) ∊ PS implies that l=r is in PE,
(iii) for all l=r in E-BE l contains at least one operation symbol of OP-BOP, but no operation symbols of POP-{true,false},
(iv) E-BE is linear, terminating and contextually normalizing,
(v) all critical pairs of E-BE into E-BE, all parallel critical pairs of BE into E-BE and all recursive critical pairs of E-BE into BE are contextually (E-BE)-convergent.

(Note also the "Boolean assumptions" at the beginning of section 4.) □

References

/1/ ADJ: J.A. Goguen, J.W. Thatcher, E.G. Wagner: An Initial Algebra Approach to the Specification, Correctness and Implementation of Abstract Data Types, in R.T. Yeh, ed., Current Trends in Programming Methodology, Vol. 4, Prentice-Hall (1978), 80-149

/2/ ADJ: J.W. Thatcher, E.G. Wagner, J.B. Wright: Data Type Specification: Parameterization and the Power of Specification Techniques, ACM Transactions on Programming Languages and Systems 4 (1982), 711-732

/3/ ADJ: H. Ehrig, H.-J. Kreowski, J.W. Thatcher, E.G. Wagner, J.B. Wright: Parameter Passing in Algebraic Specification Languages, Proc. Workshop in Program Specification, Springer LNCS 134 (1982), 322-369

/4/ G. Birkhoff: On the Structure of Abstract Algebras, Proc. Cambridge Phil. Soc. 31 (1935), 433-454

/5/ N. Dershowitz: Orderings for Term-Rewriting Systems, Theoretical Computer Science 17 (1982), 279-301

/6/ H. Ganzinger: Parameterized Specifications: Parameter Passing and Implementation with Respect to Observability, TOPLAS 5,3 (1983), 318-354

/7/ G. Huet, D.C. Oppen: Equations and Rewrite Rules: A Survey, in: R.V. Book, ed., Formal Language Theory: Perspectives and Open Problems, Academic Press (1980)

/8/ J.-P. Jouannaud: Confluent and Coherent Equational Term Rewriting Systems. Application to Proofs in Abstract Data Types, Proc. Coll. on Trees in Algebra and Programming, Springer LNCS 159 (1983), 269-283

/9/ P. Padawitz: Correctness, Completeness, and Consistency of Equational Data Type Specifications, Ph.D.thesis, Technische Universität Berlin (1983)

/10/ P. Padawitz: Towards a Proof Theory of Parameterized Specifications, Proc. Symp. on Semantics of Data Types, Springer LNCS 173 (1984), 375-391

/11/ H. Perdix: Proprietes Church-Rosser de Systemes de Reecriture Equationnels ayant la Propriete de Terminaison faible, Proc. Symp. on Theoretical Aspects of Computer Science, Springer LNCS 166 (1984), 97-108

/12/ J.L. Remy: Proving Conditional Identities by Equational Case Reasoning, Rewriting and Normalization, Research Report, Nancy (1983)

ON THE PARAMETERIZED ALGEBRAIC SPECIFICATION OF CONCURRENT SYSTEMS(*)

E.ASTESIANO(1), G.F.MASCARI(2), G.REGGIO(1), M.WIRSING(3)

(1) Istituto di Matematica, Università di Genova, Italy

(2) Collaboratore esterno, Istituto Applicazioni del Calcolo C.N.R. Roma, Italy

(3) Fakultät für Mathematik und Informatik, Universität Passau, FRG

Abstract. A technique for specifying concurrent systems is shown, that uses the algebraic specification language ASL. A system is algebraically specified as a transition system and a concurrent system is the result of composing systems by three basic operations: synchronization, parallel composition and monitoring. These operations are schematically described using the parameterization concept of ASL and they are in the same time examples for the power of ASL since they cannot be formally specified in other specification languages. Each particular synchronization, parallel composition or monitoring is defined by instantiating on appropriate parameters a unique specification, which produces a transition system out of an input transition system. By combining the three operations we obtain a formal support for a methodology of hierarchical and modular specification of concurrent systems. Moreover it is shown that combining tools for defining semantics in ASL with the above parameterized schema provides a standard way for giving to a concurrent system a variety of semantics depending on observability constraints.

Introduction. It is well known that abstraction is a major problem in specification and to cope with that various brands of abstract data type techniques have been developed, now generalized to allow specifications of programming languages (cf. e.g.[BW1]). When the task is the specification of concurrent systems there is a need of combining a good level of abstraction with the use of some specific structures, serving both as a guide in specifying and as a tool for deriving properties.

The main aim of this paper is precisely to connect an abstract data type approach to a view of concurrency based on some very fundamental parameterized operations and structures. In a sense we try to lay a bridge between the purely translational approach (specify a system by translating into a basic language/model, like CCS or CSP) and the approach which considers a concurrent system simply as an abstract data type [BW2]. Note that our approach is more semantics than syntax oriented; we can see many points of contacts with the aims of Mosses' ASA theory [Mo].

Basically we view a concurrent system as a transition system obtained by composing other transition systems, representing the component processes and we decompose the description of a system into three basic operations of composition: synchronization, parallel composition and monitoring.

Our approach presents three main technical features.

(*) Work partially supported by CNR ITALY PFI-CNET, by a grant MPI and by the ESPRIT project METEOR.

- We use the algebraic specification language ASL [W],which provides tools for the description of parameterized abstract structures, not available in other specification languages: apart from specifications also other types of (high-level) objects such as signatures or sets of formulas can be passed as parameters. That way we define static structures (states of systems), dynamic structures (transition systems), composition of systems (the operations above).

- In particular we can define specific synchronization, parallel composition or monitoring mechanisms as instantiations on appropriate parameters of a unique operation (hierarchical construction) which derives a system out of an input system.

- Every specification has a precise mathematical semantics, defined by ASL. Due to the special form of the axioms in our hierarchical construction we can show that the algebraic semantics corresponds to an abstract operational semantics. Moreover, by using the ASL "observe" construct, we show how to give a variety of semantics depending on what we want to observe in a system; that is essential in concurrency, where now it is recognized that no single semantic equivalence can capture the variety of meanings while preserving a sufficient level of abstraction. .

Our hierarchical and parameterized approach is the formal support for a precise methodology of modular specification. For example, we can show that the methodology underlying the SMoLCS approach [AMMR],[AR] developed in project CNET can be expressed by describing systems via a specification consisting of the nested instantiations of our composition operations.

In section 1 we present an ASL description of hierarchical transition systems; in section 2 we define the basic concurrent schemata for synchronization, parallel composition and monitoring; in section 3 we discuss semantic issues.

1. HIERARCHICAL TRANSITION SYSTEMS.

The concurrent schemata of the following section are based on an operational viewpoint of communicating processes, which is formally expressed by describing processes with the help of (hierarchically structural) transition systems. Therefore we describe "hierarchical transition systems".

Note that we use partial algebra semantics,with minimally defined models;for these and other basic algebraic notions see /BW2/. For a description of ASL we refer to [SW] or [W], although we hope it will be possible to understand the principal ideas of our schemata with the help of the informal explanations of ASL in this section.

1.1 Transition systems.

Usually, a transition system such as a term rewriting system is understood as a binary relation between terms, the so-called "single step transition relation", and this relation is defined by a set of rewriting rules. For the application to concurrency this notion is extended in two ways (cf. e.g. [P1]). Instead of a binary relation a ternary one is used to express the rewriting of terms (here called "states") as well as to indicate further information such as output or synchronization of data (here called "flags"); moreover instead of simple rewriting rules, conditional axioms are used where the premises of the axioms express applicability conditions.

Algebraically we say that a <u>conditional transition system</u> is based on two specifications STATE and FLAG of data structures with the sorts state and flag. Then a ternary boolean function symbol

$$. \overset{\cdot}{\Longrightarrow} . : \text{state} \times \text{flag} \times \text{state} \longrightarrow \text{bool}$$

is defined by axioms E_T of the form cond $\supset$ s $\overset{f}{\Longrightarrow}$ s'=true where cond is an applicability condition, consisting usually of a conjunction of equations. Usually we write s $\overset{f}{\Longrightarrow}$ s' instead of s $\overset{f}{\Longrightarrow}$ s' = true.

In an algebraic style this can be presented as the following specification $\overline{TS}$:

<u>enrich</u> BOOL + STATE + FLAG <u>by</u>
<u>opns</u> $. \overset{\cdot}{\Longrightarrow} .$: state × flag × state ⟶ bool
<u>axioms</u> E_T

In the body of $\overline{TS}$ we use the ASL-operator "+" which builds the sum of two specifications T+T'. Formally the signature of T+T' is the union Σ_+ of the signatures of T and T' and a Σ_+-algebra is a model of T+T' iff its restriction to the signature of T is a model of T and also its restriction to the signature of T' is a model of T'. The other specification operator <u>enrich</u> T <u>by</u> <u>sorts</u> S <u>opns</u> F <u>axioms</u> E corresponds to the enrich-operation of CLEAR and can be derived using"+". The signature of this expression is the signature of T extended by the sorts in S and the function symbols in F. An algebra of this extended signature is a model if it satisfies the axioms E and its restriction to the signature of T is a model of T.

More generally, in ASL we get a function which has the specifications STATE and FLAG and a set of formulas (the conditional axioms) as parameters and which yields a specification (of the transition system) as result. In the following ASL-specification TS of a schema for conditional transition systems the reserved word <u>specfunct</u> expresses that a function with a specification as result is defined. Parameters are written similarly to typed λ-calculus : " λ<u>m</u> x:b. " indicates that x is a formal parameter of mode (or type) <u>m</u> which is subject to the parameter restriction b. There we have a restriction of the third parameter, the set of formulas E_T, which is an application of a function from the domain <u>formulas</u> of sets of formulas into booleans (<u>boolfunct</u>) saying that every element of E_T must have the syntactic form (*).

(*) cond $\supset$ (s $\overset{f}{\Longrightarrow}$ s')=true

Moreover in the following by writing T<s> (where T is a specification expression and s a sort) we mean that s is considered the main sort of T. Similarly <u>specfunct</u> F ≡λ<u>spec</u> T.e is an abbreviation of <u>specfunct</u> F ≡λ <u>spec</u> T,<u>sort</u> t:t ∈sorts(T).e and, correspondingly, the instantiations F(A) and F(A<s>) are abbreviations of F(A,a) and F(A,s).

<u>specfunct</u> TS ≡λ<u>spec</u> STATE,FLAG,
<u>formulas</u> E_T : E_T is"cond $\supset$ (s $\overset{f}{\Longrightarrow}$ s')=true" . $\overline{TS}$

where <u>boolfunct</u> .is"cond $\supset$ (s $\overset{f}{\Longrightarrow}$ s')=true" ≡λ<u>formulas</u> E .
∀<u>formula</u> e : e ∈ E <u>implies</u> is $\supset$(e) <u>and</u> is=(son(2,e)) <u>and</u>
root(son(1,son(2,e))) = $. \overset{\cdot}{\Longrightarrow} .$: state × flag × state ⟶ bool <u>and</u>
root(son(2,son(2,e))) = true : ⟶ bool

For any schema ψ of formulas (above ψ is the schema (*)) the function .is" ψ" analyses whether every formula in a set of formulas has the syntactic form ψ

by means of the tree-like operators root, son and isα (where isα(e)=true iff root(e)= α). In the following the exact definition is left to the reader.
The specification function TS provides a syntactic schema for transition systems. No semantic properties are required. By instantiating TS with actual parameters we get a specification satisfying the requirements for signature and form of axioms.

Example CL. We illustrate the definitions by a small concurrent language CL. For a more significant application of the method to a real project see [AMRZ]. A program of CL consists of a set of processes. The processes can cooperate by means of a CSP-like communication mechanism (statements ? and !) and by means of a shared memory assigning to a local variable the content of a shared variable and viceversa (statements read and write). Moreover a process can create other processes (by the statement start) and there is a construct (or) for nondeterministic choice. A process can either terminate normally, when it has completed the execution of its body, or fail, when it tries to communicate with a terminated process. The program execution terminates when all processes are terminated (either normally or failing). To every program will be given an input-output semantics (see section 3.3) where the outputs are the states of the shared memory. For lack of space we only give the fragments of the full specification which are related to the basic concepts of the paper; it should be clear that all omitted details can be easily defined in ASL.
We start with the basic transition system PBTS which specifies the behaviour of the processes:

$$\underline{\text{spec}}\ \text{PBTS} \equiv \text{TS(PS,PFLAG,PACT)}[\longrightarrow / \Longrightarrow].$$

T[$\longrightarrow/\Longrightarrow$] denotes the renaming of the operation symbol "$\Longrightarrow$" by "$\longrightarrow$" in the signature of T and all models of T (formally this is an application of the specification operation derive, see e.g. [SW]).
The states of PBTS (specification PS) are triples of process identifiers (PID), statements (specification STAT with the zeroary function symbol Λ for the null statement) and local memories (MEM), plus a special element nil, needed for specifying creation and termination of processes. The local memory MEM is represented by finite maps from variable identifiers (VID) to values (specification VAL with the zeroary function symbol vl_o), the parametric specification MAP is given in the appendix.

$$\underline{\text{spec}}\ \text{PS} \equiv \underline{\text{enrich}}\ \text{PROD(PID,STAT,MEM)}\ [ps/prod]\ \underline{\text{by}}$$
$$\underline{\text{opns}}\ \text{nil} : \longrightarrow ps$$
$$\underline{\text{spec}}\ \text{MEM} \equiv \text{MAP(VID,VAL)}[mem/map]$$

The flags of PBTS (specification PFLAG) are in correspondence with the concurrent statements of CL, exception made for τ, which corresponds to the sequential statements (Milner's internal action).
The set of formulas describing the process actions PACT is the union of CACT (concurrent actions) and SACT (sequential actions). We do not specify SACT, whose formulas have the form:

$$\text{cond} \supset \langle p,\text{seqstat},m\rangle \overset{\tau}{\Longrightarrow} \langle p,\Lambda,m'\rangle.$$

Note that a complete specification of the sequential statements can be given separately from CACT. We now give the specification of CACT that will be essential for the rest of the example.
Formula C2 completely defines the global nondeterministic choice statement, for the operator "or" is commutative (st_1 or st_2 = st_2 or st_1). Note that a process in a state <p,st,m> ,where st is an input statement (? or read) can perform an infinite number of actions, one for every possible value which can be received (formulas C3 and C5). "empty" and " .[./.]"are function symbols of the parametric specification MAP defining respectively the map with empty domain and the usual substitution operation . Formula C7 denotes the capability of a process of creating a new process named p_1, with local variables $v_1,\ldots,v_n$ (with initial value vl_o) and the body given

by the statement st. Formula C8 defines the capability of being created, which is represented by a transition of the null process (nil) in the initial state of the created process and is denoted by the flag new(...).

<u>formulas</u> CACT ≡

$$\{C1\ \langle p,st,m\rangle \xRightarrow{f} \langle p,\Lambda,m'\rangle \supset \langle p,st;st_1,m\rangle \xRightarrow{f} \langle p,st_1,m'\rangle,$$

$$C2\ \langle p,st_1,m\rangle \xRightarrow{f} \langle p,st_1',m'\rangle \supset \langle p,st_1 \text{ or } st_2,m\rangle \xRightarrow{f} \langle p,st_1',m'\rangle,$$

$$C3\ \langle p,v?p',m\rangle \xRightarrow{?(p,vl,p')} \langle p,\Lambda,m[vl/v]\rangle,$$

$$C4\ \langle p,v!p',m\rangle \xRightarrow{!(p,m(v),p')} \langle p,\Lambda,m\rangle,$$

$$C5\ \langle p,read(v,sv),m\rangle \xRightarrow{read(vl,sv)} \langle p,\Lambda,m[vl/v]\rangle,$$

$$C6\ \langle p,write(v,sv),m\rangle \xRightarrow{write(m(v),sv)} \langle p,\Lambda,m\rangle\ \} \cup$$

$$\{C7\ \langle p,start(p_1,\{v_1,\dots,v_n\},st),m\rangle \xRightarrow{start(p_1,\{v_1,\dots,v_n\},st)} \langle p,\Lambda,m\rangle,$$

$$C8\ nil \xRightarrow{new(p,\{v_1,\dots,v_n\},st)} \langle p,st,empty[vl_o/v_1]\dots[vl_o/v_n]\rangle \mid n \geq 0\} \cup$$

$$\{C9\ \langle p,\Lambda,m\rangle \xRightarrow{end(p)} nil\} \cup$$

$$\{C10\ \langle p,st,m\rangle \xRightarrow{fail(p,p')} nil \mid \underline{term}\ st \in \{v?p',v!p'\}\ \}$$

<u>1.2 Hierarchical systems.</u> A <u>hierarchical transition system</u> refines the concept of transition system defining a new transition system HTS based on a "basic transition system" BTS with sorts bstate (basic states), bflag (basic flags) and transition operation $\Longrightarrow$. HT is based on specification STATE and FLAG and its relationship to BTS is expressed in the axioms E_H of form $cond \wedge \bigwedge_j (bs_j \xRightarrow{bt_j} bs_j') \supset s \xrightarrow{f} s'$.

Thus HT is simply the sum of the basic transition system BTS and of another transition system i.e. of an appropriate instantiation of TS:

$$BTS + TS(STATE,\ FLAG,\ E_H).$$

Similarly to the specification of transition systems, a schema for hierarchical transition systems is specified as a function with a specification as result. It has as parameters the basic transition system BTS and the parameters STATE,FLAG,E_H for the new system.

If srt is a sort then ≪srt≫ indicates the trivial specification in which there is just one sort and no operations nor axioms. An expression of the form <u>specfunct</u> F≡ λ<u>spec</u> G:X... where X is a specification indicates that in every instantiation of F the actual parameter instantiating G must satisfy the specification X.

<u>specfunct</u> HTS≡ λ<u>spec</u> BTS : TS(≪bstate≫,≪bflag≫,∅)[⟶/⟹],
<u>spec</u> STATE,FLAG,
<u>formulas</u> E_H : E_H is "$cond \wedge (\bigwedge_{j=1}^{n} bs_j \xRightarrow{bf_j} bs_j') \supset s \xrightarrow{f} s'$ ". HT

2. CONCURRENT SCHEMATA.

In this section transition system schemata for concurrent processes are established w.r.t. three different semantic features:

- synchronization
- parallelism (without synchronization)
- monitoring.

These schemata will all be special hierarchical transition systems; i.e. particular instantiations of HTS. To get a description of some particular parallel programming language (or system) one can combine them in various ways using the operators of ASL, in particular the function application.

2.1 Synchronization. A synchronous transition system $\overline{\overline{STS}}$ can be seen as a hierarchical transition system and thus as an instantiation ("makro-expansion") of the schema HTS where:

- states of the basic system are combined to multisets (mset stands for multiset) of states, that, together with an information part, will form the states SST of the resulting system according to some conditions expressed in E_{SST};
- the specification SYNC of the flags of the new system is done in two steps: the set of the possible flags of the new system is given as a parameter SFLAG, then the admissible new flags are filtered out by a function

  $$iss : sflag \times mset(flag) \times inf \longrightarrow bool$$

  denoting that an sflag is the (nondeterministic) result of the synchronization of a multiset of flags;
- for the transformation of the information in a transition from one state to another we need a function

  $$sit : sflag \times inf \longrightarrow inf$$

  which gives the new information part;
- the axioms E_{SYNC} define the functions iss and sit;
- with iss and sit the axioms for a synchronized system are given by a fixed set of axioms E_S.

In the following specification these properties are formally expressed. In contrast to the specification of transition systems the synchronization schema involves semantic properties such as the axiom set E_S specifying the result of a synchronization and the multiset structure of the first component of STATE. To describe the admissible states as pairs of multisets of basic states and an information part the parameterized specifications MSET of finite multisets and PRODCOND of pairs of objects are used which are defined in the appendix.

$$\underline{spec}\ \overline{\overline{STS}} \equiv HTS(BTS, SST, SYNC\langle sflag\rangle, E_S)$$

where $SST = PRODCOND(MSET(\langle\!\langle bstate\rangle\!\rangle), INF, E_{SST})$ [sstate/prod]

$SYNC = \underline{enrich}\ MSET(\langle\!\langle bflag\rangle\!\rangle) + SFLAG + INF\ \underline{by}$

$\underline{opns}$ $iss : sflag \times mset(bflag) \times inf \longrightarrow bool$

$sit : sflag \times inf \longrightarrow inf$

$\underline{axioms}$ E_{SYNC}

$$E_S = \{iss(sf, bf_1|\ldots|bf_n, i) = true \wedge \Big(\bigwedge_{j=1}^{n} bs_j \xrightarrow{bf_j} bs'_j\Big) \supset$$

$$\langle bs_1|\ldots|bs_n, i\rangle \overset{sf}{\Longrightarrow} \langle bs'_1|\ldots|bs'_n, sit(sf,i)\rangle \mid n \geq 1\}$$

Note that in the resulting system $s \overset{f}{\Longrightarrow} s'$ iff f is the result of a synchronization, and also that, for n=1, also actions (or action capabilities) of a unique agent can be seen as a result of a synchronization. Creation of processes can be handled at this step, by synchronizing the creating process with a null state process evolving into the initial state of the new process (see the example in what follows). A synchronized action can influence the information.

By considering the unspecified identifiers of STS as formal parameters one gets the synchronization schema STS:

$\underline{specfunct}$ STS $\equiv \lambda$ $\underline{spec}$ BTS : TS($\langle\!\langle$bstate$\rangle\!\rangle$,$\langle\!\langle$bflag$\rangle\!\rangle$,$\emptyset$)[$\longrightarrow$/$\Longrightarrow$],
$\underline{spec}$ INF,SFLAG,
$\underline{formulas}$ E_{SST}, E_{SYNC} . $\overline{\overline{STS}}$

Example CL. The synchronous interactions between the processes of a CL program are described by CLSTS.

$$\underline{spec}\ CLSTS \equiv STS(PBTS, CLINF, PFLAG, E_{CLSST}, E_{CLSYNC})$$

The information needed for synchronization is described as a cartesian product of shared memory states (SMEM) and sets of process identifiers (TERM, representing the names of the terminated processes).

spec CLINF ≡ PROD(SMEM,TERM)[inf/prod].

The shared memory states are defined as finite mappings from shared variable identifiers (SVID) to values (VAL):

spec SMEM ≡ MAP(SVID,VAL)[smem/map] .

spec TERM ≡ SET(PID)[term/set(pid)]

The flags of the synchronous actions are the same as for process actions (the actual parameter for SFLAG is PFLAG).

The axioms E_{CLSST} ensure that a process is not contemporaneously present and terminated and that nil is the neutral element for the parallel composition of processes.

formulas E_{CLSST} ≡ {¬D(<{<p,st,m>}|mps,<sm,{p} ∪tr>>),<{nil}|mps,i>=<mps,i>}.

The synchronous actions are defined by the following set of axioms.

formulas E_{CLSYNC} ≡

{S1 iss(τ ,mfs,i)=true | term mfs ∈ {τ},{!(p,vl,p_1)}|{?(p_1,vl,p)},
{start(p,vs,st)}|{new(p,vs,st)}} } ∪

{S2 sit(τ ,i)=i,
S3 iss(read(vl,sv),{read(vl,sv)},<sm,tr>) = (sm(sv)=vl),
S4 sit(read(vl,sv),i)=i,
S5 iss(write(vl,sv),{write(vl,sv)},i)= true,
S6 sit(write(vl,sv),<sm,tr>) = <sm [vl/sv] ,tr>,
S7 iss(end(p), {end(p)},i) = true,
S8 sit(end(p), <sm,tr>)=<sm,tr ∪{p}>,
S9 iss(end(p), {fail(p,p_1)},<sm,tr>) = (p_1 ∈ tr)}.

The axiom schema S1 expresses handshaking communication, creation of new processes and the fact that every internal process action is understood as synchronous action. The process action capabilities of form read(vl,sv) are accepted as synchronous actions only if the value of sv in the actual state of the shared memory is vl (S3). The synchronous actions with flags write(vl,sv) update the value of sv transforming the first component of the information part (S6). When a process terminates (normally or by failing) its name is added to the second component of the information part (S8). The process capabilities of the form fail(p,p_1) become synchronous actions only when the process p_1 is terminated (S9).

2.2 Parallel composition. Parallel transition systems $\overline{PTS}$ are hierarchical transition systems, similar to synchronized systems, but with some important differences:

- the states and the flags are the same as in the basic system: the states consist of a product of an information part and of a multiset of some component states; this is expressed by a parameter restriction on the basic transition system BTS;
- the transitions of the system include those of the basic system, hence the basic transition is also called "⟹". The basic transitions are augmented by a binary composition expressed by the semigroup operation // on flags, and an axiom E_P specifying the particular transition resulting from the parallel composition;
- a function pit represents the transformation of the information resulting from composing the two component transformations.

The basic intuitive idea behind $\overline{PTS}$ is that the actions are composed to form a new action, corresponding to a parallel execution of the component actions. The formulas E_{PAR} specify the function pit and the definedness of the composition operation.

spec $\overline{PTS}$ ≡ HTS(BTS,<<bstate>>,PAR<flag>,E_P)

where

$$\begin{array}{lll} PAR = \underline{sorts} & flag, inf & \\ \quad \underline{opns} & /\!/ : flag \times flag & \longrightarrow flag \\ & pit : flag \times inf & \longrightarrow inf \\ \quad \underline{axioms} & \{f_1 /\!/ f_2 = f_2 /\!/ f_1 \, , \, (f_1 /\!/ f_2) /\!/ f_3 = f_1 /\!/ (f_2 /\!/ f_3)\} \cup E_{PAR} \text{ and} & \end{array}$$

$$E_P = \{D(f_1 /\!/ f_2) \wedge (\bigwedge_{j=1}^{2} \langle ms_j, i\rangle \xRightarrow{f_j} \langle ms'_j, i_j\rangle) \supset \langle ms_1 | ms_2, i\rangle \xRightarrow{f_1 /\!/ f_2} \langle ms'_1 | ms'_2, pit(f_1 /\!/ f_2, i)\rangle\}$$

Similarly to synchronization we get the following schema for parallel composition:

$$\underline{specfunct}\ PTS \equiv$$
$$\lambda\, \underline{spec}\ BTS : TS(PRODCOND(MSET(\langle\!\langle state \rangle\!\rangle), \langle\!\langle inf \rangle\!\rangle, \emptyset)[bstate/prod], \langle\!\langle flag \rangle\!\rangle, \emptyset),$$
$$\underline{formulas}\ E_{PAR}\ .\ \overline{PTS}$$

Note that for an instantiation of PTS the specification of FLAG (which is a subspecification of BTS) must admit models which are not term generated models, since otherwise the introduction of the function symbol $/\!/$ could cause inconsistencies in PTS.

Example CL. The parallel composition between the synchronous actions of the CL processes is defined by the formulas E_{CLPAR}.

$$\underline{spec}\ CLPTS \equiv PTS(CLSTS, E_{CLPAR}).$$

$$\underline{formulas}\ E_{CLPAR} \equiv$$
$$\{P1\ D(write(vl,sv) /\!/ write(vl',sv')) \Leftrightarrow (sv \neq sv')\} \cup$$
$$\{P2\ D(f) \supset D(f /\!/ f') \mid \underline{term}\ f' \in \{read(vl,sv), end(p), \tau\}\} \cup$$
$$\{P3\ sit(f,i) = i' \supset pit(f,i) = i',$$
$$P4\ pit(f_1 /\!/ f_2, i) = pit(f_1, pit(f_2, i))\}.$$

P1 and P2 forbid two contemporaneous writings on the same shared variable while allow contemporaneous readings. It can be easily proved that $pit(f_1, pit(f_2, i))$ is equal to $pit(f_2, pit(f_1, i))$ whenever $D(f_1 /\!/ f_2)$ is true, so the specification is consistent.

Different assumptions on the mutual exclusion of the accesses to the shared variables can be expressed modularly by changing P1 and P2. For example

$$\{P'_1\ D(f_1 /\!/ f_2) \Leftrightarrow (sv \neq sv') \mid \underline{term}\ f_1 \in \{write(vl,sv), read(vl,sv)\}, \underline{term}\ f_2 \in \{write(vl',sv'), read(vl',sv')\}\} \cup$$
$$\{P'_2\ D(f) \supset D(f /\!/ f') \mid \underline{term}\ f' \in \{end(p), \tau\}\}$$

forbid contemporaneous accesses to a single shared variable.

2.3 Monitoring. Monitored systems arise from a basic transition system by restricting the possible transitions to those that are allowed by the monitor. The restriction is achieved by using (in a specification MON) the function

$$mon : bflag \times bstate \times minf \longrightarrow mflag$$

which depends also on some information in MINF. This information is updated in a transition by the function

$$mit : bflag \times minf \longrightarrow minf.$$

A basic transition with flag f, starting from a basic state bs is allowed by the monitor according to the monitor information mi, when mon(f,bs,mi) is defined; the resulting value is in MFLAG and represents the interaction of that transition with the external world (see E_M).

The states, MSTATE, of a monitored system are specified by a conditional product of the basic states of BTS and a monitor information part MINF.

We usually assume as further constraint (which is not specified in the following) that mon, for any fixed f and mi, depends only on the multiset of all flags of the transitions starting in a state.

spec $\overline{HTS}$ ≡ HTS(BTS,MSTATE,MON<mflag>,E_M)

where

MSTATE = PRODCOND(≪bstate≫,MINF,E_{MST})[mstate/prod],

MON = enrich BTS + MFLAG by

opns mon : bflag × bstate × minf ⟶ mflag

mit : bflag × minf ⟶ minf

axioms E_{MON} and

E_M={mon(bf,<$\bar{s}$,i>,mi)=mf ∧ <s,i> $\xrightarrow{bf}$ <s',i'> ∧ s ⊆ $\bar{s}$ ⊃

<<$\bar{s}$,i>,mi> $\xRightarrow{mf}$ <<($\bar{s}$-s)| s',i'>,mit(bf,mi)> }

Intuitively monitoring can express some kind of centralized control (e.g. interleaving mode vs. parallel mode, fairness constraints and so on) or the result of some global external observation (cutting capabilities needing synchronization when the system is isolated, converting flags to some external behaviour).

By abstracting from unspecified identifiers we get the following schema:

specfunct MTS ≡

λ spec BTS : TS(PRODCOND(MSET(≪state≫),≪inf≫,∅)[bstate/prod],≪bflag≫,∅)[⟶/⟹],

spec MINF,MFLAG,

formulas E_{MST},E_{MON}. $\overline{MTS}$

Example CL. We specify an interleaving mode for the CL processes; only one action resulting from a synchronization is allowed at each step.

spec CLMTS ≡ MTS(CLPTS,NULL,NULL,E_{CLMST},E_{CLMON})

No monitor informations are needed; so the actual parameter for MINF is the specification NULL, which has just one sort, one zeroary function symbol, no axioms and such that every model has just one element (that can be obtained by using the construct reachable of ASL).

Every state of CLPTS becomes a state of CLMTS:

formulas E_{CLMST} ≡ {D(sst) ⊃ D(<sst,null>)}.

Since we are interested in an input-output semantics of CL we do not care about the flags of the system CLMTS; then the actual parameter for MFLAG is NULL.

formulas E_{CLMON} ≡

{M1 mon(f,sst,null)=null| term f ∈ {τ,read(vl,sv),write(vl,sv),end(p)} } ∪

{M2 mit(f,null) = null}.

These formulas define the wanted interleaving mode. Note that mon(f_1//f_2,...) is undefined in minimally defined models (then the contemporaneous execution of several synchronous actions is prohibited).

Due to the modularity of our methodology, by choosing a different axiom schema M1 we could define different modes for the CL processes. For example, for a parallel mode where all parallel actions are allowed, we use the axiom

M1' mon(f,sst,null) = null ;

while for a parallel mode, where, when a process fails, the program execution stops, we use the axiom

M1" (∄ ps ∈ mps ps $\xRightarrow{fail(p,p')}$ nil) ⊃ mon(f,<mps,i>,null) = null .

2.4 Iterated hierarchical systems. Finally we give two examples illustrating how the previous specifications can be combined. Structured Linear Concurrent Systems SLCS can be obtained by instantiating PTS with STS.

spec $\overline{SLCS}$ ≡ PTS(STS(BTS,INF,SFLAG,E_{SST},E_{SYNC}),E_{PAR})

By λ-abstraction we get

<u>specfunct</u> SLCS ≡ λ <u>spec</u> BTS : TS(≪state≫,≪flag≫,∅)[⟶/⟹],
<u>spec</u> INF,SFLAG,
<u>formulas</u> $E_{SST}, E_{SYNC}, E_{PAR}$. $\overline{SLCS}$

Structured Monitored Linear Concurrent Systems SMoLCS are obtained by instantiating MTS with SLCS (see [AMMR],[AR]) and then by applying λ-abstraction.

<u>specfunct</u> SMoLCS ≡ λ <u>spec</u> BTS : TS(≪state≫,≪flag≫,∅)[⟶/⟹],
<u>spec</u> INF,SFLAG,MINF,MFLAG,
<u>formulas</u> $E_{SST}, E_{SYNC}, E_{PAR}, E_{MST}, E_{MON}$.
MTS(SLCS(BTS,INF,SFLAG,$E_{SST}, E_{SYNC}, E_{PAR}$)[⟹/⟹],
MINF,MFLAG,E_{MST}, E_{MON})

<u>Example CL.</u> CLMTS is an example of SMoLCS system; indeed
CLMTS = SMoLCS(PBTS,CLINF,PFLAG,NULL,NULL,$E_{CLSST}, E_{CLSYNC}, E_{CLPAR}, E_{CLMST}, E_{CLMON}$).

3. SEMANTICS.

In this paragraph we consider three kinds of semantics for transition systems by analysing the hierarchical specifications $\overline{TS}$ (see 1.1), $\overline{TREE\text{-}SEM}$ and $\overline{OBS\text{-}SEM}$.
The structure of the models of hierarchical specifications has been studied in the literature (cf. e.g. [BW2],[WPPDB]) and we will apply these results in this section. For notions of the theory of partial abstract types such as "minimally defined models","hierarchy-persistency","partial completeness","monomorphic type" we refer to [BW1] and [WPPDB].

3.1 Operational semantics.

The specification $\overline{TS}$ permits already to consider an operational semantics in which intermediate states and flags are observable.
In order to analyse the semantics of the specification of the transition system $\overline{TS}$ of the previous section we fix specifications STATE, FLAG and the set of formulas E_T and make the following technical assumptions.

- The specification FLAG is monomorphic, that is all the models of FLAG are isomorphic.
- The axioms of STATE are positive conditional formulas (see [BW2]) of the form :

$$\bigwedge_{i=1}^{n} D(p_i) \wedge \bigwedge_{j=1}^{k} (D(r_j) \wedge r_j = q_j) \supset \bigwedge_{h=1}^{m} e_h$$

  where e_h is of the form D(t) or $t_1 = t_2$; then STATE has an initial model I_{STATE} [BW2].

- The preconditions "cond" in any axiom cond ⊃ t $\overset{f}{\Longrightarrow}$ t' are of the form $\bigwedge_i (D(u_i) \wedge u_i = v_i)$ where u_i is of sort bool or flag.

Then we can consider $\overline{TS}$ as hierarchical specification with primitive sorts bool and flag and the sorts of STATE as non-primitive sorts.
If A is a specification, then Sig(A) indicates the signature of A; if Σ is a signature, srt and srt' are sorts, then $W_\Sigma|_{srt}$ and $W_\Sigma(\{srt'\ x\})|_{srt}$ indicate the set of the ground terms on Σ of sort srt and the set of the terms on Σ of sort srt with at most one variable x of sort srt'.

Proposition 1.

There exists an initial model $I_{\overline{TS}}$ of $\overline{TS}$ such that for every $s,s' \in W_{Sig(\overline{TS})}|_{state}$:

(1) $I_{\overline{TS}} \models s=s'$ iff $STATE \vdash s=s'$ or $(\overline{TS} \not\vdash D(s)$ and $\overline{TS} \not\vdash D(s'))$,

(2) $I_{\overline{TS}} \models D(s)$ iff $\overline{TS} \vdash D(s)$,

(3) $I_{\overline{TS}} \models s \xrightarrow{f} s'$ iff $\overline{TS} \vdash s \xrightarrow{f} s'$,

(4) $I_{\overline{TS}} \models \neg D(s \xrightarrow{f} s')$ iff $\overline{TS} \not\vdash s \xrightarrow{f} s'$.

There exists a weakly terminal model $Z_{\overline{TS}}$ of $\overline{TS}$ such that

for every $s,s' \in W_{Sig(\overline{TS})}|_{state}$:

(1) $Z_{\overline{TS}} \models s=s'$ iff (for every $c \in W_{Sig(\overline{TS})}(\{state\ x\})|_{state}$, $\bar{s} \in W_{Sig(\overline{TS})}|_{state}$, $f \in W_{Sig(\overline{TS})}|_{flag}$:

$$\overline{TS} \vdash c[s] \xrightarrow{f} \bar{s} \text{ iff } \overline{TS} \vdash c[s'] \xrightarrow{f} \bar{s})$$
or $(\overline{TS} \not\vdash D(s)$ and $\overline{TS} \not\vdash D(s'))$,

(2) $Z_{\overline{TS}} \models D(s)$ iff $\overline{TS} \vdash D(s)$,

(3) $Z_{\overline{TS}} \models s \xrightarrow{f} s'$ iff $\overline{TS} \vdash s \xrightarrow{f} s'$,

(4) $Z_{\overline{TS}} \models \neg D(s \xrightarrow{f} s')$ iff $\overline{TS} \not\vdash s \xrightarrow{f} s'$.

<u>Proof idea.</u> From the form of the axioms and our assumption it is easy to see that $\overline{TS}$ is hierarchy consistent and partially complete (cf. e.g. [BW2]). Then Proposition 1 and corollary 1 of [BW2] ensure the existence of initial algebras with properties (1)-(3).
Proposition 2 and corollary 4 of [BW2] ensure the existence of weakly terminal algebras with properties (1)-(3). Property (4) follows from the form of the axioms E_T which implies that $(s \overset{f}{\Rightarrow} s') = false$ can never be derived. □

Hence the provable equality between states depends only on the axioms of states and not on those of the transition system. The weakly terminal models represent an operational equality, since two states s and s' are equal in $Z_{\overline{TS}}$ iff all transitions involving s are also possible for s' and viceversa.

<u>3.2. Tree semantics.</u> We generalize now the "single step" transition relation ⇒ to its tree closure that is we consider the unordered trees of possible finite derivations from a given state.

First we define unordered finite trees with arcs labelled by flags and leaves labelled by states. The specification TRFLST is based on FLAG and STATE and defines finite unordered trees using a specification MSET(<<trflst>>) (cf. appendix) of multisets over some (auxiliary) sort trflst. The terms of sorts mset(trflst) and trflst are built by mutual recursive use of the operations on these sorts.

<u>spec</u> TRFLST ≡

<u>reachable</u>

<u>enrich</u> FLAG + STATE + MSET(<<trflst>>)

<u>by opns</u> tr: state → trflst

.o.: flag x mset(trflst) → trflst

sel1: trflst → flag, sel2: trflst → mset(trflst)

<u>axioms</u> {sel1(f o mst) = f, sel2(f o mst) = mst, D(tr(s))}

The ASL-construct <u>reachable</u> T restricts the models of T to the term generated ones.

The trees of flags and states are represented by terms of sort mset(trflst), precisely: trees of depth 0 by terms of form {tr(s)}, while trees of depth greater than 0

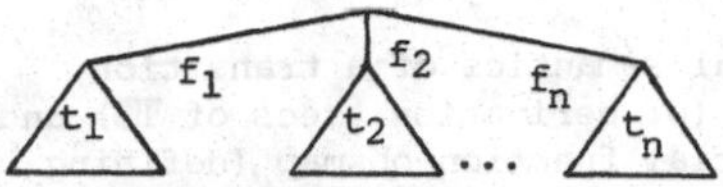

are represented by terms of form $\{f_1 \circ mst_1\}|\ldots|\{f_n \circ mst_n\}$ where for i=1,...,n mst_i represents t_i.

In the following we will write t instead of {t}.

By considering STATE and FLAG as parameters and NAT as primitive one easily proves that TRFLST is persistent wrt these specifications. Moreover, if they are monomorphic, then TRFLST is monomorphic.

By the specification $\overline{\text{TREE-SEM}}$ we give the tree semantics of a transition system $\overline{\text{TS}}$ by introducing a multistep relation $\Rightarrow$ which associates to every state a family of (unordered) trees in TREEFLST.

<u>spec</u> $\overline{\text{TREE-SEM}}$ ≡

<u>enrich</u> $\overline{\text{TS}}$ + TREEFLST <u>by</u>

<u>opns</u> $. \Rightarrow .$: state × mset(trflst) ⟶ bool

<u>axioms</u> $\{s \Rightarrow tr(s)\} \cup$

$$\{\bigwedge_{i=1}^{n} (s \xrightarrow{f_i} s_i \wedge s_i \Rightarrow mst_i) \supset s \Rightarrow f_1 \circ mst_1 |\ldots| f_n \circ mst_n \mid n \geq 1\}.$$

Since TREEFLST is a persistent extension of $\overline{\text{TS}}$ and the new axioms of $\overline{\text{TREE-SEM}}$ define just $\Rightarrow$ in a partially complete way, $\overline{\text{TREE-SEM}}$ is a persistent extension of $\overline{\text{TS}}$ and we get the following properties (by a proof similar to the one of prop. 1, (2)-(3)).

<u>Proposition 2.</u>

There exists an initial algebra $I_{\overline{\text{TREE-SEM}}}$ and a weakly terminal algebra $Z_{\overline{\text{TREE-SEM}}}$ which both satisfy the following properties:

for $A \in \{I_{\overline{\text{TREE-SEM}}}, Z_{\overline{\text{TREE-SEM}}}\}$ and for all ground terms s,s' of sort state and mst of sort mset(trflst) :

(1) $A \models D(s)$ iff $\overline{\text{TS}} \vdash D(s)$,

(2) $A \models s \xrightarrow{f} s'$ iff $\overline{\text{TS}} \vdash s \xrightarrow{f} s'$,

(3) $A \models s \Rightarrow mst$ iff $\overline{\text{TREE-SEM}} \vdash s \Rightarrow mst$.

Therefore the initial and weakly terminal algebras of $\overline{\text{TREE-SEM}}$ are just extensions of those of TS: $\overline{\text{TREE-SEM}}$ is not more abstract than $\overline{\text{TS}}$.

<u>3.3 Observational semantics.</u> For the (mathematical) observational semantics of concurrent transition systems several abstractions are possible: one may be interested in observing the input-output relation of a program or the stream of communications the program is capable to produce. In the following we formalize input-output semantics and strong equivalence semantics (in the sense of Milner [M1],[M2]) for transition systems in the framework of ASL, as a particular case of a general observational semantics of transition systems,by means of the the <u>observe</u> ASL-construct. This operator enlarges the class of the models of a specification T to all those algebras which behave like a model of T on some fixed set of observable terms; more formally, for any fixed set W of terms and for any specification T, the semantics of "<u>observe</u> T <u>by</u> W" is the set of all (classes of isomorphic) algebras A on the signature of T,s.t. there exists an algebra A_o, which is a model of T and is W-equivalent to A, i.e. such that there exist surjective assignments $v:X \rightarrow A$ and $v_o:X \rightarrow A_o$

such that for all $t,t' \in W$

$A,v \models t=t' \Leftrightarrow A_o,v_o \models t=t'$ and $A,v \models D(t) \Leftrightarrow A_o,v_o \models D(t)$.

To get the specification $\overline{\text{OBS-SEM}}$ of the observational semantics of a transition system $\overline{\text{TS}}$ we enrich first the specification $\overline{\text{TREESEM}}$ (of derivation trees of $\overline{\text{TS}}$) and a specification OBS of the observable part by a partial function obsmap (defining the observable part of a tree by means of a set of formulas E_{SEM}) and by a transition relation $\boxed{\text{obs}}\rangle$ (defining the observable transitions). Then we forget all sorts and operation symbols not involving $\boxed{\text{obs}}\rangle$ by means of the derive-construct (in particular we forget flags and the transition relations $\Rightarrow$ and $\sqsupset\!\!\rangle$ which is expressed by an injection inSig(...) of the observable signature into the signature of TREE-SEM + OBS) and abstract from these algebras by means of observe

<u>spec</u> OBS-SEM ≡ observe derive

<u>enrich</u> $\overline{\text{TREE SEM}}$ + OBS <u>by</u>

<u>opns</u> obsmap : mset(trflst) $\longrightarrow$ obs

. $\boxed{\text{obs}}\rangle$. : state × obs $\longrightarrow$ bool

<u>axioms</u> E_{SEM} ∪

$\{(D(\text{obsmap}(mst)) \wedge s \sqsupset\!\!\rangle mst) \supset s \boxed{\text{obs}}\rangle \text{obsmap}(mst)\}$

<u>by</u> inSig(Sig(STATE)∪ Sig(OBS)∪ { $\boxed{\text{obs}}\rangle$}) <u>by</u> $\{s \boxed{\text{obs}}\rangle x \mid s \in W_{Sig(STATE)}|_{state} \wedge x \in X_{obs}\}$

In the following we assume that the enrichment of $\overline{\text{TREE-SEM}}$ + OBS is hierarchy-persistent. In the minimally defined models of OBS-SEM a state is defined only if it is already defined in STATE and two states are equivalent if they cannot be distinguished using the relation $\boxed{\text{obs}}\rangle$.

<u>Proposition 3.</u>

Let OBS be a partially complete specification which is monomorphic w.r.t. fixed isomorphism classes of primitive models and assume that the enrichment of $\overline{\text{TREE-SEM}}$ + OBS is hierarchy-persistent. Then $\overline{\text{OBS-SEM}}$ has an initial algebra $I_{\overline{\text{OBS-SEM}}}$ and a weakly terminal algebra $Z_{\overline{\text{OBS-SEM}}}$ with the following properties.

Let $A \in \{I_{\overline{\text{OBS-SEM}}}, Z_{\overline{\text{OBS-SEM}}}\}$, s and o be ground terms respectively of sort state and obs:

(1) $A \models D(s)$ iff STATE $\vdash D(s)$,

(2) $A \models s \boxed{\text{obs}}\rangle o$ iff $\overline{\text{OBS-SEM}} \vdash s \boxed{\text{obs}}\rangle o$.

Moreover the equality in $I_{\overline{\text{OBS-SEM}}}$ and $Z_{\overline{\text{OBS-SEM}}}$ is defined for states s and s' with $\overline{\text{OBS-SEM}} \vdash D(s) \wedge D(s')$ as follows:

$I_{\overline{\text{OBS-SEM}}} \models s=s'$ iff s and s' are syntactically the same term,

$Z_{\overline{\text{OBS-SEM}}} \models s=s'$ iff for all ground terms o of sort obs and all contexts $c[x] \in W_{Sig(\overline{\text{OBS-SEM}})}(\{\text{state } x\})|_{state}$:

$$\overline{\text{OBS-SEM}} \vdash c[s] \boxed{\text{obs}}\rangle o \text{ iff } \overline{\text{OBS-SEM}} \vdash c[s'] \boxed{\text{obs}}\rangle o.$$

<u>Proof idea.</u> $\overline{\text{OBS-SEM}}$ can be considered as an equational specification over BOOL where the only axioms are of the form s $\boxed{\text{obs}}\rangle$ o such that s $\boxed{\text{obs}}\rangle$ o is an axiom iff s $\boxed{\text{obs}}\rangle$ o holds in all models of the enrichment of $\overline{\text{TREE-SEM}}$ + OBS. Then everything follows from the theorems in [BW2].

□

By considering $\overline{\text{TS}}$, OBS and E_{SEM} as parameter we get the following schema:

<u>specfunct</u> OBS-SEM ≡ λ <u>spec</u> $\overline{\text{TS}}$, OBS, <u>formulas</u> E_{SEM}. $\overline{\text{OBS-SEM}}$

Different semantics of transition systems can be obtained by instantiating the parameterized specification OBS-SEM with different sets of formulas E_{SEM}.
The specifications IO-SEM and SE-SEM define respectively the input-output and strong equivalence ([M1],[M2]) semantics.

Input-output semantics. For the IO-semantics we need the notion of final (or normal) state and of observable result of a final state which will be represented by the operations

normal : state $\longrightarrow$ bool
res : state $\longrightarrow$ result

where result is some observable sort.

specfunct IO-SEM $\equiv \lambda$ spec T : TS(≪state≫,≪flag≫,∅),
spec RESULT : C .
OBS-SEM(T,SET(RESULT),E_{IO})

where C is the requirement specification

sorts state,result,bool
opns normal : state $\longrightarrow$ bool
res : state $\longrightarrow$ result
axioms normal(s) ⊃ D(res(s)) and where

E_{IO} = {normal(s)=true ⊃ obsmap(tr(s))={res(s)} ,
normal(s)=false ⊃ obsmap(tr(s))=∅ ,
, $s \overset{f}{\Longrightarrow} s'$ ⊃ normal(s)=false } ∪
{obsmap(f_1 o mst_1 |...|f_n o mst_n)=obsmap(mst_1) ∪... ∪obsmap(mst_n) | n ≥1}

Then in any weakly terminal model $Z_{\overline{IO}}$ of an instantiation $\overline{\text{IO-SEM}}$, two defined states s and s' are equal iff they lead to the same result in any context c:

$Z_{\overline{IO}}$ ⊨ s=s' iff for all ground terms r of sort result
for all contexts c [x] ∈ $W_{Sig(\overline{\text{IO-SEM}})}$({state x}) $|_{state}$:
$\overline{\text{IO-SEM}}$ ⊢ c [s] obs r iff $\overline{\text{IO-SEM}}$ ⊢ c [s'] obs r.

Termination is ignored in this specification. It could be added e.g. by introducing a function stop : state $\longrightarrow$ bool.

Example CL. To define an IO semantics for the language CL we consider the transition system CLMTS (defined in section 2.3). First to every program of CL a state of CLMTS is associated by a "representation" function R defined by:

R(program shared var sv_1,...,sv_n $proc_1$‖...‖$proc_m$ end program) =
<<PR($proc_1$)|...|PR($proc_m$),<empty [vl_o/sv_1] ... [vl_o/sv_n],∅>>,null>

PR(process p var v_1,...,v_k stat end process) =
<p,stat,empty [vl_o/v_1] ... [vl_o/v_k] >.

Then we get an IO semantics for CLMTS by instantiating the parameterized specification IO-SEM with CLMTS and a specification CLRES of admissible results. These are chosen to be the final states of the shared memory:CLRES is SMEM + CLMTS enriched by operations "normal" and "res" defined by normal(<<mps,i>,null>)=(mps=∅) and res(<<mps,<sm,tr>>,null>) = sm.

Then we define two programs $prog_1$ and $prog_2$ of CL to be equivalent iff R($prog_1$) and R($prog_2$) coincide in the weakly terminal models of IO-SEM(CLMTS,CLRES). That means that $prog_1$ and $prog_2$ are equivalent iff for every context c [x] of sort state c [R($prog_1$)] and c [R($prog_2$)] yield the same results. But the form of R($prog_i$) implies that proper contexts c [x] of R($prog_i$) do not exist. Hence $prog_1$ and $prog_2$ are equivalent iff R($prog_1$) and R($prog_2$) yield the same set of results.

Strong equivalence semantics. For the specification of strong equivalence semantics we need the specification TREEFL of trees of flags which are the same as trees of flags and states in TREEFLST (where states are the leaves of unordered trees) except that now flags are the labels of leaves and ares.
Given a tree of flags and states t in TREE-SEM, t^- indicates the associated tree of flags (i.e. without its leaves). Then we observe t^- from a state s

s $\boxed{obs}\!\!>$ t^- iff s $\boxed{\ }\!\!>$ t.

spec $\overline{\text{SE-SEM}}$ ≡ OBS-SEM($\overline{\text{TS}}$, TREEFL, E_{SE})

where

$$E_{SE}=\{obsmap(f_1 \circ tr(s_1)\,|\ldots|\,f_m \circ tr(s_m)\,|\,f_{m+1} \circ mst_{m+1}\,|\ldots|\,f_{m+n} \circ mst_{m+n})=$$
$$tr(f_1)\,|\ldots|\,tr(f_n)\,|\,f_{m+1} \circ obsmap(mst_{m+1})\,|\ldots|\,f_{n+m} \circ obsmap(mst_{m+n})\ \ |n+m \geq 1\}.$$

As expected in the weakly terminal models $Z_{\overline{SE}}$ of the specification $\overline{\text{SE-SEM}}$, defined states s and s' are equal iff in all contexts they accept the same communication trees.

$Z_{\overline{SE}} \models$ s=s' iff for all ground terms mtf of sort mset(trfl),
for all contexts $c[x] \in W_{Sig(\overline{SE-SEM})}(\{state\ x\})\big|_{state}$:
$\overline{\text{SE-SEM}} \vdash c[s]$ $\boxed{obs}\!\!>$ mtf iff $\overline{\text{SE-SEM}} \vdash c[s']$ $\boxed{obs}\!\!>$ mtf.

Concluding remarks. In this paper we have mainly intended to provide a precise formal support for a methodology of specification of concurrent systems, supporting abstraction and modularity. Some work has already been done for giving properties of richer subtheories, for example relating different execution modes (interleaving vs. parallel); but much work has still to be done in that direction.
An interesting topic could be to expose the exact relationship between our approach, the SOS methodology [P], Mosses' ASA's [Mo] and the algebraic approach to concurrency in [BW1]. Here we give only some hints.
In the SOS methodology transition systems are specified using derivation rules which correspond to axioms of the form

$$\text{(o)}\qquad \bigwedge_i s_i \xrightarrow{f_i} s_i' = true \supset s \xrightarrow{f} s' = true\ .$$

One or more such derivations are associated to every construct of a programming language yielding a term model which is an initial model for the axioms (o) in our sense. The main methodological difference to our approach seems to be that Plotkin's approach is more oriented to the syntax of programming languages and that no semantic equations between concurrent constructs are postulated.
In that way our approach resembles more to Mosses' Abstract Semantic Algebras [Mo] where equational algebraic specifications are associated to many concepts of programming languages. Methodologically Mosses does not specify transition systems but gives equations which hold in denotational semantics; as technical consequence the underlying semantics of specifications are initial models of continuous algebras.
From the technical point of view our approach is strongly related to the algebraic specification of CSP in [BW1]. We have applied the techniques of [BW1] (such as transition systems, partial abstract types, the analysis of their models and the consideration of different abstractions from operational semantics) to the SMoLCS methodology of operational specification of concurrent systems. This was made possible by the use of the specification language ASL in which our general specification schemata of concurrent transition systems as well as the transition from operational semantics to more abstract mathematical semantics could be formulated.

REFERENCES

[AMMR] E.Astesiano,F.Mazzanti,U.Montanari,G.Reggio: Design and formal specification of a family of distributed architectures. Quaderni CNET n. 88,ETS Pisa,1983.

[AMRZ] E.Astesiano,F.Mazzanti,G.Reggio,E.Zucca: Formal specification of a concurrent architecture in a real project. To appear in Proc. of ICS'85 ACM (International Computing Symposium),1985.

[AR] E.Astesiano,G.Reggio: A unifying viewpoint for the constructive specification of cooperation, concurrency and distribution. Quaderni CNET n.115, ETS PISA ,1983.

[BaWo] F.L.Bauer,H.Wossner: Algorithmic language and program development, Springer Verlag Berlin, 1982.

[BW1] M.Broy,M.Wirsing: On the algebraic specification of finitary infinite communicating sequential processes. In Proc. IFIP TC2 Working Conference on "Formal Description of Programming Concepts II", Garmisch 1982.

[BW2] M.Broy,M.Wirsing: Partial abstract types. Acta Informatica 18, 47-64, 1982.

[BW3] M.Broy,M.Wirsing: Generalized heterogeneous algebras and partial interpretations. In Proc. CAAP 83, Lecture Notes in Computer Science n. 159, Springer Verlag Berlin, 1983.

[M1] R.Milner: A calculus of communicating systems. Lecture Notes in Computer Science n.92, Springer Verlag Berlin, 1980.

[M2] R.Milner: Calculi for synchrony and asynchrony. TCS 25,267-310, 1983.

[Mo] P.Mosses: A basic abstract semantic algebra. In Proc. Semantics of Types, Lecture Notes in Computer Science,Springer Verlag,Berlin, 1984.

[P] G.Plotkin: A structural approach to operational semantics. Lecture notes, Aarhus University, 1981.

[P1] G.Plotkin: An operational semantics for CSP. In Proc. IFIP TC2 Working Conference on "Formal Description of Programming Concepts II", Garmisch 1982.

[SW] D.T.Sannella,M.Wirsing: A kernel language for algebraic specification and implementation. In Proc. Int. Conf. on Foundations of Computation Theory, Borgholm Sweden, Lecture Notes in Computer Science n.158,1983.

[W] M.Wirsing: Structured algebraic specifications: a kernel language. Habilitation thesis, Technische University Munchen, 1983.

[WPPDB] M.Wirsing,P.Pepper,H.Partsch,W.Dosch,M.Broy: On the hierarchies of abstract types. Acta Informatica 20,1-33, 1983.

Appendix.

```
specfunct PRODCOND ≡ λ spec A,B,formulas COND .
                     reachable enrich A+B by sorts prod
                                             opns <.,.> :a×b ⟶ prod
                                             axioms COND

specfunct MAP ≡ λspec IND,VAL.
   reachable enrich IND+VAL by
      sort map  opns empty:               ⟶ map   axioms{i≠j ⊃ m[v/j](i)=m(i),
                     .(.):map×ind         ⟶ val          m[v/i](i)=v,
                     .[./.]:map×val×ind   ⟶ map          ¬D(empty(i))}
```

```
specfunct MSET ≡ λspec ELEM.
          enrich ELEM+NAT by
   sort mset(elem)
   opns {.}: elem                        → mset(elem)
     .-.,.|.: mset(elem)× mset(elem)     → mset(elem)
        card: mset(elem) x elem → nat, ∅: → mset(elem)
   axioms { ms|(ms'|ms'') = (ms|ms') |ms'', ms|ms' = ms'|ms,
            card({e},e) = 1,
            e ≠ e' ⊃ card({e},e') = 0,
            card(ms|ms',e) = card(ms,e) + card(ms',e),
            ms-(ms'|ms'') = (ms-ms') - ms'',
            (ms|{e}) - {e} = ms,
            e ≠ e' ⊃ (ms|{e}) - {e'} = (ms-{e'})|{e},
            ∅|ms = ms, ms - ∅ = ms, ∅ - ms = ∅, card(∅,e) = 0}
```

THE SEMANTICS OF SHARED SUBMODULES SPECIFICATIONS*

E.K. Blum and F. Parisi-Presicce
Department of Mathematics
University of Southern California
Los Angeles, CA 90089-1113

ABSTRACT. After reviewing the concept of module specification with import and export interfaces introduced by H. Ehrig for the modular development of software systems, precise definitions of submodule and union of modules specifications are given along with some basic results on their compatibility and semantics. The notion of amalgamated sum is used for the semantics of unions of modules and some connections are made with parametrized specifications. The results are restricted to the basic algebraic case.

1. INTRODUCTION

In [2], a new algebraic specification concept, called a "module", was introduced for the modularization of software systems. A module is an abstract data type equipped with an import interface and an export interface: the import interface represents the operations available (and previously specified) inside the module while the export interface consists of the operations available to the user of the module. The two interfaces are combined in the body of the module, whose operations not in the export interface are considered "hidden". The interfaces are allowed to share a common parameter part.

The focus in [2] is on operations of composition of modules and actualization of parameterized modules and their interaction. It is obvious that these operations do not suffice for the construction of more complicated modules from simpler ones and that , in practical languages such as Ada, the union operation is an important one. The mere formation of the union of two

*This research was supported in part by the National Science Foundation under Grant MCS 82-03666.

disjoint modules poses no algebraic difficulties, nor does the formation of the union of modules which share a common part, provided we are willing to duplicate this common part. In many situations, however, having two copies of a common part leads to difficulties in using the union module in a wider context requiring importing or exporting the common part. Therefore, we need a kind of union which allows us to avoid such duplication.

In this paper, we describe, in increasing order of generality, the union of modules which share the parameter part, a subparameter part and finally a complete submodule. This last situation requires the precise definition of submodule specification and its semantics, which are given in Section 4. The results, in Section 5, relating the semantics of the union module to those of its components are described in terms of amalgamated sums of algebras. This notion is introduced in Section 2, where we present a constructive definition of the amalgamated sum $A1 +_{A0} A2$ of two algebras A1 and A2 with respect to a third algebra A0. The amalgamated sum is then related to pullback constructs in the category of SPEC-algebra categories and to pushouts in a new category <u>UAlg</u>. In Section 3, we briefly illustrate the connection between amalgamated sums and loose semantics of standard parameter passing in parameterized data types. Further developments pointing the way to an algebra of modules are outlined in Section 6.

2. AMALGAMATED SUM

In this section, we introduce the concept of <u>amalgamated sum</u> and relate its constructive definition to other well known constructions in the framework of category theory. Let SPECi, i = 0,1,2,3, be algebraic specifications and **SPECi** the corresponding categories of SPECi-algebras. Also let f_i: SPEC0 → SPECi and g_i: SPECi → SPEC3, for i = 1,2, be specification morphisms and Ui and Vi the forgetful functors associated with the specification morphisms g_i and f_i respectively, such that (1) is a pushout diagram in the category of algebraic specifications and (2) is a pullback diagram in the category of SPEC-algebras categories and forgetful functors.

$$\begin{array}{ccc} SPEC0 & \xrightarrow{f1} & SPEC1 \\ {\scriptstyle f2}\downarrow & (1) & \downarrow{\scriptstyle g1} \\ SPEC2 & \xrightarrow[g2]{} & SPEC3 \end{array} \quad . \quad \begin{array}{ccc} \mathbf{SPEC0} & \xleftarrow{V1} & \mathbf{SPEC1} \\ {\scriptstyle V2}\uparrow & (2) & \uparrow{\scriptstyle U1} \\ \mathbf{SPEC2} & \xleftarrow[U2]{} & \mathbf{SPEC3} \end{array}$$

2.1 Definition (Amalgamated Sum)

Given algebras $Ai \in$ **SPECi** for $i = 0,1,2$ with the property that $V1(A1) = A0 = V2(A2)$, the amalgamated sum $A1+_{A0}A2$ of A1 and A2 with respect to A0 is the unique SPEC3-algebra A3 satisfying the following conditions on sorts $s \in S3$ and operators $\sigma \in \Sigma 3$

$$(A3)_s = \begin{cases} A1_{s1} & \text{if } g1(s1) = s \\ \\ A2_{s2} & \text{if } g2(s2) = s \end{cases}$$

$$\sigma_{A3} = \begin{cases} \sigma1_{A1} & \text{if } g1(\sigma1) = \sigma \\ \\ \sigma2_{A2} & \text{if } g2(\sigma2) = \sigma \quad . \end{cases}$$

The algebra A3 is well-defined, since SPEC3 is the pushout of SPEC1 and SPEC2. Thus if $s \in S3$, then $g1(s1) = s$ for some $s1 \in S1$ or $g2(s2) = s$ for some $s2 \in S2$ and if both are true, then there exists $s0 \in S0$ such that $fi(s0) = si$, $i = 1,2$ in which case $(A1)_{s1} = (A0)_{s0} = (A2)_{s2}$. A similar argument shows that σ_{A3} is also well defined and that, in fact, $A1+_{A0}A2$ is a SPEC3-algebra.

The amalgamated sum $A1+_{A0}A2$ can also be defined implicitly in terms of the pullback diagram of the SPECi-algebra categories.

2.2 Lemma (Pullback Property)

Given $Ai \in$ **SPECi**, $i = 0,1,2$, with $V1(A1) = A0 = V2(A2)$, the amalgamated sum $A3 = A1+_{A0}A2$ is the unique SPEC3-algebra such that $U1(A3) = A1$, $U2(A3) = A2$ and if $B \in$ **SPEC** for an algebraic specification SPEC with (forgetful) functors Fi: **SPEC** $\rightarrow$ **SPECi**, $i = 1,2$, $V1 \cdot F1 = V2 \cdot F2$ and $Fi(B) = Ai$, then there exists a unique (forgetful) functor F: **SPEC** $\rightarrow$ **SPEC3** such that $Fi = Ui \cdot F$, $i = 1,2$, and $F(B) = A3$.

2.3 Corollary

SPEC3 = **SPEC1** $+_{\text{SPEC0}}$ **SPEC2** = $\{A1 +_{A0} A2$: $Ai \in$ **SPECi**, $V1(A1) = A0 = V2(A2)\}$

The amalgamated sum $A1 +_{A0} A2$ can also be viewed as the pushout of A1 and A2 w.r.t. A0 in the appropriate category. In order to define this category, we need the notion of "generalized homomorphism".

2.4 Definition (Generalized Homomorphism)

Let SPECi be algebraic specifications and Ai be SPECi-algebras for $i = 0,1$. A generalized homomorphism from (A0, SPEC0) to (A1, SPEC1), denoted by (h,f): (A0, SPEC0) $\to$ (A1, SPEC1) is a pair of functions (h,f) where f: SPEC0 $\to$ SPEC1 is a specification morphism and h is a family $h_s: A0_s \to A1_{f(s)}$ of functions indexed by the set S0 of sorts in SPEC0 such that, for every σ: s1x ... xsn $\to$ s in $\Sigma 0$, the following diagram commutes:

$$\begin{array}{ccc} A0_{s1} \times \dots \times A0_{sn} & \xrightarrow{\sigma_{A0}} & A0_s \\ \downarrow {\scriptstyle h_{s1} \times \dots \times h_{sn}} & & \downarrow {\scriptstyle h_s} \\ A1_{f(s1)} \times \dots \times A1_{f(sn)} & \xrightarrow{f(\sigma)_{A1}} & A1_{f(s)} \end{array}$$

Given f: SPEC0 $\to$ SPEC1, we denote by $GENHOM_f(A0,A1)$ the set of all functions h such that (h,f): (A0, SPEC0) $\to$ (A1, SPEC1) is a generalized homomorphism.

Notice that any h $\in GENHOM_f(A0,A1)$ is also a SPEC0-morphism from A0 to $V_f(A1)$ since $h_s: A0_s \to A1_{f(s)} = V_f(A1)_s$ and $\sigma_{V_f(A1)}$ is defined as $f(\sigma)_{A1}$. Conversely, if h $\in$ **SPEC0**$(A0,V_f(A1))$, then $h_s: A0_s \to V_f(A1)_s = A1_{f(s)}$ and the above diagram commutes since V_f is a functor. We have just proved the first part of the following result.

2.5 Proposition

$GENHOM_f(A0,A1) \cong$ **SPEC0**$(A0,V_f(A1)) \cong$ **SPEC1**$(F_f(A0),A1)$.

The second isomorphism follows from general properties of the forgetful functor V_f and its left adjoint, the free functor F_f.

2.6 Definition

Given generalized homomorphisms (h0,f0): (A0,SPEC0) $\to$ (A1,SPEC1) and (h1,f1): (A1,SPEC1) $\to$ (A2,SPEC2), the composition is given by the pair (h1 $\cdot$ h0, f1 $\cdot$ f0) and it is clear, either directly from Definition 2.4 or using Proposition 2.5, that it is again a generalized homomorphism. It is also clear that the composition of generalized homomorphisms, when defined, is associative. We denote by UAlg the category with objects the pairs (A, SPEC) with A $\in$ **SPEC** and morphisms the generalized morphisms (h,f): (A0,SPEC0) $\to$ (A1,SPEC1). We can now state the main result of this section.

2.7 Proposition

Given specification morphisms fi: SPEC0 → SPECi and SPECi-algebras Ai, i = 0,1,2, satisfying V1(A1) = A0 = V2(A2), the amalgamated sum $A1+_{A0}A2$ is the pushout of A1 and A2 w.r.t. A0, that is, the diagram

$$\begin{array}{ccc} (A0,SPEC0) & \xrightarrow{(h1,f1)} & (A1,SPEC1) \\ {\scriptstyle (h2,f2)}\downarrow & & \downarrow{\scriptstyle (k1,g1)} \\ (A2,SPEC2) & \xrightarrow[(k2,g2)]{} & (A1+_{A0}A2,SPEC3) \end{array}$$

is a pushout diagram in the category UAlg, where hi and ki are the obvious inclusion maps satisfying $hi_s(A0_s) = Ai_{fi(s)}$ and $ki_{si}(Ai) = (A1+_{A0}A2)_{gi(si)}$.

Remark Notice that if SPEC0 = ϕ in diagram (1), then SPEC3 is the disjoint union of SPEC1 and SPEC2 and every algebra A3 ∈ **SPEC3** is the disjoint union of an algebra A1 ∈ **SPEC1** and an algebra A2 ∈ **SPEC2**.

3. PARAMETERIZED DATA TYPES AND AMALGAMATED SUMS

The notion of parameterized data type is an important one in the hierarchical design of large programming systems. While several authors ([1,4,6]) have considered the problems of abstract data type specifications and implementations, the problem of parameter passing has not received as much attention ([3,5]). Here, we look at amalgamated sums of algebras as a "constructive" parameter passing technique. We take the following definition from [5].

3.1 Definition (Parameterized Data Type)

A parameterized data type PDT = (SPEC0, SPEC1, T) consists of two algebraic specifications SPEC0 and SPEC1 with SPEC0 ⊆ SPEC1 (componentwise) and a functor T: **SPEC0** → **SPEC1** which is assumed to be strongly persistent, i.e. V(T(A)) = A for every A ∈ **SPEC0**, where V is the forgetful functor associated with the inclusion specification morphism j: SPEC0 → SPEC1.

In the case of initial algebra semantics [5,6], the functor T is taken to be the free functor F: **SPEC0** → **SPEC1**. In order to pass an actual parameter specification SPEC2 for the parameter part SPEC0 of a parameterized specification PSPEC = (SPEC0,SPEC1), a "parameter passing" morphism h: SPEC0 → SPEC2 is specified and a new specification SPEC3 is constructed as in the following pushout diagram:

$$\begin{array}{ccc} SPEC0 & \xrightarrow{j} & SPEC1 \\ {\scriptstyle h}\downarrow & & \downarrow{\scriptstyle h'} \\ SPEC2 & \xrightarrow{j'} & SPEC3 \end{array}$$

(In [5], the construction of the new specification is explicit).
The semantics of this standard (i.e. non-parameterized) parameter passing is taken to be $(F, T_{SPEC2}, T_{SPEC3})$, where T_{SPECi} is the initial algebra in **SPECi** and F: **SPEC0** → **SPEC1** is the free functor of the parameterized specification PSPEC. The assumption that F be strongly persistent is then sufficient to guarantee the <u>semantical conditions:</u>

1) actual parameter protection: $V_{j'}(T_{SPEC3}) = T_{SPEC2}$
2) passing compatibility: $F(V_h(T_{SPEC2})) = V_{h'}(T_{SPEC3})$.

In the loose semantics case, the result of passing a SPEC2-algebra as actual parameter can be expressed as an amalgamated sum of algebras. Let again PSPEC, SPEC2, h: SPEC0 → SPEC2 and SPEC3 be given as above. Let $A2 \in$ **SPEC2** and define $A0 = V_h(A2)$. Then $A1 = T(A0)$ is a SPEC1-algebra with the property that $V(A1) = A0$ (by strong persistency of T). Since A0, A1 and A2 satisfy the assumptions of Definition 2.1, we can define their amalgamated sum $A3 = A1 +_{A0} A2$. Then $A3 = T(V_h(A2)) +_{V_h(A2)} A2$ is the result of passing A2 to the parameterized data type PDT = (SPEC0,SPEC1,T). It is easy now to check that by the properties of the amalgamated sum, similar semantical conditions are satisfied. The actual parameter A2 is protected since

$$V_{j'}(A3) = V_{j'}(T(A0) +_{A0} A2) = A2$$

and the parameter passing in "compatible", i.e. it reflects the behavior of the functor T, since $T(V_h(A2)) = T(A0) = A1 = V_{h'}(A3)$. Hence **SPEC1** $+_{\mathbf{SPEC0}}$ **SPEC2** (see Corollary 2.3) can be taken as the loose semantics of (standard parameter) passing SPEC2 for SPEC0 in PDT.

4. MODULE AND SUBMODULE SPECIFICATIONS

In this section, we first review the basic notions of module specification with import and export interfaces as introduced by Ehrig ([2]) briefly mentioning the operations of composition and actualization and their semantics. We then introduce the notion of submodule specification and semantics to be used in the next section in the context of unions of modules sharing a common part.

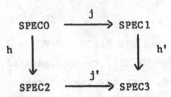

4.1 Definition (Module Specification)

A module specification MOD consists of four algebraic specifications PAR, IMP, EXP, BOD along with specification morphisms e, s, i and v (e and s injective) making the following diagram commute:

$$
\begin{array}{ccc}
\text{PAR} & \xrightarrow{e} & \text{EXP} \\
{\scriptstyle i}\downarrow & & \downarrow{\scriptstyle v} \\
\text{IMP} & \xrightarrow{s} & \text{BOD}
\end{array}
$$

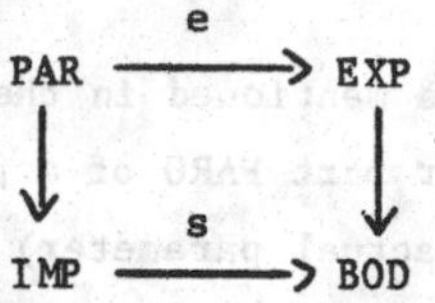

IMP and EXP are the import and export interfaces, respectively, and PAR is the parameter part shared by IMP and EXP. We will assume that e and s are actually inclusions.

4.2 Definition (Semantics of Modules)

Given a module specification MOD as in Definition 4.1, denote by V_s, V_v and V_e the forgetful functors induced by s, v and e, respectively, and by FREE: IMP → BOD the free functor associated with V_s.

The (unrestricted) semantics SEM of MOD is the functor

$$SEM = V_v \cdot FREE: IMP \to EXP.$$

The restriction semantics RSEM of MOD is the functor

$$RSEM = R \cdot SEM: IMP \to EXP$$

where, for $A \in$ EXP, $R(A) = \cap\{B \in EXP: B \subseteq A, V_e(B) = V_e(A)\}$.

Assumptions Using the unrestricted semantics SEM, we will assume that FREE is strongly persistent (i.e. $V_s \cdot$ FREE is the identity on IMP). Using RSEM, we will add the assumption that FREE preserves injective homomorphisms. For a discussion of the interpretation of both definitions, see [2].

In the composition of two modules the import interface of one module is "matched" with the export interface of the other one.

4.3 Definition (Composition of Modules)

Given two modules specifications MODi = (PARi, EXPi, IMPi, BODi) with a specification morphism h: IMP1 → EXP2, the composition of MOD1 and MOD2 w.r.t. h, denoted by MOD2 $\cdot_h$ MOD1, is the module specification MOD3 = (PAR3, EXP1, IMP2, BOD3) with PAR3 and BOD3 defined as in the diagram

$$
\begin{array}{ccccc}
\text{PAR3} & \longrightarrow & \text{PAR1} & \longrightarrow & \text{EXP1} \\
\downarrow & {\scriptstyle (1)} & \downarrow & & \downarrow \\
\text{PAR2} & \longrightarrow & \text{IMP1} & \longrightarrow & \text{BOD1} \\
\downarrow & & \downarrow & {\scriptstyle (2)} & \downarrow \\
\text{IMP2} & \longrightarrow & \text{EXP2} & & \\
 & & \downarrow & & \downarrow \\
 & & \text{BOD2} & \longrightarrow & \text{BOD3}
\end{array}
$$

where (1) and (2) are a pullback and a pushout diagram respectively.

4.4 Theorem (Semantics of Composition)

i) $SEM3 = SEM1 \circ V_h \circ SEM2$

ii) If h: IMP1 → EXP2 is "parameter consistent", i.e. there exists p: PAR1 → PAR2 such that $e_2 \circ p = h \circ i_1$, then $RSEM3 = RSEM1 \circ V_h \circ RSEM2$.

The other operation on modules mentioned in the introduction is that of actualization, where the parameter part PAR0 of a parametrized module MOD0 is replaced by a specification ACT (actual parameter) to yield a parameterless module specification. The actualization of MOD0 by ACT w.r.t a specification morphism h: PAR0 → ACT is the parameterless module $ACT_h(MOD0) = (\phi, EXP, IMP, BOD)$ where EXP (resp. IMP) is the pushout of ACT and EXP0 (resp. IMP0) w.r.t. PAR0 and BOD is obtained by "gluing" IMP and BOD0. For the precise definition and results dealing with the induced semantics of actualization and compatibility properties of composition and actualization, see [2].

We now introduce the concept of submodule specification. As in [2], we restrict our attention to the basic algebraic case, without logical or algebraic constraints on the interfaces.

4.5 Definition (Submodule Specification)

Given two module specifications MODi = (PARi, EXPi, IMPi, BODi) for i = 1,2, MOD1 is a submodule specification of MOD2 if there exist four specification morphisms m_p: PAR1 → PAR2, m_e: EXP1 → EXP2, m_i: IMP1 → IMP2 and m_b: BOD1 → BOD2 such that the following four diagrams commute:

$$\begin{array}{ccc} PAR1 & \xrightarrow{e_1} & EXP1 \\ {\scriptstyle m_p}\downarrow & (1) & \downarrow{\scriptstyle m_e} \\ PAR2 & \xrightarrow[e_2]{} & EXP2 \end{array} \qquad \begin{array}{ccc} PAR1 & \xrightarrow{i_1} & IMP1 \\ {\scriptstyle m_p}\downarrow & (2) & \downarrow{\scriptstyle m_i} \\ PAR2 & \xrightarrow[i_2]{} & IMP2 \end{array} \qquad \begin{array}{ccc} IMP1 & \xrightarrow{s_1} & BOD1 \\ {\scriptstyle m_i}\downarrow & (3) & \downarrow{\scriptstyle m_b} \\ IMP2 & \xrightarrow[s_2]{} & BOD2 \end{array} \qquad \begin{array}{ccc} EXP1 & \xrightarrow{v_1} & BOD1 \\ {\scriptstyle m_e}\downarrow & (4) & \downarrow{\scriptstyle m_b} \\ EXP2 & \xrightarrow[v_2]{} & BOD2 \end{array}$$

Assumptions We have already assumed that, in module specifications, the free functor FREEi: **IMPi** → **BODi** is strongly persistent (when using unrestricted semantics) or strongly conservative (with restriction semantics). In the case of submodule specification, we will add the condition that the free functors FREE1 and FREE2 commute with the vertical forgetful functors of diagram 3, i.e. $V_{m_b} \circ FREE2 = FREE1 \circ V_{m_i}$. This formalizes our intuitive notion that, for MOD1 to be a submodule of MOD2, the free construction in MOD1 should reflect the free construction in MOD2. When using the restriction semantics, we will also add the assumption that, for every EXP2-algebra A, $V_{m_e}(R2(A)) = R1(V_{m_e}(A))$. These assumptions are sufficient to relate the semantics of MOD1 and MOD2.

4.6 Proposition (Submodule Semantics)

Given module specifications MOD1 and MOD2 with MOD1 a submodule specification of MOD2 and the above assumptions on the behavior of the forgetful functors, we have

i) $V_{m_e} \cdot SEM2 = SEM1 \cdot V_{m_i}$

ii) $V_{m_e} \cdot RSEM2 = RSEM1 \cdot V_{m_i}$

where SEMi, RSEMi: **IMPi → EXPi**, V_{m_e}: **EXP2 → EXP1** and V_{m_i}: **IMP2 → IMP1.**

Remarks If in Definition 4.5 we take PAR1 = PAR2, then the notion of MOD1 being a submodule specification of MOD2 is equivalent to MOD2 being a "refinement" of MOD1 with the additional specification morphism m_b: BOD1 → BOD2 (see [2] sec. 4.5). In view of Proposition 4.6, our assumptions on submodule specifications imply Ehrig's notions of "correct" and "R-correct" refinements. If in addition we take IMP1 = IMP2 and diagram (4) as a pushout, then we obtain a special case of an "extension" of module specification as in 4.6 of [2].

5. UNION OF MODULES WITH SHARED SUBMODULES

As mentioned already in the Introduction, composition and actualization are but two of the operations that can be used to build up complex modules from simpler ones. Another possible construction, allowed, for example, in Ada, is that of a union of two (or more) modules. A larger module can be obtained whose import and export interfaces are formed by combining the import and export interfaces of the component modules, respectively. The simplest possible combination is that of a union of two disjoint modules or, equivalently, of two modules which share a common part, may it be a submodule or just part of an interface, and we are willing to duplicate that part in the composite module. There are instances, however, where two modules share a common part, say the parameter part, which should not be duplicated since PAR is intended to be instantiated, at a later stage in the development, with the same actual parameter. This is the situation we analyze next.

5.1 Definition (Union of Modules with Shared Parameter)

The union of two modules specifications MODi = (PAR, EXPi, IMPi, BODi) for i = 1,2, which share the parameter part PAR, is denoted by MOD1 $+_{PAR}$ MOD2 and is the module specification MOD3 = (PAR, EXP3, IMP3, BOD3) where the last three specifications are given by the pushout diagrams:

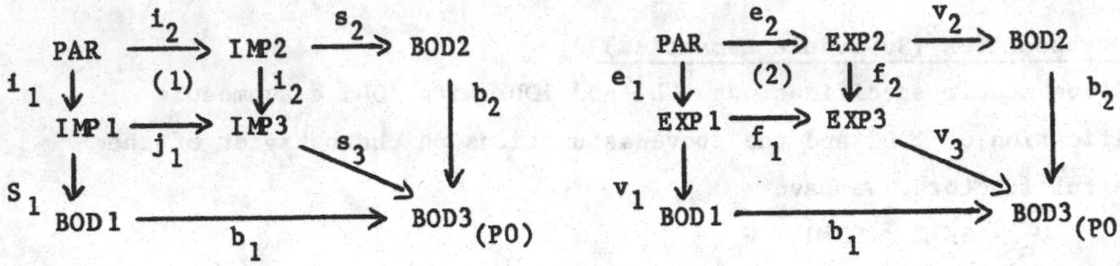

By definition of MOD1 and MOD2, the two outer diagrams are the same and they define BOD3 as a pushout w.r.t. PAR. Since (1) and (2) are pushouts, s_3 and v_3 exist and are unique. They are also injective and $v_3 \cdot e_3 = b_2 \cdot v_2 \cdot e_2 = b_2 \cdot s_2 \cdot i_2 = s_3 \cdot i_3$.

The following Lemma is needed to prove Theorem 5.3.

5.2 Lemma

Given the diagram

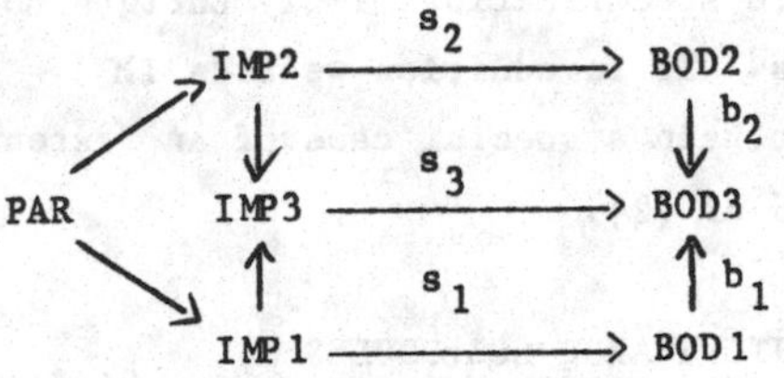

with IMP3 and BOD3 as in Definition 5.1, let Vi be the forgetful functor associated with si and Fi be the corresponding free functor. Define $F = F1 +_{PAR} F2$: IMP3 → BOD3 by letting $F(I1 +_P I2) = F1(I1) +_P F2(I2)$. Then F = FREE3 is the free functor associated with s_3 and if F1 and F2 are strongly persistent (resp. conservative), then so is F.

5.3 Theorem (Semantics of Union with Shared Parameter)

Given the module specification $MOD3 = MOD1 +_{PAR} MOD2$ as in Definition 5.1, its semantics are given by

i) $SEM3 = SEM1 +_{PAR} SEM2$

ii) $RSEM3 = RSEM1 +_{PAR} RSEM2$

where $(SEM1 +_{PAR} SEM2)(I1 +_P I2) = SEM1(I1) +_P SEM2(I2)$

and $RSEM1 +_{PAR} RSEM2$ is defined similarly.

The next situation we consider is that of a union of two module specifications MODi = (PARi, EXPi, IMPi, BODi) for i = 1,2 where PAR1 and PAR2 share a common subparameter part PAR0 which should not be duplicated in the union. The sharing of this common subparameter is indicated by two specification morphisms pi: PAR0 → PARi for i = 1,2.

5.4 Definition (Union of Modules with Shared Subparameter)

Given module specifications MODi = (PARi, EXPi, IMPi, BODi) for i = 1,2 and a specification PAR0 with specification morphisms pi: PAR0 → PARi, the union MOD1 $+_{PAR0}$MOD2 of MOD1 and MOD2 w.r.t. PAR0 is the module specification MOD3 = (PAR3, EXP3, IMP3, BOD3) where PAR3 is defined as the pushout of PAR1 and PAR2 w.r.t. PAR0, and EXP3, IMP3 and MOD3 are obtained as in Definition 5.1 with PAR0 replacing PAR, e.g.

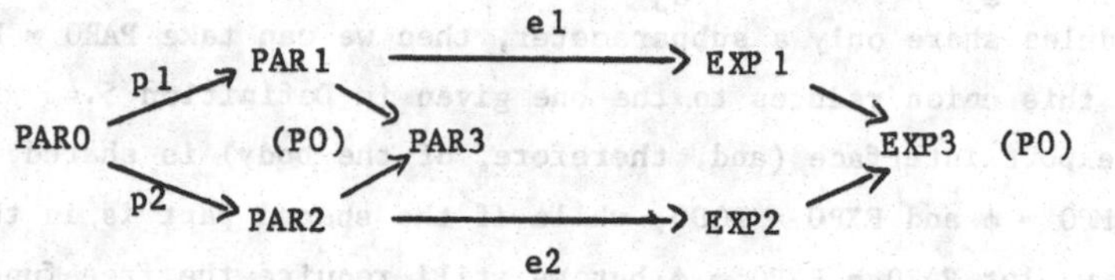

(3)

If PAR0 = PAR1 = PAR2, we are back to the case of Shared Parameter, while if PAR0 = ϕ we have disjoint union. The same arguments can be used to show that the diagram of MOD3 commutes and that FREE3: IMP3 → BOD3 is nothing more than FREE1 $+_{PAR0}$FREE2 and is again strongly persistent (conservative) whenever FREE1 and FREE2 are.

5.5 Theorem (Semantics of Union with Shared Subparameters)

The unrestricted and restriction semantics SEM3 and RSEM3, respectively, of MOD3 = MOD1 $+_{PAR0}$MOD2 as in Definition 5.4 are given by

i) SEM3 = SEM1 $+_{PAR0}$SEM2 and

ii) RSEM3 = RSEM1 $+_{PAR0}$RSEM2.

The only situations considered so for are those involving union of either disjoint modules or module sharing part or all of the parameter part. In the more general situation, two modules to be combined can share part (or all) of the import and/or export interfaces, and therefore part (or all) of the body.

5.6 Definition (Union of Modules with Shared Submodule)

Given a submodule MOD0 = (PAR0, EXP0, IMP0, BOD0) of two module specifications MODj = (PARj, EXPj, IMPj, BODj) for j = 1,2 with specification morphisms m_{pj}: PAR0 → PARj, m_{ej}: EXP0 → EXPj, m_{ij}: IMP0 → IMPj, m_{bj}: BOD0 → BODj for j = 1,2 as in Definition 4.5, the union of MOD1 and MOD2 with shared MOD0, denoted by $MOD1 +_{MOD0} MOD2$, is the module specification MOD3 = (PAR3, EXP3, IMP3, BOD3) where each of its specifications is given as a pushout of the corresponding specifications in MOD0, MOD1 and MOD2 with the appropriate specification morphisms.

Remark In this definition of union, the parts shared by MOD1 and MOD2 are required to form a submodule of both MOD1 and MOD2. According to our assumptions in Definition 4.5, this implies not only that FREE0: **IMP0** → **BOD0** is strongly persistent (or conservative) but also that $V_{m_{bj}} \cdot FREEj = FREE0 \cdot V_{m_{ij}}$ for j = 1,2 and that $V_{m_{ej}} \cdot Rj = R0 \cdot V_{m_{ej}}$ for j = 1,2.

If the two modules share only a subparameter, then we can take PAR0 = EXP0 = IMP0 = BOD0 and this union reduces to the one given in Definition 5.4. If only part of the export interface (and, therefore, of the body) is shared, we can take PAR0 = IMP0 = ϕ and EXP0 = BOD0, while if the shared part is in the import interface, we let PAR0 = EXP0 = ϕ but we still require the free functor from **IMP0** to **BOD0** to be strongly persistent (or conservative).

The semantics of the union of two modules with a shared submodule behaves exactly as we expect it (or hope for it) to behave.

5.7 Theorem (Semantics of Union with Shared Submodule)

The semantics SEM3 of the union module specification $MOD3 = MOD1 +_{MOD0} MOD2$ is the amalgamated sum of the semantics of MOD1 and MOD2 w.r.t. the semantics of MOD0, i.e. SEM3 is uniquely defined by $SEM3 = SEM1 +_{SEM0} SEM2$.

Proof Let Vj: **BODj** → **EXPj** denote the forgetful functor associated with the specification morphism vj: EXPj → BODj. We first prove that $V3 = V1 +_{V0} V2$, i.e. $V3(B1 +_{B0} B2) = V1(B1) +_{V0(B0)} V2(B2)$, where Bj ∈ **BODj** for j = 0,1,2 and $B1 +_{B0} B2 \in$ **BOD3**. Since MOD0 is a submodule of both MOD1 and MOD2, $V0 \cdot V_{m_{bj}} = V_{m_{ej}} \cdot Vj$ or, equivalently, $V0(B0) = Vj(Bj)_{EXP0}$ for j = 1,2. Then $(V3(B1 +_{B0} B2))_{EXPj} = ((B1 +_{B0} B2)_{EXP3})_{EXPj} = Bj_{EXPj} = Vj(Bj)$ for j = 0,1,2 and therefore $V3(B1 +_{B0} B2) = V1(B1) +_{V0(B0)} V2(B2)$ by uniqueness of the amalgamated sum (Lemma 2.2 or Proposition 2.7). We now show that, if we define F3: $\mathbf{IMP1} +_{\mathbf{IMP0}} \mathbf{IMP2} \to \mathbf{BOD1} +_{\mathbf{BOD0}} \mathbf{BOD2}$ by $F3(I1 +_{I0} I2) = F1(I1) +_{F0(I0)} F2(I2)$, where Ij ∈ **IMPj** and Fj is the free functor from **IMPj** to **BODj** for j = 0,1,2, then F3 is the free functor from **IMP3** to **BOD3** and is strongly persistent if F0, F1 and F2 are. To this extent, let

$f3 = f1 +_{f0} f2\colon I1 +_{I0} I2 \to V_{s3}(B1 +_{B0} B2) = V_{s1}(B1) + _{V_{s0}(B0)} V_{s2}(B0)$

be an IMP3-morphism. Since Fj is the free functor, there exists a unique BODj-morphism $\overline{fj}\colon Fj(Ij) \to Bj$ making the diagram

$$\begin{array}{ccc} Ij & \xrightarrow{\;fj\;} & V_{sj}(Bj) \\ {\scriptstyle \eta j}\downarrow & \nearrow{\scriptstyle V_{sj}(\overline{fj})} & \\ V_{sj}(Fj(Ij)) & & \end{array}$$

commute.

Then the BOD3-morphism $\overline{f1} +_{\overline{f0}} \overline{f2}$ makes the diagram

$$\begin{array}{ccc} I1 +_{I0} I2 & \xrightarrow{\;f3\;} & V_{s3}(B1 +_{B0} B2) \\ \downarrow & \nearrow{\scriptstyle V_{s3}(\overline{f1} +_{\overline{f0}} \overline{f2})} & \\ V_{s3}(F1(I1) +_{F0(I0)} F2(I2)) & & \end{array}$$

commute. Furthermore, F3 is strongly persistent since

$V_{s3}(F1(I1) +_{F0(I0)} F2(I2)) = V_{s1}(F1(I1)) + _{V_{s0}(F0(I0))} V_{s2}(F2(I2))$

$= I1 +_{I0} I2$ if F0, F1 and F2 are strongly persistent.

Finally, $SEM3 = V3 \cdot F3 = (V1 +_{V0} V2) \cdot (F1 +_{F0} F2) = (V1 \cdot F1) +_{(V0 \cdot F0)} (V2 \cdot F2) =$ $= SEM1 +_{SEM0} SEM2$.

5.8 Theorem (Restriction Semantics)

The restriction semantics RSEM3 of $MOD3 = MOD1 +_{MOD0} MOD2$ is uniquely given by $RSEM3 = RSEM1 +_{RSEM0} RSEM2$.

Proof Since $RSEM3 = R3 \cdot SEM3$ the result will be established as soon as we show that $R3 = R1 +_{R0} R2$.

First notice that, for $Ej \in$ **EXP**j, j = 0,1,2, $R1(E1) +_{R0(E0)} R2(E2) \subseteq E1 +_{E0} E2$ and that $(R1(E1) +_{R0(E0)} R2(E2))_{PAR3} = R1(E1)_{PAR1} +_{R0(E0)_{PAR0}} R2(E2)_{PAR2} = E1_{PAR1} +_{E0_{PAR0}} E2_{PAR2} = (E1 +_{E0} E2)_{PAR3}$ and hence $R3(E1 +_{E0} E2) \subseteq R1(E1) +_{R0(E0)} R2(E2)$.

(Notice that we use here the assumption made in Definition 4.5 that $V_{m_{ej}}(Rj(Ej)) = R0(V_{m_{e0}}(Ej)) = R0(E0)$ for j = 1,2).

On the other hand, if $R3(E1 +_{E0} E2) = \overline{E1} +_{\overline{E0}} \overline{E2}$ with $\overline{Ej} \subseteq Ej$, $\overline{Ej}_{PARj} = Ej_{PARj}$

and $R3(E1 +_{E0} E2)_{PARj} = \overline{Ej}$, then $Rj(Ej) \subseteq \overline{Ej}$ and therefore

$R1(E1) +_{R0(E0)} R2(E2) \subseteq \overline{E1} +_{\overline{E0}} \overline{E2}$. Hence $R3(E1 +_{E0} E2) = R1(E1) +_{R0(E0)} R2(E2)$.

6. CONCLUSION AND FURTHER DEVELOPMENTS

Let us first give a short summary of the main constructions and results of this paper. In Section 4, after reviewing the basic concepts of module specification and semantics (as in [2]), we have introduced the concept of a submodule M of module M', imposed some (natural) restrictions on the connecting specification morphisms and related the semantics of M and M'. A precise notion of submodule is not only worthy of independent investigation, but also important for a precise treatment of the union of modules which share common parts. Different possible unions of modules have been presented (in Section 5) in increasing degree of difficulty, from the simple case of shared parameter to the most general one of union of modules with shared submodules. Both the unrestricted and the restriction semantics of the union modules have been shown to relate in a natural way to the semantics of its components. In discussing the semantics of the union of modules, we have made use of the notion of amalgamated sum of algebras, whose basic definition and properties have been introduced in Section 2. Connections between amalgamated sums and parametrized data types have been briefly touched upon in Section 3, where parameter passing has been formulated from a constructive point of view in both the initial and loose semantics cases.

Several questions arise from the developments in this paper and in [2] and are currently under investigation. Among the results that will be presented in full details in forthcoming papers, are some compatibility conditions on union and composition of modules that guarantee distributivity properties of these two operations, such as $(M1 + M2) \cdot M3 = (M1 \cdot M3) +_{M3\phi} (M2 \cdot M3)$ where $M3\phi$ is the submodule of M3 given by $M3\phi = (\phi, \phi, IMP3, BOD3)$. Similar results can also be obtained with unions of modules with shared submodules. Composition on the left, e.g. $M1 \cdot (M2 +_{M0} M3)$, seems to be more complicated, but we have some encouraging preliminary results of a "pseudo-distributive" nature. Results of this type are a prerequisite to a comprehensive development of an algebra of modules. The possibility of partial composition, i.e. the matching of only part of the import interface of a module with the export interface of another one, is also under investigation as a first step toward the construction of complex modules using a "recursive-like" interaction of simpler ones. The compatibility of the operations of union and actualization has been investigated in [7]. An example of shared submodules can be constructed from the example in [2]. This will be given in an expanded version of this paper.

Acknowledgements. The results in this paper are part of ongoing research being conducted jointly with H. Ehrig, Technische Universitat Berlin, and initiated during his visit to the University of Southern California in the Spring 1984. A more comprehensive account of this research will be presented in a forthcoming joint paper. Many thanks to Cynthia Summerville for fast and accurate typing.

References

[1] Blum, E.K., Parisi-Presicce, F., Implementation of Data Types by Algebraic Methods, J. Comput. System Sci. 27, 2 (Oct. 1983) 304-330.

[2] Ehrig, H., An Algebraic Specification Concept for Modules, Draft Version, Techn. Report No. 84-02, TU Berlin, FB 20, March 1984.

[3] Ehrig, H., Kreowski, H.-J., Compatibility of Parameter Passing and Implementation of Parameterized Data Types, Theoret. Comp. Sci. 27(1983) 255-286.

[4] Ehrig, H., Kreowski, H.-J., Mahr, B., Padawitz, P., Algebraic Implementation of Abstract Data Types, Theoret. Comp. Sci. 20(1982) 209-264.

[5] Ehrig, H., Kreowski, H.-J., Thatcher, J.W., Wagner, E.G., Wright, J.B., Parameter Passing in Algebraic Specification Languages, Proc. Aarhus Workshop on Prog. Spec., 1981, LNCS 134(1982) 322-369.

[6] Goguen, J.A., Thatcher, J.W., Wagner, E.G., An Initial Algebra Approach to the Specification, Correctness and Implementation of Abstract Data Types, Current Trends in Prog. Method., IV: Data Structuring (R.T. Yeh, Ed.) Prentice-Hall, NJ (1978) 80-149.

[7] Parisi-Presicce, F., The operations of union and actualization of module specifications are compatible, Extended Abstract, Univ. of Southern California, September 1984.

Why Horn Formulas Matter in Computer Science: Initial Structures and Generic Examples
(extended abstract)

J.A. Makowsky

Department of Computer Science
Technion - Israel Institute of Technology
Haifa 32000, Israel
and
Institut für Informatik
Swiss Federal Institute of Technology
CH-8092 Zürich, Switzerland

Abstract: We introduce the notion of *generic examples* as a unifying principle for various phenomena in computer science such as *initial structures* in the area of abstract data types and *Armstrong relations* in the area of data bases. Generic examples are also useful in defining the semantics of logig programming, in the formal theory of program testing and in complexity theory. We characterize initial structures in terms of their generic properties and give a syntactic characterization of first order theories *admitting initial structures*. The latter can be used to explain why Horn formulas have gained a predominant role in various areas of computer science.

1. Introduction

Verification by example has always been alternative to formal deduction. Historically, in mathematics, it usually also preceeded the development of formal deduction methods. The Babylonians "knew" that $(x-y)^2 = x^2 + 2xy + y^2$ but they did not have a notational system which allowed them carry out a formal, i.e. algebraic, proof. Instead they wrote $(3+5)^2 = 3^2 + 2x3x5 + 5^2$, from which they immediately concluded all the other instances of the general formula. The choice of the particular instance $x=3, y=5$ is important here. It is clear why $x=1, y=2$ would confuse the matter, and we informally describe an appropriate choice of an instance as the finding of a "generic" example. The art of finding "generic" examples has been pushed to the extreme in Euclidean plane geometry, where we convince ourselves of many theorems by just drawing *one* picture of a non-degenerate case. The generalization of this approach to other areas of reaoning is usually highly non-trivial. In algebraic geometry, for example, a satisfactory definition of "generic points" was only found in this century.

In computer science one is often concerned with the specification and analysis of algorithms and programs. Methods for formal specification and verification of programs have been developed intensively without leaving too much impact on the *practical* programmers. These methods are all very much in the spirit of formal deduction. The use of "generic" examples can be observed occasionally with various degrees of explicitness. Strassen [Str74] and his school have used the generic points of algebraic geometry with considerable success to obtain lower bounds in algebraic complexity theory. Recent work in the mathematical foundation of program testing, as presented in the survey edited by B. Chandrasekaran and S. Radicchi [CR81], focus around various notions of "generic" input. In data base theory, W. Armstrong [Ar74] has introducted a kind of "generic" relation for functional dependencies and R. Fagin has investigated the possibilities of generalizing this for implicational dependencies [Fa82]. Last but not least there is M. Zloof's approach to data base query languages where queries are specified by giving "generic" examples, an approach he most recently generalized to operate more complex systems in office automation [Zl82]. It is not surprising that *specification and verification by example* is more appealing to the computer engineer than formal deduction. A look at Euclidean geometry can be revealing again: People involved in surveying and drawing plans have, in general, very little use for formal deductions Euclidean style, but are very much aware of the role of the "generic" non-generate configurations.

The purpose of this paper is to introduce some variations of notions of "*genericity*" which arise in abstract specification of data structures, in relational data bases and in logic programming. What these three areas of computer science have in common, is the use of *first order logic* as its basic specification language. In each of these areas *Horn formulas* play an important role. In algebraic specification of abstract data structures one first used pure *equational logic* with the semantics of *initial structures* as a specification language (hence algebraic). Later one felt the need to extend this to *conditional equations*, which are *universal Horn formulas* without relation symbols. In relational data bases various specification languages where introduced, such as the arrow notation between finite sets of attribute names, to express functional and multivalued dependencies. It was soon realized by R. Fagin, C. Beeri and others, that implicational dependencies, which are Horn formulas without function symbols, could capture all the previously considered cases (cf. [Fa82]). In logic programming Horn formulas are used both as a specification and a programming language because, as R. Kowalski put it, the allow a *procedural* interpretation (cf. [Ko79]).

From a semantic point of view all these approaches can be described in a similar way: Instead of thinking of arbitrary models (= algebraic structures, relations or first order structures resp.) one only considers a *restricted* class of structures (= initial structures, Armstrong relations, minimal Herbrand universes). These restricted classes have all in common some sort of genericity, $\exists^+$-*genericity*, which we shall describe below. Intuitively $\exists^+$-genericity captures rather well the notion of a *generic example*, here of a $\exists^+$-generic structure satisfying the required specifications.

Various attempts exist in the literature to explain why Horn formulas are the *right* class of formulas to be used in the respective context. B. Mahr and J. Makowsky [MM83] prove that under certain assumptions for the semantics of algebraic specifications, conditional equations form the largest specification language satisfying these assumptions. J. Makowsky and M. Vardi [MV84] characterize various classes of data base dependencies in terms of preservation properties under operations on relations which come from data manipulation. In logic programming,, it was shown by Tarnlund [Tarn77] that Horn logic is enough to program every recursive function, a result, stated in slightly different form in a different context, proven already by S. Aanderaa and independently by E. Börger. For an excellent survey see [Bö84].

Our main result in this paper is a characterization of Horn formulas in terms of the existence of $\exists^+$-generic structures. It simultaneously extends and unifies results of [MM83], [Ma84] and [VM84] and remedies objections raised to [MM83] by A. Tarlecki. It states that a first order theory T admits initial ($= \exists^+$-generic) models iff there is a set of definable partial functions such that adding functions to the vocabulary of T gives us a theory T_1, which is equivalent to a universal Horn theory. Additionally, if T is finite, this set of definable partial functions can be chosen to be finite, too. In other words, if we want to define a semantic over $\exists^+$-generic structures only, we can, without loss of generality, confine our specification language to universal first order Horn formulas. Without loss of generality can mean two things: Given a specification which is not a set of universal Horn formulas, then either we can find an equivalent set of universal Horn formulas or we have chosen the basic vocabulary (set of basic symbols) wrongly, and then the theorem tells us that there is a unambiguous way to correct this.

In detail the paper is organized as follows:

In section 2 we introduce the concepts of *A*-genericity and $\exists^+$-genericity and relate these definitions to initiality. We state the basic definability theorem for initial models; we characterize initial term models as *A*-generic models and initial models as $\exists^+$-generic pseudo-term models.
In section 3 we characterize first order theories which admit initial term models as the universal Horn theories. This theorem was already proved in [MM83].
In section 4 we establish the intersection property of first order theories admitting $\exists^+$-generic structures and review some classical model theoretic results on first order theories with the intersection property. From this we get that theories admitting $\exists^+$-generic models can always be axiomatized by universal-existential sentences. We also state a theorem of M. Rabin [Ra60], which characterizes first order theories with the intersection property.
In section 5 we state an analogue of Rabin's theorem to obtain our main result. We show that a first order theory admits initial models iff it is a partially functional $\forall\exists$-Horn theory.
No proofs are presented in this extended abstract, since the editors of these proceedings put a severe space limit on the papers to be presented. We hope that the complete paper will appear soon elsewhere.
In section 6 we state some conclusions.

The reader familiar with the introduction to model theory by G. Kreisel and J.L. Krivine will realize how much, in spirit, this work is influenced by chapter 6 of [KK66]. I am indebted also to A. Tarlecki for his remarks in our correspondence concerning [MM83], to S. Shelah, who suggested theorem 2.10 and to B. Mahr for the discussions around [MM83].

2. Initial models and genericity

In this section we deal with first order languages *with equality*. *Vocabularies (= similarity types)* are allowed to be *many-sorted* and may include *function symbols, relation symbols* and *constant symbols*. Vocabularies are denoted by τ,σ. A τ*-structure* **A** is a collection of *universes* (= sets) $A_1,...,A_n$, for each sort in τ one, together with interpretations for all the function, relation and constant symbols in τ. τ*-terms, atomic formulas* and τ*-formulas* are defined as usual. If T is a set of τ-formulas, φ is a τ-formula and **A** is a τ-structure we write $\mathbf{A}\models T$ if the universal closure of all the formulas φ in T are true in **A**. We write $T\models\varphi$ if in every τ-structure **A** such that $\mathbf{A}\models T$ we also have that $\mathbf{A}\models\varphi$. We call sets of τ-formulas *theories* and formulas *without free variables* also τ-sentences. We call τ-structures also *models* and denote by *Mod(T)* the class of t-structures **A** such that **A**$\models$*T*.

2.1. Defintions:

(i) Let K be a class of τ-structures closed under isomorphisms and **A** $\in$ **K**. We say that **A** *is initial in* **K** *(is an initial model for* **K***)* if for every structure **B** $\in$ **K** there is a *unique* homomorphism h_B:**A** $\rightarrow$ **B**.

(ii) If **K** is of the form *Mod(T)*, where T is some first order theory, we also say that **A** is *initial for T*.

(iii) A τ-structure **A** is a *term model (reachable model)* if for every $a \in$ **A** there is τ-term t such that its interpretation **A**(t) in **A** is the element a.

Next we introduce the concept of *generic* structures for first order theories and relate it to initial structures.x

2.2. Definitions: Let **K** be a class of τ-structures closed under isomorphisms and **A**$\in$**K**. Let Σ be a set of first order sentences (i.e. formulas without free variables).

(i) We say that **A** *is generic* in **K** *for* Σ if *for* every $\varphi \in \Sigma$ we have that **A**$\models\varphi$ iff **B**$\models\varphi$ for every **B**$\in$**K**.

(ii) If Σ is the set of atomic τ-formulas we say *A-generic* instead of generic for Σ.

(iii) Let $\exists^+$ be the set of τ-formulas of the form $\exists x \bigwedge^n \varphi_i$ with each φ_i an atomic formula, $x=(x_1,x_2,...,x_n)$; and $\exists$ be the set of τ-formulas of the form $\exists x \psi(x)$ with ψ quantifier free.

(iv) If Σ is the set of $\exists^+$-sentences we say $\exists^+$*-generic* instead of generic for Σ.

2.3. Remarks:

(i) If $\Sigma_0 \subset \Sigma$ and **A** is Σ-generic then **A** is also Σ_0-generic.

(ii) If **A** is a *A*-generic term model then **A** is an initial term model.

2.4. Theorem ($\exists^+$-genericity):

Let **K** be a class of τ-structures closed under isomorphisms and $\mathbf{A}_I$ be initial for **K**. Then $\mathbf{A}_I$ is $\exists^+$-generic.

2.5. Corollary: Let **K** be a class of τ-structures closed under isomorphisms and **A**$\in$**K**. Then **A** is an initial term model for **K** iff **A** is *A*-generic.

2.6. Definitions: Let T be a set of τ-sentences, let **A** be a τ-structure and $a \in A$ be an element of the universe of **A**.

(i) a is *definable over* **A** if there is a τ-formula $\varphi_a(x)$ with x the only free variable of φ such that $\mathbf{A} \models \varphi_a(a)$ and if $\mathbf{A} \models \varphi_a(b)$ for any $b \in A$ then $\mathbf{A} \models a = b$. We call φ_a the defining formula of a.

(ii) a is $\exists^+$*-definable* ($\exists$*-definable, atomically definable*) over **A** if a is definable over **A** and the defining formula is an $\exists^+$-formula ($\exists$-formula, conjunction of atomic formulas).

(iii) a is *definable over* T if there is a τ-formula $\varphi_a(x)$ with x the only free variable of φ_a such that $\mathbf{A} \models \varphi_a(a)$ and $T \models \forall x \forall y (\varphi_a(x) \wedge \varphi_a(v) \Rightarrow x = y)$.

(iv) a is $\exists^+$*-definable* ($\exists$*-definable, atomically definable*) over T if a is definable over T and the defining formula is an $\exists^+$-formula ($\exists$-formula, conjunction of atomic formulas).

(v) We say that $\mathbf{A} \models T$ is a *pseudo term model of* T if every element $a \in A$ is $\exists^+$-definable.

The following theorem shows that in an initial model of a theory T is always a pseudo term model.

2.7. Theorem ($\exists^+$-definability): Let T be a first order theory and let $\mathbf{A}_I$ be an initial model of T. Then every $a \in A$ is definable over T by a $\exists^+$-formula φ_a. In other words $\mathbf{A}_I$ is a pseudo term model.

We now are in a position to characterize initial models as pseudo term models which are $\exists^+$-generic.

2.8. Theorem: Let T be a first order theory and let **A** be a model of T. Then **A** is initial (for T) iff **A** is a $\exists^+$-generic pseudo-term model.

3. Characterizing first order theories which admit initial term models

In this section we characterize first order theories which admit initial term models. Such a characterization was first given in [MM83], based on a theorem due to Mal'cev [Mal56]. In [Mal56] there is a minor mistake as pointed out by [Mo59], which propagated not [MM83] in as far as one had to assume that every first order theory admitting initial term models also has a trivial model. In the full version of this paper, which had been submitted to this conference, this mistake was corrected. A. Tarlecki had found a different way to avoid this difficulty in [Tar84].

3.1. Definitions: Let **K** be a class of τ-structures closed under isomorphisms.

(i) We say that **K** *admits initial term models* if for every σ and for every set Δ of atomic (variable free) σ-sentences either Δ has no model in **K** or there is an initial term model in **K** which satisfies Δ.

(ii) We say that **K** *strongly admits term models* if for every σ and for every set Δ of atomic or negated atomic (variable free) σ-sentences either Δ has no model in **K** or there is an initial term model in **K** which satisfies Δ.

(iii) We say that **K** is *closed under substructures* if whenever $\mathbf{A} \in \mathbf{K}$ and $\mathbf{B} \subset \mathbf{A}$ is a substructure of **A** then $\mathbf{B} \in \mathbf{K}$.

(iv) Let T be a first order theory. We say that T is *preserved under substructures* if whenever $\mathbf{A} \models T$ and $\mathbf{B} \subset \mathbf{A}$ is a substructure of **A** then $\mathbf{B} \models T$.

We first state that classes of structures admitting initial term models are closed under substructures. This theorem was inspired by [Mal56] and first stated in [MM83].

3.2. Theorem: (Mahr-Makowsky) Let **K** be a class of τ-structures closed under isomorphisms.

(i) If **K** strongly admits term models, then **K** is closed under substructures.

(ii) If **K** admits initial term models, then **K** is closed under substructures.

Classes of structures closed under isomophisms and substructures were characterized by Tarski [Ta52].

3.3. Theorem: (Tarksi) Let T be a first order theory which is preserved under substructures. Then T is equivalent to a universal theory $T_\forall$. Additionally, if T is finite so is $T_\forall$.

3.4. Corollary: Let T be a first oder theory which admits term models. Then T is equivalent to a universal theory $T_\forall$. Additionally, if T is finite so is $T_\forall$.

The next theorem characterizes first order theories which admit initial term models. A similar theorem was proved in [MM83].

3.5. Theorem: Let T be a first order theory which admits initial term models. Then T is equivalent to a universal Horn theory T_H. Additionally, if T is finite so is T_H.

3.6. Definitions: Let **K** be a class of τ-structures closed under isomorphisms.
(i) We say tat **K** is *closed under products* if whenever $\mathbf{A}_i \in \mathbf{K}(i \in I)$ is a family of τ-strucrues then $\Pi\, \mathbf{A}_i \in \mathrm{K}$.
(ii) Let T be a first order theory. We say that T is *preserved under proudcts* if whenever $\mathbf{A}_i \models T$, $(i \in I)$ is a family of τ-structures then $\Pi\, \mathbf{A}_i \models T$.

The following characterization of universal Horn formulas is due to McKinsey [McK43].

3.7. Theorem: (McKinsey) A first order theory T is equivalent to a universal Horn theory iff T is closed under products and substructures.

This theorem can be used to get the following theorem.

3.8. Theorem: Let T be a universal Horn theory (over avocabulary τ containing at least one constant symbol). Then T admits initial term models.

We can now collect the results of this section into one theorem:

3.9. Theorem: For a first order theory T the following are equivalent:
(i) T admits initial term models;
(ii) T is preserved under products and substructures;
(iii) T is equivalent to a universal Horn theory.

4. First order theories with the Intersection Property

In this section we show that first order theories admitting initial models have the intersection property. This will allow us to use a theorem of M. Rabin [Ra60, Ra62], here heorem 4.7, together with the result of sections 2 and 3, to characterize first order theories vhich admit initial models.

Convention: Let $x=x_0,...x_n$ and $y=y_0,...,y_n$. We write in this and the following sections $\exists! x\varphi(x)$ for the conjunction of the two formulas $\exists x\varphi(x)$ and $\forall x\forall y(\varphi(x) \wedge_{i=0}^{i=n} x_i=y_i)$.

4.1. Definitions:

(i) Let **K** be a class of structures closed under isomorphisms. **K** is said to have the *Intersection Property* if for every $\mathbf{A},\mathbf{A}_1,\mathbf{A}_2 \in \mathbf{K}$ with $\mathbf{A}_i \subset \mathbf{A} (i=1,2)$ the intersection $\mathbf{A}_1 \cap \mathbf{A}_2 \in \mathbf{K}$.

(ii) A first order theory T is said to have the *Intersection Property* if $Mod(T)$ has the Intersection Property.

(iii) A first order theory T is said to have the *Infinite Intersection Property* if with every family $\mathbf{A}_i$ $(i \in I)$ such that $\mathbf{A}_i \models T$ also $\bigcap_{i \in I} \mathbf{A}_i \models T$.

4.2. Theorem: (Robinson) A first order theory T has the Intersection Property iff T has the Infinite Intersection Property.

4.3. Example: If T is preserved under substructures then T has the Intersection Property. Therefore, by theorem 3.3, every universal theory has the Intersection Property.

4.4. Theorem: Let T be a first oder theory which admits initial models. Then T has the Intersection Property.

First order theories which ave the Intersection Property have been studied in the early days of model theory by A. Robinson and M. Rabin [Rab62,Rob51].

4.5. Theorem: (A. Robinson, Chang and Los-Suszko, cf. [Ro63]).
Let T be a first order theory with the Intersection Property. Then T is equivalent to a set of $\forall\exists$-sentences.

The above theorem is a condensed form of two theorems: Robinson's theorem asserts that a theory with the Intersection Property is a theory which is preserved under unions of chains; and the theorem of Chang and, independently, of Los and Suszko asserts that a theory which is preserved under union of chains is equivalent to a set of $\forall\exists$-sentences.

4.6. Corollary: Let T be a theory which admits initial models. Then T is equivalent to a set of $\forall\exists$-sentences.

The next theorem, due to Rabin [Rab60], characterizes theories with the Intersection Property. For this let $\varphi(x,y)$ be a first order formula with free variables $x=x_1,x_2,...,x_t$ and $y=y_1,y_2,...y_n$ and let k be a natural number. We denote by $N(k,y,\varphi(x,y))$ the first order formula which says that there are exactly k different n-tuples satisfying the formula $\varphi(x,y)$.

4.7. Theorem: (Rabin [Rab60]) A necessary and sufficient condition for a first order theory T to have the Intersection Property is that for every $\forall\exists$-sentence $\forall x \exists y \varphi(x,y)$ which is a consequence of T, there exist two sequences of quantifier-free formulas

$$\sigma_1(x,u),\sigma_2(x,u),...,\sigma_\mu(x,u) \text{ and } \theta_1(x,y,z),\theta_2(x,y,z),....,\theta_\mu(x,y,z)$$

and a sequence of natural numbers $k_1,k_2,...,k_\mu$ such that

$$T \models \forall x(\forall u \sigma_1(x,u) \vee \forall u \sigma_2(x,u) \vee ... \vee \forall u \sigma_\mu(x,u))$$

and for $1 \leq i \leq \mu$

$$T \models \forall x(\forall u \sigma_i(x,u) \Rightarrow N(k_i, y \exists z \varphi(x,y) \wedge \theta_i(x,y,z))).$$

In the next section we want to give a similar characterization for first order theories admitting initial models. Our next goal is to show the existence of initial models for certain theories which have the Intersection Property and which are preserved under products. For this we need some more definitions.

4.8. Definitions: Let T be a first order theory with the Intersection Property. A model $\mathbf{A}_0$ of T is a *core model* if there is no proper submodel $\mathbf{B} \subset \mathbf{A}_0$ such that $\mathbf{B} \models T$. If $\mathbf{A}$ is a model of T and $\mathbf{A}_0 \subset \mathbf{A}$, $\mathbf{A}_0 \models T$ is a core model we say that $\mathbf{A}_0$ is a *T-core of* $\mathbf{A}$.

4.9. Lemma: Let T be a first order theory with the Intersection Property. Then every model $\mathbf{A}$ of T has a T-core $\mathbf{A}_0$.

4.10. Proposition: Let T be a first order theory with the Intersection Property. Then every core model of T is an $\exists$-term model.

4.11. Definition: A first order theory T is *pseudo algebraic* if T is preserved under products, has the Intersection Property and if every core model of T is a pseudo term model.

4.12. Theorem: A pseudo algebraic first order theory T has an initial model A_I.

A converse of theorem 4.12 will proved in the next section.

5. Charcterizing first order theories which admit initial models

The purpose of this section is to characterize first order theories which admit initial models. We first want to show that such a theory is equivalent to an $\forall\exists$-Horn theory.

5.1. Theorem: Let T be a first order theory which admits initial models. Then:

(i) T is equivalent to an $\forall\exists$-Horn theory $T_{\forall\exists H}$.

(ii) If T is finite, so is $T_{\forall\exists H}$.

Next we want to state an analogue of Rabin's theorem (theorem 4.7) for theories which admit initial models.

5.2. Theorem: Let T be a first order theory which admits initial models. Then for every $\forall\exists$-sentence $\forall x\exists y\varphi(xy)$ which is a consequence of T, there exist two sequences of formulas

$$\sigma_1(x,u),\sigma_2(x,u),\ldots,\sigma_\mu(x,u) \text{ and } \theta_1(x,y,z),\theta_2(x,y,z),\ldots,\theta_\mu(x,y,z),$$

where σ_i are quantifier free formulas and θ_i are $\exists^+$-formulas, such that

$$T\models\forall x(\forall u\sigma_1(x,u)\vee\forall u\sigma_2(x,u)\vee\ldots\vee\forall u\sigma_\mu(x,u))$$

and for $1\leq i\leq\mu$

$$T\models\forall x(\forall u\sigma_i(x,u)\Rightarrow\exists!y(\exists z(\varphi(x,y)\wedge\theta_i(x,y,z)))$$

5.3. Definition: We call a first order theory which satisfies the conclusion of theorem 5.2

partially functional. This is justified since theorem 5.2 says that every $\forall\exists$-formula which is a consequence of T can be Skolemized with finitely many partial functions.

5.4. Corollary: Let T be a first order theory which admits initial models. Then every $\exists$-term model **A** is a pseudo term model.

We need another well known result from model theory, see e.g. ([CK73]):

5.5. Theorem: Let T be an $\forall\exists$-Horn theory. Then T is preserved under products.

Putting everything together we obtain:

5.6. Theorem: (Main theorem) Let T be a first order theory. The following are equivalent:
(i) T admits initial models;
(ii) T Is equivalent to a partially functional $\forall\exists$-Horn theory.
(iii) T is pseudo algebraic.

6. Conclusions

We have given a characterization of universal Horn theories in terms of the existence of initial, or equivalently, A-generic term models (theorem 3.9) and a characterization of partially functional $\forall\exists$-Horn theories in terms of the existence of initial, or equivalently $\exists^+$-generic pseudo term models (theorem 5.6). The latter essentially says that a first order theory which admits initial models which are not term models does so by oversight: The vocabulary (similarity type) was badly chosen, such as not to allow that all elements are denoted by some term. This can be almost remedied: Either by adding definable partial Skolem functions or by allowing pseudo terms, i.e. elements uniquely definable by $\exists^+$-formulas.

The paper also sheds more light into the question why in [ADJ75] initial structures were proposed as the framework for abstract data types. We have given in theorem 2.13 a characterization of initial structures as $\exists^+$-generic pseudo term models. For somebody not familiar with category theory this may be more appealing since it relates directly to or concept of verification by example. However, this characterization has also its technical merits for it provides the missing link between the category theoretic concept and the model theoretic tools needed to prove 5.9.

Last but not least we have yet added another explanation as to why Horn formulas play such an important role in various branches of computer science. We have shown that universal Horn theories (partially functional ∀∃-theories) are exactly the framework in which the notion of a generic example can be applied. This should prevent other researchers from trying to generalize Logic Programming or the semantics abstract data types to larger classes of first order formulas. If it has to be generalized then the direction chosen by R.M. Burstall and J.A. Goguen in [GB84] seems to be much more appropriate.

References

[ADJ75] Goguen, J.A., Thatcher, J.W., Wagner, E.G. and Wright, J.A.; Abstract data types as initial algebras and the correctness of data representations, Proc. of Conf. on Computer Graphics, Pattern Recognition and Data Structures, 1975, pp. 89-93.

[Ar74] Armstrong, W.W.; Dependency structures of database relationships Proc. IFIP 74, North Holland 1974, pp. 580-583.

[Bo84] Börger, E.; Decision problems in predicate logic, in: Logic Colloquium '82, G. Lolli, G. Longo and Am Marcja eds., North Holland 1984, pp. 263-302.

[CK73] Chang, C.C. and Keisler, H.J.; *Model Theory*, North Holland 1973.

[CR81] Chandrasekaran, B. and Radicchi, S.; *Computer Program Testing*, North Holland 1981.

[Fa82] Fagin, R.; Horn Clauses and data base dependencies, J.ACM vol. 29 (1982) pp. 252-285.

[GB84] Goguen, J.A. and Burstall, R.M., Introducing institutions, in Proc. of Logic Programming Workshop, E. Clarke ed., Lecture Notes of Computer Science 1984, to appear.

[KK66] Kreisel, G. and Krivine, J.L.; Elements de logique mathematique, Dunod 1966.

[Ko79] Kowalski,R.; *Logic for Problem solving*, North Holland 1979.

[McK43] McKinsey, J.C.C.; The decision problem for some classes of sentences without quantifiers, Journal of Smb. Logic vol. 8 (1943) pp. 61-76.

[Mal56] Mal'cev, A.; Quasi primitive classes of abstract algebras, in: The Metamathematics of algebraic systems, collected papers of A.I. Mal'cev, North Holland 1971, pp. 27-31.

[MM83] Mahr, B. and Makowsky, J.A.; Characterizing specification languages which admit initial semantics, Proc. of 8th CAAP, Springer LNCS vol. 159 (1983) pp. 300-316.

[Mv84] Makowsky, J.A. and Vardi, M.Y.; On the expressive power of data dependencies, submitted 1984.

[Ma84] Makowsky, J.A.; Model theoretic issues in Computer Science, Part I: Relational data bases and abstract data types, in: Logic Colloquium '82, G. Lolli, G. Longo and A. Marcja eds., North Holland 1984, pp. 303-344.

[Mo59] Mostowski, A.; Review of [Mal56], Journal of Symb. Logic vol. 24 (1959) p. 57.

[Ra60] Rabin, M.; Characterization of convex systems of axioms, Notices AMS, Abstract 571-65 (1960) p. 505.

[Ra62] Rabin M.; Classes of models and sets of sentences with the intersection property, Ann.Fac.Sci. Universite de Clermont, vol. 7.1 (1962) pp. 39-53.

[Rob63] Robinson, A.; *Introduction to model theory and the Metamathematics of algebra*, North Holland 1963.

[Sm84] Smorynski, C.; Lectures on non-standard models of arithmetic, in Logic Colloquium '82, Lolli, G. Longo, G. and Marcja, A. eds., North Holland 1984, pp. 1-70.

[St74] Strassen, V.; Polynomials with rational coefficients which are hard to compute, SIAM J.Comput. vol. 3.2 (1974) pp. 128-149.

[Ta52] Tarski, A.; Some notions and methods on the borderline of algebra and metamathematics, Proc.Int.Congr.of Mathe., Cambridge, MA vol. 1 (1952) pp. 705-720.

[Tar84] Tarlecki, A.; Free constructions in abstract algebraic institutions, draft February 1984.

[Tarn77] Tarnlund, S.A.; Horn clause computability, BIT vol. 17 (1977) pp. 215-226.

[Zi82] Zloof, M.M.; Office-by-example: A business language that unifies data and word processing and electronic mail, IBM Syst.J. vol. 21.3 (1982) pp. 272-304.

ON THE IMPLEMENTATION OF ABSTRACT DATA TYPES
BY PROGRAMMING LANGUAGE CONSTRUCTS

Axel Poigné
Dept. of Computing
Imperial College
London SW7 2BZ

Josef Voss
Abt. Informatik
Universitaet Dortmund
Postfach 500500
D-4600 Dortmund 50

ABSTRACT. Implementations of abstract data types are defined via an enrichment of the target type. We suggest to use an extended typed λ-calculus for such an enrichment in order to meet the conceptual requirement that an implementation has to bring us closer to a (functional) program. Composability of implementations is investigated, the main theorem being that composition of correct implementations is correct if terminating programs are implemented by terminating programs. Moreover we provide syntactical criteria to guarantee correctness of composition.

0. INTRODUCTION

The concept of abstract data types (ADTs) has pushed forward the investigations for a systematic and formal software design. The given problem is made precise as a set of data with operations on it. The way from the problem to a first exact description is often difficult and beyond formal methods. But a lot of work has been spent on ADTs and algebraic specifications and their relationship to programming languages in the last years. On the one hand the theory is involved in structuring large ADTs resp. specifications (parameterization), on the other hand the stepwise refinement of non-algorithmic specifications in direction of a higher order programming language (implementations) is investigated.

This paper is about implementation. There are two points of view how to deal with this subject: a purely semantical reasoning as in the embracing work of Lipeck [7], or a syntax-oriented reasoning on specifications, especially algebraic ones with an initial algebra semantics in mind. The latter approach is taken by several authors like Ehrig, Kreowski, ADJ-group, Ehrich, Ganzinger,... Their recent work investigates convenient correctness criteria, compatibility of parameterizations and implementations, and extensions to wider classes of specification techniques.

We join the latter approach and consider yet another notion of implementation of algebraic specifications. There are two conceptual requirements we attach importance to:

1. An implementation of SPEC0 by SPEC1 has to bring us closer to a program for SPEC0.
2. There has to be a natural way to compose implementations syntactically such that the correctness criteria are preserved.

It has been Ganzinger [6] who used the word 'program' for certain enrichments that may be used for implementations. They were characterized by some semantical conditions but do not look like programs at all. We extend his approach and specify enrichments as programs where new sorts are introduced by domain equations, and new operators are introduced as λ-terms. In that, we restrict enrichments to special, often appearing patterns like products, sums, tree etc. for sorts, and to operator definitions using case-distinction and recursion.

In chapter 1 we introduce an extended typed λ-calculus over a base specification to denote our programs. Several properties of the calculus are investigated which will be used for our main theorem that composition of correct implementations is correct if terminating programs are implemented by terminating programs. The termination condition may be natural from a programmer's point of view but the difficulty of the proof seems to be rather surprising. This may cast some light on the difficulty to find sufficient but not constraining conditions to ensure correctness of composition of the more general notion of implementation used in abstract data type theory.

We assume that the reader is familiar with abstract data type theory. For the introduction of signatures, terms etc. we informally use sorted sets $M = (M_i | i \in I)$. A *signature* is a pair (S,Σ) with a set S of sorts and an $S^* \times S$-sorted set Σ of operators. We use $\sigma: w \to s$ to denote operators with arity w and coarity s. Variables are created from a fixed countable set X by indexing. $x{:}s$ is a variable of sort s. If possible sort indices are omitted. $T_\Sigma(Y)$ denotes the S-sorted set of *λ-terms with variables* Y being defined as usual (Y is a S-sorted set of variables). Syntactical equality is denoted by $\equiv$. Substitution is defined as usual. A *Σ-equation* is a pair (t,t') with $t,t' \in T_\Sigma(Y)$ of the same type. A *specification* is a tripel (S,Σ,E) with a signature (S,Σ) and a set E of Σ-equations. The congruence on $T_\Sigma(Y)$ generated by the equations E is denoted by $=$.

We use the following notation, similar to CLEAR [2]:

```
ZFOURO = sorts Z4
         ops 0,1,2,3: → Z4
             +,-,*: Z4 Z4 →  Z4
             pred, suc: Z4 →  Z4
         eqns 0 * 1 = 0    0 + 1 = 1    0 - 1 = 3   pred(0) = 3   suc(0) = 1
              0 * 2 = 0    0 + 2 = 2    0 - 2 = 2       :             :
                  :            :            :       pred(3) = 2   suc(3) = 0
              3 * 2 = 2    3 + 2 = 1    3 - 2 = 1
              3 * 3 = 1    3 + 3 = 2    3 - 3 = 0

ZFOUR1 = enrich ZFOURO by sorts -
                          opns -
                          eqns (z * 2)* 2 = 0
```

(These specifications of $\mathbb{Z}/4$ which will be used below, but do not constitute what we consider to be a typical specification). Apart from the notation we will as well use the nomenclature of [4].

1. PROGRAMS OVER A SPECIFICATION

1.1 MOTIVATION

Enrichments of specifications occur if we construct complex specifications out of smaller ones, or in implementations as studied by Ehrig et al. [3,4]: SPEC = (S,Σ,E) is extended to SPEC' = SPEC + (S',Σ',E') where the added part (S',Σ',E') need not to be a specification. The partition is used to structure the specification. Thus SPEC and SPEC' should depend on each other in an easy way. There are different notions to catch this semantically:

1. *Consistency*: No identifications of old constants: $t,t' \in T_\Sigma(\emptyset)$, $t =_{E+E'} t' \Rightarrow t =_E t'$.
2. *Completeness*: No new constants on old sorts: $t \in T_{\Sigma+\Sigma'}(\emptyset), s\in S \Rightarrow \exists t' \in T_\Sigma(\emptyset)$: $t =_{E+E'} t'$.
3. *Persistency*: 1. and 2. hold for terms with variables of S-sorts.

Consistency and completeness together guarantee the protection of the SPEC-part in the enrichment. Persistency is stronger in that the introduction of so-called derivors is allowed on S-sorts only. To check wether one of the conditions holds is difficult, in general undecidable.

One observes that over and over again the same constructs are used for enrichments. The new sorts represent lists, trees etc, the added equations are (often primitive) recursive schemes to define the new operators. The restriction to those standard constructs yields a syntactical notion of enrichment which is transparent and of course not exactly equivalent to the semantic ones above. The constructs to be permitted will be both, expressable in higher order programming languages and definable by algebraic specifications. One can state that the operators to be defined are recursive programs in an applicative language with recursive data structures, which uses the given specification, resp. the thereby defined ADT, as a kernel of basic data structures, and functional procedures on them. For new operators a λ-notation will be introduced which underlines the affinity to languages like LISP. The defining equations are replaced by rewrite rules of a typed λ-calculus.

We may lead a longer discussion about the definite choice of constructs to be used: What is typical for higher order PL's and algebraic specifications? What constructs are at least needed? Therefore the following choice is somewhat arbitrary. Some nice properties to be proved below may justify it.

The language Λ contains the following elements:

1. *Products*: In higher PL's products appear as records or classes. In algebraic specifications we write

```
PROD = enrich SPEC by sorts prod-a-b
                 ops   p: prod-a-b → a
                       q: prod-a-b → b
                    pair: a b → prod-a-b
                 eqns p(pair(x,y)) = x,  q(pair(x,y)) = y
                      pair(p(z),q(z)) = z
```

2. *Sums*: Variant records in PASCAL and subclasses in SIMULA correspond to sums. In a specification we write

$$\text{SUM} = \text{enrich SPEC by sorts } \underline{\text{sum-a-b}} \quad \text{ops } \text{inl}: \underline{a} \to \underline{\text{sum-a-b}}, \quad \text{inr}: \underline{b} \to \underline{\text{sum-a-b}}$$

Besides the embeddings we need a means to define functions on the sum by case distinction. In PASCAL we have the case statement. For specifications we use

$$\text{SUMgh} = \text{enrich SUM by ops } f: \underline{\text{sum-a-b}} \to \underline{c} \quad \text{eqns } f(\text{inl}(x)) = g(x),\ f(\text{inr}(y)) = h(y)$$

where we presume that $\underline{a},\underline{b},\underline{c}$, $g:\underline{a} \to \underline{c}, h:\underline{b} \to \underline{c}$ are in SPEC.

3. *Recursive Types*: In PASCAL we can describe recursive data structures using recursive schemes of records and variant records. If a PL does not allow this a controlled use of pointers can help. As a means for the description of recursive types we introduce domain equations, for example

$$\text{tree} = 1 + (\text{tree} \times \text{entry} \times \text{entry}) \qquad \begin{array}{l} \text{expr} = \text{term} \times \text{operator} \times \text{term} \\ \text{term} = \text{identifier} + \text{expr} \end{array}$$

Again entry, operator and identifier are given sorts. Senseless schemes like $d = d \times d$ are excluded.

4. *Recursion*: This is the essential construct which, in combination with the case distinction, allows to write non-trivial programs, but brings along the problems of non-termination. In specifications recursive schemes are those definition schemes which for each new operator symbol σ have one equation with $\sigma(x_o,\ldots,x_{n-1})$ on the left and an arbitrary right hand side. In our language we will use a fixpoint operator to denote recursive operator definitions.

1.2 RECURSIVE TYPES OVER BASE SORTS

For the set S of a given specification (S,Σ,E) we construct products, sums and recursive types as congruence classes of sorts terms over S:

Let DTn(S) denote sort terms with type variables $d_0,\ldots,d_{n-1}$ constructed as the smallest set such that

1. $S \subseteq DTo(S)$, $1 \in DTo(S)$
2. $DTn(S) \subseteq DTm(S)$ for $n \leq m$
3. $d_i \in DTi(S)$ $i \geq 1$
4. $t,t' \in DTn(S) \Rightarrow t + t', t \times t' \in DTn(S)$.

Now take the recursive type scheme

$$\begin{array}{c} d_o = t_o(d_o,\ldots,d_{n-1}) \\ \vdots \\ d_{n-1} = t_{n-1}(d_o,\ldots,d_{n-1}) \end{array}$$

with n variables and n equations. We introduce names for the n solutions of this scheme by $D^i_n(t_o,\ldots,t_{n-1})$, $i \in \underline{n}:=\{0,\ldots,n-1\}$. We get arbitrarily nested schemes if we regard these solutions as new constants. Thus the defition of DTn(S) is completed by the line

5. $t_o,\ldots,t_{n-1} \in DTn(S) \Rightarrow D^i_n(t_o,\ldots,t_{n-1}) \in DTo(S)$

To obtain the $D_n^i(t_o,\dots,t_{n-1})$ as the solution of the respective recursive scheme the type terms are to be factorized by the least equivalence relation $\sim$ containing

1. $D_n^i(t_o,\dots,t_{n-1}) \sim t_i[d_o \leftarrow D_n^o(t_o,\dots,t_{n-1}),\dots,d_{n-1} \leftarrow D_n^{n-1}(t_o,\dots,t_{n-1})]$ for $i \in \underline{n}$
2. $t_i \sim t_i',\ i \in \underline{2} \Rightarrow t_o + t_1 \sim t_o' + t_1'$, $t_o \times t_1 \sim t_o' \times t_1'$
3. $t_i \sim t_i',\ i \in \underline{n} \Rightarrow D_n^j(t_o,\dots,t_{n-1}) \sim D_n^j(t_o',\dots,t_{n-1}')$ for $j \in \underline{n}$

In fact, we state that $\sim$ is a congruence. Hence the operators $\_+\_$, $\_\times\_$ and $D_n^i(\dots)$ are well defined on equivalence classes.

It is reasonable to restrict our attention to *acceptable types*, i.e. types whith non-empty solution ($d = d + d$ is not useful). We define acceptable types to be those types t such that $t \twoheadrightarrow 1$ with regard to (where $\twoheadrightarrow$ is the refl.,trans.,substitutive & compatible closure of $\rightarrow$).

$s \rightarrow 1$ for $s \in S$

$t \rightarrow 1$ or $t' \rightarrow 1 \Rightarrow t + t' \rightarrow 1$

$t \rightarrow 1$ and $t' \rightarrow 1 \Rightarrow t \times t' \rightarrow 1$

$D_n^i(t_o,\dots,t_{n-1}) \rightarrow t_i[d_i \leftarrow D_n^i(t_o,\dots t_{n-1}), i \in \underline{n}]$.

Remarks: 1. The whole calculus in the rest of the paper is essentially the same if we do not restrict to acceptable types, but the proof technique has to be extended sometimes. For instance the following property will be used:

2. The maximal decomposition of an acceptable type into products is finite.

Definition: Let $ATn(S) \subseteq DTn(S)$ denote the set of acceptable types with at most n variables.

The set of *base types* over S is given by $BType(S) := ATo(S)_{/\sim}$

The set of (*higher order*) *types* is defined to be the smallest set Type(S) with

1. $BType(S) \subseteq Type(S)$
2. $t, t' \in Type(S)$, t or $t' \notin BType(S) \Rightarrow t + t', t \times t' \in Type(S)$
3. $t, t' \in Type(S) \Rightarrow t \twoheadrightarrow t' \in Type(S)$.

1.3 THE PROGRAMMING LANGUAGE Λ

Compound operations will be denoted by Λ-terms. The set Type(S) is used to type the terms where S is a given set of sorts. The set FV(t) of free variables of a Λ-term t is defined simultaneously:

For a given signature (S,Σ) and a set of variable names X we define the language $\Lambda_\Sigma(X)$ or for short Λ, to be the smallest Type(S)-sorted set with

1. $x \in X,\ s \in Type(S) \Rightarrow x{:}s \in \Lambda_s$ — $FV(x{:}s) := \{x{:}s\}$
2. $\emptyset \in \Lambda_1$ — $FV(\emptyset) := \emptyset$
3. $t_i \in \Lambda_{s_i},\ i \in \underline{n},\ \sigma{:}s_o\dots s_{n-1} \rightarrow s \in \Sigma \Rightarrow \sigma(t_o,\dots,t_{n-1}) \in \Lambda_s$ — $FV(\sigma(t_o,\dots,t_{n-1})) := \bigcup_{i \in \underline{n}} FV(t_i)$
4. $inl_{s,s'} \in \Lambda_{s \rightarrow s+s'}$, $inr_{s,s'} \in \Lambda_{s' \rightarrow s+s'}$ — $FV(inl_{s,s'}) := FV(inr_{s,s'}) := \emptyset$

5. $p_{s,s'} \in \Lambda_{s\times s' \to s}$, $q_{s,s'} \in \Lambda_{s\times s' \to s'}$ $\qquad FV(p_{s,s'}) := FV(q_{s,s'}) := \emptyset$

6. $t_i \in \Lambda_{s_i}$, $i \in \underline{2}$ => $\langle t_o, t_1\rangle \in \Lambda_{s_o \times s_1}$ $\qquad FV(\langle t_o,t_1\rangle) := FV(t_o) \cup FV(t_1)$

7. $t, t' \in \Lambda_{s''}$ => case $x{:}s.t$, $y{:}s'.t'$ esac $\in \Lambda_{s+s' \to s''}$ $\qquad FV(\text{case } x{:}s.t, y{:}s'.t' \text{ esac}) := FV(t)\backslash\{x{:}s\} \cup FV(t')\backslash\{y{:}s'\}$

8. $t \in \Lambda_{s'}$ => $\lambda x{:}s.t \in \Lambda_{s \to s'}$ $\qquad FV(\lambda x{:}s.t) := FV(t)\backslash\{x{:}s\}$

9. $t \in \Lambda_{s \to s'}$, $t' \in \Lambda_s$ => $(t\ t') \in \Lambda_{s'}$ $\qquad FV(t\ t') := FV(t) \cup FV(t')$

10. $Y_s \in \Lambda_{(s\to s)\to s}$ $\qquad FV(Y_s) := \emptyset$

Substitution is defined as usual in λ-calculus. We consider terms modulo α-conversion [1]. For convenience indices are omitted if the typing is obvious from the context. As standardization we assume that $\emptyset$ is the only term of type 1.

Examples: For better readability a more general notation is allowed in that we use manyfold sums and products, more than one parameter for abstractions and case-statements, omit brackets, and write t(t') instead of (t t').

We define some 'functions' on non-empty lists and trees over a sort entry:

list = entry + (entry × list)
tree = entry + (entry × tree) + (tree × entry) + (tree × entry × tree)

Attaching to the left side of a list is given by

latt ≡ λe,l.inr <e,l>

For attaching to the right side we write a recursive program

ratt ≡ Y(λf.λl,e.case ee.inr<ee,inl(e)>, ee,ll.inr<ee,f<ll,e>> esac (l))

Other functions on lists and trees are

conc ≡ Y(λf.λl,l'.case e.ratt(l,e), e,l".f(ratt(l,e),l") esac (l'))

inorder ≡ Y(λf.λb.case e.inl(e), e,b'.inr<e,f(b')>, b',e.ratt(f(b'),e),
b',e,b".conc(f(b'),inr<e,f(b")>) esac (b)) .

If we add some syntactical sugar - for instance replacing fixpoint operators by recursive procedures and using type declarations - we would get a more or less standard procedural language. But for proof theoretic reasons we prefer the more clumsy Λ-notation.

1.4 REDUCTIONS ON Λ

We use a reduction system to define the operational semantics of our language. It should be remarked that an equivalent algebraic semantics can be defined [8]. Hence all arguments hold in a purely algebraic framework compatible with abstract data type theory. But in proofs we heavily rely on operational properties.

We define a special notion of reduction denoted by YβηE. The equations E of the underlying specification (S,Σ,E) are understood as rewriting rules from left to right. We assume the following restrictions on E:

(E1) $(t,t') \in E$ => $FV(t') \subseteq FV(t)$

(E2) $\overline{E}$ is Church-Rosser ($\overline{E}$ trans.,refl.,substitutive & compatible closure of E).

(E3) $(t,t') \varepsilon E \Rightarrow x \varepsilon FV(t)$ appears only once in t

(E4) $(t,t') \varepsilon E \Rightarrow$ t is not a variable.

In addition we require that $T_\Sigma(\emptyset)_s$ is non empty for all $s \varepsilon S$.

Definition: $\rightarrow$ is the smallest (Type(S)-sorted) relation on Λ such that

E 1. $(t,t') \varepsilon E,\ t_i \varepsilon \Lambda_\Sigma(X)_{s_i}, i \varepsilon \underline{n} \Rightarrow t[x_i{:}s_i \leftarrow t_i, i\varepsilon\underline{n}] \rightarrow t'[x_i{:}s_i \leftarrow t_i, i\varepsilon\underline{n}]$

β
2. $(\lambda x{:}s.t)t' \rightarrow t[x{:}s \leftarrow t']$
3. case x:s.t,x':s'.t' esac (inl t") $\rightarrow$ t[x:s $\leftarrow$ t"]
   case x:s.t,x':s'.t' esac (inr t") $\rightarrow$ t[x':s' $\leftarrow$ t"]
4. p <t,t'> $\rightarrow$ t   q <t,t'> $\rightarrow$ t'

Y 5. (Yt) $\rightarrow$ t(Yt)

η
6. $\lambda x{:}s.(tx) \rightarrow t$ if $x{:}s \notin FV(t)$, $\lambda x{:}1.(t\emptyset) \rightarrow t$
7. <pt,qt> $\rightarrow$ t

We use $\twoheadrightarrow$ to denote the reflexive, transitive and compatible (with the structure) closure of $\rightarrow$. If we refer to a specific subset of rules we index by $\rightarrow_{Y\beta E}$, $\rightarrow_E$, $\rightarrow_6$,...

Example: According to the generalized notations for terms we have generalized reductions.
For the constant T ≡ in2<e1,in4<in1(e3),e2,in1(e4)>> where e1,...,e4 are given constants of type entry, we reduce the term 'inorder(T)':

inorder(T) → (λf.λb.case...esac(b))(inorder)(T)
→ λb.case e.in1(e), e,b".in2<e,inorder(b")>, b'e.ratt(inorder(b'),e), b',e,b".conc(inorder(b'),in2<e,inorder(b")> esac (b)(T)
↠ case...esac (in2<e1,in4<in1(e3),e2,in1(e4)>>)
↠ in2<e1,inorder(in4<in1(e3),e2,in1(e4)>)>
↠ in2<e1,conc(inorder(in1(e3),in2<e2,inorder(in1(e4))>)>
↠ in2<e1,conc(in1(e3),in2<e2,in1(e4)>)>
↠ in2<e1,in2<e3,in2<e2,in1e4)>>>

Remarks on the use of the equations E:

1. The essential use of the equations in the calculus is as 'stop-equations'. Beside β-reductions E-reductions are able to eliminate Y's and thereby stop a recursive calculation.
2. It is not realistic to require (E1)-(E4) to hold for all specifications. But we can take the following point of view: The use as stop-equations is a kind of error recovery. Like other authors we may distinguish a certain subset of E to be allowed for this purpose. Only these special equations may be used outside $T_\Sigma(X)$ in that arbitrary terms (especially those with Y's) are substituted for the variables. Then we require (E1)-(E4) only to hold for this subset.

As usual Church-Rosser property (CR), weakly Church-Rosser (WCR), finiteness of reductions are introduced. For illustration of such properties we use commuting diagram [1]. Unbroken lines stand for given reductions, dashed lines indicate when existence of reductions is claimed.

1.5 PROPERTIES OF THE CALCULUS

In this section we consider several properties of reductions which will be useful in proofs on composition of implementations. Because of lack of space only proof ideas can be given. For full proofs the reader is refered to [9] (or to [8] for an extended version).

Terms of special interest are those which are equivalent to a Y-free term, especially those of base types.

Definition: $t \in \Lambda_\Sigma(X)$ is called *terminating* $:\Leftrightarrow \exists t' \in \Lambda_\Sigma(X)$: no Y occurs in t' & $t \simeq t'$
Closed terms of base types are called *base constants* (BC).
($\simeq$ is the symmetric closure of $\twoheadrightarrow$)

Next we distinguish a class of Y-free BCs of a very simple form.

Definition: BCs only built up by inl,inr, <_,_>, $\emptyset$ and $T_\Sigma(\emptyset)$ are called *normal forms* (NF).

Facts: All base types have normal forms. Here we use that $T_\Sigma(\emptyset)_s$ is non empty, and the restriction to acceptable sort terms.
It is undecidable wether a term is terminating or not.

Next we will state some properties of the calculus in general, and then some special results about terminating BCs. It is essential for the proofs that there is no mixing-up of types, especially that a function space is not a product or a sum, and that the requirements (E1)-(E4) hold for equations. The proofs are restricted to terms of acceptable types but the proofs can be extended (compare [8]).

1.1 *Proposition*: β-reductions are finite. YβE is CR.

For the first statement we adapt the proof of Gandy [5]. For the Church-Rosser property use generalized (simultaneous) 1-step-reductions $\rightarrowtriangle$ as in [1]. To avoid difficulties with E-reductions we generalize E-redices to maximal connected Σ-parts of a term and regard $\twoheadrightarrow_E$-reductions as one step.

1.2 *Proposition*: η-reductions can be shifted to the end: $t \twoheadrightarrow_{Y\beta\eta E} t' \Rightarrow \exists t'': t \twoheadrightarrow_{Y\beta E} t'' \twoheadrightarrow_\eta t'$
Introduce a generalized 1-step-reduction $\rightarrowtriangle_\eta$. Then check $t \rightarrowtriangle_\eta t' \rightarrow_{Y\beta E} t''$ implies $t \twoheadrightarrow_{Y\beta E} t''' \rightarrowtriangle_\eta t''$ by case distinction on the origin of the YβE-redex in t.

1.3 *Proposition*:

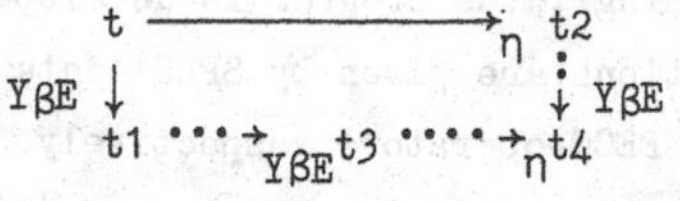

By a more complex redex marking we can show which yields the result by a diagram chase together with 1.1 and 1.2.

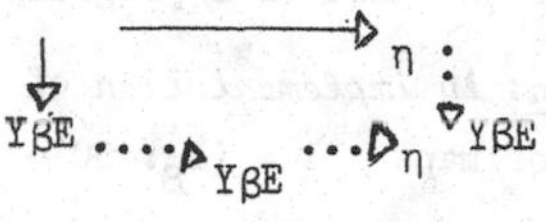

1.4 *Proposition*: βη is finite CR)

1.5 *Proposition*: t terminating $\Rightarrow \exists t'$: t' Y-free and $t \twoheadrightarrow_{Y\beta E} t'$.

We have a Y-free tn and reductions $t \equiv t0 \twoheadleftarrow t1 \twoheadrightarrow t2 \twoheadleftarrow \ldots \twoheadrightarrow tn$. With 1.1 and 1.3 we stepwise construct shorter chains beginning at the right side. We do not need η-reductions as they (by 1.3 at the end of a reduction) cannot eliminate Y's.

1.6 *Proposition*: t terminating BC $\Rightarrow \exists t'$: t' in NF and $t \twoheadrightarrow_{Y\beta E} t'$

Because of 1.5 it is sufficient to show that Y-free BCs are β-reducable to a term without Y, ,case,p,q. Now by a case distinction prove that a Y-free BC contains a β-redex as long as it contains a λ,case,p or q.

1.7 *Proposition*: t terminating BC $\Rightarrow \exists t'$: t' in NF and $t \twoheadrightarrow_{Y\beta} t'' \twoheadrightarrow_E t'$.

We have to show that in the situation $t \rightarrow_E t1 \twoheadrightarrow_{Y\beta} t2 \twoheadrightarrow_E t3$ with t3 in NF, E-reductions can be shifted to the right. If t3 is in normal form then we can assume that Y -reductions are of such a form that they consist only of $-\!\!\triangleright_Y$ and $-\!\!\triangleright_\beta$ steps of maximal breadth. Such reductions treat syntactical equivalent subterms of t1 in the same way. Then the E- and the Y -reductions are exchangeable ($\twoheadrightarrow_E$ may become a $-\!\!\triangleright_E$ -step).

(1.8 *Proposition*: t,t' in NF, $t \simeq t' \Rightarrow t =_E t'$)

1.4 is used for the proof of 1.9 which we do not need for the following. But 1.8 proves that the enrichment of $T_\Sigma(X)$ to $\Lambda_\Sigma(X)$ is consistent. But there are new constants on S-sorts (at least $Y(\lambda x{:}s.x)$, hence the enrichment is not complete nor persistent.

2. IMPLEMENTATIONS

2.1 MOTIVATION AND DEFINITION

The notion of implementation makes the idea of stepwise refinement precise. Program development is the construction of a hierarchy of specification levels with decreasing abstractness. An implementation builds a bridge between two neighbouring levels with the aim to come closer to a program. If we have two specifications SPEC0 and SPEC1, an implementation of SPEC0 by SPEC1 should preserve correctness of SPEC0-programs. We might call this idea 'relative programming'; the program is developed on the SPEC0-level but run on the SPEC1-level. Implementations are given by SPEC1-data structures and SPEC programs implementing SPEC0-sorts and SPEC0-operators respectively. This proceeding see to capture the task of a programmer who has to write a program realizing a data type.

Definition: An *implementation* of SPEC0 = (S0,Σ0,E0) by SPEC1 =(S1,Σ1,E1) is given by a pair of maps $I = (I_S{:}\ S0 \rightarrow ATo(S1),\ I_\Sigma{:}\ \Sigma0 \rightarrow \Lambda_{\Sigma1})$ s.t. $I_\Sigma(\sigma) \in \Lambda_{s_o \times \ldots \times s_{n-1} \rightarrow s}$ if $\sigma{:}s_o \ldots s_{n-1} \rightarrow s \in \Sigma$. We extend I_Σ to all terms in $T_\Sigma(X)$ by

$$I_\Sigma(x{:}s) := x{:}\tilde{I}_S(s) \qquad I_\Sigma(\sigma(t_0,\ldots,t_{n-1})) := T_\Sigma(\sigma) < I_\Sigma(t_o),\ldots,I_\Sigma(t_{n-1}) >$$

where $\tilde{I}_S := \Pi \circ I_S$ with the factorization Π: ATo(S1) $\rightarrow$ BType(S1).

Example: Assume that we have a standard specification of stacks and arrays [3,4]. Stacks are implemented by arrays plus a pointer as follows

ARRAY impl STACK by

sorts $\underline{stack} = \underline{array} \times \underline{nat}$

ops push = $\lambda s{:}\underline{stack}, n{:}\underline{nat}.\langle add(p(s),suc(q(s)),n),suc(q(s))\rangle$
pop = $\lambda s{:}\underline{stack}.\langle p(s),pred(q(s))\rangle$
empty = $\langle nil,0\rangle$
top = $\lambda s{:}\underline{stack}.\ p(s)[q(s)]$

(the notation hopefully is self-explaining. Sorts and operators which are implemented identically are omitted).

The given syntactical definition has to be completed by semantical constraints which express the correctness of an implementation. We of course intend that the given example is correct. The example illustrates two features of the notion of correctness to be defined:

1. We allow manifold representation of data. An element of type stack may be represented by different elements of type $\underline{array} \times \underline{nat}$, especially by arrays which differ in components above the pointer.
2. Not all elements of the implementing data type are used. In the example arrays with non-trivial entries under inder 0 are not used to represent stacks.

Definition: I is called *correct* iff

1. I is *consistent* :<=> $I_\Sigma(t) \simeq I_\Sigma(t')$ implies $t =_{E0} t'$ for all $t,t' \in T_{\Sigma 0}(\emptyset)_s$, $s \in S0$
2. I is *terminating*:<=> $I_\Sigma(t)$ is terminating for all $t \in T_{\Sigma 0}(\emptyset)$.

There is a close connection to the notion of correctness in the work of EKP [3], especially to their 'term version'. Consistency corresponds to their RI-correctness, and preservation of termination to OP-completeness. EKP add the requirement that the SPEC1-part remains unchanged in SORTIMPL. In our approach we have to examine what happens to T_Σ-terms in Λ_Σ. Property 1.9 guarantees that there are no additional identifications on $T_\Sigma(\emptyset)$-terms. On the other hand the only new terms on S-sorts that are not equivalent to $T_\Sigma(\emptyset)$-terms are non-terminating ones. But those we regard as error-programs which should not be used for implementations. In a more recent version [4] EKP restrict their SORTIMPL to special patterns which describe exactly those types over S which can be defined by recursive domain equation schemes in our approach. The equations EKP allow in their OPIMPL specification to implement operators are much more general than our recursive programs.

2.2 COMPOSABILITY

The most important property expected to hold for implementations is the composability of the single steps to one large implementation which then yields a computable program for every operator of the very first specification level.

We want to compose implementations I1 of SPEC0 by SPEC1 and I2 of SPEC1 by SPEC2.

Syntactically we intend the following: A $\Sigma 0$-operator $\sigma: w \to s$ has the SPEC1-implementation $I1(\sigma)$. Now replace all $\Sigma 1$-symbols in $I1(\sigma)$ by their SPEC2-implementations under I2. We obtain a SPEC2-program which is the implementation of σ in SPEC2.

For this purpose we have to extend I2 to all terms of $\Lambda_{\Sigma 1}(X)$. The composition of this extension with I1 then yields the implementation of SPEC0 by SPEC1.

Let $I = (I_S, I_\Sigma)$ be an implementation of SPEC0 by SPEC1. We extend I_S in the obvious way to I_S: ATo(S0) $\to$ ATo(S1). This defines a mapping I_S: BType(S0) $\to$ BType(S1) (using congruence properties) which finally extends to I_S^*:Type(S0) $\to$ Type(S1).
We extend I_Σ to $I_\Sigma^*: \Lambda_{\Sigma 1}(X) \to \Lambda_{\Sigma 2}(X)$ by

$I_\Sigma^*(x:s) := x:I_S^*(s) \qquad I_\Sigma^*(\emptyset) := \emptyset$

$I_\Sigma^*(\sigma(t_o,\ldots,t_{n-1})) := I_\Sigma(\sigma)\langle I_\Sigma^*(t_o),\ldots,I_\Sigma^*(t_{n-1})\rangle$

$I_\Sigma^*(p_{s,s'}) := p_{I_S^*(s),I_S^*(s')}$ and similar for q, inl,inr

$I_\Sigma^*(\lambda x:s.t) := \lambda x:I_S^*(s).I_\Sigma^*(t)$

and so on preserving the structure of programs.

Definition: For given implementations I1 of SPEC0 by SPEC1 and I2 of SPEC1 by SPEC2 the *syntactical composition* I2.I1 is defined by $I2.I1_S := I2_S^* \circ I1_S$ and $I2.I1_\Sigma := I2_\Sigma^* \circ I1_\Sigma$.

Fact: I2.I1 is an implementation of SPEC0 by SPEC2.

It should be noted that our notion of composition of implementations is different to that of EKP [4] as there SPEC0 is implemented using SPEC1 as a hidden part of the composed implementation while in our approach the intermediate specification disappears

Questions: Is the composition of correct implementations correct again?

Do the consistency- and termination-conditions still hold if extended to all terminating BCs?

The answers are in general negative. The following examples shows that the composition is not necessary terminating:

The program p over stacks

$p \equiv Y(\lambda f.\lambda s.pop(push(empty,top(f(s)))))(empty)$

is a constant of type stack and is equivalent to empty. But its implementation $I_\Sigma^*(p)$ as a program over arrays with pointers is not terminating:

$I_\Sigma^*(p) \equiv Y(\lambda f.\lambda s.\langle add(p\langle nil,0\rangle,..\rangle)(\langle nil,0\rangle)$

$\twoheadrightarrow$ $p1 \equiv Y(\lambda f.\lambda s.\langle add(nil,suc(0),p(f(s))[q(f(s))]),0\rangle)$

$\twoheadrightarrow$ $\langle add(nil,suc(0),p(p1)[q(p1)]),0\rangle$

$\twoheadrightarrow$ ...

We never get rid of p1 and the Y's in it.

But we can prove the following

2.1 *Main Theorem*: If I1 and I2 are correct implementations and I2.I1 is terminating then I2.I1 is correct.

We outline the idea of the proof. We use $\simeq$ and $\twoheadrightarrow$ ambiguously for reductions with regard to E1 and E2-equations.

Let $t,t' \in T_{\Sigma 0}(\emptyset)$ with $I2.I1(t) \simeq I2.I1(t')$ both terminating. As I1 is correct, I1(t) and I1(t') are terminating BCs. We have to show that they are equivalent in $\Lambda_{\Sigma 1}(X)$. Then the consistence of I1 yields $t \simeq t'$. Therefore it is suffient to prove

<u>Claim</u>: If I2 is correct, $t,t' \in \Lambda_{\Sigma 1}(X)$ are terminating BCs, $I2(t) \simeq I2(t')$ both terminating, then $t \simeq t'$.

The proof of the claim takes several steps where we use the properties of the calculus stated in chapter 1. For convenience we use I instead of I2 (Remark: The claim states that correctness extends to terminating programs which are implemented by a terminating program with regard to an arbitrary correct implementation).

<u>Step 1</u>: We can assume that all terms are *well-formed* (wf) in that Y's only occur in the form Y(t) (idea: replace all Y's by $YY \equiv Y(\lambda F.\lambda f.f(F(f)))$).

<u>Step 2</u>: In terminating terms we can add arbitrary many Y-reductions:

$$t \twoheadrightarrow t' \twoheadrightarrow t'' \qquad t \twoheadrightarrow_Y t''' \twoheadrightarrow t''$$

where t'' is Y-free.

<u>Step 3</u>: Terminating computations with wf BCs have without restriction of generality the following form: $t \twoheadrightarrow_Y t' \twoheadrightarrow_\beta t'' \twoheadrightarrow_E t'''$ such that (i) t'' is maximal β-reduced, and t''' is in NF.

<u>Step 4</u>: We can synchronize the reduction of t and I(t) as follows:

$$t \twoheadrightarrow_Y t' \twoheadrightarrow_\beta t'' \twoheadrightarrow_{E1} t'''$$
$$I(t) \twoheadrightarrow_Y I(t') \twoheadrightarrow_\beta I(t'') \twoheadrightarrow_{Y\beta\eta E2} t^{IV} \quad \text{with}$$

1. $t \twoheadrightarrow t'''$ is given as in step 3, t''' is in NF.
2. There are no Y-reductions of those Y's inherited from t in $I(t'') \twoheadrightarrow t^{IV}$.
3. t'' and I(t'') have the following form

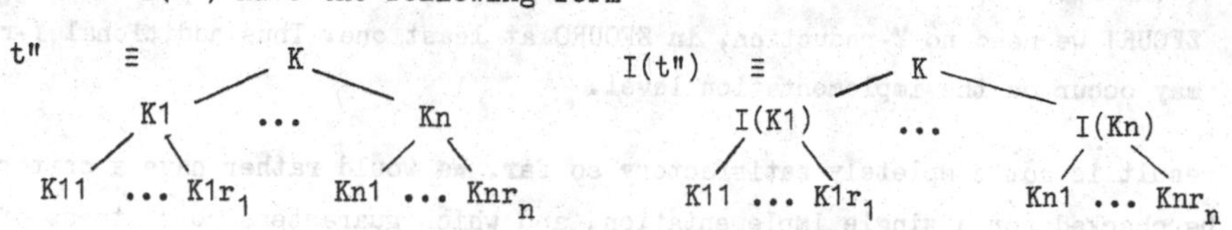

where a) the Ki's only contain $\Sigma 1$-operators, K contains no $\Sigma 1$-operators (hence only Λ-operators inl,inr,<_,_>), and the Kij have a Λ-operator in the root.

b) The terms Kij are no more β-reducible (thus contain a Y)

c) Up to sort indices the terms K and Kij are the same in t'' and I(t'')

d) If we replace the subterms Kij of Ki by suitably typed variables, resulting term being $Ki(\vec{x})$, $Ki(\vec{x})$ is E1-reducible to a $T_{\Sigma 1}(\emptyset)$ term.

<u>Step 5</u>: In the above situation the terms $I(Ki(\vec{x}))$ terminate. The reason is that the Y's in the Kij's inherited from t'' need not to be reduced in $I(t'') \twoheadrightarrow t'$ and that

there are no β-redexes in the Kij's, hence the Kij's do not interfere with the computations on the I(Ki)'s. Therefore we may replace the Kij's by arbitrary sub-terms of appropiate sort.

Step 6: - The final argument:

We take the situation of the claim and construct for both s and t the synchronized computations. We have

```
I(s")   ≡        Ks                    I(t")   ≡        Kt
              /     \                              /      \
      I(Ks1) ... I(Ksn)                    I(Kt1) ... I(Ktn)
       /\        /\                         /\         /\
```

I(s") and I(t") are reducible to the same term, and Ks ≡ Kt ≡: K (by step 4,3a). Hence

$$I(s) \simeq I(t) \simeq \overset{K}{\underset{t1 \ \cdots \ tn}{\diagup \diagdown}} \qquad \text{for some } ti \text{ in NF (use 1.6).}$$

Now replace, as sketched in step 5, the subtrees Ksij and Ktij by type matching terms of $T_{\Sigma O}(\emptyset)$ (which exists for any sort by general assumtion on SPEC, compare section 1.4). We obtain terms $\bar{s}$ and $\bar{t}$ of the form

$$\bar{s} \equiv \overset{K}{\underset{\bar{s1} \ \cdots \ \bar{sn}}{\diagup \diagdown}} \qquad \qquad \bar{t} \equiv \overset{K}{\underset{\bar{t1} \ \cdots \ \bar{tn}}{\diagup \diagdown}}$$

where $\bar{s1}, \bar{t1} \in T_{\Sigma O}(\emptyset)$. As the Ki(x) terminate we have $s \simeq \bar{s}$ and $t \simeq \bar{t}$. To show $s \simeq t$ we have to prove the equivalence of $\bar{s1}$ and $\bar{t1}$:

$I(\bar{s1}) \simeq I(Ksi) \simeq ti \simeq I(Kti) \simeq I(\bar{t1})$. The consistency of I gives us $\bar{s1} \simeq \bar{t1}$.

This completes the proof of the main theorem.

Example: (Compare step 4) ZFOURO implements ZFOUR1 (cf. introduction). The constant $(Y(\lambda x.3)(1)*2)* 2$ is reducible to 0 with regard to ZFOURO and ZFOUR1. But in ZFOUR1 we need no Y-reduction, in ZFOURO at least one. Thus additional Y-reductions may occur on the implementation level.

The result is not completely satisfactory so far. We would rather have a criterion that can be checked for a single implementation, and which guarantees correctness of composition. The proof of 2.1 gives a hint: It is sufficient to require that if a term t does not depend on one of its arguments x, then the implementation of t does not depend on x as well.

Definition: t is called I-*representative* of t' iff there exists a t" such that t" ≃ t' and t ≃ I(t").

An implementation is called *strong* iff for some I-representative of t' ∈ $T_{\Sigma O}(\emptyset)$ there exists a term t" with t ≃ t" and FV(t") = ∅ .

Remark: I is strong if SPECO has only equations with FV(t) = FV(t').

Proposition: Strongness is preserved by composition.

2.2 *Theorem*: If I2 is strong and correct, and I1 is correct then I2.I1 is correct.

We only have to show that I2.I1 is terminating. Take a terminating BC $t \varepsilon \Lambda_{\Sigma 1}(X)$. Then there exists a computation $t \twoheadrightarrow t'' \twoheadrightarrow_{E1} t'''$ such that t''' is in NF, and t'' has the form

t'' ≡ 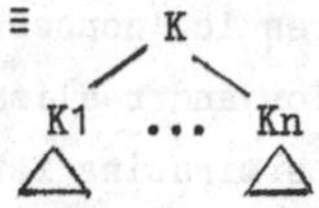

with $Ki(\vec{x}) \twoheadrightarrow ti \; \varepsilon \; T_{\Sigma 1}(\emptyset)$

and a computation $I2(t) \twoheadrightarrow I2(t'') \equiv$ 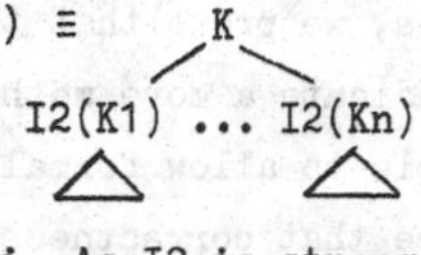

$I2(Ki(\vec{x}))$ is a I2-representative of ti. As I2 is strong I2(Ki) does not depend on its subtrees. Again we replace the subtrees by terms from $T_{\Sigma 1}(\emptyset)$. Let the resulting terms be $\overline{ti}$. As I2 is terminating $I2(\overline{ti})$ is terminating and equivalent to I2(Ki). Then the whole term I2(t) is terminating.

Example: The implementation of stacks by arrays is not strong as
$t \equiv pop(push(empty,n)) = empty$ does not depend on n. But its implementation
$I(t) \equiv \langle p\langle add(p\langle nil,0\rangle, suc(q\langle nil,0\rangle),n),... \quad \approx \langle add(nil,suc(0),n),0\rangle$
depends on n. We can do better and change the implementation of pop to
$pop \equiv \lambda s{:}\underline{stack}.\langle add(p(s),q(s),0),pred(q(s))\rangle$,
in that we erase the entry on the top and replace the pointer. Now the implementation is strong (assuming that all entries of the nil-array are 0's).

Remark: Strongness rules out a phenomenon well known in programming: If boolean expressions of a programming language are implemented evaluation strategies are used like "To evaluate 'x and y', first evaluate x. If x evaluates to 'false' then 'x and y' evaluates to false. If x evaluates to 'true' evaluate y". Different evaluation strategies of this kind yield different results with regard to non-termination. As evaluation strategies may be expressed by equations ("false and y = y") choosing different sets of equations to characterize the same (initial) algebra may change the intensional character with regard to 'infinitary' or 'non-terminating' terms. Strongness states that the intensional character of the equations is to be preserved to a certain extend.

CONCLUDING REMARKS AND OUTLOOK

1. An implementation step makes a part of a specification more computable. A typical situation is that SPEC1 is an enrichment of SPEC0, and the enrichment is to be implemented by programs. Considering parameterized data types may support an analysis of such a situation.

2. Other programming language constructs may be added. In [8] we add a fixed boolean sort and if-then-else-fi-constructs for any type with the restriction that any SPEC-term of sort boolean is equivalent to 'true' or 'false'. We then obtain similar results without any restrictions on the equations E of the base specification.

3.As already pointed out our notion of composition is different to that of EKP [4] which to our opinion is somewhat counter intuitive. For instance there the identical implementation of SPECO by SPECO not always is a unit with regard to composition.
4. At a first sight there seems to be little connection to the work of Lipeck [7] but in fact the extension of a data type by recursive data structures is 'conservative' in terms of [7(4.12)] which guarantees compatibility of construction and realization steps [7(4.11)]. Now the termination condition allows to reduce any terminating term to a normal form or, with other words, we prove that the respective functor is conservative.
5. The observation of 4. may indicate a more methodological aspect: The syntax of an implementation should be flexible to allow formalizations close to the given problem. This freedon has the consequence that correctnesss proofs (that functors are conservative) are more complicated. The situation is well known from programming languages.
6. (Added when preparing this version) There seems to be a close connection to [10] where recursive schemes are used as 'programs'. If we add (as in [8]) a fixed boolean sort and if_then_else_fi-operators our notion of programs seem to cover that of [10] (apart from the semantical side conditions). The in [10] indicated conditions for correctness of compositions seems to be a semantic counterpart to strongness. The connection needs further investigation.

REFERENCES

[1] Barendregt,H.: The Lambda Calculus, North-Holland 1981

[2] Burstall,R.M., Goguen,J.A.: Putting Theories Together to Make Specifications, Proc. of 1977 IJCAI MIT Cambridge, 1977

[3] Ehrig,H., Kreowski,H.J., Padawitz,P.: Algebraic Implementation of Abstract Data Types. Concept,Syntax,Semantics and Correctness, Proc. ICALP'80, LNCS 85, 1980

[4] Ehrig,H., Kreowski,H.J., Mahr,B., Padawitz,P.: Algebraic Implementation of Abstract Data Types, TCS 20, 1982

[5] Gandy,R.O.: Proofs of Strong Normalisation, In: To H.B.Curry: Essays on Combinatory Logic, Lambda Calculus and Formalism, Academic Press 1980

[6] Ganzinger,H.: Parameterized Specifications: Parameter Passing and Optimizing Implementation, Bericht Nr. TUM-I8110, TU Muenchen 1981

[7] Lipeck,U.: Ein algebraischer Kalkuel fuer einen strukturierten Entwurf von Datenabstraktionen, Dissertation, Ber. Nr.148, Abt. Informatik, Universitaet Dortmund, 1983

[8] Poigné,A., Voss,J.: Programs over Algebraic Specifications - On the Implementation of Abstract Data Types, Ber. Nr. 171, Abt. Informatik, Uni Dortmund, 1983

[9] Voss,J.: Programme ueber algebraischen Spezifikationen - Zur Implementierung von Abstrakten Datentypen, Diplomarbeit,Abt. Informatik, Uni Dortmund, 1983

[10] Blum,E.K., Parisi-Presicce,F.: Implementation of Data Types by Algebraic Methods, JCSS 27, 1983

A LISP COMPILER FOR FP LANGUAGE AND ITS PROOF VIA ALGEBRAIC SEMANTICS

C. CHOPPY, G. GUIHO, S. KAPLAN

Laboratoire de Recherche en Informatique
Université de Paris-Sud
Bâtiment 490
91405 Orsay - Cedex, FRANCE

INTRODUCTION

Since Backus pioneer paper [BACKUS 78], much work has been devoted to the study of the Functional Programming (FP) approach. As a main reason for this success, FP environment is characterized by a clean algebraic framework for reliable program design. The general purpose of this paper is to describe a LISP computation system for a FP language and to provide a FP algebraic semantics in order to prove the correctness of the system.

Part I provides a modelization of FP algebra of programs within the framework of abstract data types. The purpose of this part is to describe and define FP in a totally algebraic way (which strongly relies on the algebraic structures of FP).

Part II is the description of an actual interactive FP computation system. FP expressions are compiled into LISP code that is evaluated by the LISP interpreter. The system includes some efficient mechanisms, such as rapid error transmission (*strictness* in FP sense), simple optimization for function composition.

Part III sketches the proof of the compiler w.r.t. semantics given in part I. More precisely, we show how to perform a complete proof (which consists in proving the *axioms* given in the first part - assuming reasonable lemmas about the LISP environment), and conduct typical axioms demonstrations.

We assume the reader has basic knowledge about Backus Functional Programming systems. This paper is a short version of [CHOPPY et al. 83].

I - ALGEBRAIC SEMANTICS OF FP LANGUAGE

I.1. - INTRODUCTION

The algebraic behaviour of FP languages has been considered as an essential asset, in particular compared with more classical programming languages. This is investigated in [BACKUS 78], where the set of FP programs is viewed as an algebra, and some properties (*theorems*) of this algebra are stated, as :

$$\forall\ f,g,h\quad [f, g] \circ h = [f \circ h, g \circ h].$$

In order to consider this aspect more systematically, the formalism of abstract data types [ADJ 78, ZILLES 79,...] is particularly well suited. In this part, we provide an extensive modelization of FP environment in that framework. This gives a mathematical semantics for FP language, according to which it is proven (part III) that the FP compiler described in part II is correct.
Conversely, there has been several attempts to modelize programming languages with abstract data types [GOGUEN et al. 79, GAUDEL 80, WIRSING et al. 81, BROY et al. 80, ...]. It clearly happened to be easier for FP, due to its strong natural algebraic structure.

I.2 - DEFINITION OF THE ABSTRACT TYPE

The general structure of the type is summarized as follows :

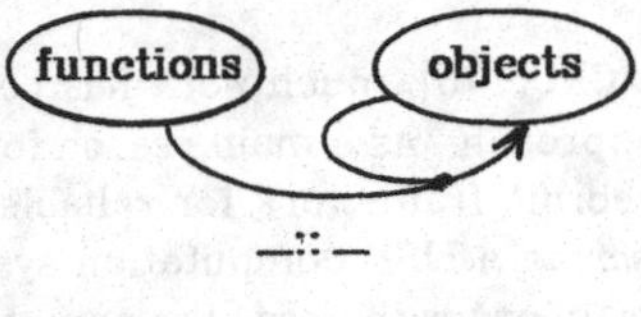

Figure 1 : FP type structure

[this denotes a function _:: _: **functions** x **objects** → **objects** , the "_" symbol standing for the position of the arguments.]
We now develop more precisely the structure of the two previous sorts.

OBJECTS AND LISTS OF OBJECTS

The structure of the sorts involved in the definition of FP base objects is the following :

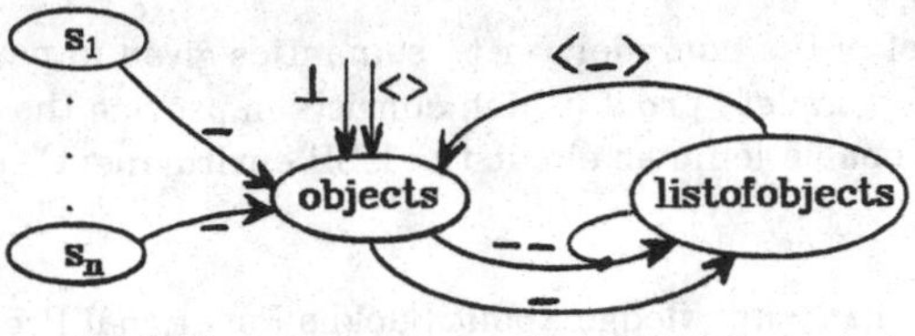

Figure 2 : General structure of objects sort

• The s_i are predefined sorts of disjoint atomic objects, with s_1 and s_2 being for instance the usual **boolean** and **integer** sorts.
We will require the existence of "equality predicates" for all s_i :

$\_?\ EQ_i ?\ \_ : s_i \times s_i \to s_1 =$ **boolean**

that are sufficiently complete and hierarchically consistent with respect to **boolean** (cf. e.g. [GUTTAG et al. 78]).
The s_i are any types algebraically specified (i.e. with axioms and signatures). We will suppose that axioms about s_i (and $\_?\ EQ_i ?\ \_$) may be interpreted as canonical rewrite rule systems.

• To keep Backus syntax for lists, the following set of constructors for the sort **listofobjects** is needed :

$\_$: **objects** → **listofobjects**,
$\_\,\_$: **objects** x **listofobjects** → **listofobjects**

To ensure the construction of list to be a strict operation, the following equations "between constructors" are given, similar to classical "error propagation" axioms :

$< \_\,\_(\bot, l) > = \bot \qquad < \_\,\_(o, \_(\bot)) > = \bot$
$< \_(\bot) > = \bot \qquad <\bot> = \bot$

On the whole, there are exactly four kinds of constructors for the sort **objects** :
the constant $\bot$,
the constant <> corresponding to the object "empty-list",
the coercions $\_$: $\mathbf{s_i}$ → **objects** yielding atomic FP-objects, and
the operator < $\_$> transforming a **listofobjects** into an **object**.
From now on, we present set of axioms that are sufficiently complete w.r.t. this family of constructors.

We now define a *definedness* predicate : $?\bot?\ \_$: **objects** → **bool**, with the (sufficiently complete set of) axioms :

For $i \in [1..n]$, $\forall x \in \mathbf{s_i}$, $?\bot?(\_(x)) = ff$
$?\bot?(\bot) = tt$
$\forall o \in$ **objects** , $\forall l \in$ **listofobjects**
$?\bot?(<>) = ff \qquad ?\bot?(<\_(o)>) = ?\bot?(o)$
$?\bot?(<\_\,\_(o,l)>) = ?\bot?(o)$ OR $?\bot?(<l>)$

This achieves the description of the **FP base objects**. It is tedious, though easy to check the following facts :
- semantics of the $\mathbf{s_i}$ is not altered by the new axioms (more precisely, the global specification is sufficiently complete w.r.t. the former specifications of the $\mathbf{s_i}$.
- semantics of the **objects** and **listofobjects** part is correct w.r.t. the usual FP model.
- we still have a runnable (i.e. confluent and noetherian) rewrite rule system (when the specifications of the $\mathbf{s_i}$ are so themselves).

THE "FUNCTIONS" SORT

FP functions are elements of the sort **functions**. We consider successively the three classical classes of functions : primitive functions (becoming constants in our formalization), functions obtained by application of combinators (that are operators of the algebra with non null arity), and recursively defined functions .

In FP, functions need to be **strict**. We thus adopt the following general axiom :

$\forall f \in$ **functions**, $\quad f :: \bot = \bot$

• *Primitive functions*
We shall give the algebraic specification of the primitive functions head and tail ; the other functions (selectors, identity, reverse, ...) would be modelized in the same way (cf. [CHOPPY et al. 83]).

Head and tail functions

We define the two "constants" head and tail in the following way :

$\forall x \in$ **objects** , $\forall l \in$ **listofobjects**

$?\bot?(<x\,l>) = ff \implies$ head :: $<x\,l> = x$

$?\bot?(<x\,l>) = ff \implies$ tail :: $<x\,l> = <l>$

$\forall l \in$ **listofobjects**

$?\bot?(<l>) = ff \implies$ head :: $<l> = l$

$?\bot?(<l>) = ff \implies$ tail :: $<l> = <>$

head :: $<> = \bot$ tail :: $<> = \bot$

$\forall x \in \mathbf{s_i}$ (for $i \in [1..n]$) head :: $\_x = \bot$ tail :: $\_x = \bot$

• *Functionals (combinators)*

Combinators, in our framework, are just operators of range **functions**, and of a non-null arity.

Composition

We have _o_ : **functions** x **functions** → **functions**, and :

$\forall f,g \in$ **functions**, $\forall x \in$ **objects** (f o g) :: x = f :: (g :: x)

Construction

To modelize the FP-construction of functions

(which associates to $f_1,\ldots,f_n$ the function $[f_1,\ldots,f_n]$, s.t:

$$[f_1,\ldots,f_n] :: x = < f_1 :: x,\ldots,f_n :: x >)$$

we apply the pattern we used to construct lists of objects introducing a new sort **listoffunctions**, and new operators "_ _", "_", [], and [_].

Apply to all

Let alpha : **functions** → **functions** with

$\forall f \in$ **functions** alpha f :: <> = <>

$\forall f \in$ **functions**, $\forall x \in$ **objects**, $\forall l \in$ **listofobjects**

alpha f :: $<x\,l> = <(f :: x)\ (\text{alpha } f :: <l>)>$

$\forall f \in$ **functions**, $\forall t \in \mathbf{s_i}$ (for $i \in [1..n]$) alpha f :: $\_t = \bot$

• *Recursively defined functions*

For recursively defined functions, a new sort **identifiers** is given, with a coercion from **identifiers** into **functions**. Because there is no notion of environment, and thus no global binding between function identifiers and their possible semantics, we use the following axiom :

$\forall f \in$ **identifiers**, $\forall x \in$ **objects**, f :: x = $\bot$

(Applying the name of a function without body to an object gives bottom.)

We now introduce the fixpoint operator:

fix _ ≡ _ : **identifiers** x **functions** → **functions** ,

taking a name and a body, and producing a (well-defined) function.

Let us call *context* any term $T \in T_{S,\Sigma}[X]$ of range **functions**, X being a variable of sort **functions** too. For instance,

$$\mathbf{T}_{FACT}[X] = \text{eg0} \rightarrow 1 \text{ or mult o [id , X o sub1]}$$

is a context.

We now adopt the following *meta*-axiom :

for any context T, $\forall id \in$ **identifiers**, $\forall f \in$ **functions**

[fix id ≡ T[id]] :: x = T [fix id ≡ T[id]] :: x

It just states that we can **unfold** the definition of a function.

Example :
Let t_{FACT} be fix idfunct $\equiv$ $\mathbf{T_{FACT}}$ [idfunct] ; one proves by induction on n=0 that

$$t_{FACT} :: n = n!$$

NOTE :
In the meta-axiom hereabove, the expression " fix id $\equiv$ **T**[id] " must be substituted to the formal argument of **T**. We shall suppose that the substitution is total (i.e. performed at every occurence in **T**). This is discussed in the next part.

I.3 - INITIAL ALGEBRA SEMANTICS

Having modelized FP with abstract data types (ADTs), we have provided a mathematical semantics for FP (which is used in the last part to prove the correctness of the compiler). Actually, it is often considered that the semantics of an abstract data type is its initial model (when it exists). This can been done here, the initial model being :

$$\hat{I} = T_{S,\Sigma} / \equiv_{\{Ax\}},$$

where {Ax} is the set of (positive) axioms previously given.

This approach naturally provides a ***least fixed point semantics*** for recursively defined functions, in the following sense.
Suppose a classical recursive equation scheme f $\equiv$ **F**[f] is given, where **F** is a continuous functional.
Let f_0 be the least fixed point of this equation. f_0 may be partial. Let $x \in DOM(f_0)$.
It is possible, accordingly to [BACKUS 81b], to "lift" (which means approximately "to translate into FP") the definition of **F** , yielding a corresponding Φ , which is clearly a context in our sense. We then have :

THEOREM

$$(\text{fix id} \equiv \Phi[\text{id}]) :: \text{lift}[x] = \text{lift}[f_0(x)] \quad \textbf{in } \hat{I}.$$

This means that in $\hat{I}$, (fix id $\equiv$ **T**[id]) :: x ***computes*** the least fixed point of **T** (considered as the ***lifted*** of a classical recursive functional)
The proof of the theorem is clear, using the fact that successively applying the meta-axiom to a scheme corresponds to its computation through the ***full-substitution computation*** rule, which is safe (i.e. computes the least fixed point [MANNA 74]).

Notes :
- As shown in [ADJ 77], initial algebra semantics for an abstract data type is naturally equivalent to denotational semantics. Nevertheless, when applied to our previous modelization, this does not provide denotational semantics for FP ***considered as a language***. On the other hand, it is possible to directly deduce denotational semantics from our framework in the following way :
  - the domain of the objects is just the **objects** sort of the initial algebra $\hat{I}$, ordered by $x < y$ iff $x = \text{class}_{\hat{I}}(\perp)$
  - FP functions are interpreted by their action in $\hat{I}$, through their application by $(\_ :: \_)^{\hat{I}}$. Functionals are treated analogously.
- The axioms that have been given may be interpreted as rewrite rules, including the meta-

axiom (applied by full-substition). This provides a confluent system, but which clearly does not always terminate ; for instance, the meta-axiom may be applied infinitely often to any operator fixpoint definition.
Nevertheless, it is a simple way of interpreting FP languages, which is correct w.r.t. to the previous initial algebra semantics (according to classical results on rewriting).

II. - LISP Computation of FP

This FP computation system is available both in MULTICS/MACLISP/EMACS and in UNIX/FRANZLISP/WINNIE [AMAR 83] environments (i.e. user interface is done through full-page editor). It makes use of a grammar generator and of a specialized parser [VOISIN 84].

The system is given the definition of the FP language that we deal with ; semantic attributes are attached to each FP symbol :

- the type of the symbol, used during the parsing process (which uses strong typing)
- a LISP expression representing semantics for the symbol, and used during the code generation step.

The computation itself simply works by parsing a given FP expression, realizing semantic attribute attachment (i.e. the FP expression is compiled into LISP code), and evaluating the resulting tree by mere LISP evaluation process.
This compiling process being simple, it will be shortly described with an example ; we shall give more details on features concerning strictness and function definition.

II.1 - EXAMPLE

In order to provide some feeling for the system overall use and processing, and its evaluation mechanism, let us take an example : the figure (next page) is the resulting parse tree (after semantic attachment) of the FP expression :

head o tail :: < a b c >

Let us comment on this figure :

- the "_" signs in the operator name indicate the argument positions ("mixfixed" operators are allowed), therefore coercion is denoted by an operator name reduced to "_" ; in the operator semantics, the arguments are referred to by "_1", "_2",... (this is a macro provided by the system)
- semantics of the application "_:: _" : FPapply is application preserving strictness (see next paragraph), unparse translates a LISP expression into FP syntax
- semantics of _o _, which is simplified for sake of readability, actually involves beta-reduction
- primitive function semantics are expressed by LISP lambda expressions ; see for instance, semantics of head and tail : fpcar and fpcdr return (by "throw" mechanism) bottom when the argument is an atom
- objects are atoms (a, b, c,..., bottom, also booleans and integers via coercion) and sequences of atoms constructed by means of a sort "listofobjects" and operators : <> (empty sequence), <_>, _ _, _.

To compute this FP expression, one has to generate the code (i.e. the S-expression) corresponding to it, by traversing the tree, and evaluate, through the LISP interpreter, the resulting S-expression.

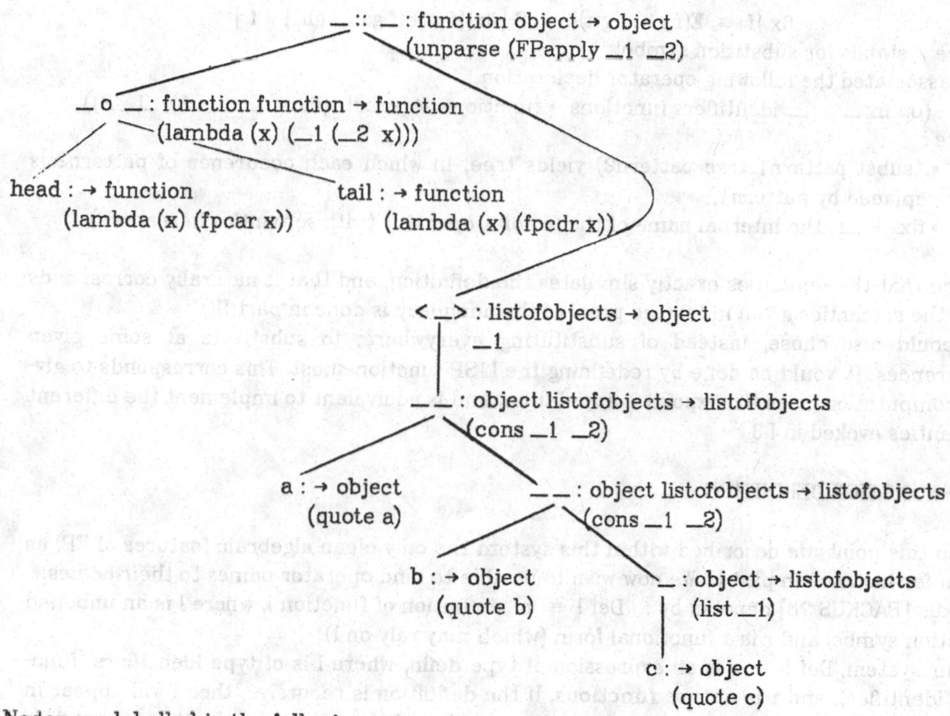

Nodes are labelled in the following way :

operator-name : operator-profile
operator-semantics

Figure 3 : Parse tree of head o tail :: < a b c >

II.2 - STRICTNESS

A rough way for implementing strictness would be to implement a ?⊥? predicate (as described in I.2) and have the application and the functionals use it. There are essentially two cases to consider :

(i) a mere application of a function to bottom :

head :: ⊥ = ⊥

(ii) bottom coming up at some point while evaluating an expression :

head o tail o head :: <a l> = ⊥

where a ∈ objects, l ∈ listofobjects. In this case, one wants bottom to "bubble up" through the several steps of evaluation. The LISP catch-throw mechanism provides an elegant way to by-pass evaluation steps. Along with the two cases considered above, catch-throws are used in the system :

(i) at the parsing level, the corresponding throw coming from the bottom semantics

(ii) at the application level, the corresponding throw coming from the primitive function semantics

II.3 - FIXPOINT SEMANTICS

Backus fixpoint definition is :

$$\text{fix}\,\{f \equiv \mathbf{E}(f\,g1 \ldots gn)\} = E\,[\text{fix}\,\{f \equiv \mathbf{E}(f\,g1 \ldots gn)\} \,/\, f\,]$$

where / stands for substition symbol.
It is associated the following operator declaration :

(op fix _ ≡ _: identifiers functions → functions // (subst '(fix_≡ _ _1 _2) _1 _2))

where :

- (subst pattern1 tree pattern2) yields tree, in which each occurence of pattern2 is replaced by pattern1,
- fix_≡ _is the internal name of the fixpoint operator (with semantics hereabove).

Notice that the semantics exactly simulates the definition, and that it naturally corresponds with the semantics given in I.3 (the proof of the adequacy is done in part III).
We could also chose, instead of substituting everywhere, to substitute at some given occurences. It would be done by redefining the LISP function subst. This corresponds to giving computation rules for fixpoint computation, and is equivalent to implement the different semantics evoked in I.3.

II.4 - FUNCTION DEFINITION

Up to this point, we described within this system the only clean algebraic features of FP, as given for instance in part I. We now wish to be able to bind operator names to their bodies. Backus [BACKUS 78] denotes by : Def l ≡ r definition of function l, where l is an unbound function symbol and r is a functional form (which may rely on l).
In our system, Def l ≡ r is an expression of type **defin**, where l is of type **identifiers** (function identifier), and r is of type **functions**. If the definition is recursive, then l will appear in the right hand side r in "call l", call being an operator with an argument l of type identifiers and a result l of type functions having the semantics of the function associated to the identifiers l.
The declaration of this operator is :

(op Def _ ≡ _: identifiers functions → defin // (eval '(op _1 : → functions // _2)))

Parsing and evaluating the expression Def l ≡ <body> causes the following actions :
- at a pre-parsing level :

declaration of l as an identifiers : (op l : identifiers) with default semantics (quote l). This pre-parsing action is somewhat inelegant, for it is generates side-effects during the parsing process; it allows the user to create new names without having to introduce the corresponding declaration at the grammar building level (in particular, one has to get out of FP environment to do this).
- when evaluating Def l ≡ <body> , the declaration :

(op l : → functions // (quote <body>)) is evaluated, thus realizing the binding.

In conclusion, the system described above for FP computation is efficient and works within a nice user interface. Its caracteristics are : the use of a strongly typed parser and fast and elegant processing for strictness and fixpoint computation.

III. - PROOF OF THE COMPILER

In this section, we prove that the FP computation system described in the second part is *valid* w.r.t. the semantics given in the first part. Actually, it is shown that *the FP compiler* (in its LISP environment) *is a model* [an algebra] *of the abstract data type defined in part I.*

In the following, the compiler (within its LISP environement) will be denoted by Ξ, and the abstract data type FP by $\overline{FP}$.
In order to prove that Ξ is a $\overline{FP}$-algebra, one has to
1) assign to each operator of $\overline{FP}$'s signature its *interpretation* in Ξ.
This is the easy part : each operator is interpreted as described in the second section, with its LISP action. For instance, the interpretation $(\_::\_)_\Xi$ of the operator :

_::_: functions x objects -> objects

is the function associating to its arguments _1 and _2 the LISP-result : (unparse (catch (apply _1 _2))) [cf. part II].

2) Now, Ξ is viewed as a sig($\overline{FP}$)-algebra. To prove that it is a $\overline{FP}$-algebra, it still has to be shown that Ξ satisfies the axioms of $\overline{FP}$, given in part I.
To do so, we shall avoid giving a formal description of the LISP system itself (the correctness of which would have to be proved) and then of the compiler Ξ, and finally prove the correctness of this formal model w.r.t. the algebraic semantics. It would be a large work, which is not absolutely necessary for our purpose.
The proofs are performed, assuming some reasonable *axioms* about the system, and some natural *deduction rules*. For instance, one will usually agree with the following axiom :

(car (cons a b)) $=_\Xi$ a (LISP-axiom)

provided that the evaluation of *a* yields no side-effect.

III.1. - THE PROOFS

In this part, we prove that Ξ satisfies the axioms of part I, i.e.

- the axioms concerning the object sorts. It is easy to check them in our implementation, and this wil not be developed here
- the general axiom : $f :: \perp = \perp$ (proven below)
- the axioms defining the behaviour of FP "constants" (in our sense) and "combinators". We will just prove one of them, the other proofs being similar
- the meta-axiom concerned with the definition of recursively defined functions (proved below).

Strictness of functions
The axiom to be proven is : $\forall f \in$ **functions** $f :: \perp = \perp$ **in** Ξ
or : $\forall f \in$ **functions** $(f :: \perp)_\Xi = (\perp)_\Xi$.
But : $(f :: \perp)_\Xi = f_\Xi ::_\Xi \perp_\Xi$ = (unparse (catch (apply 'f$_\Xi$ $\perp_\Xi$))) .
Since : $\perp_\Xi$ = (throw 'bottom) ,
then : $(f :: \perp)_\Xi$ = (unparse 'bottom) = 'bottom $= \perp_\Xi$ QED

Proof of the correction of "head"
The three axioms defining head are :

- head :: <> = $\perp$
- $\forall x \in \mathbf{s}_i$ (for i $\in$ [1..n]) head :: _x = $\perp$
- $\forall x \in$ **objects**, $\forall y \in$ **listofobjects**
  $?\perp?(\langle x\ y \rangle) = ff \implies$ head :: < x y > = x

1) head :: <> = $\perp$
(head :: <>)$_\Xi$ = (unparse (catch (apply 'head$_\Xi$ <>$_\Xi$)))
= (unparse (catch (apply 'fpcar nil)))

But recall that :
fpcar = (lambda (x) (cond ((atom x) (throw 'bottom))) (t (car x))))
So : (head :: <>)$_{\Xi}$ = bottom = $\perp_{\Xi}$

2) head :: _x = ⊥
Let x be from any $\mathbf{s}_i$. We have : (head :: _x)$_{\Xi}$ = (unparse (catch (fpcar [_x]$_{\Xi}$)))
Since : (atom [_x]$_{\Xi}$) = t , therefore : (head :: _x)$_{\Xi}$ = bottom = $\perp_{\Xi}$

3) ?⊥?(<x y>) = ff ⟹ *head :: <x y> = x*
(head :: <x y>)$_{\Xi}$ = (unparse (catch (fpcar <x y>$_{\Xi}$)))
Now, recall that :

<x y> = < _ >
x : **objects** y : **listofobjects**

and that : (< _>)$_{\Xi}$ = (lambda (x) x) and (_ _)$_{\Xi}$ = (lambda (x y) (cons x y))
So : (head :: <x y>)$_{\Xi}$ = (unparse (catch (fpcar (cons x$_{\Xi}$ y$_{\Xi}$))))
Suppose the following condition holds :

?⊥?(<x y>)$_{\Xi}$ = ff$_{\Xi}$

or ?⊥?$_{\Xi}$(cons x$_{\Xi}$ y$_{\Xi}$) = nil
Note that the exact meaning of ?⊥?$_{\Xi}$ has not yet been defined. We shall suppose that it returns "t" whenever the Ξ-evaluation of its argument yields "bottom", and "nil" otherwise. Actually, it is a so-called *hidden function*, that does not physically exist in Ξ nor needs to be implemented, but that has to be supplied for the proofs.
Here, because ?⊥?$_{\Xi}$(cons x$_{\Xi}$ y$_{\Xi}$) = nil , we can apply the LISP-theorem :
(fpcar (cons a b)) = a when no side effect occurs.
Hence : ?⊥?$_{\Xi}$(cons x$_{\Xi}$ y$_{\Xi}$) = nil ⟹ (head :: <x y>)$_{\Xi}$ = x$_{\Xi}$

QED

This achieves the proof that head in the system Ξ is correct w.r.t. $\overline{FP}$ algebraic semantics.

Proof of the meta-axiom
To prove : [fix id ≡ **T**[id]] :: x = **T**[fix id ≡ **T**[id]] :: x .
Recall that :
(fix_≡ _)$_{\Xi}$ = (lambda (name body) (subst '(fix_≡ _$_{\Xi}$ name body) name body))
where (subst x y z) is the result of substituting x for y everywhere in z.
Hence :
([fix id ≡ **T**[id]] :: x)$_{\Xi}$ =
(catch (unparse (apply (subst '(fix_≡ _$_{\Xi}$ id$_{\Xi}$ **T**$_{\Xi}$[id$_{\Xi}$]) x$_{\Xi}$))) =
(catch (unparse (apply **T**$_{\Xi}$ [(fix_≡ _$_{\Xi}$ id$_{\Xi}$ **T**$_{\Xi}$[id$_{\Xi}$]] x$_{\Xi}$))) =
(catch (unparse (apply (**T**[fix id ≡ **T**[id]])$_{\Xi}$ x$_{\Xi}$))) =
(**T**[[fix id ≡ **T**[id]] :: x)$_{\Xi}$

QED

Thus, the compiler Ξ satisfies the meta-axiom. Note that the unfolding is performed in Ξ at every occurence of the identifier, as specified in the standard interpretation of the meta-axiom ("full-substitution" strategy). Another choice would be immediatly implemented by changing the action of the LISP-function "subst".
This achieves the proof of the computation system (recalling that the missing demonstrations are identical, in their principle, to the previous ones).

Remark : More precisely, the compiler belongs to the class $\overline{FP}$-alg* of the $\overline{FP}$-algebras satisfying :

$$tt \neq ff .$$

We may notice that the terminal model T^*_{FP} is characterized by the property of extensional equivalence : two objects f and g of sort **functions** are equal in T^*_{FP} iff $\forall x \in$ **objects**, f :: x = g :: x . This is because the only *observator* acting on the sort **functions** is the application (_:: _), and that its range is the sort **objects** which is monomorphic in $\overline{FP}$-alg* (this stems from the completeness of the predicates _?EQ?$_i$_ w.r.t. the **booleans**).

• Now, suppose we decide to consider that two **functions** in Ξ are equal if they are extensionally equivalent **in** Ξ, then : **Ξ is the terminal model of** T^*_{FP} .

Proof

We just have to prove that, if f,g ∈ $T_{S,\Sigma}$ of sort **functions**

$$f =_{ext} g \quad \text{iff} \quad f_\Xi =_{ext} g_\Xi$$

This is a simple consequence of the fact that the mechanism for computing recursive functions in Ξ *and* the formal evaluation process are similar, having both the least fixed point semantics property, and acting both by full-substition.

Note that getting extensional equivalence between functions through the consideration of terminal models is quite classical [KAMIN 80, WIRSING et al. 81].

• On the other hand, we could consider that two **functions** in Ξ are equal if they have the same definition (LISP genuine notion of equality). Then, we now claim that :

Ξ is the initial model of T_{FP} ,

Proof

We just need to prove that, if f,g ∈ **functions**

$$f_\Xi = g_\Xi \quad \text{iff} \quad \{Ax\} \models f = g$$

⟹ If f and g are (LISP-)equal, they must have exactly the same definition. Then, f = g , and, a fortiori, {Ax} ⊨ f = g.

⟸ If {Ax} ⊨ f = g , then, in fact, f = g. This is because there is no equalities between objects of **functions** in {Ax}. QED

In conclusion, Ξ may be considered as the initial or the terminal model of *FP**, according to which philosophical meaning we agree to give to the notion of equality between functions in Ξ.

But anyhow, the choice corresponds exactly to the different notions of "meanings" we gave to the abstract data type describing FP.

Thus, we showed that Ξ is a correct FP compiler , with respect to the abstract type FP , **and** to the semantics we agreed to give to the type.

CONCLUSION

The aim of this paper was to describe a FP computation system, and to prove it.

This led to consider FP semantics using two different approaches : an axiomatic approach using the abstract data type theory, and a lambda-calculus and operational approach using a LISP environment. A connection was then established between them proving the validity of the latter w.r.t. the former.

In the first part, a complete formalization for a FP environment is given in the framework of algebraic abstract data types. This led straightforwardly to different semantics for FP. In particular, a thorough treatment for fixpoint definition of functions, parameterized by evaluation mechanisms was provided.

In the second part, a FP computation system is described within LISP environment. The organization of the system remained close to FP philosophy, even for the very conception of the compiler : for instance, FP functions are implemented as functions constants. An original way of implementing strictness is also provided.

Finally, this system is proven to be a model of the FP type. Notice that, the proof really dealt with the physical aspect of the compiler (not trying to further modelize it).

Ackowledgments
We want to thank Frèderic VOISIN for help in using and interfacing the parser with the compiler. We thank the members of the ASSPRO project for helpfull discussions.

We thank John WILLIAMS for fruitfull discussion and remarks.

REFERENCES

[ADJ 78]
Goguen, J.A., J.W. Thatcher and E.G. Wagner, An initial algebra approach to the specification, correctness, and implementation of abstract data types, in Current Trends in Programming Methodology, vol. IV Data Structuring, R. Yeh ed., Prentice-Hall, 1978.

[AMAR 83]
Amar, P., Winnie : un editeur de textes multifenetres extensible, Actes des Journees BIGRE, Le Cap d'Agde, France, 1983.

[BACKUS 78]
Backus, J., Can programming be liberated from the von Neumann style? A functional style and its algebra of programs, C.A.C.M., **21**, 8, pp. 613-639, August 1978.

[BACKUS 81a]
Backus, J., The algebra of functional programs : function level reasoning, linear equations and extended definitions, Proc. Int. Coll. on the Formalization of programming concepts, L.N.C.S. No 107, Peniscola, 1981.

[BACKUS 81b]
Backus, J., Function level programs as mathematical objects, Proc. of the 1981 Conf. on Functional Programming Languages and Computer Architecture, Wentworth-by-the-Sea, New Hampshire, October 1981.

[BIDOIT 81]
Bidoit, M., Une méthode de présentation des types abstraits : applications, Thèse de 3eme cycle, Université Paris-Sud, Juin 1981.

[BROY et al. 80]
Broy, M., M. Wirsing, Algebraic definitions of a functional programming language and its semantical models, Institute fur Informatik der TU

[CHOPPY et al. 83]

Choppy, C., G. Guiho, S. Kaplan, Algebraic semantics for FP langguages, a lisp compiler and its proof, Rapport LRI N° 133, Orsay, 1983.

[GAUDEL 80]

Gaudel, M.C., Génération et preuve de compilateurs basées sur une sémantique formelle des langages de programmation, Thèse d'Etat, Nancy, 1980.

[GOGUEN et al. 79]

Goguen, J.A., J.J. Tardo, An introduction to OBJ : a language for writing and testing formal algebraic program specifications, Specifications of Reliable Software Conf. Proc., Cambridge MA, April 1979.

[GUTTAG et al. 78]

Guttag, J., J. Horning, The algebraic specification of abstract data types, Acta Informatica, 10, pp. 27-52, 1978.

[GUTTAG et al. 81]

Guttag, J., J. Horning, and J. Williams, FP with data abstraction and strong typing, Proc. of the 1981 Conf. on Functional Programming Languages and Computer Architecture, Wentworth-by-the-Sea, New Hampshire, October 1981.

[HUET et al. 80]

Huet, G., J.M. Hullot, Proof by induction in equational theories with constructors, 21st IEEE Symp. on Foundations of Computer Science, 1980.

[KAMIN 80]

Kamin, S., Final data type specifications : a new data type specification method, 7th ACM Symp. on Principles of Programmong Languages, Las Vegas, 1980.

[KAPLAN 83]

Kaplan, S., Un langage de spécification de types abstraits algébriques, Thèse de 3eme cycle, Orsay, Février 1983.

[MANNA 74]

Manna, Z., Mathematical theory of computation, Mc Graw Hill, 1974.

[VOISIN 84]

Voisin, F., CIGALE : Construction Interactive de Grammaire et Analyse Libérale d'expressions, Thèse de 3eme cycle, Université d'Orsay, France.

[WILLIAMS 80]

Williams, J.H., On the development of the algebra of functional programs, Report RJ2983, I.B.M. Research Laboratory, San Jose, 1980.

[WILLIAMS 81]

Williams, J.H., Notes on the FP style of functional programing, Lecture Notes for the Course "FP and its applications", Newcastle-upon-Tyne, July 1981.

[WIRSING et al. 81]

Wirsing, M., M. Broy, An analysis of semantic models for algebraic specifications, Marktoberdof Summer School on Theor. Found. of Progr. Methodology, 1981.

[ZILLES 79]

Zilles, S.N., An introduction to data algebras, L.N.C.S. No 86, Springer Verlag, 1979.

AUTHOR INDEX